VOLUME EIGHTY SEVEN

ADVANCES IN MARINE BIOLOGY

Population Dynamics of the Reef Crisis

ADVANCES IN MARINE BIOLOGY

Series Editor

CHARLES SHEPPARD
Emeritus Professor, School of Life Sciences
University of Warwick, U.K.
Charles.Sheppard@Warwick.ac.uk

VOLUME EIGHTY SEVEN

ADVANCES IN MARINE BIOLOGY

Population Dynamics of the Reef Crisis

Edited by

BERNHARD M. RIEGL

Department of Marine and Environmental Sciences,
Halmos College of Arts and Sciences,
Nova Southeastern University,
Dania Beach, FL, United States

ELSEVIER

ACADEMIC PRESS

An imprint of Elsevier

Academic Press is an imprint of Elsevier
125 London Wall, London, EC2Y 5AS, United Kingdom
The Boulevard, Langford Lane, Kidlington, Oxford OX5 1GB, United Kingdom
525 B Street, Suite 1650, San Diego, CA 92101, United States
50 Hampshire Street, 5th Floor, Cambridge, MA 02139, United States

First edition 2020

ISBN: 978-0-12-821529-6
ISSN: 0065-2881

For information on all Academic Press publications
visit our website at https://www.elsevier.com/books-and-journals

Publisher: Zoe Kruze
Acquisitions Editor: Jason Mitchell
Editorial Project Manager: Leticia M. Lima
Production Project Manager: Denny Mansingh
Cover Designer: Christian J. Bilbow

Typeset by SPi Global, India

Dedication

This collection of studies is dedicated to Richard E. Dodge.

In recognition of his impact on coral reef research as scientist, teacher, mentor, administrator and friend.

Those who interacted with him benefitted greatly from his guidance, thank him for his service to science and academia, and wish him much enjoyment on his continuing, hopefully now more relaxing, path.

Contributors to Volume 87

Esmeralda Alcantar
Scripps Institution of Oceanography, UC San Diego, La Jolla, CA, United States

Renata Alitto
Instituto de Biologia, Universidade Estadual de Campinas, São Paulo, Brazil

Richard B. Aronson
Department of Ocean Engineering and Marine Sciences, Florida Institute of Technology, Melbourne, FL, United States

Radhouane Ben-Hamadou
Department of Biological and Environmental Sciences, Qatar University, Doha, Qatar

Lorenzo Bramanti
CNRS-Sorbonne Université, Laboratoire d'Ecogéochimie des Environnements Benthiques, LECOB, Observatoire Océanologique de Banyuls sur Mer, Banyuls sur Mer, France

Brandon Brulé
Department of Marine and Environmental Sciences, Halmos College of Arts and Sciences, Nova Southeastern University, Dania Beach, FL, United States

Edmo Campos
Department of Biology, Chemistry and Environmental Sciences, College of Arts and Sciences, American University of Sharjah, Sharjah, United Arab Emirates; Instituto Oceanografico, Universidade de São Paulo, São Paulo, Brazil

Geórgenes Cavalcante
Department of Biology, Chemistry and Environmental Sciences, College of Arts and Sciences, American University of Sharjah, Sharjah, United Arab Emirates; Instituto de Ciências Atmosféricas (ICAT), Universidade Federal de Alagoas, Maceió, Brazil

Kendall S. Chancellor
Scripps Institution of Oceanography, UC San Diego, La Jolla, CA, United States

Ana B. Christensen
Biology Department, Lamar University, Beaumont, TX, United States

Isabelle M. Côté
Department of Biological Sciences, Simon Fraser University, Burnaby, BC, Canada

Kyle Dettloff
National Oceanographic and Atmospheric Administration, National Marine Fisheries Service, Southeast Fisheries Science Center, Miami, FL, United States

Joshua Dominguez
Undergraduate Marine and Atmospheric Science Program, Rosenstiel School of Marine and Atmospheric Science, University of Miami, Coral Gables, FL, United States

Peter J. Edmunds
Department of Biology, California State University, Northridge, CA, United States

Clinton B. Edwards
Scripps Institution of Oceanography, UC San Diego, La Jolla, CA, United States

Joshua S. Feingold
Department of Marine and Environmental Sciences, Halmos College of Arts and Sciences, Nova Southeastern University, Dania Beach, FL, United States

Michael D. Fox
Woods Hole Oceanographic Institution, Woods Hole, MA, United States

Toby A. Gardner
Stockholm Environment Institute, Stockholm, Sweden

Jennifer A. Gill
Centre for Ecology, Evolution and Conservation, School of Biological Sciences, University of East Anglia, Norwich, United Kingdom

Phillip Gillette
Department of Marine Biology and Ecology, Rosenstiel School of Marine and Atmospheric Science, University of Miami, Miami, FL, United States

David S. Gilliam
Halmos College of Arts and Sciences, Nova Southeastern University, Dania Beach, FL, United States

Peter W. Glynn
Department of Marine Biology and Ecology, Rosenstiel School of Marine and Atmospheric Science, University of Miami, Miami, FL, United States

Elizabeth A. Goergen
Department of Biological and Environmental Sciences, Qatar University, Doha, Qatar

Julie P. Hawkins
Environment Department, University of York, York, United Kingdom

Edwin A. Hernández-Delgado
Sociedad Ambiente Marino; Department of Environmental Sciences and Center for Applied Tropical Ecology and Conservation, Applied Marine Ecology Laboratory, University of Puerto Rico, San Juan, Puerto Rico

Walter C. Jaap
Lithophyte Research LLC, Saint Petersburg, FL, United States

Matthew Johnston
Department of Marine and Environmental Sciences, Halmos College of Arts and Sciences, Nova Southeastern University, Dania Beach, FL, United States

Howard R. Lasker
Department of Environment and Sustainability and Department of Geology, University at Buffalo, Buffalo, NY, United States

Zi Wei Lim
Experimental Marine Ecology Laboratory, Department of Biological Sciences, National University of Singapore, Singapore, Singapore

Yossi Loya
School of Zoology, George S. Wise Faculty of Life Sciences, Tel-Aviv University, Tel-Aviv, Israel

Kathleen Semon Lunz
National Operations Center, Bureau of Land Management, Denver, CO, United States

Nicolas Martinez
Undergraduate Marine and Atmospheric Science Program, Rosenstiel School of Marine and Atmospheric Science, University of Miami, Coral Gables, FL, United States

Priscilla Martínez
Instituto Nazca de Investigaciones Marinas, Salinas, Ecuador

S.A. Matthews
Australian Research Council Centre of Excellence for Coral Reef Studies, James Cook University; Australian Institute of Marine Science, Townsville, QLD, Australia

Tim R. McClanahan
Wildlife Conservation Society, Marine Programs, Bronx, NY, United States

Melanie D. McField
Smithsonian Marine Station, Ft Pierce, FL, United States

Mike McWilliam
Hawai'I Institute of Marine Biology, University of Hawai'I at Manoa, Kaneohe, HI, United States

C. Mellin
The Environment Institute and School of Biological Sciences, University of Adelaide, Adelaide, SA, Australia

Alex E. Mercado-Molina
Sociedad Ambiente Marino, San Juan, Puerto Rico

Holly V. Moeller
Department of Ecology, Evolution and Marine Biology, University of California, Santa Barbara, Santa Barbara, CA, United States

Jonas Mortensen
DHI A/S. Agern Allé 5, Hørsholm, Denmark

Thaddeus J.T. Murdoch
Murdoch Marine Ltd., Warwick, WKBX, Bermuda

Maggy M. Nugues
EPHE, Laboratoire d'Excellence "CORAIL", PSL Research University, UPVD, CNRS, USR, Perpignan, France

Gaia Pavoni
Visual Computing Lab, Istituto di Scienza e Tecnologie dell'Informazione "A. Faedo", Consiglio Nazionale delle Ricerche, Pisa, Italy

Nicole E. Pedersen
Scripps Institution of Oceanography, UC San Diego, La Jolla, CA, United States

Vid Petrovic
Department of Computer Science and Engineering, UC San Diego, La Jolla, CA, United States

Chiara Pisapia
Department of Biology, California State University, Northridge, CA, United States; Department of Ocean Science and Hong Kong Branch of the Southern Marine Science and Engineering Guangdong Laboratory (Guangzhou), The Hong Kong University of Science and Technology, Kowloon, Hong Kong

Federico Ponchio
Visual Computing Lab, Istituto di Scienza e Tecnologie dell'Informazione "A. Faedo", Consiglio Nazionale delle Ricerche, Pisa, Italy

Rian Prasetia
Experimental Marine Ecology Laboratory, Department of Biological Sciences, National University of Singapore, Singapore, Singapore

Morgan S. Pratchett
Australian Research Council Centre of Excellence for Coral Reef Studies, James Cook University, Townsville, QLD, Australia

William F. Precht
Marine and Coastal Programs, Dial Cordy and Associates, Miami, FL, United States

Pedro Range
Department of Biological and Environmental Sciences, Qatar University, Doha, Qatar

Bernhard M. Riegl
Department of Marine and Environmental Sciences, Halmos College of Arts and Sciences, Nova Southeastern University, Dania Beach, FL, United States

Fernando Rivera
Instituto Nazca de Investigaciones Marinas, Salinas, Ecuador

Callum M. Roberts
Environment Department, University of York, York, United Kingdom

Randi D. Rotjan
Department of Biology, Boston University, Boston, MA, United States

Alberto M. Sabat
University of Puerto Rico, San Juan, Puerto Rico

Stuart A. Sandin
Scripps Institution of Oceanography, UC San Diego, La Jolla, CA, United States

Christiane K. Schelten
GEOMAR, Helmholtz Centre for Ocean Research Kiel, Kiel, Germany

Tom Shlesinger
Institute for Global Ecology, Florida Institute of Technology, Melbourne, FL, United States

K. Shoemaker
Department of Natural Resources and Environmental Science, University of Nevada Reno, Reno, NV, United States

Brenna Stallings
Department of Biology, Boston University, Boston, MA, United States

Sascha C.C. Steiner
Cooperación Alemana al Desarrollo (GIZ), Quito, Ecuador; Institute for Tropical Marine Ecology (ITME) Inc., Roseau, Dominica

Christopher J. Sullivan
Scripps Institution of Oceanography, UC San Diego, La Jolla, CA, United States

Aaron Teo
Experimental Marine Ecology Laboratory, Department of Biological Sciences, National University of Singapore, Singapore, Singapore

Peter A. Todd
Experimental Marine Ecology Laboratory, Department of Biological Sciences, National University of Singapore, Singapore, Singapore

Georgios Tsounis
Department of Biology, California State University, Northridge, CA, United States

Filipe Vieira
Department of Biology, Chemistry and Environmental Sciences, College of Arts and Sciences, American University of Sharjah, Sharjah, United Arab Emirates

Andrew R. Watkinson
Living with Environmental Change, School of Environmental Sciences, University of East Anglia, Norwich, United Kingdom

Christopher D. Wells
Department of Geology, University at Buffalo, State University of New York, Buffalo, NY, United States

Brian J. Zgliczynski
Scripps Institution of Oceanography, UC San Diego, La Jolla, CA, United States

Contents

Series Contents for Last Fifteen Years*,†

*The full list of contents for volumes 1–37 can be found in volume 38
†The full list of contents for volumes 38–43 can be found in volume 78

Cristián J. Monaco and Brian Helmuth. Tipping Points, Thresholds and the Keystone Role of Physiology in Marine Climate Change Research. pp. 123–160.

David A. Ritz, Alistair J. Hobday, John C. Montgomery and Ashley J.W. Ward. Social Aggregation in the Pelagic Zone with Special Reference to Fish and Invertebrates. pp. 161–228.

Volume 61, 2012.

Gert Wörheide, Martin Dohrmann, Dirk Erpenbeck, Claire Larroux, Manuel Maldonado, Oliver Voigt, Carole Borchiellini and Denis Lavrov. Deep Phylogeny and Evolution of Sponges (Phylum Porifera). pp. 1–78.

Paco Cárdenas, Thierry Pérez and Nicole Boury-Esnault. Sponge Systematics Facing New Challenges. pp. 79–210.

Klaus Rützler. The Role of Sponges in the Mesoamerican Barrier-Reef Ecosystem, Belize. pp. 211–272.

Janie Wulff. Ecological Interactions and the Distribution, Abundance, and Diversity of Sponges. pp. 273–344.

Maria J. Uriz and Xavier Turon. Sponge Ecology in the Molecular Era. pp. 345–410.

Volume 62, 2012.

Sally P. Leys and April Hill. The Physiology and Molecular Biology of Sponge Tissues. pp. 1–56.

Robert W. Thacker and Christopher J. Freeman. Sponge–Microbe Symbioses: Recent Advances and New Directions. pp. 57–112.

Manuel Maldonado, Marta Ribes and Fleur C. van Duyl. Nutrient Fluxes Through Sponges: Biology, Budgets, and Ecological Implications. pp. 113–182.

Grégory Genta-Jouve and Olivier P. Thomas. Sponge Chemical Diversity: From Biosynthetic Pathways to Ecological Roles. pp. 183–230.

Xiaohong Wang, Heinz C. Schröder, Matthias Wiens, Ute Schloßmacher and Werner E. G. Müller. Biosilica: Molecular Biology, Biochemistry and Function in Demosponges as well as its Applied Aspects for Tissue Engineering. pp. 231–272.

Klaske J. Schippers, Detmer Sipkema, Ronald Osinga, Hauke Smidt, Shirley A. Pomponi, Dirk E. Martens and René H. Wijffels. Cultivation of Sponges, Sponge Cells and Symbionts: Achievements and Future Prospects. pp. 273–338.

Volume 63, 2012.

Michael Stat, Andrew C. Baker, David G. Bourne, Adrienne M. S. Correa, Zac Forsman, Megan J. Huggett, Xavier Pochon, Derek Skillings, Robert J. Toonen, Madeleine J. H. van Oppen, and Ruth D. Gates. Molecular Delineation of Species in the Coral Holobiont. pp. 1–66.

Daniel Wagner, Daniel G. Luck, and Robert J. Toonen. The Biology and Ecology of Black Corals (Cnidaria: Anthozoa: Hexacorallia: Antipatharia). pp. 67–132.

Cathy H. Lucas, William M. Graham, and Chad Widmer. Jellyfish Life Histories: Role of Polyps in Forming and Maintaining Scyphomedusa Populations. pp. 133–196.

T. Aran Mooney, Maya Yamato, and Brian K. Branstetter. Hearing in Cetaceans: From Natural History to Experimental Biology. pp. 197–246.

Volume 64, 2013.

Dale Tshudy. Systematics and Position of *Nephrops* Among the Lobsters. pp. 1–26.

Mark P. Johnson, Colm Lordan, and Anne Marie Power. Habitat and Ecology of *Nephrops norvegicus*. pp. 27–64.

Emi Katoh, Valerio Sbragaglia, Jacopo Aguzzi, and Thomas Breithaupt. Sensory Biology and Behaviour of *Nephrops norvegicus*. pp. 65–106.

Edward Gaten, Steve Moss, and Magnus L. Johnson. The Reniform Reflecting Superposition Compound Eyes of *Nephrops norvegicus*: Optics, Susceptibility to Light-Induced Damage, Electrophysiology and a Ray Tracing Model. pp. 107–148.

Susanne P. Eriksson, Bodil Hernroth, and Susanne P. Baden. Stress Biology and Immunology in *Nephrops norvegicus*. pp. 149–200.

Adam Powell and Susanne P. Eriksson. Reproduction: Life Cycle, Larvae and Larviculture. pp. 201–246.

Anette Ungfors, Ewen Bell, Magnus L. Johnson, Daniel Cowing, Nicola C. Dobson, Ralf Bublitz, and Jane Sandell. *Nephrops* Fisheries in European Waters. pp. 247–314.

Volume 65, 2013.

Isobel S.M. Bloor, Martin J. Attrill, and Emma L. Jackson. A Review of the Factors Influencing Spawning, Early Life Stage Survival and Recruitment Variability in the Common Cuttlefish (*Sepia officinalis*). pp. 1–66.

Dianna K. Padilla and Monique M. Savedo. A Systematic Review of Phenotypic Plasticity in Marine Invertebrate and Plant Systems. pp. 67–120.

Leif K. Rasmuson. The Biology, Ecology and Fishery of the Dungeness crab, *Cancer magister*. pp. 121–174.

Volume 66, 2013.

Lisa-ann Gershwin, Anthony J. Richardson, Kenneth D. Winkel, Peter J. Fenner, John Lippmann, Russell Hore, Griselda Avila-Soria, David Brewer, Rudy J. Kloser, Andy Steven, and Scott Condie. Biology and Ecology of Irukandji Jellyfish (Cnidaria: Cubozoa). pp. 1–86.

April M. H. Blakeslee, Amy E. Fowler, and Carolyn L. Keogh. Marine Invasions and Parasite Escape: Updates and New Perspectives. pp. 87–170.

Michael P. Russell. Echinoderm Responses to Variation in Salinity. pp. 171–212.

Daniela M. Ceccarelli, A. David McKinnon, Serge Andréfouët, Valerie Allain, Jock Young, Daniel C. Gledhill, Adrian Flynn, Nicholas J. Bax, Robin Beaman, Philippe Borsa, Richard Brinkman, Rodrigo H. Bustamante, Robert Campbell, Mike Cappo, Sophie Cravatte, Stéphanie D'Agata, Catherine M. Dichmont, Piers K. Dunstan, Cécile Dupouy, Graham Edgar, Richard Farman, Miles Furnas, Claire Garrigue, Trevor Hutton, Michel Kulbicki, Yves Letourneur, Dhugal Lindsay, Christophe Menkes, David Mouillot, Valeriano Parravicini, Claude Payri, Bernard Pelletier, Bertrand Richer de Forges, Ken Ridgway, Martine Rodier, Sarah Samadi, David Schoeman, Tim Skewes, Steven Swearer, Laurent Vigliola, Laurent Wantiez, Alan Williams, Ashley Williams, and Anthony J. Richardson. The Coral Sea: Physical Environment, Ecosystem Status and Biodiversity Assets. pp. 213–290.

Volume 67, 2014.

Erica A.G. Vidal, Roger Villanueva, José P. Andrade, Ian G. Gleadall, José Iglesias, Noussithé Koueta, Carlos Rosas, Susumu Segawa, Bret Grasse, Rita M. Franco-Santos, Caroline B. Albertin, Claudia Caamal-Monsreal, Maria E. Chimal, Eric Edsinger-Gonzales, Pedro Gallardo, Charles Le Pabic, Cristina Pascual, Katina Roumbedakis, and James Wood. Cephalopod Culture: Current Status of Main Biological Models and Research Priorities. pp. 1–98.

Paul G.K. Rodhouse, Graham J. Pierce, Owen C. Nichols, Warwick H.H. Sauer, Alexander I. Arkhipkin, Vladimir V. Laptikhovsky, Marek R. Lipiński, Jorge E. Ramos, Michaël Gras, Hideaki Kidokoro, Kazuhiro Sadayasu, João Pereira, Evgenia Lefkaditou, Cristina Pita, Maria Gasalla, Manuel Haimovici, Mitsuo Sakai, and Nicola Downey. Environmental Effects on Cephalopod Population Dynamics: Implications for Management of Fisheries. pp. 99–234.

Henk-Jan T. Hoving, José A.A. Perez, Kathrin Bolstad, Heather Braid, Aaron B. Evans, Dirk Fuchs, Heather Judkins, Jesse T. Kelly, José E.A.R. Marian, Ryuta Nakajima, Uwe Piatkowski, Amanda Reid, Michael Vecchione, and José C.C. Xavier. The Study of Deep-Sea Cephalopods. pp. 235–362.

Jean-Paul Robin, Michael Roberts, Lou Zeidberg, Isobel Bloor, Almendra Rodriguez, Felipe Briceño, Nicola Downey, Maite Mascaró, Mike Navarro, Angel Guerra, Jennifer Hofmeister, Diogo D. Barcellos, Silvia A.P. Lourenço, Clyde F.E. Roper, Natalie A. Moltschaniwskyj, Corey P. Green, and Jennifer Mather. Transitions During Cephalopod Life History: The Role of Habitat, Environment, Functional Morphology and Behaviour. pp. 363–440.

Volume 68, 2014.

Paul K.S. Shin, Siu Gin Cheung, Tsui Yun Tsang, and Ho Yin Wai. Ecology of Artificial Reefs in the Subtropics. pp. 1–64.

Hrafnkell Eiríksson. Reproductive Biology of Female Norway Lobster, *Nephrops norvegicus* (Linnaeus, 1758) Leach, in Icelandic Waters During the Period 1960–2010: Comparative Overview of Distribution Areas in the Northeast Atlantic and the Mediterranean. pp. 65–210.

Volume 69, 2014.

Ray Hilborn. Introduction to Marine Managed Areas. pp. 1–14.

Philip N. Trathan, Martin A. Collins, Susie M. Grant, Mark Belchier, David K.A. Barnes, Judith Brown, and Iain J. Staniland. The South Georgia and the South Sandwich Islands MPA: Protecting A Biodiverse Oceanic Island Chain Situated in the Flow of the Antarctic Circumpolar Current. pp. 15–78.

Richard P. Dunne, Nicholas V.C. Polunin, Peter H. Sand, and Magnus L. Johnson. The Creation of the Chagos Marine Protected Area: A Fisheries Perspective. pp. 79–128.

Michelle T. Schärer-Umpierre, Daniel Mateos-Molina, Richard Appeldoorn, Ivonne Bejarano, Edwin A. Hernández-Delgado, Richard S. Nemeth, Michael I. Nemeth, Manuel Valdés-Pizzini, and Tyler B. Smith. Marine Managed Areas and Associated Fisheries in the US Caribbean. pp. 129–152.

Alan M. Friedlander, Kostantinos A. Stamoulis, John N. Kittinger, Jeffrey C. Drazen, and Brian N. Tissot. Understanding the Scale of Marine Protection in Hawai'i: From Community-Based Management to the Remote Northwestern Hawaiian Islands. pp. 153–204.

Louis W. Botsford, J. Wilson White, Mark H. Carr, and Jennifer E. Caselle. Marine Protected Area Networks in California, USA. pp. 205–252.
Bob Kearney and Graham Farebrother. Inadequate Evaluation and Management of Threats in Australia's Marine Parks, Including the Great Barrier Reef, Misdirect Marine Conservation. pp. 253–288.
Randi Rotjan, Regen Jamieson, Ben Carr, Les Kaufman, Sangeeta Mangubhai, David Obura, Ray Pierce, Betarim Rimon, Bud Ris, Stuart Sandin, Peter Shelley, U. Rashid Sumaila, Sue Taei, Heather Tausig, Tukabu Teroroko, Simon Thorrold, Brooke Wikgren, Teuea Toatu, and Greg Stone. Establishment, Management, and Maintenance of the Phoenix Islands Protected Area. pp. 289–324.
Alex J. Caveen, Clare Fitzsimmons, Margherita Pieraccini, Euan Dunn, Christopher J. Sweeting, Magnus L. Johnson, Helen Bloomfield, Estelle V. Jones, Paula Lightfoot, Tim S. Gray, Selina M. Stead, and Nicholas V. C. Polunin. Diverging Strategies to Planning an Ecologically Coherent Network of MPAs in the North Sea: The Roles of Advocacy, Evidence and Pragmatism in the Face of Uncertaintya. pp. 325–370.
Carlo Pipitone, Fabio Badalamenti, Tomás Vega Fernández, and Giovanni D'Anna. Spatial Management of Fisheries in the Mediterranean Sea: Problematic Issues and a Few Success Stories. pp. 371–402.

Volume 70, 2015.
Alex D. Rogers, Christopher Yesson, and Pippa Gravestock. A Biophysical and Economic Profile of South Georgia and the South Sandwich Islands as Potential Large-Scale Antarctic Protected Areas. pp. 1–286.

Volume 71, 2015.
Ricardo Calado and Miguel Costa Leal. Trophic Ecology of Benthic Marine Invertebrates with Bi-Phasic Life Cycles: What Are We Still Missing? pp. 1–70.
Jesse M.A. van der Grient and Alex D. Rogers. Body Size Versus Depth: Regional and Taxonomical Variation in Deep-Sea Meio- and Macrofaunal Organisms. pp. 71–108.
Lorena Basso, Maite Vázquez-Luis, José R. García-March, Salud Deudero, Elvira Alvarez, Nardo Vicente, Carlos M. Duarte, and Iris E. Hendriks. The Pen Shell, *Pinna nobilis*: A Review of Population Status and Recommended Research Priorities in the Mediterranean Sea. pp. 109–160.

Volume 72, 2015.

Thomas A. Jefferson and Barbara E. Curry. Humpback Dolphins: A Brief Introduction to the Genus *Sousa*. pp. 1–16.

Sarah Piwetz, David Lundquist, and Bernd Würsig. Humpback Dolphin (Genus *Sousa*) Behavioural Responses to Human Activities. pp. 17–46.

Tim Collins. Re-assessment of the Conservation Status of the Atlantic Humpback Dolphin, *Sousa teuszii* (Kükenthal, 1892) Using the IUCN Red List Criteria. pp. 47–78.

Caroline R. Weir and Tim Collins. A Review of the Geographical Distribution and Habitat of the Atlantic Humpback Dolphin (*Sousa teuszii*). pp. 79–118.

Gill T Braulik, Ken Findlay, Salvatore Cerchio, and Robert Baldwin. Assessment of the Conservation Status of the Indian Ocean Humpback Dolphin (*Sousa plumbea*) Using the IUCN Red List Criteria. pp. 119–142.

Stephanie Plön, Victor G. Cockcroft, and William P. Froneman. The Natural History and Conservation of Indian Ocean Humpback Dolphins (*Sousa plumbea*) in South African Waters. pp. 143–162.

Salvatore Cerchio, Norbert Andrianarivelo, and Boris Andrianantenaina. Ecology and Conservation Status of Indian Ocean Humpback Dolphins (*Sousa plumbea*) in Madagascar. pp. 163–200.

Muhammad Shoaib Kiani and Koen Van Waerebeek. A Review of the Status of the Indian Ocean Humpback Dolphin *Sousa plumbea* in Pakistan. pp. 201–228.

Dipani Sutaria, Divya Panicker, Ketki Jog, Mihir Sule, Rahul Muralidharan, and Isha Bopardikar. Humpback Dolphins (Genus *Sousa*) in India: An Overview of Status and Conservation Issues. pp. 229–256.

Volume 73, 2016.

Thomas A. Jefferson and Brian D. Smith. Re-assessment of the Conservation Status of the Indo-Pacific Humpback Dolphin (*Sousa chinensis*) Using the IUCN Red List Criteria. pp. 1–26.

Leszek Karczmarski, Shiang-Lin Huang, Carmen K. M. Or, Duan Gui, Stephen C. Y. Chan, Wenzhi Lin, Lindsay Porter, Wai-Ho Wong, Ruiqiang Zheng, Yuen-Wa Ho, Scott Y. S. Chui, Angelico Jose C. Tiongson, Yaqian Mo, Wei-Lun Chang, John H. W. Kwok, Ricky W. K. Tang, Andy T. L. Lee, Sze-Wing Yiu, Mark Keith, Glenn Gailey, and Yuping Wu. Humpback Dolphins in Hong Kong and the Pearl River Delta: Status, Threats and Conservation Challenges. pp. 27–64.

Bernd Würsig, E.C.M. Parsons, Sarah Piwetz, and Lindsay Porter. The Behavioural Ecology of Indo-Pacific Humpback Dolphins in Hong Kong. pp. 65–90.

John Y. Wang, Kimberly N. Riehl, Michelle N. Klein, Shiva Javdan, Jordan M. Hoffman, Sarah Z. Dungan, Lauren E. Dares, and Claryana Araújo-Wang. Biology and Conservation of the Taiwanese Humpback Dolphin, *Sousa chinensis taiwanensis*. pp. 91–118.

Bingyao Chen, Xinrong Xu, Thomas A. Jefferson, Paula A. Olson, Qiurong Qin, Hongke Zhang, Liwen He, and Guang Yang. Conservation Status of the Indo-Pacific Humpback Dolphin (*Sousa chinensis*) in the Northern Beibu Gulf, China. pp. 119–140.

Gianna Minton, Anna Norliza Zulkifli Poh, Cindy Peter, Lindsay Porter, and Danielle Kreb. Indo-Pacific Humpback Dolphins in Borneo: A Review of Current Knowledge with Emphasis on Sarawak. pp. 141–156.

Guido J. Parra and Daniele Cagnazzi. Conservation Status of the Australian Humpback Dolphin (*Sousa sahulensis*) Using the IUCN Red List Criteria. pp. 157–192.

Daniella M. Hanf, Tim Hunt, and Guido J. Parra. Humpback Dolphins of Western Australia: A Review of Current Knowledge and Recommendations for Future Management. pp. 193–218.

Isabel Beasley, Maria Jedensjö, Gede Mahendra Wijaya, Jim Anamiato, Benjamin Kahn, and Danielle Kreb. Observations on Australian Humpback Dolphins (*Sousa sahulensis*) in Waters of the Pacific Islands and New Guinea. pp. 219–272.

Alexander M. Brown, Lars Bejder, Guido J. Parra, Daniele Cagnazzi, Tim Hunt, Jennifer L. Smith, and Simon J. Allen. Sexual Dimorphism and Geographic Variation in Dorsal Fin Features of Australian Humpback Dolphins, *Sousa sahulensis*. pp. 273–314.

Volume 74, 2016.

J. Salinger, A.J. Hobday, R.J. Matear, T.J. O'Kane, J.S. Risbey, P. Dunstan, J.P. Eveson, E.A. Fulton, M. Feng, É.E. Plagányi, E.S. Poloczanska, A.G. Marshall, and P.A. Thompson. Decadal-Scale Forecasting of Climate Drivers for Marine Applications. pp. 1–68.

S.A. Foo and M. Byrne. Acclimatization and Adaptive Capacity of Marine Species in a Changing Ocean. pp. 69–116.

N.D. Gallo and L.A. Levin. Fish Ecology and Evolution in the World's Oxygen Minimum Zones and Implications of Ocean Deoxygenation. pp. 117–198.

R.J. Olson, J.W. Young, F. Ménard, M. Potier, V. Allain, N. Goñi, J.M. Logan, and F. Galván-Magaña. Bioenergetics, Trophic Ecology, and Niche Separation of Tunas. pp. 199–344.

Volume 75, 2016.

G. Notarbartolo di Sciara. Marine Mammals in the Mediterranean Sea: An Overview. pp. 1–36.

L. Rendell and A. Frantzis. Mediterranean Sperm Whales, *Physeter macrocephalus*: The Precarious State of a Lost Tribe. pp. 37–74.

G. Notarbartolo di Sciara, M. Castellote, J.-N. Druon, and S. Panigada. Fin Whales, *Balaenoptera physalus*: At Home in a Changing Mediterranean Sea?. pp. 75–102.

M. Podestà, A. Azzellino, A. Cañadas, A. Frantzis, A. Moulins, M. Rosso, P. Tepsich, and C. Lanfredi. Cuvier's Beaked Whale, *Ziphius cavirostris*, Distribution and Occurrence in the Mediterranean Sea: High-Use Areas and Conservation Threats. pp. 103–140.

R. Esteban, P. Verborgh, P. Gauffier, D. Alarcón, J.M. Salazar-Sierra, J. Giménez, A.D. Foote, and R. de Stephanis. Conservation Status of Killer Whales, *Orcinus orca*, in the Strait of Gibraltar. pp. 141–172.

P. Verborgh, P. Gauffier, R. Esteban, J. Giménez, A. Cañadas, J.M. Salazar-Sierra, and R. de Stephanis. Conservation Status of Long-Finned Pilot Whales, *Globicephala melas*, in the Mediterranean Sea. pp. 173–204.

A. Azzellino, S. Airoldi, S. Gaspari, C. Lanfredi, A. Moulins, M. Podestà, M. Rosso, and P. Tepsich. Risso's Dolphin, *Grampus griseus*, in the Western Ligurian Sea: Trends in Population Size and Habitat Use. pp. 205–232.

D. Kerem, O. Goffman, M. Elasar, N. Hadar, A. Scheinin, and T. Lewis. The Rough-Toothed Dolphin, *Steno bredanensis*, in the Eastern Mediterranean Sea: A Relict Population?. pp. 233–258.

J. Gonzalvo, G. Lauriano, P.S. Hammond, K.A. Viaud-Martinez, M.C. Fossi, A. Natoli, and L. Marsili. The Gulf of Ambracia's Common Bottlenose Dolphins, *Tursiops truncatus*: A Highly Dense and yet Threatened Population. pp. 259–296.

G. Bearzi, S. Bonizzoni, N.L. Santostasi, N.B. Furey, L. Eddy, V.D. Valavanis, and O. Gimenez. Dolphins in a Scaled-Down Mediterranean: The Gulf of Corinth's Odontocetes. pp. 297–332.

M.C. Fontaine. Harbour Porpoises, *Phocoena phocoena*, in the Mediterranean Sea and Adjacent Regions: Biogeographic Relicts of the Last Glacial Period. pp. 333–358.

G. Notarbartolo di Sciara and S. Kotomatas. Are Mediterranean Monk Seals, *Monachus monachus*, Being Left to Save Themselves from Extinction?. pp. 359–386.

T. Scovazzi. The International Legal Framework for Marine Mammal Conservation in the Mediterranean Sea. pp. 387–416.

Volume 76, 2017.

K.S. Meyer. Islands in a Sea of Mud: Insights From Terrestrial Island Theory for Community Assembly on Insular Marine Substrata. pp. 1–40.

E.M. Montgomery, J.-F. Hamel, and A. Mercier. Patterns and Drivers of Egg Pigment Intensity and Colour Diversity in the Ocean: A Meta-Analysis of Phylum Echinodermata. pp. 41–104.

F. Espinosa and G.A. Rivera-Ingraham. Biological Conservation of Giant Limpets: The Implications of Large Size. pp. 105–156.

L. Yebra, T. Kobari, A.R. Sastri, F. Gusmão, and S. Hernández-León. Advances in Biochemical Indices of Zooplankton Production. pp. 157–240.

Volume 77, 2017.

Dayv Lowry and Shawn E. Larson. Introduction to Northeast Pacific Shark Biology, Ecology, and Conservation. pp. 1–8.

David A. Ebert, Jennifer S. Bigman, and Julia M. Lawson. Biodiversity, Life History, and Conservation of Northeastern Pacific Chondrichthyans pp. 9–78.

Shawn E. Larson, Toby S. Daly-Engel, and Nicole M. Phillips. Review of Current Conservation Genetic Analyses of Northeast Pacific Sharks. pp. 79–110.

Joseph J. Bizzarro, Aaron B. Carlisle, Wade D. Smith, and Enric Cortés. Diet Composition and Trophic Ecology of Northeast Pacific Ocean Sharks. pp. 111–148.

Jonathan C.P. Reum, Gregory D. Williams, and Chris J. Harvey. Stable Isotope Applications for Understanding Shark Ecology in the Northeast Pacific Ocean. pp. 149–178.

Mary E. Matta, Cindy A. Tribuzio, David A. Ebert, Kenneth J. Goldman, and Christopher M. Gburski. Age and Growth of Elasmobranchs and Applications to Fisheries Management and Conservation in the Northeast Pacific Ocean. pp. 179–220.

Volume 78, 2017.
Shawn E. Larson and Dayv Lowry. Introduction to Northeast Pacific Shark Biology, Research, and Conservation, Part B. pp. 1–8.
Jackie King, Gordon A. McFarlane, Vladlena Gertseva, Jason Gasper, Sean Matson, and Cindy A. Tribuzio. Shark Interactions With Directed and Incidental Fisheries in the Northeast Pacific Ocean: Historic and Current Encounters, and Challenges for Shark Conservation. pp. 9–44.
Dovi Kacev, Timothy J. Sippel, Michael J. Kinney, Sebastián A. Pardo, and Christopher G. Mull. An Introduction to Modelling Abundance and Life History Parameters in Shark Populations. pp. 45–88.
Michael Grassmann, Bryan McNeil, and Jim Wharton. Sharks in Captivity: The Role of Husbandry, Breeding, Education, and Citizen Science in Shark Conservation. pp. 89–120.
Peter A. Mieras, Chris Harvey-Clark, Michael Bear, Gina Hodgin, and Boone Hodgin. The Economy of Shark Conservation in the Northeast Pacific: The Role of Ecotourism and Citizen Science. pp. 121–154.
Dayv Lowry. Conclusions: The Future of Shark Management and Conservation in the Northeast Pacific Ocean. pp. 155–164.

Volume 79, 2018.
Patricia L.M. Lee, Gail Schofield, Rebecca I. Haughey, Antonios D. Mazaris, and Graeme C. Hays. A Review of Patterns of Multiple Paternity Across Sea Turtle Rookeries. pp. 1–32.
Johanne Vad, Georgios Kazanidis, Lea-Anne Henry, Daniel O.B. Jones, Ole S. Tendal, Sabine Christiansen, Theodore B. Henry, and J. Murray Roberts. Potential Impacts of Offshore Oil and Gas Activities on Deep-Sea Sponges and the Habitats They Form. pp. 33–60.
Gianmarco Ingrosso, Marco Abbiati, Fabio Badalamenti, Giorgio Bavestrello, Genuario Belmonte, Rita Cannas, Lisandro Benedetti-Cecchi, Marco Bertolino, Stanislao Bevilacqua, Carlo Nike Bianchi, Marzia Bo, Elisa Boscari, Frine Cardone, Riccardo Cattaneo Vietti, Alessandro Cau, Carlo Cerrano, Renato Chemello, Giovanni Chimienti, Leonardo Congiu, Giuseppe Corriero, Federica Costantini, Francesco De Leo, Luigia Donnarumma, Annalisa Falace, Simonetta Fraschetti, Adriana Giangrande, Maria Flavia Gravina, Giuseppe Guarnieri, Francesco Mastrototaro, Marco Milazzo, Carla Morri, Luigi Musco, Laura Pezzolesi, Stefano Piraino, Fiorella Prada, Massimo Ponti, Fabio Rindi, Giovanni Fulvio Russo, Roberto Sandulli, Adriana Villamor, Lorenzo Zane, and Ferdinando Boero. Mediterranean Bioconstructions Along the Italian Coast. pp. 61–136.
Alex D. Rogers. The Biology of Seamounts: 25 Years on. pp. 137–224.

Volume 80, 2018.

Michael H. Schleyer and Sean N. Porter. Drivers of Soft and Stony Coral Community Distribution on the High-Latitude Coral Reefs of South Africa. pp. 1–56.

Sara González-Delgado and José Carlos Hernández. The Importance of Natural Acidified Systems in the Study of Ocean Acidification: What Have We Learned? pp. 57–100.

Volume 81, 2018.

Melissa C. Cook, Adam May, Lucas Kohl, Geert Van Biesen, Christopher C. Parrish, and Penny L. Morrill. The Potential Impact of Hydrocarbons on Mussels in Port au Port Bay, Newfoundland pp. 1–22.

Yunwen Tao, Bing Chen, Baiyu (Helen) Zhang, Zhiwen (Joy) Zhu, and Qinhong Cai. Occurrence, Impact, Analysis and Treatment of Metformin and Guanylurea in Coastal Aquatic Environments of Canada, USA and Europe pp. 23–58.

Gong Zhang, Chun Yang, Mariam Serhan, Graeme Koivu, Zeyu Yang, Bruce Hollebone, Patrick Lambert, and Carl E. Brown. Characterization of Nitrogen-Containing Polycyclic Aromatic Heterocycles in Crude Oils and Refined Petroleum Products. pp. 59–96.

Arpana Rani Datta, Qiao Kang, Bing Chen, and Xudong Ye. Fate and Transport Modelling of Emerging Pollutants from Watersheds to Oceans: A Review. pp. 97–128.

Meijin Du, Wenwen Gu, Xixi Li, Fuqiang Fan, and Yu Li. Modification of Hexachlorobenzene to Molecules with Lower Long-Range Transport Potentials Using 3D-QSAR Models with a Full Factor Experimental Design pp. 129–166.

Weiyun Lin, Xixi Li, Min Yang, Kenneth Lee, Bing Chen, and Baiyu (Helen) Zhang. Brominated Flame Retardants, Microplastics, and Biocides in the Marine Environment: Recent Updates of Occurrence, Analysis, and Impacts. pp. 167–212.

Xixi Li, Zhenhua Chu, Jiawen Yang, Minghao Li, Meijin Du, Xiaohui Zhao, Zhiwen (Joy) Zhu, and Yu Li. Synthetic Musks: A Class of Commercial Fragrance Additives in Personal Care Products (PCPs) Causing Concern as Emerging Contaminants. pp. 213–280.

Volume 82, 2019.

Junaid S. Khan, Jennifer F. Provencher, Mark R. Forbes, Mark L. Mallory, Camille Lebarbenchon, and Karen D. McCoy. Parasites of seabirds: A survey of effects and ecological implications. pp. 1–50.

Charlotte Havermans, Holger Auel, Wilhelm Hagen, Christoph Held, Natalie S. Ensor, and Geraint A. Tarling. Predatory zooplankton on the move: *Themisto* amphipods in high-latitude marine pelagic food webs. pp. 51–92.

Angel Borja, Guillem Chust, and Iñigo Muxika. Forever young: The successful story of a marine biotic index. pp. 93–128.

Ferdinando Boero, Francesco De Leo, Simonetta Fraschetti, and Gianmarco Ingrosso. The Cells of Ecosystem Functioning: Towards a holistic vision of marine space. pp. 129–154.

Volume 83, 2019.

Dayv Lowry and Shawn Larson. Introduction: The sharks of Pacific Mexico and their conservation: Why should we care?. pp. 1–10.

Luz Erandi Saldaña-Ruiz, Emiliano García-Rodríguez, Juan Carlos Pérez-Jiménez, Javier Tovar-Ávila, and Emmanuel Rivera-Téllez. Biodiversity and conservation of sharks in Pacific Mexico. pp. 11–60.

Felipe Galván-Magaña, José Leonardo Castillo-Geniz, Mauricio Hoyos-Padilla, James Ketchum, A. Peter Klimley, Sergio Ramírez-Amaro, Yassir Eden Torres-Rojas, and Javier Tovar-Avila. Shark ecology, the role of the apex predator and current conservation status. pp. 61–114.

Jonathan Sandoval-Castillo. Conservation genetics of elasmobranchs of the Mexican Pacific Coast, trends and perspectives. pp. 115–158.

Volume 84, 2019.

Brian Morton and Fabrizio Marcondes Machado. Predatory marine bivalves: A review. pp. 1–98.

Volume 85, 2020.

James T. Ketchum, Mauricio Hoyos-Padilla, Alejandro Aldana-Moreno, Kathryn Ayres, Felipe Galván-Magaña, Alex Hearn, Frida Lara-Lizardi, Gador Muntaner-López, Miquel Grau, Abel Trejo-Ramírez, Darren A. Whitehead, and A. Peter Klimley. Shark movement patterns in the Mexican Pacific: A conservation and management perspective. pp. 1–38.

Oscar Sosa-Nishizaki, Emiliano García-Rodríguez, Christian D. Morales-Portillo, Juan C. Pérez-Jiménez, M. del Carmen Rodríguez-Medrano, Joseph J. Bizzarro, and José Leonardo Castillo-Géniz. Fisheries interactions and the challenges for target and nontargeted take on shark conservation in the Mexican Pacific. pp. 39–70.

A.M. Cisneros-Montemayor, E.E. Becerril-García, O. Berdeja-Zavala, and A. Ayala-Bocos. Shark ecotourism in Mexico: Scientific research, conservation, and contribution to a Blue Economy. pp. 71–92.

Oscar Sosa-Nishizaki, Felipe Galván-Magaña, Shawn E. Larson, and Dayv Lowry. Conclusions: Do we eat them or watch them, or both? Challenges for conservation of sharks in Mexico and the NEP. pp. 93–102.

Volume 86, 2020.

David J. Vance and Peter C. Rothlisberg. The biology and ecology of the banana prawns: *Penaeus merguiensis* de Man and *P. indicus* H. Milne Edwards. pp. 1–140.

Jeff A. Eble, Toby S. Daly-Engel, Joseph D. DiBattista, Adam Koziol, and Michelle R. Gaither. Marine environmental DNA: Approaches, applications, and opportunities. pp. 141–170.

The population dynamics of the coral reef crisis—Prologue

Bernhard M. Riegl
Department of Marine and Environmental Science, Halmos College of Arts and Sciences, Nova Southeastern University, Dania Beach, FL, United States

The ongoing coral reef crisis is beyond debate and is evidenced in an extensive body of research studies. Severe declines of biodiversity, along with fears of a sixth mass extinction, have not only occurred in coral reefs during the past few decades, but also in virtually all major ecosystems on Planet Earth (Barnosky et al., 2011). The primary driving force behind these changes, within the aptly named Anthropocene, is defined by human population dynamics and alterations of Earth surface processes (Cabellos et al., 2015; Hughes et al., 2017).

Thus, to introduce this volume's relevance, Chapter 1 by Riegl and Glynn outlines human population dynamics in the context of metapopulation theory to derive from first principles the likely impacts on coral reefs. The burgeoning human population requires space and resources. The resultant dynamics leads inexorably to extinctions, as is demonstrated by the differential equation expressions of theory. Evidence points to a difficult future for coral reefs—whether humankind manages requisite natural resources judiciously or not. Chapter 2 by Pisapia and coauthors explores the future of coral sizes based on matrix population models derived from decades of meticulous monitoring. The presently observed stresses will force corals overall into smaller colony size classes—future coral reefs will consist of smaller and often fewer colonies and likely will lose their Methuselah giants. This is clearly shown in the ergodic properties of the matrices; the model code is provided as a supplement to allow readers to explore this evolving trend for themselves. Mercado-Molina and coauthors in Chapter 3 also use a matrix population model, but specifically to explore the impacts of a coral disease. They show the remarkably severe effects of epizootics and quantify a quasiextinction threshold for the Caribbean coral *Acropora cervicornis* in Puerto Rico. The importance of epizootics on the status of basin-wide Caribbean populations is also investigated by Precht and coauthors in Chapter 12. Goergen and coauthors in

Chapter 4 investigate *A. cervicornis* dynamics at the colony-level at several sites in South Florida. These chapters together demonstrate an image of surprisingly dynamic populations of *A. cervicornis* and raise concern for their future existence that can be quantified. Chapter 5, by Prasetia and coauthors, and Chapter 6, by Feingold and Brulé, track the population status of mushroom corals, unattached single-polyp corals, respectively, in a highly stressed urban habitat in Singapore, and a largely pristine habitat in the Galápagos Islands. Prasetia and coauthors document some species losses over the past decades and different population dynamics on reef slopes and reef flats based on polyp size distributions, which exhibited quite uniform growth rates. Overall, they suggest that mushroom coral populations appear stable on the stressed urban reefs of Singapore. This is not the case for the Galápagos populations monitored by Feingold and Brulé, who witnessed overall population declines and local extinctions over the past 3 decades. In Chapter 7, Sandin and coauthors further explore such discrepancies in dynamics among closely related species across an ocean basin. They quantify variability in growth and survivorship in two common coral taxa (*Pocillopora* and *Porites*), one of the branching colony morphology and the other massive, across a west-east Pacific basin gradient. Surprising local variability in the tracked life history dynamics is observed. This chapter is also a good demonstration of how the latest photogrammetry techniques can be employed to improve the evaluation of population dynamics. While the previous chapters consider primarily the postsettlement dynamics of corals, Chapter 8 by Cavalcante and coauthors presents a model of larval dispersal in Persian/Arabian Gulf corals drawing attention to the overarching importance of recruitment processes. They show that both long-distance, basin-wide larval transport as well as local retention are likely to be influential, with the latter occurring far more frequently. Thus, local oceanographic conditions seem to have a disproportionate effect on coral larval transport processes.

Coral reefs consist not only of reef-building corals but also of multitudes of associated species that are also part and parcel of the reef crisis. The population dynamics of corals in the Indo-Pacific region is strongly influenced by outbreaks of predatory organisms, such as the Crown-of-Thorns Starfish (CoTS, *Acanthaster solaris*). In Chapters 9 and 10, Matthews and coauthors treat key questions regarding this important relationship. In Chapter 9, a spatial model is developed based on water quality and larval connectivity, predicting where CoTS outbreaks are most likely to occur. In Chapter 10 a metacommunity model is developed, based on coral and CoTS population dynamics, to explain the outcomes of predator–prey interactions with a view towards management. The latter chapter illustrates how the dynamics of several populations contribute towards the dynamics of entire communities. McClanahan's Chapter 11 quantifies links between coral dynamics and environmental stressors, such as fisheries, geomorphological setting and sea temperature distributions. He finds that the second to fourth moments of temperature distributions are particularly useful tools to separate regions of greater vs lesser stress that result in different impacts on specific coral life histories, thus allowing for an understanding of certain coral community response types in East Africa. In Chapter 12, Precht and coauthors also examine community dynamics, which is the sum of all population processes, across the Caribbean region. Their extensive metaanalysis reveals several shifts from corals to algae and back again, suggesting that more shifts occurred than would be expected by chance alone. They build an argument for the existence of a regional driver of such multiple changes, based primarily on coral diseases that deplete populations of formerly dominant species.

As quantified in Chapter 12, significant recent changes have occurred in Caribbean coral reefs, and Chapter 13 by Lasker and coworkers provides a detailed review of how and where octocoral 'forests' have replaced former hexacoral (scleractinian) 'gardens of stone'. This chapter provides an in-depth overview of faunal replacement on Caribbean reefs over the past decades driven by the processes described in Chapter 12. In Chapter 14, Steiner and coauthors describe patterns of octocoral populations, communities and connectivity in the Eastern Tropical Pacific (Ecuador), a region not yet adequately researched. While Chapter 13 shows Caribbean octocorals to be mostly mixotrophic, Chapter 14 identifies those in the Eastern Tropical Pacific as mostly heterotrophic. Communities are characterized by relatively few species, populations are in lower abundances than in the Caribbean and are not generally associated with reef-building corals. Connectivity patterns among populations across the Eastern Tropical Pacific suggest that mainland Ecuador is a diversity and dispersal hub for octocorals.

The final Chapter 15 by Glynn and coworkers presents research on an often-overlooked aspect of the reef crisis, namely that of introduced species and their impacts. They describe an invasive Indo-Pacific brittle star (*Ophiothela mirabilis*) that now occurs from southern Brazil northwards to the Caribbean Sea and southeast Florida, present as an epizoite on octocorals and other coral reef-associated species. Their exceedingly high abundances and competitive abilities have raised concerns of potentially negative effects on their hosts and native, cooccurring epizoites.

Ranging widely among taxa, processes and regions, this collection of studies provides a concentrated wealth of new information to inform, inspire, and thus further the study of population dynamics on coral reefs.

References

Barnosky, A.D., Matzke, N., Tomiya, S., Woga, G.O.U., Swartz, B., 2011. Has the Earth's sixth mass extinction already arrived? Nature 471, 51–57.

Cabellos, G., Ehrlich, P.R., Barnosky, A.D., Garcia, A., Pringle, R.M., Palmer, T.M., 2015. Accelerated modern human-induced species losses: entering the sixth mass extinction. Sci. Adv. 1 (5). https://doi.org/10.1126/sciadv.1400253.

Hughes, T.P., Barnes, M.L., Bellwood, D.R., Cinner, J.E., Cumming, G.S., et al., 2017. Coral reefs in the Anthropocene. Nature 546, 82–90.

CHAPTER ONE

Population dynamics of the reef crisis: Consequences of the growing human population

Bernhard M. Riegl[a,*] and Peter W. Glynn[b]
[a]Department of Marine and Environmental Sciences, Halmos College of Arts and Sciences, Nova Southeastern University, Dania Beach, FL, United States
[b]Department of Marine Biology and Ecology, Rosenstiel School of Marine and Atmospheric Science, University of Miami, Miami, FL, United States
[*]Corresponding author: e-mail address: rieglb@nova.edu

Contents

Abstract

An unequivocal link exists between human population density and environmental degradation, both in the near field (local impacts) and far field (impacts due to teleconnections). Human population is most widely predicted to reach 9–11 billion by 2100, when the demographic transition is expected in all but a handful of countries. Strongest population growth is in the tropics, where coral reefs face dense human population and concomitant heavy usage. In most countries, >50% will be urbanized but growth of rural population and need for food in urban centres will not alleviate pressure on reef resources. Aquaculture will alleviate some fishing pressure, but still utilizes reef surface and is also destructive. Denser coastal populations and greater wealth will lead to reef degradation by coastal construction. Denser populations inland will lead to more runoff and siltation. Effects of human perturbations can be explored with metapopulation theory since they translate to increases in patch-mortality and decreases in patch-colonization (=regeneration). All such changes will result in a habitat with overall fewer settled patches, so fewer live reefs. If rescue effects are included, bifurcations in system dynamics will allow for many empty patches and, depending on system state relative to stable and unstable equilibria, a part-empty system may either trend towards stability at higher patch occupancy or extinction. Thus, unless the disturbance history is known, it may be difficult to assess the direction of system trajectory—making management difficult. If habitat is decreased by destruction, rescue effects

Advances in Marine Biology, Volume 87
ISSN 0065-2881
https://doi.org/10.1016/bs.amb.2020.07.004

become even more important as extinction-debt, accumulated by efficient competitors with weaker dispersal ability, is realized. Easily visible trends in human population dynamics combined with well-established and tested ecological theory give a clear, intuitive, yet quantifiable guide to the severity of survival challenges faced by coral reefs. Management challenges and required actions can be clearly shown and, contrary to frequent claims, no scientific ambiguity exists with regards to the serious threat posed to coral reefs by humankind's continued numerical increase.

1. Introduction

Spectacular losses have occurred in coral reefs over the past decades, following unprecedented environmental disturbances. This has resulted in stark reductions of biomass in many critical ecosystem components such as fishes, invertebrates including corals, which are quintessential ecosystem engineers. Coral reef ecosystems now stand at a crossroads (Hubbard et al., 2016) with some doubt whether matters can improve or whether inexorable decline will lead to their disappearance as functioning ecosystems (Pandolfi et al., 2005). Recognition of crises has been consistent in the research community from early on (Brown, 1987; Glynn, 1984; Hughes, 1994; Jackson et al., 2001; Sale, 2008; Wilkinson, 1992) and a rich literature debates local and global effects. Justifiably, climate change has been a focus (Buddemeier, 1992; Gattuso et al., 2015; Hoegh-Guldberg and Bruno, 2010; Hughes et al., 2017; Williams and Bunkley-Williams, 1990). Some pioneering population biologists recognized humankind's expansion as a fundamental cause of environmental degradation (Ehrlich, 1971; Hardin, 1968; Meadows et al., 1972). While under specific circumstances some beneficial effects of increased human population density on the management of remnant natural resources can be realized, there is general consensus that human population size is the main driver of environmental degradation (Boyd and Slaymaker, 2000; Bradshaw and Brook, 2014; Cinner et al., 2020; De Sherbinin et al., 2007; Duarte et al., 2020; IPCC, 2013; Johnson, 2000; Malakoff, 2011; Mora and Sale, 2011; Sale, 2011; Sale et al., 2014; Weber and Sciubba, 2019).

The International Coral Reef Symposia regularly convened reef scientists over the past 50 years and these proceedings indicate how coral reef disturbances were recognized. While 9% of all papers described local human-made impacts in 1969, 24.4% and 26.3% of all papers addressed the issue in 1988 and 1992, including mostly local but also widespread and combined stressor effects. During the 1980s the first widespread coral bleaching and mortality events were observed, coincident with El Niño Southern Oscillation ocean warming and indicative of global scale disturbances (Brown, 1987; Glynn, 1983, 1984; Williams and Bunkley-Williams, 1990). From 1996 to 2016, the proportion of studies dedicated to local and widespread stresses affecting coral reefs have predominated, underscoring the concern of rapid decline and likely loss of coral reef ecosystems at the cost of burgeoning human population growth and associated resource demands (Fig. 1).

That humans cause significant problems for coral reefs is not new information, it is well-documented in the literature, and failures to act have been pointed out (Mumby and Steneck, 2008; Risk, 2014; Sale, 2008, 2011; Sheppard, 2016). Despite progress in conservation, remarkably little concrete action or directed serious discussion is apparent with regards to some of the overarching issues, such as the need to manage human population growth for society's sake and to mitigate against associated impacts on natural ecosystems. Often a perceived lack of scientific evidence or clarity is a

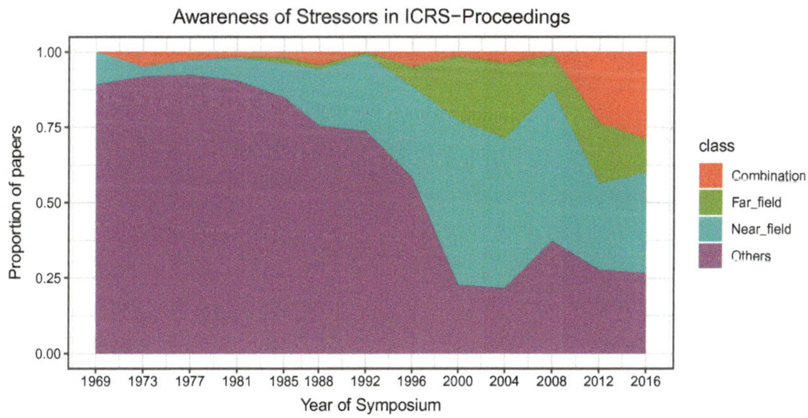

Fig. 1 Increase in concern and studies relating to coral reef degradation, protection, and rehabilitation over a 47-year period, based on publications in the Proceedings of the International Coral Reef Symposia. Threats described in the papers were categorized as acting in the near field and far field (see main text). "Combination" refers to near- and far-field stressors acting together, either additively or synergistically. The number of publications in the proceedings varied: Mandapam Camp, 1969 ($n = 37$); Great Barrier Reef, 1973 (123); Miami, 1977 (183); Manila, 1981 (276); Tahiti, 1985 (424); Townsville, 1988 (262); Guam, 1992 (434); Panama, 1996 (340); Bali, 2000 (202); Okinawa, 2004 (290); Ft. Lauderdale, 2008 (292); Cairns, 2012 (176); Hawaii, 2016 (45) (accessed at http://www.reefbase.org/resource_center/publication/icrs.aspx, and http://coralreefs.org/publications/icrs-2016-proceedings/).

front argument for delaying action or necessary, frequently painful, discussion. However, science cannot be blamed for either overlooking the issues or a lack of clarity (Fig. 1). Among innumerable case-studies, a rich body of ecological theory exists that can be employed to derive from first principles the consequences of human-mediated habitat degradation and destruction (e.g., Hardin, 1968). The present paper emphasizes that almost any aspect of theoretical or applied ecology can be used to draw quantifiable and dynamically predictive conclusions about impacts on the dynamics of ecosystems as caused by human populations. This contribution employs metapopulation dynamics (Hanski, 1999) as a theoretical framework, which is well-established and has been demonstrated as helpful to understanding the dynamics and conservation needs of coral reefs and other marine ecosystems (Kritzer and Sale, 2006).

Initially, key aspects of human population dynamics will be presented from which we will derive, applying the above noted well-established concepts from population ecology, the consequences for coral reef ecosystems. This information will be neither new nor surprising, nor will it answer all questions, but may remind us that the scientific framework for rigorous discussions about the effects of human population expansion on ecological resources has been in place for a long time in many sub-disciplines.

2. Humans: Their numbers, distribution, and impacts

With regards to human influences, coral reefs are faced by threats in two basic categories: (1) those acting near-field and (2) those acting far-field (Riegl et al., 2009). Near-field stresses are due to local phenomena such as volcanic eruptions, flash-flooding, unusually low tides, or oxygen depletion of waters by anomalous local conditions. They also include point-source chemical and light pollution (Shlesinger and Loya, 2019), local overfishing, destructive fishing, and other direct impacts by nearby human populations. While local, often seemingly contained and apparently of only small-scale significance, intense utilization of the world's tropical shorelines makes these impacts ubiquitous. Ever fewer exceptional and remote reef systems are spared (Gilmour et al., 2013; Glynn et al., 2015). Far field stressors impact large geographical regions and can act via teleconnections. Environmental changes by modification of climate and ocean chemistry fall into this category. Global warming, sea level rise and ocean acidification are much-discussed topics, and questions arise on how coral reefs can remain resilient to such changes, especially when combined with near-field stressors.

As in most ecosystems, degradation of coral reefs coincides, correlates with, or can be demonstrated to be caused by human activities (Bradshaw and Brook, 2014). Humans and coral reefs have coexisted for millennia, but human population densities and resultant impacts have reached unsustainable levels in many places. The IPCC's (2013) unequivocal conclusion that human activity has caused climate change shows humankind's responsibility for most coral reef degradation by either near- or far-field stresses (Bruno et al., 2019; Carpenter et al., 2008; Cinner et al., 2020; Duarte et al., 2020; Hughes et al., 2003; Jackson et al., 2001; Mora and Sale, 2011; Sale, 2011).

The present human population is ~7.8 billion people and ~40% live in tropical countries, many of which have coral reefs. About 80% live within 200 km, and 25% within 50 km of a coastline, resulting in 20% of Earth's population living on just 7% of its land area (Sale et al., 2014). The largest urban conglomerations are also located in the tropics, many at or near the coast, in which live 55% of Indonesians, 48% of Filipinos, 75% of Malaysians, 77% of Cubans, 83% of Bahamians, 40% of Maldivians, 53% of Seychellois and the list goes on (Ritchie and Roser, 2020a). Near- and far-field impacts by dense coastal human populations, often near or adjacent to coral reefs, are abundantly demonstrated in the literature (e.g., Cinner et al., 2020; Jackson et al., 2014; Sale et al., 2011; Sheppard et al., 2010).

Predictions of future human population growth vary and depend on assumptions of societal progress in gender equality, education, and governance (Caldwell, 1979; GBD 2017 Population and Fertility Collaborators, 2018; Lutz, 2010; Speringer et al., 2019). Most models assume that by 2100, long after the >2% annual growth-peak in the 1960s (Lutz et al., 2019), the demographic transition, where natality balances mortality (Fig. 2B; Galor, 2011), will be approached and human population will reach between ~11 billion (UN "medium projection") and 8–9 billion (SSP1, various CEPAM, WIC; Lutz et al., 2019). The UN "high projection", however, suggests an unabated increase to >20 billion. Even if a significant mortality event were to affect humans (e.g., 0.5 billion deaths), the population in 2100 could still reach ~10 billion (Bradshaw and Brook, 2014). At present, an annual net 85 million newly-born humans enter the world's population.

Human societies have been prolific and successful. Never have more people existed or have there been more affluent economies. Never has a smaller share of the world's population lived in abject poverty than over the past decades (Fig. 2; Roser and Ortiz-Ospina, 2020). More people than ever have access to some cash, leading to better nutrition—and therewith greater demand for food (Fig. 3). Although the number of those living in

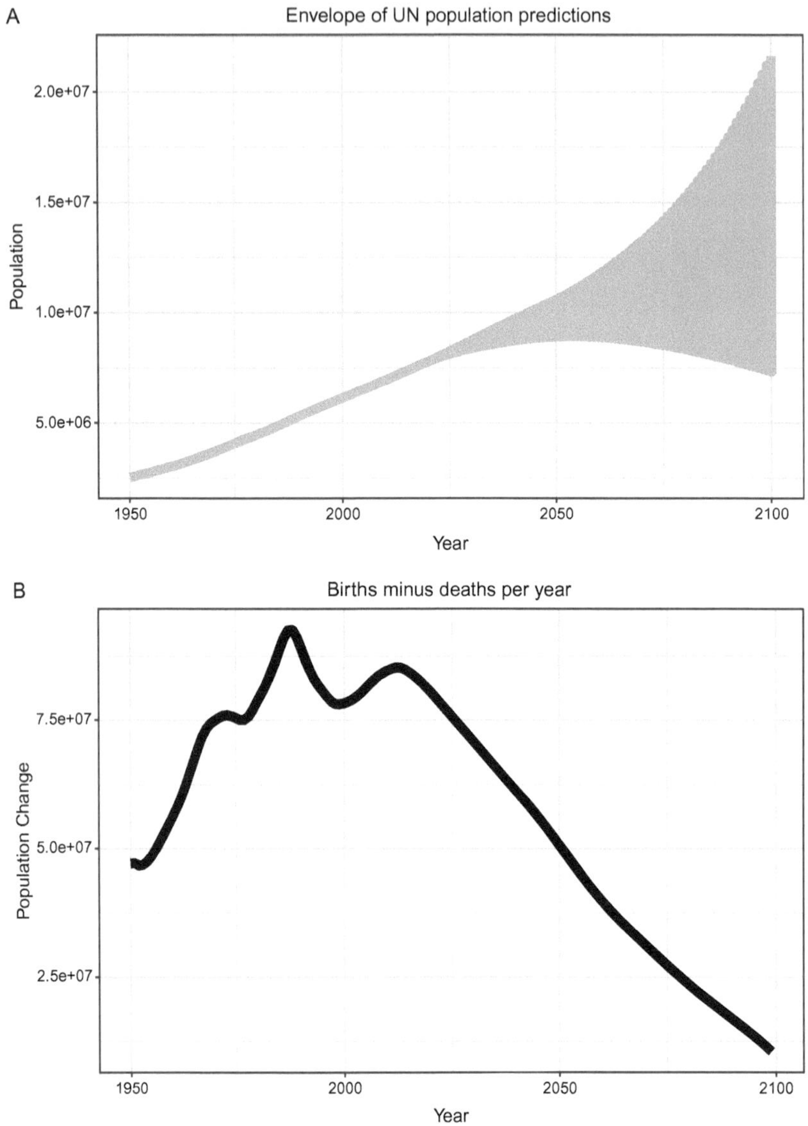

Fig. 2 (A) Human population growth, past and potential future trends. The envelope of the UN model high and low variants includes virtually all other model projections, most of which predict a population size of 8–11 billion by 2100 (Lutz et al., 2019; Roser, 2020) (B) Total population increase is expected to abate in the medium and low scenarios, leading to a decline from the present addition of ~85 million new humans per year (140 million births minus ~58 million deaths) to about 10 million per year in 2100. More than half of that new increase occurs in the tropics (Ritchie and Roser, 2020a,b; Roser, 2020). At the demographic transition, death and birth rates will balance. Whatever the scenario, human population will continue to grow over the next decades and thus exert more pressure on coral reefs. Graphs from data provided online in http://population.un.org/wpp and in Roser (2020).

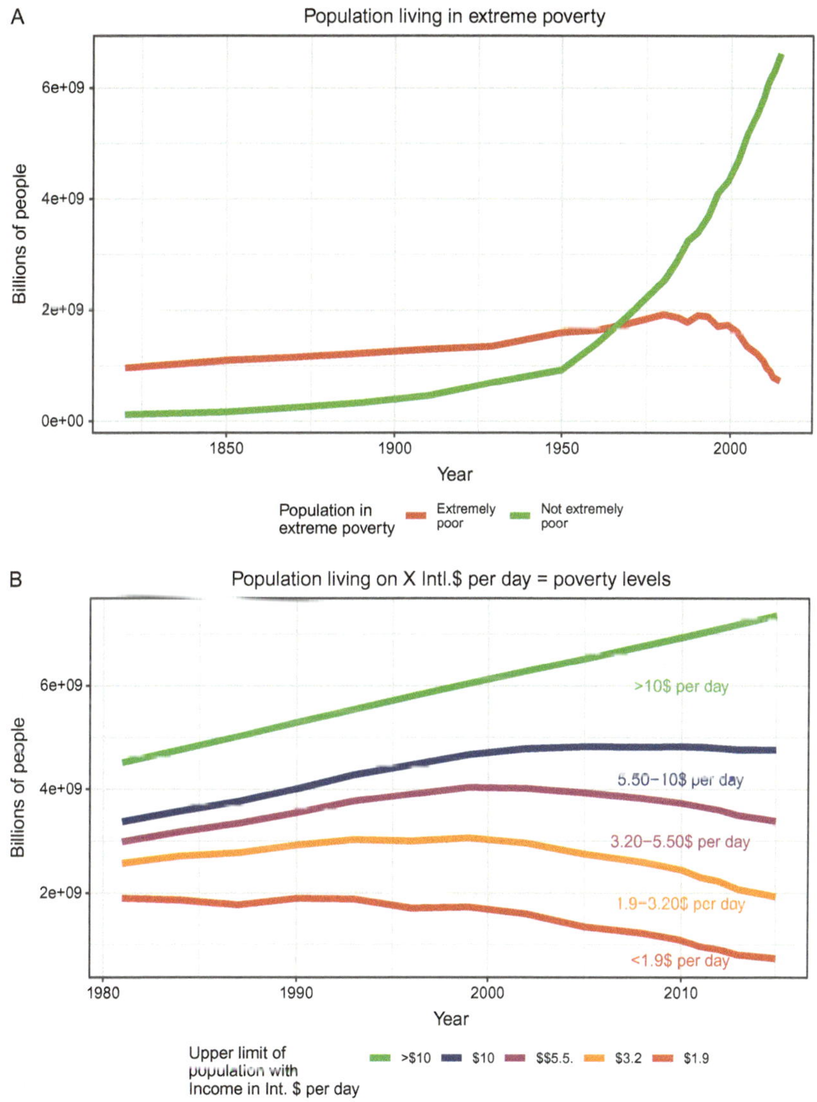

Fig. 3 (A) The total number and share of the human population living in extreme poverty (defined as earning <1 International $ per day) has lately decreased; (B) A greater share of the world's population enjoyed higher earnings than before. However, still ~2 billion people live at <3.2 Intl. $ per day, many of them in the vicinity of coral reefs, which thus represent an important means of sustenance. "Intl. $" is a hypothetical currency unit with comparable purchasing power as the US$ within the United States. *Redrawn from data in Roser, M., Ortiz-Ospina, E., 2020. Global Extreme Poverty. Published online at OurWorldInData.org. Retrieved from: https://ourworldindata.org/extreme-poverty [Online Resource].*

the most crushing poverty (<1.9 Intl. \$ per day) has declined relative to the total population, still ~700 million people are condemned to a life with essentially no money (Fig. 3B). Many are forced to live off the land or sea as much as possible. As poverty abated, wealth also increased, most notably in Eastern Asia and the USA. The world's population of high-net worth individuals (HNWI, with >1 million \$ in investable assets) has grown to ~46 million souls (Capgemini, 2020). Disposable cash drives luxury markets and ~30% of HNWI wealth is in cash. More disposable cash, also in the not "poverty-stricken" and increasingly urbanized population (Fig. 3), creates large and potent markets. An increasing range of coral reef organisms is marketable—either as a food staple for the poorer stratum of society or as novelty items and high-end consumer goods for the wealthier. While poverty will drive local extractive consumption for sustenance (near-field effect), more money available in markets for seafood or bushmeat will incentivize the abundant poor to exploit available natural resources for sale in more distant markets (far-field effect). On coral reefs this latter effect has led to over-harvesting of precious coral, sea cucumbers, shark fins, sea trout (traded both alive and dead) and wrasses (Bruckner, 2014; Rhyne et al., 2014).

An increasing share of the world's population will live in cities, but rural populations will remain dense (Ritchie and Roser, 2020a). In Indonesia, urban centres grow rapidly and the relative share of rural population declines. But that may not reduce human-reef conflict since Indonesia has on average 4.8 people per linear meter of coastline. If 55% of the population is urban, it still leaves at least 2 people per meter of coast outside the urban centres. In the Philippines, which has 2.8 people per meter coastline, rural and urban populations will grow at comparable pace, leading to expected greater pressure on reef resources (Fig. 4). In both countries, seafood consumption is important. Per capita consumption of fish in the Philippines ranged from 30 to 35 $kg\,p^{-1}\,y^{-1}$ of which ~15% was demersal and non-pelagic marine fish—mostly from coral reefs; Indonesia's seafood consumption has recently strongly increased from ~20 $kg\,p^{-1}\,y^{-1}$ in 2000 with one quarter made up by non-pelagic species (Dey, 2008) to nearly 50 $kg\,p^{-1}\,y^{-1}$ in 2020 (Fig. 4). Much of this increase was satisfied by aquaculture (Fig. 4E), which relieves wild stocks from exploitation pressure but is space-intensive and often destructive to natural marine habitats.

In a population of ~7.8 billion in 2020, there were ~7 million reef fishers (Teh et al., 2013), which by 2100 may increase to ~10 million. Since the present number of reef fishers already is unsustainable in many areas,

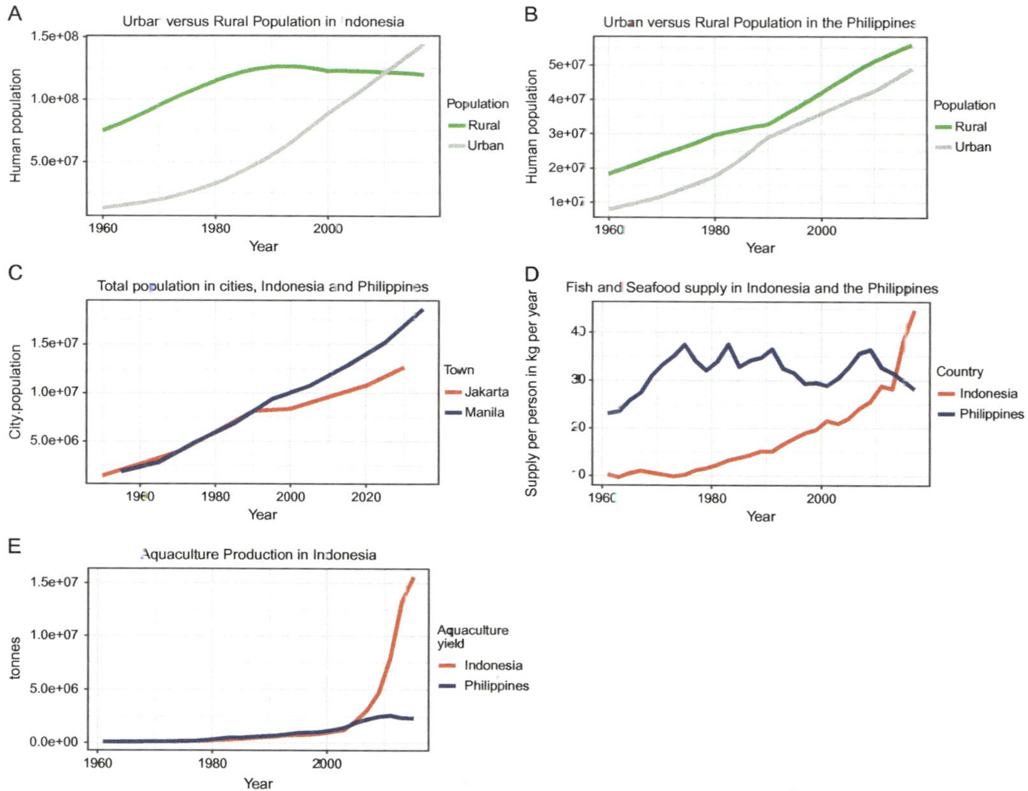

Fig. 4 (A) In Indonesia, more people live in cities than in the country since 2010, generating a powerful market of >120,000,000 people in urban centres. (B) In the Philippines, rural and urban populations grow apace, creating both powerful urban markets and strong local, rural demand for food. (C) Among the biggest urban conglomerations in the two countries, Manila (27% of Philippine urban population) is outgrowing Jakarta (7.2% of Indonesian urban population). (D) The per capita fish and seafood supply in the Philippines is fluctuating at a relatively high level with a recent minor decline, which suggests overharvest. The supply in Indonesia is exponentially increasing, which makes overexploitation of resources highly likely. (E) Aquaculture production in Indonesia has sharply increased and can therefore account for a portion of the increase in fish and seafood supply (data source for all 5 graphs: Ritchie and Roser, 2020b).

another 3 million fishers will only add to the problem of over-exploitation. Additional millions will be employed in aquaculture and may have a similarly strong impact on reefs as fishers. Given the significant market for reef-derived food, wild-caught or farmed, and the large number of rural people available for resource exploitation, managing coral reef resources will be increasingly difficult (Cinner, 2014). Rarity tends to increase price and luxury value of declining resources. Thus, future pressure on coral reef resources will increase in a vicious cycle of more money being offered to more people (fishers and middlemen) to procure more of the sought after, but rarer, items.

Not only direct impacts on reefs will increase. More people inland will translate into more watershed alteration/degradation and increased silt and pollutant transport towards the coast by rivers and canals. Negative impacts of changed hydrographic regimes and sediment loads were experienced on the Great Barrier Reef and the East African coast already with the onset of early European-style farming in the Australian and African highlands (Fleitmann et al., 2007; Roff et al., 2012). These impacts have continued and may likely become more widespread. These far-field pressures augment threats to reefs posed by the near-field pressures like fisheries.

From the above, one may expect a quantifiable link between human population density and coral reef degradation. This is illustrated by examples from the Caribbean and Persian/Arabian Gulf (Fig. 5) where coral cover has declined almost in-step with human population density. The situation is not much different in other places, e.g., the Philippines (Magdaong et al., 2014). While coincidence is not correlation (or causation), there is a suspicion of linkage suggesting a shortfall in management capacity in these regions to control near field impacts from their human population. But even in remote and isolated regions, such as Australia's Great Barrier Reef, the distant reefs and islands of Oceania and the few isolated reef-fringed continental shorelines that still exist, coral decline is evident (Hughes et al., 2018). This suggests a management shortfall of far field effects that relate to basin-wide phenomena. Different from management of near field stressors, local action is useless in the battle against far field stressors like climate change. Caused by the sum of the world's economic activity, pollutant sources that damage reefs often originate at different latitudes on different continents and thus complicate efforts to save a far-off coral reef. While difficult to manage, some of the impacts could be mitigated by efficiently addressing near-field stresses at the local scale, but also that is rarely accomplished.

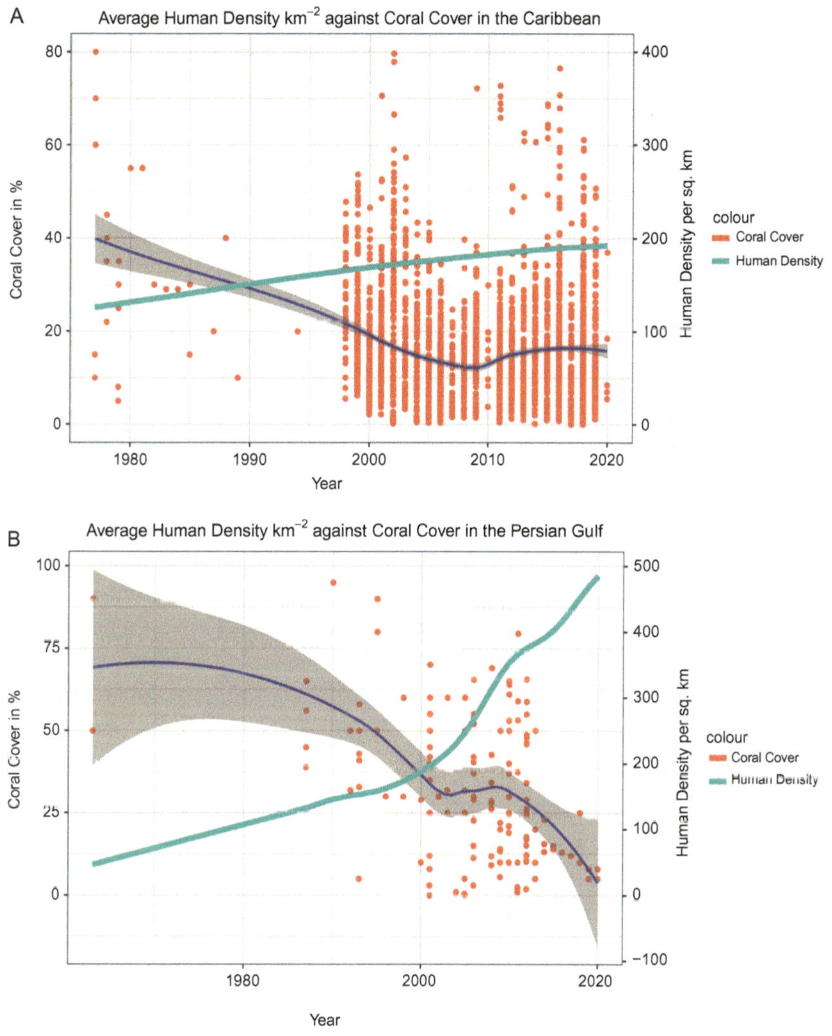

Fig. 5 Trajectory of human population density against coral cover in two well-studied regions where the coral declines are well known (A) the Caribbean and (B) the Persian/Arabian Gulf. Both regions have witnessed coral declines in-step with increased human population density (coral cover data from: Jackson et al., 2014; AGRRA Data Explorer, 2020; Riegl et al., 2018; Human density data from: population.un.org/wpp; Roser, 2020). Red dots = reported coral cover, blue and grey line = locally estimated scatterplot smoother (LOESS, degree = 2) with 95% confidence interval; green line = actual human density in persons km^{-2}.

Where reefs are within easy reach of humans, especially if impoverished, over-exploitation and degradation are often observed. Some of the highest birth rates are found within the tropical reef belt (Somalia with 5.9 children per woman in 2019; Comores with 4.4, Eritrea 3.9, Tanzania 4.8; Mozambique 5; Kenya 3.6; Madagascar 4, Papua New Guinea 3.5; Roser, 2020). Consequently, these countries have some of the lowest median ages (Roser, 2020), which combine with some of the world's highest poverty levels. Reef fisheries, reef mining, and aquaculture are physically demanding activities and mostly an occupation of younger people. Steepest increases in numbers of young people that could take up fishing will occur in East Africa and the Western Indian Ocean. But Africans consume less fish (e.g., Kenya with $2.8\,kg\,y^{-1}$ of which only 4% are marine, FAO, 2020), have fewer fishers (1.5 million; Teh et al., 2013) and are strongly urbanized (Ritchie and Roser, 2020b). In Asia, slower population growth (Malaysia 1.98; Philippines 2.84; Indonesia 2.3 children per woman in 2019; Roser, 2020) is offset by higher per capita fish consumption, much of which is reef-derived (Fig. 3D) and a greater number of fishers (3.35 million, Teh et al., 2013). Therefore, neither slower population growth resulting in fewer new fishers in Asia nor lower consumption of seafood in the burgeoning populations bordering the W-Indian Ocean are likely to alleviate pressure on coral reef resources.

Climate change, clearly a far field driver, has been amply demonstrated to be potentially lethal for coral reef ecosystems via increased heat-induced coral bleaching and mortality, displacement of vagile fauna, or changes in ocean alkalinity that depress calcification of reef organisms and cause behavioural changes in reef fishes (Baker et al., 2008; Bruno et al., 2019; Gattuso et al., 2015; Glynn et al., 2014; Hoegh-Guldberg and Bruno, 2010; Hughes et al., 2018; Pauly, 2010). Increased heat and acidity, alone or in synergy, are far-field drivers likely to influence near-field stresses. If average sizes and latitudinal distribution of fishes changed (Pauly, 2010), then local exploitation dynamics would also change (Cheung et al., 2013). Since climate change is primarily driven by CO_2 emissions, it is directly linked to the size of the human population producing energy, modifying vegetation, and consuming raw materials (IPCC, 2013). Increase in wealth will also translate to ~4 billion motor vehicles by 2050 (World Energy Council, 2011). Even if electrically powered, this electricity must still be produced. A larger and wealthier population will demand more space, more resources and more power. That suggests rather slow reductions, if any at all, in emissions—even with serious political and economic commitment.

It is evident that human populations impact coral reefs and will constrain the latter's persistence with large areas becoming unsuitable for reef development. Human impacts on coral reefs cause (1) increased probability of mortality of individual populations as well as of entire reefs due to direct exploitation, immediate impacts by pollution, siltation, etc.; (2) decreased probability of recolonization, if larval supply or fitness decline, if degraded habitats are incapable of regeneration, or if only a few areas remain to be resettled; (3) loss of habitat by degradation to a point where regeneration is impossible (due to continuous pollution, dredge-and-fill, etc.). Effects of such dynamics can be investigated with the tools of population ecology. These methods are widely published, well-known, well-tested and provide clear indications of what effects human population and perturbation will have on coral reefs.

3. Coral population dynamics in response to human depredation

Besides the urgent conservation imperative, coral reefs are prime systems for study and verification of constructs from the theory of population dynamics—especially regarding the effects of habitat degradation and destruction. Human and natural impacts restrict the area occupied by coral reefs, with some areas degraded for extended periods or low to nil coral cover due to near- and far-field disturbances. Ecological theory can, from first principles, provide guidance regarding expected effects of (1) increased coral mortality, (2) retarded recovery, and (3) habitat destruction.

Coral reefs are typically discrete patches (structures) with well-defined boundaries, mostly surrounded by habitat that can be traversed, but is not suitable for settlement. Reef spacing and connectivity via dispersal (migrants or offspring) occur over a variety of spatial and temporal scales dependent on currents and the biological potential for dispersal. Thus, patch area and isolation effects can be studied, and paradigms such as island biogeography and metapopulation dynamics can be invoked to derive effects expected from human population perturbations (Hanski, 1999; Kritzer and Sale, 2006; Lande, 1993; Mumby and Dytham, 2006; Tilman et al., 1994).

Evidence for net loss of coral habitat is conclusive (Jackson et al., 2014; McClenachan et al., 2017; Sale et al., 2011; Sheppard et al., 2010) as is the potential for extinction of many coral reef species (Carpenter et al., 2008). But coral reef organisms live not only in reef frameworks and cavities, they occupy a wide depth zone in deeper water (the mesophotic zone; Loya et al., 2019) and occur in non-reefal habitats. This may afford them some escape habitats (refuge or refugium; Cacciapaglia and vanWoesik, 2015; Smith et al., 2017; Bongaerts and Smith, 2019) from local impacts, if physically more removed from coastal human populations, and even from sea warming and UV-radiation that cause coral bleaching and mortality of associated fauna (Glynn, 1996; Table 1). Significant living biota that can aid in resettlement of denuded areas exists outside the typical structural coral reef setting (Fig. 6) as well as within traditional reefal habitats (Table 1; Glynn, 1996).

While human impacts reduce habitat by degradation, biota in still functioning escape habitats may assist regeneration. If patches in typical reef habitat are eliminated, they might be recolonized from other reefs or non-reef habitats. If patches are equivalent and the involved species can disperse, their dynamics can be explored by metapopulation theory (Hanski, 1999; Kritzer and Sale, 2006). Biota must be able to occasionally move among patches and across gradients by recruitment (if sessile) or by displacement (if vagile) (Figs. 6C, 7A). In a matrix of habitat patches (Hanski, 1999; Levins, 1969), impacts are represented by changed rates of patch extinction and colonization (parameters c or e, blue and red in Fig. 7B–E), or even by destruction of habitat (see further below). Equilibria with fewer or more occupied patches can arise with changed c or e (Fig. 7B, C), but environmental degradation will result in fewer occupied patches, i.e., fewer living coral reefs (Fig. 7B–E).

Metapopulation theory is concerned primarily with fluctuations in occupancy of patches and is used here for its heuristic value to demonstrate obvious results of human pressure on the general availability of reef habitat. As presented here, few assumptions are made regarding the variable abundance of biota making up the complex and diverse organismic communities of each patch. Also, subtleties of degradation that may lead to changes in the abundance of species are not addressed. But effects of wholesale habitat availability and destruction, modulated by changes in only two general and easily interpreted parameters that can be directly related to human activities, can be conveniently addressed and illustrated.

These models capture different reproductive and dispersal strategies well. For highly fecund organisms that saturate the environment with propagules, the proportion of occupied patches is not important ("propagule rain"; Gotelli, 1991; Fig. 7C). Few occupied patches, e.g., only few surviving reefs, can still allow high (re)settlement rates, which only decline as the environment fills up—while propagules always remain available (Fig. 7C). Many turf and macroalgae, for example, will settle onto any available space and

Table 1 Evidence for escape habitats from near- and far-field degradation drivers. Escape can be due to remoteness from coastal human populations, inaccessibility, economic unattractiveness, or buffering from increased ocean heat-content. Such habitats could potentially serve to maintain connectivity with shallow reefs and might be helpful to replenish shallow reefs, or to act as refuges. Their presence and relative size help to justify the assumption of rescue effects in metapopulation models. The table focuses on corals as the key ecosystem engineers of coral reefs.

Area	Reference	Non-reef escape habitat	Reefal escape habitat
Andaman Sea	Wall et al. (2014), Buerger et al. (2016)		Upwelling by internal waves causes significant bleaching relief
Galápagos	Glynn et al. (2018), Riegl et al. (2019a,b)	Only a single coral framework reef, all others incipient reefs or coral communities	Upwelling by internal waves can suppress bleaching in reefs and incipient reefs
Great Barrier Reef	Harris et al. (2013), Hock et al. (2017), Bridge et al. (2019)	Wide areas of deep reefs on the shelf and shelf-edge that support similar species as shallow reefs; some areas well-connected by low risk of bleaching	
Red Sea, Safaga Bay, Egypt	Riegl and Piller (2003)	20 times more coral habitat in 10–30 m than in <10 m	
Red Sea, Saudi Arabia	Bruckner et al. (2012), Rowlands et al. (2012)	Large areas of non-reefal habitat in shallow and deep water	
Red Sea, Israel	Eyal et al. (2019) Shlesinger et al. (2018)	Many shallow coral species extend into mesophotic zones	Asynchrony between shallow and deep coral spawning plus depressed reproduction in deep habitats make rescue of shallow habitats unlikely
SE South Africa	Porter and Schleyer (2017), Schleyer and Porter (2018), Riegl and Piller (2003)	Most of the reefal fauna >20 m depth	Frequent upwelling suppresses bleaching

Continued

Table 1 Evidence for escape habitats from near- and far-field degradation drivers. Escape can be due to remoteness from coastal human populations, inaccessibility, economic unattractiveness, or buffering from increased ocean heat-content. Such habitats could potentially serve to maintain connectivity with shallow reefs and might be helpful to replenish shallow reefs, or to act as refuges. Their presence and relative size help to justify the assumption of rescue effects in metapopulation models. The table focuses on corals as the key ecosystem engineers of coral reefs.—cont'd

Area	Reference	Non-reef escape habitat	Reefal escape habitat
SE Asia	McManus (1997)	Thousands of sq. km of coral-bearing sediments leads to assumption of large coral populations outside reefs	
Florida Keys and Gulf of Mexico, USA	Hine et al. (2008), Jaap et al. (2008), Schmahl et al. (2008)	Deeper coral communities in Tortugas and Marquesas Islands and Flower Garden Banks consist of comparable species as shallow reefs in Florida Keys; but also unique, deep assemblages occur	
US Virgin Islands	Smith et al. (2019)	At least 3 × as much coral habitat in 30–60 m as in 0–30 m, similar species as in shallow reefs with lower temperatures, but bleaching and diseases also observed	
Uva Island, Panama	Smith et al. (2017)	The slope below the reef served as a refuge for the hydrocoral *Millepora intricata*	Upwelling and cooler bottom waters protect deeper fauna from warmed surface waters
Pearl Islands, Gulf of Panama, Panama	Glynn et al. (2001)	Extensive areas of non-structural reef habitats	Seasonal upwelling periods moderating El Niño warming events
Western Australia	Moore et al. (2017)	20–80 m deep communities contain twice the abundance and 1.4 times the species diversity of shallow habitats	
South China Sea (Dongsha)	Tkachenko and Soong (2017)		Upwelling by internal waves and tides causes significant relief from bleaching

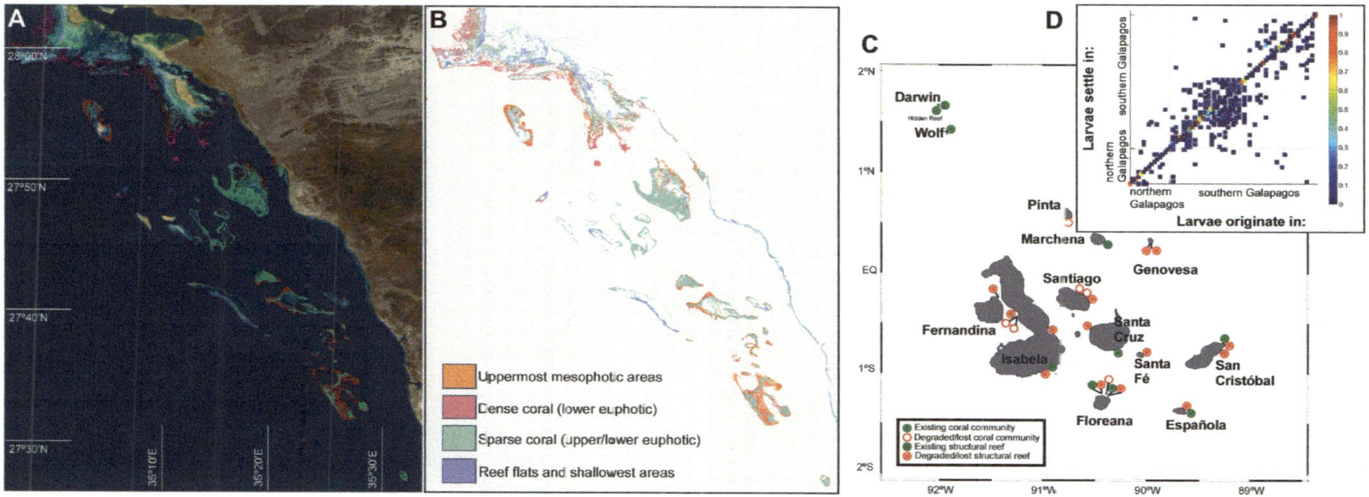

Fig. 6 (A, B) Coral reefs and coral habitats in the NE Red Sea. Fringing reefs are easily impacted by near- and far-field disturbances. Replenishment via dispersal (larvae or migrants) can occur from offshore reefs, remote from many near-field threats. Shallow reef habitats (blue and red) damaged by far-field pressures such as heat-mediated bleaching, can be replenished from deeper habitats (orange) (B) Polygons of reef type. Mesophotic reefs (>50m depth) are not mapped since they are beyond optical resolution of satellite images but they exist at many offshore banks and mainland sites. Maps based on Bruckner et al. (2012), Rowlands et al. (2012). (C) Coral habitats in the Glaápagos Islands (Glynn et al., 2018) demonstrate that reefs, incipient reef frameworks and non-reefal habitats, can form intricate mosaics and be highly connected (D) Connectivity matrix of coral patches in the Glaápagos Islands showing the dispersal of propagules among patches (Riegl et al., 2019a)—a prerequisite for metapopulations.

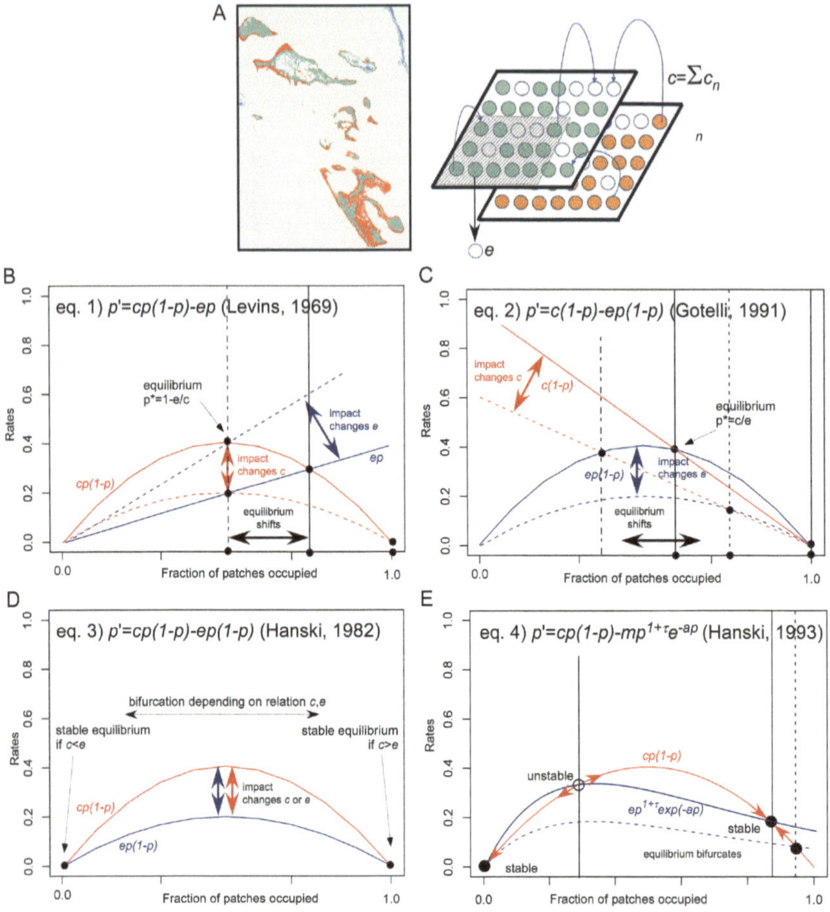

Fig. 7 Dynamics of metapopulation models—equilibria and effects of parameter changes equivalent to human impacts. Parameters are c = colonization rate, e = extinction rate, p = probability of a patch being occupied, a = area, τ = a distance scaler. (A) A reef is a patch in either a different depth layer (as shown here) or further away at the same depth and can be settled according to parameter c and will go extinct according to parameter e. Habitat can also be destroyed (striped area over the light green patches). (B) In the Levins (1969) model, colonization (red) and extinction (blue) curves intersect and a stable equilibrium exists at that point (black circle). Patches will fill up to that fraction, but if either extinction or colonization probabilities are changed, this equilibrium shifts up or down—altering how much of the environment can be settled. (C) A propagule rain exists with a linear colonization function. An abundance of propagules fills up habitat, even at lowest patch occupancy. While the colonization curve does not intercept the mortality curve, equilibrium is reached when all patches are full. Otherwise, the number of settled patches on the x-axis will decline. (D, E) illustrate rescue effects. (D) The core-satellite hypothesis posits habitat filling if colonization probability exceeds that of mortality. Here it is assumed that a large patch (such as a continent), constantly replenishes smaller peripheral patches (such as nearby islands). In (E) the rescue effect acts nonlinearly with patch size and distance. The intersection of colonization and extinction curves determines system behaviour. Stable equilibria are shown as black circles, unstable equilibria as unfilled circles; attracting versus rejecting trajectories are shown by red arrows. In (E), three stable and one unstable equilibrium exist. If disturbances push the population below (i.e., to the left of) the unstable equilibrium, extinction is the only stable outcome (Hanski, 1993).

in some reef fishes such dynamics may lead to recruitment lotteries (Sale, 1978). A reef-scape will sustain more macroalgae patches than usual if their parameter *c* is raised by eutrophication or if absence of grazers reduces *e* (Fig. 7C). If corals depend on macroalgae-free patches for settlement, the habitat will sustain a priori fewer coral patches (reduced *c* for corals). The coral's success in the landscape can therefore be reduced by patch-pre-emption, which lowers *p* (the amount of free habitat) for corals, or by reduced settlement ability (lower *c*), increased mortality (raised *e*) or any combination thereof. Algae, capable of prolific asexual and sexual reproduction (Santelices, 1990), can be described by the propagule rain (Vaz-Pinto et al., 2012) and the model captures well how fleshy algae may dominate (Arnold et al., 2010; Mumby and Dytham, 2006) but other types of models may be invoked for corals and fishes, which have more selective larvae (Glassom et al., 2003).

The Core-Satellite hypothesis (Hanski, 1982; Fig. 7C) softens this harsh dynamics by allowing rescue, which is achieved by extinction rates following a similar, but lower, quadratic function as colonisations (meaning that extinction depends on exchange between settled patches *p* and empty patches $1-p$, Fig. 7D). Migration (or, in very fecund organisms, recruitment) is possible during the same time-step as extinction. Such dynamics are most relevant for organisms such as fishes, that can move among patches and can thus rescue an emptied patch by immigration. Two equilibria exist (an unstable equilibrium at zero, and a stable one at $p=1$, when all habitat is filled) if extinction rate *m* remains lower than colonization rate *c* (Fig. 7D). If this is not so then $p=0$, the totally empty habitat, is stable. Thus, any lowering of parameter *c* or raising of *e* by impacts diminishes the entire metapopulation's chance of persistence and can lead to extinction as the only stable equilibrium. However, so long as patches persist, continuous flow of organisms into degraded patches is possible, even if those should become "sinks" (Hanski, 1999). This emphasizes the importance of escape habitats to mitigate for losses on reefs (Table 1).

The above discussions assume that all patches are equal, which is usually not the case in coral reefs. A "rescue effect" (Brown and Kodrick-Brown, 1977; Hanski, 1999), assumes that in addition to the immediate rescue of a patch by instantaneous migration (or recruitment, if that is possible), larger patches are less likely to disappear and can even more efficiently rescue neighbouring patches via immigrants—either vagile juvenile or adult stages or dispersing larvae. The shape of patch extinction functions with rescue terms can be quadratic (see above; Hanski, 1999), an upward saturating exponential (Mangel, 2006) or more complicated (Hanski, 1999, Fig. 7E). Rescue effects are important when considering the effectiveness of escape habitats as considered earlier—many of which have the potential to persist as refuges or refugia of considerable sizes (Bongaerts and Smith, 2019; Cacciapaglia and vanWoesik, 2015; Smith et al., 2017). Non-traditional reef habitats (mesophotic, coral carpets, etc.; Table 1) can be larger in extent than nearby structural reefs, which could constitute a significant rescue effect under the right circumstances. Thus, the conservation of tropical non-coral reef habitats that contain reefal biota should be given as much priority as that of structural reefs.

The shape of patch mortality curves (blue in Fig. 7), determines the number of stable and unstable equilibria. Bifurcations into alternative equilibria with very different stable patch occupancies may be observed (Fig. 7E). Importantly, some suitable patches will remain empty, even at a stable upper equilibrium (Fig. 7E) and only knowledge of perturbation history allows estimation of a metapopulation's position with regards to its equilibria (Hanski, 1999). Since either densely or sparsely occupied patches are likely to co-occur, some empty habitats (live reef fauna absent) must be expected as part of natural dynamics. This extends the possible dynamics in Fig. 7D, since the population of live reefs can fall to a lower level (more empty patches) before they must converge to the equilibrium at zero. Under a rescue-effect, wholesale metapopulation extinction is more likely with more small and empty patches but once past a threshold, the local rescue effect supports the entire metapopulation—underscoring again the importance of escape habitats (Table 1; Hanski, 1999). This has important implications when considering the cumulative effects of either near- or far-field disturbances on reefal metapopulations—many small impacts annihilating many small patches can endanger the entire metapopulation. While recolonization after impacts is possible, especially if rescue effects are considered (for example, by protected areas serving as large rescue patches), nonlinear responses make the system's path difficult to predict. And while the rescue effect of protected areas is evident, more impacts will threaten the metapopulation since cumulative impacts decrease *c* and increase *e* (Fig. 7). Thus, even without wholesale habitat destruction, the long-term success of metapopulations can be endangered.

In the preceding, only effects of reduced colonization/increased mortality were considered. Outcomes suggested loss of occupied habitat—with fewer living coral reefs a necessary consequence of reduced colonization and increased mortality. But human-induced degradation often reduces patch availability by complete habitat destruction. Introducing an explicit loss-term to Levins' (1969) model (Fig. 7B, replace $cp(1-p)$ with $cp(h-p)$; h (habitat) could also be written as *1-D, where D is the proportion of destroyed habitat*) the equilibrium modifies from $p^* = 1 - e/c$ to $p^* = h - e/c$ (where $h < 1$), one arrives at Levins' rule (Hanski, 1999), which states that the minimum requirement for metapopulation persistence is that the remaining number of habitat patches exceeds the number of suitable but empty patches prior to habitat destruction. Since healthy coral reef systems tend to occupy most available space, only a small number of unsettled patches will naturally exist. From this it follows that metapopulation persistence of reefal biota could be immediately threatened by destruction of a relatively small proportion of total available habitat. Limitations to this rule are discussed by Hanski (1999). Coral reefs are diverse and patchily distributed, so metapopulation dynamics occur across an array of species with different minimum habitat requirements. Filled and empty patches occur on a sub-reef scale, stochasticity is important in nature and not accounted for in the model, and species-specific dispersal capabilities will influence mixing among available sites. Nonetheless, while it may not lend itself as a management tool, the heuristic value of this theory in outlining the immediacy of the threats inherent in wholesale habitat loss is undebated.

In a metapopulation of several interacting species, competition can be increasingly important besides reproductive capacity (in this case p modifies to p_i, where i is the number of considered species; equation 5 in Fig. 8). While healthy corals keep algae well in check and develop competitive hierarchies among each other, once dead their habitat is quickly taken over by other benthos and coral replenishment becomes difficult or is by different species than those that dominated previously. The best competitor can also be an inferior colonizer relative to its immediate competitors. In the previous consideration of algae versus coral, no explicit differences were made other than that patch-dynamics in each taxon could be explored with a different model. The effect of a competitor would have been implicit in e and c. The metapopulation equation can be modified to capture such dynamics and even a rescue effect (Lande, 1993; Mangel, 2006; Tilman et al., 1994) to explore competition-colonization trade-offs with inverse competition-colonization capabilities. In this case, habitat loss D affects the worst colonizer (i.e., the best competitor) first in a system of single-resource competition, such as space in sunlit substrates required by corals. This suggests that extinction debt (Tilman et al., 1994), which is accumulated as a species expands in space in a community primarily due to competitive superiority but not by superior fecundity, is an important factor (Fig. 8).

eq. 5) $$p' = c_i p_i\left(1 - D - \sum_{j=1}^{i} p_j\right) - \frac{m_i p_i}{(1 + a p_i)} - \left(\sum_{j=1}^{i-1} c_j p_j p_i\right)$$

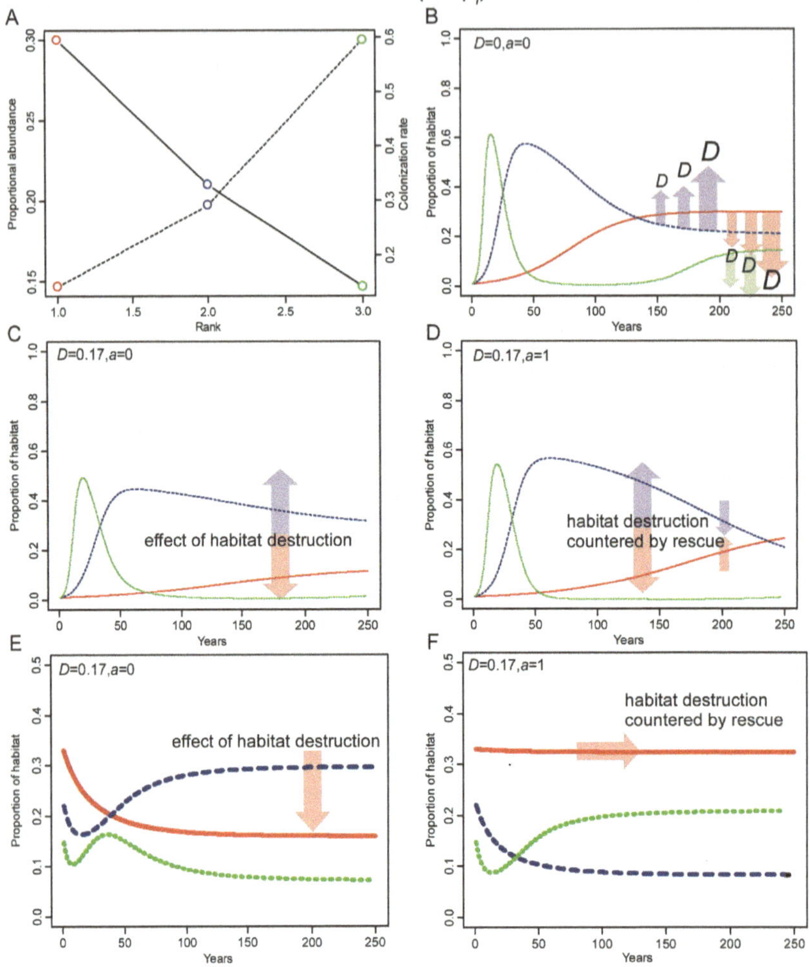

Fig. 8 Effects of habitat destruction (here set at D = 0.17, about half the space cover of the dominant competitor at equilibrium), and a possible offset by a rescue effect. Equation 5 is modified from Tilman et al. (1994) with a rescue effect added. The model is applied to three species competition, two competitively inferior good recruiters (such as *Porites astreoides* and *Undaria agaricties* in the Caribbean; blue) and a competitively dominant, but weak recruiter (such as *Acropora palmata* or *Orbicella annularis* in red). Dynamics are idealized to plausibly explain generally observed patterns and not intended to fit any specific experimental situation; model parameters S = 3 (number of competing species), pd = 0.33 (proportional abundance of dominant species), m = 0.1 (patch mortality) (A) is abundance ranking and colonization rates calculated following Tilman et al. (1994). Solid lines and left y-axis show geometric series of species abundances, broken lines and right y-axis shows inverse correlation of proportional abundance and colonization rate (B). If all habitat is undisturbed then a natural succession will occur. Habitat loss (*D*) increases the deflection of species trajectories in the direction shown by arrows. (C, E) If habitat is lost, the dominant species is disadvantaged first (red line lower, blue higher). (C) Trajectory starting with empty habitat at Year = 0, (E) trajectory starting at equilibrium level (Year = 250 in Fig. 7A) demonstrates decline of dominant species (red line); (D, F) Rescue effect reduces the decline (*a* = 1 reduces mortality by 50%). (D) Trajectory starting with empty habitat at Year = 0, (F) trajectory starting at equilibrium (Year = 250 in Fig. 7A). Due to rescue, the dominant species remains (flat red line). Unfortunately, this situation is hypothetical and has not been demonstrated to exist in Caribbean corals.

Such dynamics are recognized in some coral species, especially those that dominate (or previously dominated) in certain habitats. Corals such as *Acropora palmata* or *Orbicella annularis* dominated prominent Caribbean reef zones and are competitively superior to most other corals in these zones. They have been shown to rarely recruit and have low recruitment survival (Baums et al., 2005; Edmunds, 2015; Williams et al., 2008). Benefits from asexual colony spread by fragmentation (Foster et al., 2007) are limited to local dynamics and do not help long-distance dispersal. With suitable habitat having been decreased by bleaching events and direct destruction (see also Fig. 5), the above-mentioned coral species seem now to be paying their extinction debt. Persian Gulf *Acropora downingi* is superior in linear competitive networks and used to be a strong recruiter. But due to regional population depletion and reduced connectivity they have lately become weak colonizers (Burt and Bauman, 2019; Riegl et al., 2018). Where this species still occurs, it outcompetes others, but a combination of reliance on asexual reproduction and the now regionally reduced availability of sexual propagules suggests that due to altered metapopulation dynamics, a heavy extinction debt has occurred.

But such species can be supported by rescue effects (Fig. 8D, F), making habitats that maintain large populations for re-supply of depleted areas very important. Closer proximity will increase recruitment success (made explicit by parameter a in Fig. 7E). This suggests that deep-water patches may help rescue shallower sites (Holstein et al., 2015)—but this will only be effective if impacts are stratified by depth and if connectivity is possible (which is disputed for corals by Shlesinger et al., 2018). Large conservation areas will be vastly preferential over smaller ones, but the reality of expanding human populations needing more space will make such conservation imperatives increasingly difficult to defend.

4. Conclusion

Metapopulation theory draws attention to the effects of declining colonization rates with reduced numbers of occupied reef sites (living reef patches), the potential for rescue and resultant need to preserve large patches, and how changes in mortality and colonization schedules alter the abundance of a species or an entire assemblage across a region. It allows evaluation of species-specific effects of habitat loss depending on colonization/competition trade-offs, highlighting such important aspects as extinction debt. It also clearly points out that increasing use of space by a competitor of expanding abundance with prodigious and sustained fertility, such as the expanding human population, can only lead to reductions in the number of living coral reefs. If thresholds of minimum viable metapopulation sizes (which can be calculated via incidence functions, Hanski, 1999) are crossed, regional extinctions will follow. While the models shown here will hardly be suitable as management tools for specific regions, for example, to ask when and where tipping points will be met, they serve as quantitative tools to heuristically demonstrate the threat posed by an expanding human population to the very survival of coral reefs.

Although not spatially explicit, these models recall the critical importance of protected areas, or of landscape/seascape protection in general. Such protection can be incidental, if biota is removed from degradation pressures (Table 1) or by legal design. Until a demographic transition is achieved by humans, every year more humans will require more space and resources and will therewith consistently increase near- and far-field impacts. This unfolding scenario does not appear to leave coral reefs, or any natural ecosystem that requires undisturbed space, with many outcomes; concerted societal action will be their only hope of survival. This must include decisions as to a desirable and sustainable maximum level of the global human population, and how this can be achieved in a fair and equitable manner. Realistic policies and rapid action must follow, otherwise humankind will continue to degrade and accelerate the destruction of planet Earth's ecosystems (Sale, 2011).

Acknowledgements

Special thanks are due K. Marks for providing digests of AGRRA Data for Fig. 4 and to P. Sale, P. Mumby, S.C.C. Steiner, and K. Marks for helpful comments and review.

References

AGRRA Data Explorer, https://www.agrra.org/data-explorer/, 2020.

Arnold, S.N., Steneck, R.S., Mumby, P.J., 2010. Running the gauntlet: inhibitory effects of algal turfs on the processes of coral recruitment. Mar. Ecol. Progr. Ser. 414, 91–105.

Baker, A.C., Glynn, P.W., Riegl, B., 2008. Climate change and coral reef bleaching: an ecological assessment of long-term impacts, recovery trends and future outlook. Estuar. Coastal Shelf. Sci. 80, 435–471.

Baums, I.B., Miller, M.W., Hellberg, M.E., 2005. Regionally isolated populations of an imperiled Caribbean coral, *Acropora palmata*. Mol. Ecol. 14, 1377–1390.

Bongaerts, P., Smith, T.B., 2019. Beyond the "deep reef Refuge" hypothesis: a conceptual framework to characterize persistence at depth. In: Loya, Y., Puglise, K., Bridge, T. (Eds.), Mesophotic Coral Ecosystems. In: Coral Reefs of the World, vol. 12, pp. 881–896.

Boyd, C., Slaymaker, T., 2000. Re-examining the 'more people less erosion' hypothesis: special case or wider trend? Nat. Resour. Perspect. 63, 6.

Bradshaw, C.J.A., Brook, B.W., 2014. Human population reduction is not a quick fix for environmental problems. Proc. Nat. Acad. Sci. U. S. A 111 (46), 16610–16615.
Bridge, T.C.L., Beaman, R.J., Bongaerts, P., Muir, P., Ekins, M., Sih, T., 2019. The Great Barrier Reef and Coral Sea. In: Loya, Y., Puglise, K.A., Bridge, T.C.L. (Eds.), Mesophotic Coral Ecosystems. Coral Reefs of the World. In: vol. 12. Springer, Cham, pp. 351–386.
Brown, B.E., 1987. Worldwide death of corals—Natural cyclical events or man-made pollution? Mar. Pollut. Bull. 18, 9–13.
Brown, J.H., Kodrick-Brown, A., 1977. Turnover rates in insular biogeography: effects of immigration and extinction. Ecology 58, 445–449.
Bruckner, A.W., 2014. Advances in management of precious corals in the family Corallidae: are new measures adequate? Curr. Opin. Environ. Sust. 7, 1–8.
Bruckner, A.W., Rowlands, G.P., Riegl, B., Purkis, S.J., Williams, A., Renaud, P., 2012. Atlas of Red Sea Habitats. Khaled bin Sultan Living Oceans Foundation, Maryland. 261 pp.
Bruno, J.B., Côté, I.M., Toth, L.T., 2019. Climate change, coral loss, and the curious case of the parrotfish paradigm: why don't MPAs improve resilience? Ann. Rev. Mar. Sci. 11, 307–334.
Buddemeier, R.W., 1992. Corals, climate and conservation. In: Proceedings of the 7th International Coral Reef Symposium, 1, pp. 3–10.
Buerger, P., Schmidt, G.M., Wall, M., Held, C., Richter, C., 2016. Temperature tolerance of the coral *Porites lutea* exposed to simulated large amplitude internal waves (LAIW). J. Exp. Mar. Biol. Ecol. 471, 232–239.
Burt, J.A., Bauman, A.G., 2019. Suppressed coral settlement following mass bleaching in the southern Persian Gulf. Aquat. Ecosyst. Health Manag. https://doi.org/10.1080/14634988.2019.
Cacciapaglia, C., vanWoesik, R., 2015. Reef-coral refugia in a rapidly changing ocean. Glob. Chang. Biol. 21 (6), 2272–2282.
Caldwell, J.C., 1979. Education as a factor in mortality decline: an examination of Nigerian data. Popul. Stud. 33 (3), 395–413.
Capgemini RBC Wealth Management, 2020. World Wealth Report 2019. World wealthreport.com.
Carpenter, K.E., Abrar, M., Aeby, G., Aronson, R.B., Banks, S., et al., 2008. One third of reef-building corals face elevated extinction risk from climate change and local impacts. Science 321, 560–563.
Cheung, W.W.L., Watson, R., Pauly, D., 2013. Signature of ocean warming in global fisheries catch. Nature 497, 365–368.
Cinner, J.E., 2014. Coral reef livelihoods. Curr. Opin. Env. Sust. 7, 65–71.
Cinner, J.E., Zamborain-Mason, J., Gurney, G.G., Graham, N.A.J., MacNeil, A., et al., 2020. Meeting fisheries, ecosystem function, and biodiversity goals in a human-dominated world. Science 368, 307–311.
De Sherbinin, A., Carr, D., Cassels, S., Jiang, L., 2007. Population and evironment. Annu. Rev. Env. Resour. 32, 345–373.
Dey, M.M., 2008. Changing demand and supply structure of fish in Asia: past trends and future projection. In: Briones, R.M., Garcia, A.G. (Eds.), Poverty Reduction Through Sustainable Fisheries. ISEAS-SEARCA, pp. 40–50.
Duarte, C.M., Agusti, S., Barbier, E., Britten, G.L., Castilla, J.C., et al., 2020. Rebuilding marine life. Nature 580, 39–51.
Edmunds, P.J., 2015. A quarter-century demographic analysis of the Caribbean coral *Orbicella annularis*, and projections of population size over the next century. Limnol. Oceanogr. 60, 840–855.
Ehrlich, P.R., 1971. Impact of population growth. Science 171, 3977. https://doi.org/10.1126/science.171.3977.1212.

Eyal, G., Tamir, R., Kramer, N., Eyal-Shaham, L., Loya, Y., 2019. The Red Sea: Israel. In: Loya, Y., Puglise, K.A., Bridge, T.C.L. (Eds.), Mesophotic Coral Ecosystems. In: Coral Reefs of the World, vol. 12. Springer, pp. 199–214.

FAO, 2020. Fishery and aquaculture country profiles. Kenya (2016). Country profile fact sheet. In: FAO Fisheries and Aquaculture Department [online]. Rome. Updated 2016, http://www.fao.org/fishery.

Fleitmann, D., Dunbar, R.B., McCulloch, M., et al., 2007. East African soil erosion recorded in a 300-year old coral colony from Kenya. Geophys. Res. Lett. https://doi.org/10.1029/2006GL028525.

Foster, N.L., Baums, I.B., Mumby, P.J., 2007. Sexual vs. asexual reproduction in an ecosystem engineer: the massive coral *Montastrea annularis*. J. Anim. Ecol. 76, 384–391.

Galor, O., 2011. The demographic transition: causes and consequences. Cliometrica 6, 1–28.

Gattuso, J.P., Magnan, A., Billé, R., Cheung, W.W.L., Howes, E.L., et al., 2015. Contrasting futures for ocean and society from different anthropogenic CO_2 emission scenarios. Science 349 (6243), aac4722.

GBD 2017 Population and Fertility Collaborators, 2018. Population and fertility by age and sex for 195 countries and territories, 1950–2017: a systematic analysis for the Global Burden of Disease Study 2017. Lancet 392, 1995–2051. https://doi.org/10.1016/S0140-6736(18)32278-5.

Gilmour, J.P., Smith, L.D., Heyward, A.J., Baird, A.H., Pratchett, M.S., 2013. Recovery of an isolated coral reef system following severe disturbance. Science 340 (6128), 69–71.

Glassom, D., Zakai, D., Chadwick-Furman, N.E., 2003. Coral recruitment: a spatio-temporal analysis along the coastline of Eilat, northern Red Sea. Mar. Biol. 144, 641–651.

Glynn, P.W., 1983. Extensive "bleaching" and death of reef corals on the Pacific coast of Panama. Environ. Conserv. 10, 149–154.

Glynn, P.W., 1984. Widespread coral mortality and the 1982–83 El Niño warming event. Environ. Conserv. 11, 133–146.

Glynn, P.W., 1996. Coral reef bleaching: facts, hypotheses and implications. Glob. Chang. Biol. 2, 495–509.

Glynn, P.W., Maté, J.L., Baker, A.C., Calderon, M.O., 2001. Coral bleaching and mortality in Panama, and Ecuador during the 1997-98 El Niño-Southern Oscillation event: spatial-temporal patterns and comparisons with the 1982-1983 event. Bull. Mar. Sci. 69, 79–109.

Glynn, P.W., Enochs, I.C., Afflerbach, J.A., Brandtneris, V., Serafy, J.E., 2014. Eastern Pacific reef fish responses to coral recovery following El Niño disturbances. Mar. Ecol. Progr. Ser. 495, 233–242.

Glynn, P.W., Riegl, B., Purkis, S., Kerr, J., Smith, T., 2015. Coral reef recovery in the Galapagos: the northernmost islands (Wenman, Darwin). Coral Reefs 34, 421–436.

Glynn, P.W., Feingold, J.S., Baker, A., Banks, S., Baums, I., et al., 2018. Status of corals and coral reefs of the Galapagos Islands (Ecuador): past, present, and future. Mar. Pollut. Bull. 133, 717–733.

Gotelli, N., 1991. Metapopulation models: the rescue effect, the propagule rain, and the core-satellite hypothesis. Am. Nat. 138 (3), 768–776.

Hanski, I., 1982. Dynamics of regional distribution: the core and satellite species hypothesis. Oikos 38, 210–221.

Hanski, I., 1993. Dynamics of small mammals on islands. Ecography 16, 372–375.

Hanski, I., 1999. Metapopulation Ecology. Oxford University Press, Oxford. 313 pp.

Hardin, G., 1968. The tragedy of the commons. Science 162, 1243–1248.

Harris, P.T., Bridge, T.C., Beaman, R.J., Webster, J.M., Nichol, S.L., et al., 2013. Submerged banks in the Great Barrier Reef, Australia, greatly increase available coral reef habitat. ICES J. Mar. Sci. 70, 284–293.

Hine, A.C., Halley, R.B., Locker, S.D., Jarrett, B.D., Jaap, W.C., et al., 2008. Coral reefs, present and past, on the West Florida shelf and platform margin. In: Riegl, B., Dodge, R.E. (Eds.), Coral Reefs of the USA, Coral Reefs of the World. vol. 1, pp. 127–173.

Hock, K., Wolff, N.H., Ortiz, J.C., Condie, S.A., Anthony, K.R.N., et al., 2017. Connectivity and systemic resilience of the Great Barrier Reef. PLoS Biol. 15 (11), e2003355. https://doi.org/10.1371/journal.pbio.2003355.

Hoegh-Guldberg, O., Bruno, J.F., 2010. The impact of climate change on the world's marine ecosystems. Science 328 (5985), 1524–1528.

Holstein, D.M., Smith, T.B., Gyory, J., Paris, C.B., 2015. Fertile fathoms: deep reproductive refugia for threatened shallow corals. Sci. Rep. 5, 12407. https://doi.org/10.1038/srep12407.

Hubbard, D.K., Rogers, C.S., Lipps, J.H., Stanley Jr., G.D., 2016. Coral reefs at the crossroads. In: Coral Reefs of the World. vol. 6. Springer, Cham. 300 pp.

Hughes, T.P., 1994. Catastrophes, phase shifts, and large-scale degradation of a Caribbean coral reef. Science 265, 1547–1551.

Hughes, T.P., Baird, A.H., Bellwood, D.R., Card, M., Connelly, S.R., et al., 2003. Climate change, human impacts, and the resilience of coral reefs. Science 301 (5635), 929–933.

Hughes, T.P., Kerry, J.T., Álvarez-Noriega, M., Álvarez-Romero, J.G., Anderson, K.D., et al., 2017. Global warming and recurrent mass bleaching of corals. Nature 543 (7645), 373–377.

Hughes, T.P., Anderson, K.D., Connelly, S.R., Heron, S.F., Kerry, J.T., et al., 2018. Spatial and temporal patterns of mass bleaching of corals in the Anthropocene. Science 359 (6371), 80–83.

IPCC, 2013. AR5 Synthesis Report: Climate Change 2014. www.ipcc.ch/report.

Jaap, W.C., Szmant, A., Jaap, K., Dupont, J., Clarke, R., et al., 2008. A perspective on the biology of Florida Keys coral reefs. In: Riegl, B., Dodge, R.E. (Eds.), Coral Reefs of the USA, Coral Reefs of the World. In: vol. 1. Springer, Berlin, pp. 75–125.

Jackson, J.B.C., Kirby, M.X., Berger, W.H., Bjorndal, K.A., Botsford, L.W., et al., 2001. Historical overfishing and the recent collapse of coastal ecosystems. Science 293, 629–638.

Jackson, J.B.C., Donovan, M.K., Cramer, K.L., Lam, V.V. (Eds.), 2014. Status and Trends of Caribbean Coral Reefs: 1970–2012. Global Coral Reef Monitoring Network, IUCN, Gland, Switzerland.

Johnson, D.G., 2000. Population, food and knowledge. Am. Econ. Rev. 90 (1), 1–14.

Kritzer, J.P., Sale, P.F., 2006. Marine Metapopulations. Academic Press-Elsevier, Amsterdam. 544 pp.

Lande, R., 1993. Risks of population extinction from demographic and environmental stochasticity and random catastrophes. Am. Nat. 142, 911–927.

Levins, R., 1969. Some demographic and genetic consequences of environmental heterogeneity for biological control. Bull. Entomol. Soc. Am. 15, 237–240.

Loya, Y., Puglise, K.A., Bridge, T.C.L., 2019. Mesophotic coral ecosystems. In: Coral Reefs of the World. vol. 12. Springer. 1001 pp.

Lutz, W., 2010. Education will be at the heart of 21st century demography. In: Vienna Yearbook of Population Research. vol. 8, pp. 9–16. https://doi.org/10.1553/populationyearbook2010s9.

Lutz, W., Goujon, A., Samir, K.C., Stonawski, M., Stilianakis, N. (Eds.), 2019. Demographic and Human Capital Scenarios for the 21st Century: 2018 Assessment for 201 Countries. Publications Office of the European Union, Luxembourg, EUR 29113. https://doi.org/10.2760/41776.

Magdaong, E., Yamano, H., Masahiko, F., 2014. Development of a large-scale, long-term coral cover and disturbance database in the Philippines. In: Yahara, T., Nakashizuka, T. (Eds.), Nakano, S. Integrative Observations and Assessments, Springer-Cham, pp. 83–109.

Malakoff, D., 2011. Are more people necessarily a problem? Science 333 (6042), 544–546.
Mangel, M., 2006. The Theoretical Biologist's Toolbox. Cambridge University Press. 375 pp.
McClenachan, L., O'Connor, G., Neal, B.P., Pandolfi, J.M., Jackson, J.E.B., 2017. Ghost reefs: nautical charts document large spatial scale of coral reef loss over 240 years. Sci. Adv. 3 (9), e1603155. https://doi.org/10.1126/sciadv.1603155.
McManus, J., 1997. Tropical marine fisheries and the future of coral reefs: a brief review with emphasis on Southeast Asia. Coral Reefs 16 (Suppl), S121–S127.
Meadows, D.H., Meadows, D.L., Randers, J., Behrens III, W.W., 1972. The Limits of Growth. Potomac Associates—Universal Books. 205 pp.
Moore, C., Cappo, M., Radford, B., Heyward, A., 2017. Submerged oceanic shoals of North Western Australia are a major reservoir of marine biodiversity. Coral Reefs 37, 719–734.
Mora, C., Sale, P.F., 2011. Ongoing global biodiversity loss and the need to move beyond protected areas: a review of the technical and practical shortcomings of protected areas on land and sea. Mar. Ecol. Progr. Ser. 434, 251–266.
Mumby, P., Dytham, C., 2006. Metapopulation dynamics of hard corals. In: Kritzer, J.P., Sale, P.F. (Eds.), Marine Metapopulations. Academic Press—Elsevier, Amsterdam, pp. 157–203.
Mumby, P.J., Steneck, R.S., 2008. Coral reef management and conservation in light of rapidly evolving ecological paradigms. Trends Ecol. Evol. 23 (10), 555–563.
Pandolfi, J.M., Jackson, J.B.C., Baron, N., Bradbury, R.H., Guzman, H.M., et al., 2005. Are US coral reefs on the slippery slope to slime? Science 307 (5716), 1725–1726.
Pauly, D., 2010. Gasping fish and panting squids: oxygen, temperature and the growth of water-breathing animals. Excell. Ecol. 22, 216.
Porter, S.N., Schleyer, M.H., 2017. Long-term dynamics of a high-latitude coral reef community at Sodwana Bay, South Africa. Coral Reefs 36, 369–382.
Rhyne, A.L., Tlusty, M.F., Kaufman, L., 2014. Is sustainable exploitation of coral reefs possible? A view from the standpoint of the marine aquarium trade. Curr. Opin. Environ. Sust. 7, 101–107.
Riegl, B., Piller, W.E., 2003. Possible refugia for reefs in times of environmental stress. Int. J. Earth Sci. 92 (4), 520–531.
Riegl, B., Bruckner, A., Coles, S., Renaud, P., Dodge, R.E., 2009. Coral reefs—threats and conservation in an era of global change. Ann. N. Y Acad. Sci. 1162, 136–186.
Riegl, B., Johnston, M., Purkis, S., Howells, E., Burt, J., Steiner, S.C., Sheppard, C.R., Bauman, A., 2018. Population collapse dynamics in *Acropora downingi*, and Arabian/Persian Gulf ecosystem engineering coral, linked to rising temperature. Glob. Chang. Biol. 24 (6), 2447–2462.
Riegl, B., Johnston, M., Glynn, P.W., Keith, I., Rivera, F., et al., 2019a. Some environmental and biological determinants of coral richness, resilience and reef building in Galapagos (Ecuador). Sci. Rep. 9, 10322. www.nature.com/articles/s41598-019-46607-9.
Riegl, B., Glynn, P.W., Banks, S., Keith, I., Rivera, F., Vera-Zambrano, M., d'Angelo, C., Wiedenmann, J., 2019b. Heat attenuation and nutrient delivery by localized upwelling avoided coral bleaching mortality in northern Galapagos during 2015/16 ENSO. Coral Reefs 38, 773–785.
Risk, M.J., 2014. Assessing the effects of sediments and nutrients on coral reefs. Curr. Opin. Environ. Sust. 7, 108–117.
Ritchie, H., Roser, M., 2020a. Urbanization. Published online atOurWorldInData.org. Retrieved from, https://ourworldindata.org/urbanization. (Online Resource).
Ritchie, H., Roser, M., 2020b. Seafood Production. Published online at OurWorldInData.org. Retrieved from, https://ourworldindata.org/urbanization. (Online Resource).
Roff, G., Clark, T.R., Reymond, C.E., Zhao, J.-X., Feng, Y., et al., 2012. Paleoecological evidence of a historical collapse of corals at Pelorus Island, inshore GBR, following European settlement. Proc. R. Soc. B 280, 20122100.

Roser, M., 2020. Fertility Rate. Published online at OurWorldInData.org Retrieved from, https://ourworldindata.org/fertility-rate. (Online Resource).

Roser, M., Ortiz-Ospina, E., 2020. Global Extreme Poverty. Published online at OurWorldInData.org Retrieved from, https://ourworldindata.org/extreme-poverty. (Online Resource).

Rowlands, G., Purkis, S., Riegl, B., Metsamaa, L., Bruckner, A., Renaud, P., 2012. Satellite imaging coral reef resilience at regional scale: a case study from Saudi Arabia. Mar. Pollut. Bull. 64, 1222–1237.

Sale, P.F., 1978. Coexistence of coral reef fishes—a lottery for living space. Environ. Biol. Fishes 3, 85–102.

Sale, P.F., 2008. Management of coral reefs: where we have gone wrong and what we can do about it. Mar. Pollut. Bull. 56, 805–809.

Sale, P.F., 2011. Our Dying Planet: An ecologist's View of the Crisis we Face. University of California Press, Berkeley. 350 pp.

Sale, P.F., Feary, D.A., Burt, J.A., Bauman, A.G., Cavalcante, G.H., Drouillard, K.G., Kjerfve, B., Marquis, E., Trick, C.G., Usseglio, P., van Lavieren, H., 2011. The growing need for sustainable ecological management of marine communities of the Persian Gulf. Ambio 40, 4–17.

Sale, P.F., Agardy, T., Ainsworth, C.H., Feist, B.E., Bell, B.E., et al., 2014. Transforming management of tropical coastal seas to cope with challenges of the 21st century. Mar. Pollut. Bull. 85, 8–23.

Santelices, B., 1990. Patterns of reproduction, dispersal and recruitment in seaweeds. Oceanogr. Mar. Biol. Ann. Rev. 28, 177–276.

Schleyer, M.H., Porter, S.N., 2018. Chapter one—drivers of soft and stony coral community distribution on the high-latitude coral reefs of South Africa. Adv. Mar. Biol. 80, 1–55.

Schmahl, G.P., Hickerson, E.L., Precht, W.F., 2008. Biology and ecology of coral reefs and coral communities in the Flower Garden Banks region, Northwestern Gulf of Mexico. In: Riegl, B., Dodge, R.E. (Eds.), Coral Reefs of the USA, Coral Reefs of the World. In: vol. 1. Springer, Berlin, pp. 221–261.

Sheppard, C.R.C., 2016. Coral reefs in the Gulf are mostly dead now, but can we do anything about it? Mar. Pollut. Bull. 105 (2), 593–598.

Sheppard, C.R.C., Al-Hoseine, M., Al-Jamali, F., Al-Yamani, F., et al., 2010. The Persian/Arabian Gulf: a young sea in decline. Mar. Pollut. Bull. 60, 13–38.

Shlesinger, T., Loya, Y., 2019. Breakdown in spawning synchrony: a silent threat to coral persistence. Science 365, 1002–1007.

Shlesinger, T., Grinblat, M., Rapuano, H., Amit, T., Loya, Y., 2018. Can mesophotic reefs replenish shallow reefs? Reduced coral reproduction casts a doubt. Ecology 99, 421–437.

Smith, T.B., Mate, J., Gyory, J., 2017. Thermal refuges and refugia for stony corals in the Eastern Tropical Pacific. In: Glynn, P.W., Manzello, D.P., Enochs, I.C. (Eds.), Coral Reefs of the Eastern Tropical Pacific. In: Coral Reefs of the World, vol. 8, pp. 501–517.

Smith, T.B., Brandt, M.E., Brandtneris, V.W., Ennis, R.S., Groves, S.H., et al., 2019. The United States Virgin Islands. In: Loya, Y., Puglise, A., Bridge, T.C.L. (Eds.), Mesophotic coral ecosystems, Coral Reefs of the World. In: vol. 12. Springer, Berlin, pp. 131–147.

Speringer, M., Goujon, A., Samir, K.C., Potančoková, M., Reiter, C., et al., 2019. Global Reconstruction of Educational Attainment, 1950 to 2015: Methodology and Assessment. Vienna Institute of Demography, Vienna, Austria.

Teh, L.S.L., Teh, L.C.L., Sumaila, U.R., 2013. A global estimate of the number of coral reef fishers. PLOS One 8 (6), e65397. https://doi.org/10.1371/journal.pone.0065397.

Tilman, D., May, R.M., Lehman, C.L., Nowak, M.A., 1994. Habitat destruction and the extinction debt. Nature 371, 65–66.

Tkachenko, K.S., Soong, K., 2017. Dongscha Atoll: a potential thermal refuge for reef-building corals in the South China Sea. Mar. Env. Res. 127, 112–115.

Vaz-Pinto, F., Olabarria, C., Arenas, F., 2012. Propagule pressure and functional diversity: interactive effects on a macroalgal invasion process. Mar. Ecol. Prog. Ser. 471, 51–60.

Wall, M., Putchim, L., Schmidt, G.M., Jantzen, C., Khokiattiwong, S., Richter, C., 2014. Large-amplitude internal waves benefit corals during thermal stress. Proc. R. Soc. B 282, 20140650.

Weber, H., Sciubba, J.D., 2019. The effect of population growth on the environment: evidence from European regions. Eur. J. Popul. 35 (2), 379–402.

Wilkinson, C.R., 1992. Coral reefs of the world are facing widespread devastation: can we prevent this through sustainable management practices? In: Proceedings of the 7th International Coral Reef Symposium, Guam. vol. 1, pp. 11–21.

Williams Jr., E.H., Bunkley-Williams, L., 1990. The world-wide coral reef bleaching cycle and related sources of coral mortality. Atoll Res. Bull. 335, 1–71.

Williams, D.E., Miller, M.W., Kramer, K.L., 2008. Recruitment failure in Florida Keys *Acropora palmata*, a threatened Caribbean coral. Coral Reefs 27, 697–705.

World Energy Council, 2011. Global Transport Scenarios 2050. World Energy Council, London. ISBN: 978-0-946121-14-4.

CHAPTER TWO

Projected shifts in coral size structure in the Anthropocene

Chiara Pisapia[a,b,*], Peter J. Edmunds[a], Holly V. Moeller[c], Bernhard M. Riegl[d], Mike McWilliam[e], Christopher D. Wells[f], and Morgan S. Pratchett[g]

[a]Department of Biology, California State University, Northridge, CA, United States
[b]Department of Ocean Science and Hong Kong Branch of the Southern Marine Science and Engineering Guangdong Laboratory (Guangzhou), The Hong Kong University of Science and Technology, Kowloon, Hong Kong
[c]Department of Ecology, Evolution and Marine Biology, University of California, Santa Barbara, Santa Barbara, CA, United States
[d]Department of Marine and Environmental Sciences, Halmos College of Arts and Sciences, Nova Southeastern University, Dania Beach, FL, United States
[e]Hawai'I Institute of Marine Biology, University of Hawai'I at Manoa, Kaneohe, HI, United States
[f]Department of Geology, University at Buffalo, State University of New York, Buffalo, NY, United States
[g]Australian Research Council Centre of Excellence for Coral Reef Studies, James Cook University, Townsville, QLD, Australia
*Corresponding authors: e-mail address: pisapia@ust.hk

Contents

Abstract

Changes in the size structure of coral populations have major consequences for population dynamics and community function, yet many coral reef monitoring projects do not record this critical feature. Consequently, our understanding of current and future trajectories in coral size structure, and the demographic processes underlying these changes, is still emerging. Here, we provide a conceptual summary of the benefits to be gained from more comprehensive attention to the size of coral colonies in reef monitoring projects, and we support our argument through the use of case-history

Advances in Marine Biology, Volume 87
ISSN 0065-2881
https://doi.org/10.1016/bs.amb.2020.07.003

examples and a simplified ecological model. We neither seek to review the available empirical data, or to rigorously explore causes and implications of changes in coral size, we seek to reveal the advantages to modifying ongoing programs to embrace the information inherent in changing coral colony size. Within this framework, we evaluate and forecast the mechanics and implications of changes in the population structure of corals that are transitioning from high to low abundance, and from large to small colonies, sometimes without striking effects on planar coral cover. Using two coral reef locations that have been sampled for coral size, we use demographic data to underscore the limitations of coral cover in understanding the causes and consequences of long-term declining coral size, and abundance. A stage-structured matrix model is used to evaluate the demographic causes of declining coral colony size and abundance, particularly with respect to the risks of extinction. The model revealed differential effects of mortality, growth and fecundity on coral size distributions. It also suggested that colony rarity and declining colony size in association with partial tissue mortality and chronic declines in fecundity, can lead to a demographic bottleneck with the potential to prolong the existence of coral populations when they are characterized by mostly very small colonies. Such bottlenecks could have ecological importance if they can delay extinction and provide time for human intervention to alleviate the environmental degradation driving reductions in coral abundance.

1. Introduction

In a rapidly changing world dominated by anthropogenic disturbances, there is a great need to understand the ways through which ecosystem function is being perturbed (Doney et al., 2012; Hoegh-Guldberg and Bruno, 2010; Hughes et al., 2018). Quantifying changes in the abundance of functionally important organisms is at the frontier of efforts to address this need, and analyses of long-duration records are the central means to inform understanding of the ecological changes underway. The combination of time-series analyses with hypothesis-driven research allows ecological changes to be projected into the future (Lindenmayer et al., 2015, 2010), and to conduct proactive resource management (Flower et al., 2017; Hughes et al., 2013). Coral reefs are among the ecosystems most at risk from climate change (Hoegh-Guldberg et al., 2007; Walther et al., 2002), and while monitoring efforts are capturing the widespread, rapid, and large declines in abundance of corals (Bruno and Selig, 2007; De'ath et al., 2012), these efforts are not doing enough to identify the fundamental mechanisms causing the losses of coral to occur, or to determine whether they might reverse on a time scale of decades-to-a-century.

Decadal-scale monitoring is essential to document the dynamics of long-lived organisms like corals. Most monitoring studies report changes in live coral cover (Bruno et al., 2019; Perry et al., 2018; Stuart-Smith et al., 2018), defined as the planar percentage cover of corals, pooled among taxa. Coral cover has been used as the state variable describing coral abundance for decades (Bell and Galzin, 1984; Bruno and Selig, 2007; Hughes, 1994; Roy and Smith, 1971; Talbot, 1965; Wismer et al., 2019), and it is used to evaluate reef "condition". Arguably the coral reef crisis is so acute (Hughes et al., 2010, 2017a) that a summative metric like coral cover may be sufficient to quantify the large losses of coral that have taken place, particularly as these losses push corals to the brink of ecological extinction (Carpenter et al., 2008; McCauley et al., 2015). However, while coral cover is a good place to start for quantifying coral loss and reef degradation, this metric cannot deliver the demographic insights that are needed to understand the causes and consequences of the events now underway (Edmunds and Riegl, 2020; Hughes, 1984; Hughes and Tanner, 2000).

The focus of most coral reef monitoring on coral cover limits our ability to understand the processes driving reef degradation and perturbing reef function, and while some programs are expanding to record a wider variety of population state variables, these efforts remain nascent in implementation or interpretation. The focus of reef monitoring is shifting from measurements of coral cover to analyses of changes in recovery rates over space and time using population models that incorporate some mechanistic principles (e.g., differential equation approach with logistic and exponential growth) (Gouezo et al., 2019; MacNeil et al., 2019; Mellin et al., 2019, 2016; Ortiz et al., 2018; Osborne et al., 2017). The expansion of methodological approaches has started to increase our ability to understand reef function and to assign causality to reef degradation (Ortiz et al., 2018). Even though the importance of applying demographic approaches to coral populations has been recognized for nearly 40 years (Connell, 1973; Hughes, 1984), this suite of available approaches remain under-utilized (Edmunds and Riegl, 2020). Demography reveals mechanisms that contribute to changes in coral cover by quantifying the contributions of vital rates (recruitment, growth, and mortality) to population dynamics (Caswell, 2001). By resolving the vital rates that control population growth, and by defining their confidence limits, demography can support the testing of multiple hypotheses and strong inference (Platt, 1964) in elucidating the causes of changing coral cover. Demographic information can also substantially improve estimates of extinction risk for coral taxa (Carpenter et al., 2008), and can objectively evaluate the likelihood that depleted populations can recover, go extinct, or remain in a state of rarity (Fig. 1).

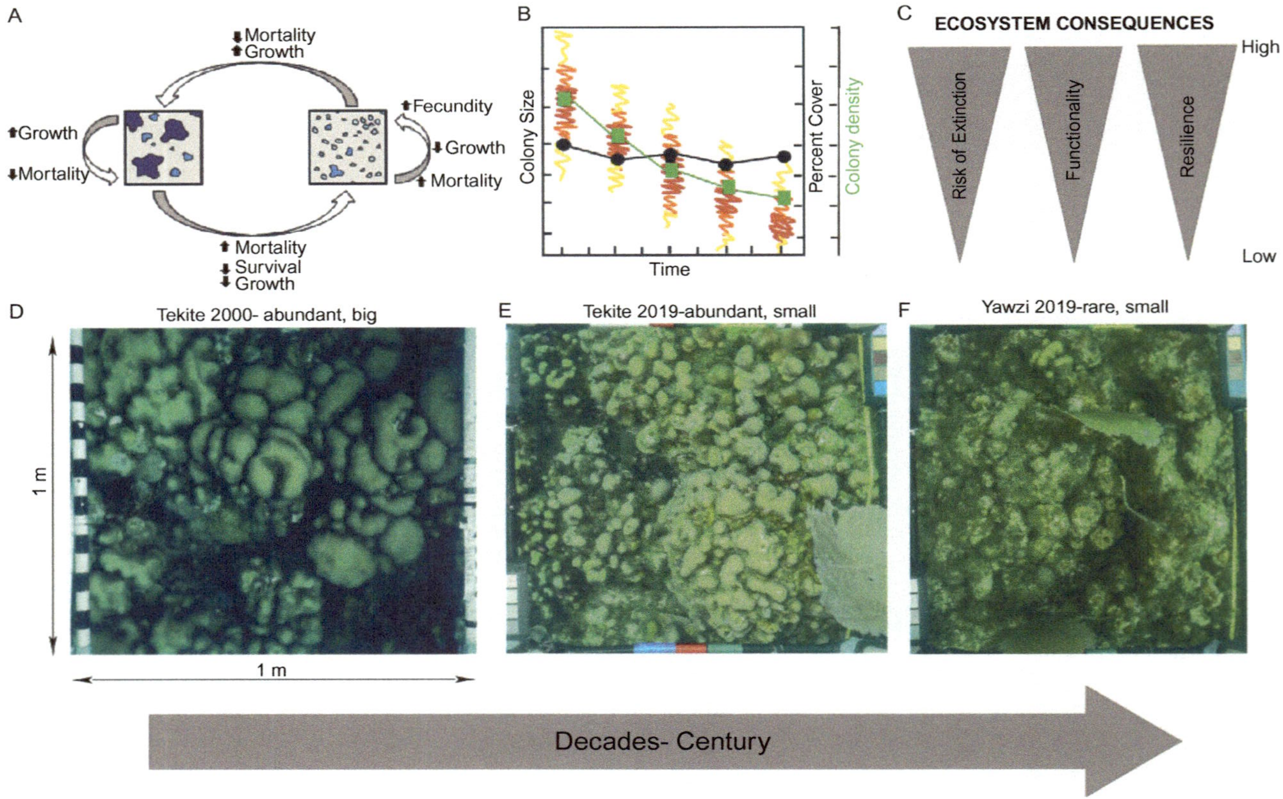

Fig. 1 See figure legend on opposite page.

Global climate change, ocean acidification, and other anthropogenic disturbances are accelerating coral mortality to favour demographic deficits in which more corals are killed than can be replaced through recruitment and growth (Hughes et al., 2017a, 2017b). Many coral reefs experience sequential disturbances that are overlapping in space and time (Ortiz et al., 2018; Wakeford et al., 2008), and sustain chronic stressors that elevate background mortality (Bythell et al., 1993; Wakeford et al., 2008). In addition to multiple disturbances, there is evidence that the growth rates of coral colonies have slowed (Cantin et al., 2010; De'ath et al., 2009; Edmunds, 2007) and rates of partial mortality have increased (Pisapia et al., 2016). Reduced growth rates, and multiple disturbances may alter the size-structure of coral populations, both due to different susceptibility of small versus large colonies, and enhanced partial mortality that reduces colony size. In some cases, larger colonies may survive a severe bleaching event (Barkley et al., 2018; Pisapia et al., 2019), but experience high rates of partial mortality and subsequent declines in colony size. These effects constraining colony growth, and reproduction, thereby degrading the capacity for population recovery following major disturbances (Hall and Hughes, 1996; Henry and Hart, 2005; Madin et al., 2020). With escalating effects of global climate change, it is expected that coral populations will have faster turnover (i.e., the rate at which individuals in the population are replaced) and will be increasingly dominated by smaller colonies (Riegl et al., 2012). There is a critical and immediate research need to better understand trends of declining size and abundance of coral colonies, and to clearly capture the demographic features of these changes.

Data on colony size with species resolution has much to contribute to the understanding of coral demography, and it is one of the easiest to measure

Fig. 1 Schematic showing the importance of colony size structure: (A) how reduction in colony size favours positive feedback on decreasing population size (colonies varying in size shown in blue); (B) hypothetical results of repeated surveys of a coral community for coral cover (black circles), colony density (green squares), or colony size-frequency structure (heat map, red = abundant, yellow = rare), as a function of time, thus revealing stable cover but declining colony abundance; (C) changes in ecosystem properties that can be associated with reduced colony size, but which remain hidden with an emphasis on coral cover as the state variable for coral community condition; (D–F) representative coral assemblages at 9–14 m depth at two sites (Tekite and Yawzi Point) in US Virgin Islands that illustrate the general transitioned (but do not convey a time series) from abundant and large colonies in 2000 (D), to abundant and small colonies at the same site in 2019 (E), and at Yawzi Point, to rare and small colonies in 2019 (F).

aspects of coral populations. Not only is colony size easy to record underwater, but the ability to acquire this information is poised to quickly expand through the application of emerging technologies to image-based data (i.e., video and still images). These technologies are approaching the capacity for automated acquisition of colony sizes from images using machine learning and artificial intelligence (Beijbom et al., 2015; Pavoni et al., 2020; Williams et al., 2019). While previous studies have identified a path for inclusion of coral demography in monitoring (e.g., Edmunds and Riegl, 2020), the present contribution conceptually addresses the causes and implications of changing size of coral colonies. Contextualized by the ongoing trends affecting the size of coral populations, ecological theory, and contemporary technological sophistication, we address four objectives. First, we summarize the ecological importance of coral colony size as it pertains to the capacity to understand the causes and consequences of changing coral abundances. Second, we use decadal-scale time series data from two case-history studies to underscore examples in which coral cover is declining, coral colonies are becoming smaller, and coral populations are transitioning to rarity. Third, we apply a generalizable stage-structured matrix model for branching and massive corals to better understand which vital rates are most influential in creating the demographic changes underway in our case example sites. By applying this model, we predict the consequences of ongoing demographic shifts for the population persistence of reef corals. Finally, we highlight the kinds of questions focusing on coral population dynamics that can be addressed with demographic approaches, and use these examples to highlight the utility of demography and its implementation in coral biology.

2. The functional implications of coral colony size

Body size has profound importance in modulating the structure and function of unitary organisms, and ultimately limits their size within bounds determined by allometric scaling (Schmidt-Nielsen and Knut, 1984). In contrast, it has long been assumed that a colonial modular body plan provides escape from these constraints (Hughes, 2005; Sebens, 1987) because colonies can increase in size while conserving module dimensions (Sebens, 1987). Colonial scleractinian corals once were though to provide a classic example of this body plan (Hughes, 2005; Jackson and Coates, 1986), and while early work indicated that at least some traits scaled allometrically with size (Jokiel and Morrissey, 1986), the notion that colony size is indeterminate and functional traits scale isometrically persisted for decades

(Hughes, 2005; Sebens, 1987). In contrast, recent work has reinforced the generality of allometry in colonial corals (Burgess et al., 2017; Edmunds and Burgess, 2016), with multiple traits, including for example fertility and growth, changing disproportionately with colony size (Burgess et al., 2017; Dornelas et al., 2017; Edmunds and Burgess, 2016). Allometric scaling in corals has important implications, because colony size varies greatly over time within the life of a single coral, and it varies greatly among colonies within a species. Moreover, if disturbances lead to overall reductions in colony size, there will be disproportionate consequences across multiple traits.

Small corals (i.e., ≤4-cm diameter) are exposed to high risks of whole-colony mortality relative to larger (i.e., adult) colonies (Edmunds and Elahi, 2007; Hughes and Tanner, 2000). In general, the probability of whole-colony mortality decreases with colony size (Madin et al., 2020), while the extent of injury (e.g., partial mortality) is expected to increase with colony size (Henry and Hart, 2005; Hughes and Jackson, 1985; Madin et al., 2014). In part, this effect reflects the limited amount of coral tissue in small colonies, which can reduce resistance to partial mortality (Henry and Hart, 2005), whereas large colonies frequently experience a wide variety of partial mortality effects (Henry and Hart, 2005; Madin et al., 2020). With high risk of mortality, small corals are subject to strong selective pressure for rapid growth (Jackson, 1977). This trend favours prioritization of food resources to tissue and skeleton, but can lead to depletion of energy reserves (e.g., storage lipids (Anthony et al., 2009)). Depletion of reserves and selective pressure for rapid growth, can push small corals into a nutritional deficit from which they may be unable to escape whole colony mortality.

Small corals are not necessarily young (Hughes and Jackson, 1980), and big colonies can undergo fission to form small colonies with relatively old tissue. Some demographic traits such as reproduction are known to revert to their younger states upon fission (Hughes et al., 1992), while other traits may persist in the daughter colonies. When older traits are retained, the resulting small, but old, corals might respond to the environment in different ways, influenced by their history (Putnam et al., 2017), thereby making them functionally different to small and young corals (Kojis et al., 1985; Szmant-Froelich, 1985). Age-related phenomena that can carry over to small fission products can include the consequences of establishing symbiosis with dinoflagellate algae, for which an initially cosmopolitan genetic assemblage (e.g., consisting of multiple symbiont genotypes in a single host) might be winnowed over time (Little et al., 2004; Mieog et al., 2009). Moreover, these effects can include the legacy of a reproductive status determined by age and colony (Kojis et al., 1985; Szmant-Froelich, 1985), and epigenetic effects mediated by parental history (Putnam et al., 2017).

Interactions between coral colonies and their fluid environment are also modulated by colony size (Madin et al., 2014; Patterson, 1992a, 1992b). For a given flow regime small corals operate at lower Reynolds numbers, while large colonies operate at higher Reynolds numbers (Patterson, 1992a, 1992b). Reynolds numbers give a measure of the ratio of inertial to viscous forces and are used to characterize the nature of the flow past a coral colony (Schlichting, 1979). As a result, colonies varying in size are affected differentially by turbulent flow and the propensity for mass transfer limitation of metabolically important gases and metabolites. Associated with these trends, flow speeds can modulate metabolic rates through the flux of metabolites between the tissue and seawater (Patterson, 1992a, 1992b). These effects vary as a function of colony size and shape, and create predictive capacity for favoured shapes and sizes of coral colonies under differing flow regimes (Patterson, 1992a, 1992b). Flow can also play strong roles in dislodging coral colonies, with the risks of dislodgement increasing with size and shapes of coral colonies (e.g., certain shaped colonies are more prone to dislodgement such as tabular and arborescent) (Madin et al., 2014). Mortality curves for these corals adopt a "bathtub-shape" with high mortality at both extremes of the colony size distribution (Madin et al., 2014).

Together, the aforementioned trends demonstrate that reductions in coral colony size and density will have profound implications that cannot be elucidated through measurements of coral cover alone (Fig. 1). Further, the multifaceted implications of a reduction in colony size at the population level indicate that the demographic consequences of this trend are likely to be complex. Reductions in colony size, for example, can support a variety of demographic outcomes, not just high risks of extinction, as has been suggested (Carpenter et al., 2008). To begin to distinguish among these possibilities, here we develop a stage-based matrix model for two functional groups of coral (massive and branching species) and compare the model output to empirical data from our case-studies. The kinds of population projections these models can produce have previously been used to evaluate the implications of decadal shifts in the colony size distributions of several corals (Edmunds and Riegl, 2020; Hughes, 1984). Specifically, we evaluate whether increased abundances of small colonies could accelerate the transition to functional extinction, due to concomitant declines in individual persistence and reproductive potential.

3. Temporal shifts in the size structure of coral populations

An increasing number of studies highlight the major changes underway in the abundance and community structure of scleractinian corals on tropical reefs (Hughes et al., 2018; Jackson et al., 2014; Pisapia et al., 2019). These changes can be attributed to multiple anthropogenic and natural disturbances (Hoegh-Guldberg et al., 2007; Hughes et al., 2010), but the gravest threats are posed by global climate change (Hoegh-Guldberg et al., 2007). In 2015–2017, extreme temperature anomalies triggered unprecedented coral mortality in all major coral reef regions throughout the world (Eakin et al., 2016, 2014). These events drastically altered the composition of coral assemblages on the Great Barrier Reef (GBR) (Hughes et al., 2018). The observed shifts in community structure relied upon the relative abundance of thermally sensitive branching corals, which exhibited severe changes in abundance (Hughes et al., 2018). Massive corals are comparatively more resilient to elevated temperatures (Hughes et al., 2018) than branching and encrusting species. Even though massive corals are more likely than branching corals to survive a bleaching event, they may experience high levels of partial mortality in response to the same stressor (Pisapia et al., 2019). However, shifts in the taxonomic composition of coral communities depend on both the frequency and severity of major disturbances such as bleaching (Pratchett et al., 2020).

Disturbances and taxonomic shifts in coral assemblages are often accompanied by changes in colony size distributions, commonly involving the selective loss of larger, reproductively mature colonies (Pisapia et al., 2019). However, changes in size-structure have been overlooked in most studies of coral bleaching, and most empirical attention has focused on measuring changes in live coral cover among taxonomic groups. Despite recognition of the importance of coral colony size, few studies have rigorously recorded colony size over time, and where sizes have been recorded, measurements rarely extend over decades (but see Kramer, 2003). A common reason for coral colony size to be measured in time-series analysis is to evaluate the effects of disturbances such as bleaching or storms (Edmunds and Elahi, 2007; Roff et al., 2014). To date, however, the colony size structure of coral communities has rarely been measured in the "recovery" phase following disturbances. This is unfortunate as such data are necessary to evaluate the extent to which the coral community returns to a holistic measure of pre-disturbance condition (Graham et al., 2015).

During the recovery phase following major disturbances, the colony size structure of coral populations could adopt a wide variety of forms, each with distinct ecological implications. In a few cases, disturbances have favoured negatively skewed frequency distributions (i.e., populations in which large colonies are more common than small colonies (Bak and Meesters, 1999; Meesters et al., 2001). More frequently, there is a progressive decline in colony size as large and old colonies are degraded through partial mortality and fission (Edmunds and Elahi, 2007). Mean colony size can also be depressed with size distributions becoming positively skewed through high rates of sexual recruitment (Bak and Meesters, 1999). In a metapopulation, persistent recruitment from distant locations to maintain abundance of small colonies, even when small corals have high rates of mortality, can continue until the source populations for the coral larvae are negatively impacted by disturbances. Distinguishing among the aforementioned demographic mechanisms of changing size structure (e.g., the effects of recruitment versus partial mortality and fission) is necessary to interpret the long-term causes and implications of disturbance and declines in coral abundance.

To illustrate temporal shifts in the size structure of coral populations, we present time-series data describing coral populations on two coral reefs, one from the Caribbean (U.S. Virgin Islands, Fig. 2A–I) and one from the Indo-Pacific (the Persian Gulf, Fig. 2J–O). Our objective in presenting these data is not to suggest that are representative of all reefs, rather to present evidence supporting our assertion that coral colonies are declining in size on at least some modern reefs. A comprehensive analysis testing for similar effects on biogeographic scales remains an important research need.

There is evidence that coral colonies are becoming smaller on the reefs we selected as case examples, although we recognize that a diversity of trends for this state variable (i.e., colony size) have been reported (e.g., Babcock, 1991; Bak and Meesters, 1999). Our data from two locations reveal consistent shifts in the size-frequency distributions of four coral species, thus favouring high abundances of small versus large colonies (Fig. 2). These trends are not unique to these species or these locations (Fong and Glynn, 2001, 1998; Hernández-Pacheco et al., 2011; Hughes and Tanner, 2000), yet as described above, congruent results do not imply congruent demographic mechanisms, and without understanding the processes driving these changes, it is impossible to project their impacts on future coral communities. While extinction remains one possible outcome of these trends, it is also possible that shrinkage, fission, and rarity could represent a pathway towards a cryptic regime shift (sensu Hughes et al., 2010).

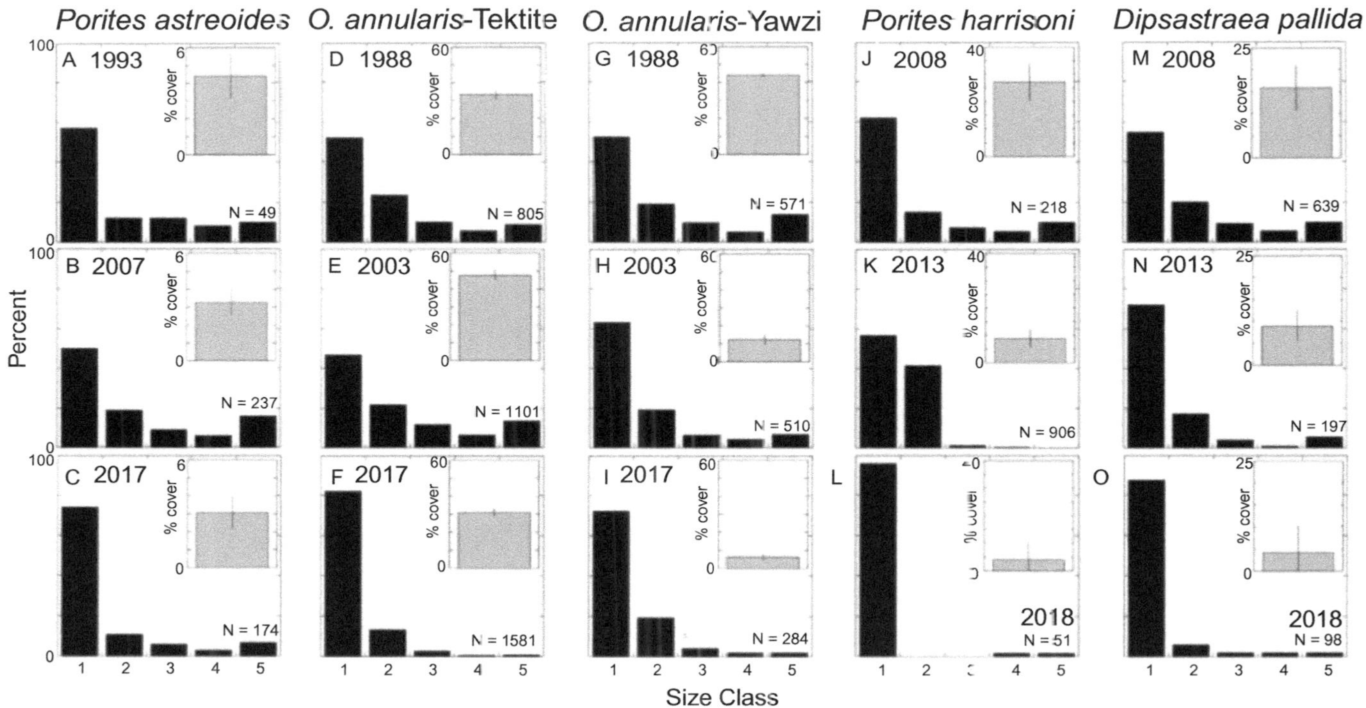

Fig. 2 Declining sizes of scleractinian colonies over a decades in the Caribbean and Persian Gulf. Bar graphs show the percentage distribution of colonies among five size classes defined by dividing the 90th percentile size range into five equal portions and adding larger colonies to size class 5. (A–C) Population structure in St. John (U.S. Virgin Islands) of *Porites astreoides* at 3–9-m depth in 1993, 2007, and July 2017 (A–C), *Orbicella annularis* at Tektite (14-m depth) in 1988, 2003, and July 2017 (D–F), and *O. annularis* at Yawzi Point (9-m depth) in 1988, 2003, and July 2017 (G–I); inset bar graphs show coral cover in the year of each size class analysis (note ordinates differ between *P. astreoides* and *O. annularis*). *P. astreoides* accounted for 16% of the coral cover throughout the study, and *O. annularis* accounted for 76% (Tektite) and 86% (Yawzi Point) of the coral cover throughout the study. Population structure in the Persian Gulf (Ras Ghanada, Abu Dhabi, UAE) for *Porites harrisoni* in 2008, 2013, and 2018 (J–L), and for *Dipsastraea pallida* in 2008, 2013, and 2018 (M–O). Data sources: (Edmunds, 2015, 2013; Edmunds and Elahi, 2007; Edmunds and Lasker, 2016; Riegl et al., 2017; Riegl and Purkis, 2015) and new data.

This shift might involve reduced ecological success of small colonies and a paucity of coral recruits, which together could favour population stabilization in a greatly reduced abundance (Fig. 2). Such an outcome would represent a vestige of the former community structure, but avoidance of extinction maintains the possibility of population recovery, and buys time for human intervention to increase the likelihood of coral population recovery.

Below we use stage-structured population matrix models to address mechanistic pathways culminating in coral populations of rare and small colonies. We project population structure into the future for both a branching and a massive coral in order to explore the consequences of the demographic trends that are underway on present day reefs (Fig. 2). These analyses are designed to highlight how a small number of processes, that can be relatively easily monitored in the field, can influence size distributions in a way that may not be captured by measuring coral cover alone.

4. Modelling changes in size distributions of coral populations

To distinguish between potential alternative demographic mechanisms that cause shifts in the size structure of coral populations, we use a stage-structured matrix population model after Caswell (2001) (Model code is available as a digital supplement, and at DOI: https://doi.org/10.5281/zenodo.3926845). Further, to account for potential differences in inherent life-history strategies (Pratchett et al., 2020), we parameterize the model for both a generalized branching coral and a generalized massive coral (Table 1). We use the models to explore the impacts of variation in three vital rates on the stable coral population size structure. First, we evaluate the role of increased whole-colony mortality, as might be caused by bleaching or disease (Hughes et al., 2017b; Weil, 2004). Second, we assess the role of declining colony growth (e.g., linear extension), which may be a consequence of increased environmental stress (Pratchett et al., 2015). Third, we explore the role of reduced fecundity, which may occur when stressed corals incur metabolic costs that reduce the resources available for spawning (Hughes et al., 2019; Szmant and Gassman, 1990).

Although we address a simplified system in which each vital rate varies independently of any other, this approach reveals the extent to which changes in each vital rate can create distinct outcomes with regard to the population size structure. By coupling changes in vital rates to their "signatures" in shifts in colony size structure, we generate hypotheses that

Table 1 Empirical data used to define the set of parameters (population reproduction, growth, partial mortality, and whole colony mortality) in the model matrix for branching and massive corals.

Model Parameters	(a) Branching	(b) Massive	Sources
Reproduction			
Minimum size for reproduction (projected planar area, cm^2)	~100	50	Hall and Hughes (1996), Stimson (1978)
Fecundity (eggs cm^{-2} y^{-1})	232	1198	Harrison and Wallace (1990)
Larval survival (maximum) (% y^{-1})	Exponential function with 50% mortality in 37 day	Exponential function with 50% mortality in 37 day	Graham et al. (2008)
Recruit survival function	Type 3; 100% at t_o, 50%[a] at t_1	Type 3; 100 at t_o, 35%[b] at t_1	Dunstan and Johnson (1998)[a], Edmunds (2000)[b]
Recruits $larva^{-1}$	0.000009	0.000009	Bramanti et al. (2015)
Growth			
Linear extension (mm y^{-1})	74	10	Pratchett et al. (2015)
Partial Mortality			
Colonies affected (% of colonies in 3 years)	65	90	Pisapia et al. (2016)
Severity (% tissue loss from each colony in 3 years)	5–30%	2–35%	Pisapia et al. (2016)
Whole colony mortality			(a) Branching
(% y^{-1} by size)	7.1% for 0.25 cm^2	6.9% for $\leq$50 cm^2	Kayal et al. (2018)
	2.0% for 3147 cm^2	4.1% for 51–150 cm^2	(b) Massive
		3.3% for 151–250 cm^2 3.4% for $\geq$250 cm^2	Edmunds (2015), Edmunds and Elahi (2007)

Whole colony mortality was calculated: (a) for branching corals for *Pocillopora* on the north shore of Moorea [Fig. 2E, in (Kayal et al., 2018)]. (b) for massive corals for *Orbicella annularis* by averaging data collected over 5 years intervals over 15 years (Yawzi, St. John) and 25 years (Tektite, St. John), divided by 5 to estimate annual rates (Edmunds, 2015; Edmunds and Elahi, 2007). However, the two taxa *Pocillopora* and *Orbicella annularis* do not represent the generic branching and massive corals in the model.

can be used to link observed changes in colony size distributions to their causal mechanisms.

A robust parameterization of a matrix model for a given coral population requires a substantial quantity of empirical data (which currently is unavailable for most corals), in part because changes in colony size structure in nature are likely to emerge from a combination of demographic mechanisms that can be statistically challenging to distinguish. Therefore, we use a generalizable parameterization for two coral types—branching and massive—grounded in empirical data but not intended to apply to a specific taxon. We seek a coarse-grain analysis with which we can illustrate the utility of the approach in aligning empirical changes in colony size distributions with mechanisms associated with specific vital rates. An emergent property of this approach is the development of benchmarks for projected population structure against which empirical population structure can be tested for goodness of fit. As we posit below, concordant size structuring likely arises from variation in specific vital rates, identification of which can support pro-active recommendations for the design of long-term monitoring and event sampling therein. In applying a stage structured matrix model we recognize that there are other demographic approaches that have been applied to scleractinian corals (Edmunds and Riegl, 2020), one of the earliest of which was a size based matrix model (Hughes, 1984). Each approach has strengths and limitations, and it is not our objective to advocate for a specific approach (or to review the methods available). Rather, we seek to shed light on the demographic mechanisms that can cause diagnostic variation in coral colony size, and explore the likelihood of local extirpation as one possible outcome of declining size and reduced abundance of coral colonies.

Matrix population models describe how individuals within a population change in state (size) over time. They can also incorporate recruitment and mortality, allowing them to capture overall population growth as well as population structure. In particular, the leading eigenvector of a matrix population model, λ, represents the population's exponential growth rate. The population grows when $\lambda > 1$, it shrinks when $\lambda < 1$ and asymptotically approaching zero, which indicates local extirpation and perhaps regional extinction (Caswell, 2001). The leading eigenvalue is also associated with an eigenvector, $\vec{x}$, which gives the population's stable size distribution. When the stable size structure of a population is reached, the numerical abundance of the population may still change while the relative contribution of each size-class to the population structure remains fixed (i.e., stable).

A stable stage distribution has been observed in natural coral populations after about a decade without disturbances (Riegl and Purkis, 2015), and serves as a useful way to explore "equilibrium dynamics" in a population. While useful for the theoretical evaluation, whether a stable stage structure is reached depends on the extent to which the population is closed (i.e., self-seeded) versus open (with recruits originating from spatially separated populations). The present analysis assumes a closed population and makes no direct reference to any export or import of larvae. Thus, the matrix model can be thought of either as encompassing the sum of all life-history parameters of several well-connected sub-populations (that themselves are isolated from other populations), or as representative of a single, primarily self-seeding locality. Either are realistic and encountered in nature.

Our matrix model describes a population of corals with a vector $\overrightarrow{x_t}$, whose entries define the number of coral colonies within each of a series of classes. Size is most easily defined as the planar surface area of living tissue (in cm^2), but the exact measurement of size is of no consequence here, and our model consists of four types of classes: Class I, diminutive corals that have not yet recruited to a reef (i.e., larvae); Class II, recruits that are new settlers to the reef (i.e., corals <1 mm diameter); Class III, a set of classes of increasing size representing juveniles that are small and sexually immature (i.e., $<\sim 100\,cm^2$ in planar area for branching corals, $< 50\,cm^2$ for massive corals); and Class IV, a set of classes of increasing size representing sexually mature adults ($>100\,cm^2$ for branching corals, $> 50\,cm^2$ for massive corals). The projection matrix $\mathbb{A}$ simulates state transitions over time:

$$\overrightarrow{x_{t+1}} = \mathbb{A} \cdot \overrightarrow{x_t} \ .$$

This projection matrix $\mathbb{A}$ is the composite of matrices representing the production of new individuals (fecundity, $\mathbb{F}$) and state transitions among existing individuals that represent growth, shrinkage, or stasis. The latter is traditionally represented as the product of the growth matrix $\mathbb{G}$ and the diagonal survival matrix $\mathbb{S}$. However, in corals, state transitions are complicated by the possibility that individuals not only increase in size but may also shrink due to partial mortality and fission (Edmunds and Elahi, 2007; Hughes and Tanner, 2000). We represent the probability of partial mortality with the diagonal matrix $\mathbb{M}$ and use two matrices $\mathbb{G}$ and $\mathbb{P}$ to represent the probabilities of state transitions to larger (via growth) and smaller (via partial mortality) size classes:

$$\mathbb{A} = \mathbb{F} + [\mathbb{G}(\mathbb{I} - \mathbb{M}) + \mathbb{P}\mathbb{M}]\mathbb{S}$$

where $\mathbb{I}$ is the identity matrix.

We use generalized functions grounded in empirical data to represent reproductive and size transitions as a function of colony size (discussed in detail in the appendix). We model the production of larvae by adults as a positive linear function of adult size (Hall and Hughes, 1996). A proportion of these larvae successfully settle on the reef as new recruits, and the remainder perish. Corals that have recruited to the reef survive with a probability that is a saturating function of colony size. This effect is driven by the high rates of mortality of small colonies (Hughes and Jackson, 1985) and the low rates of complete mortality of large colonies, augmented by a greater probability of partial mortality every year (Edmunds and Elahi, 2007; Hughes and Jackson, 1985), which favours transition to a smaller size class. We model growth as linear extension of colony margins using a mean linear extension rate of 7.5 cm y^{-1} for branching corals and 1 cm y^{-1} for massive corals (Table 1). Larger corals are also more likely to experience partial mortality (e.g., damage due to physical impacts, corallivory, etc.) because of their greater spatial extent (Edmunds and Elahi, 2007; Hughes and Jackson, 1985). When injuries are sufficiently extensive, they can cause a coral to decrease in size, thus causing a transition to a smaller size class. Thus, corals can "shrink" and transition from larger to smaller size-classes.

We begin with a baseline set of parameters (Table 1) that produce coral populations with positive growth ($\lambda = 1.11$ for branching corals, $\lambda = 1.08$ for massive corals) and, having reached a stable size structured stage, a mean colony size of 6297 cm^2 for branching corals and 67 cm^2 for massive corals. The larger branching colony would be consistent, for example, with a large plate of *Acropora hyacinthus*, and the smaller massive colony would be consistent, for example, with a hemispherical colony of *Astrae curta* (both for an Indo-Pacific reef), but taxonomic concordance is neither implied nor intended. Increases in mortality of established corals (i.e., juveniles and adults), decreases in growth rate, and reductions in fecundity all reduce the population's growth rate, with large alterations in these vital rates creating a demographic deficit in population size (i.e., where $\lambda < 1$), indicating that local population extirpation is inevitable (Figs. 3A–C). Overall, the model was less sensitive to reductions in growth rate (i.e., larger proportionate reductions were required for population growth to fall below replacement rate) than increases in mortality or decreases in fecundity. However, because our model represents a hypothetical case study, we do not view this as a definitive prediction that reductions in growth rate are of little concern. Rather, our particular scenarios indicate that, so long as growth rates are sufficient for a few colonies to reach reproductive size, populations can be

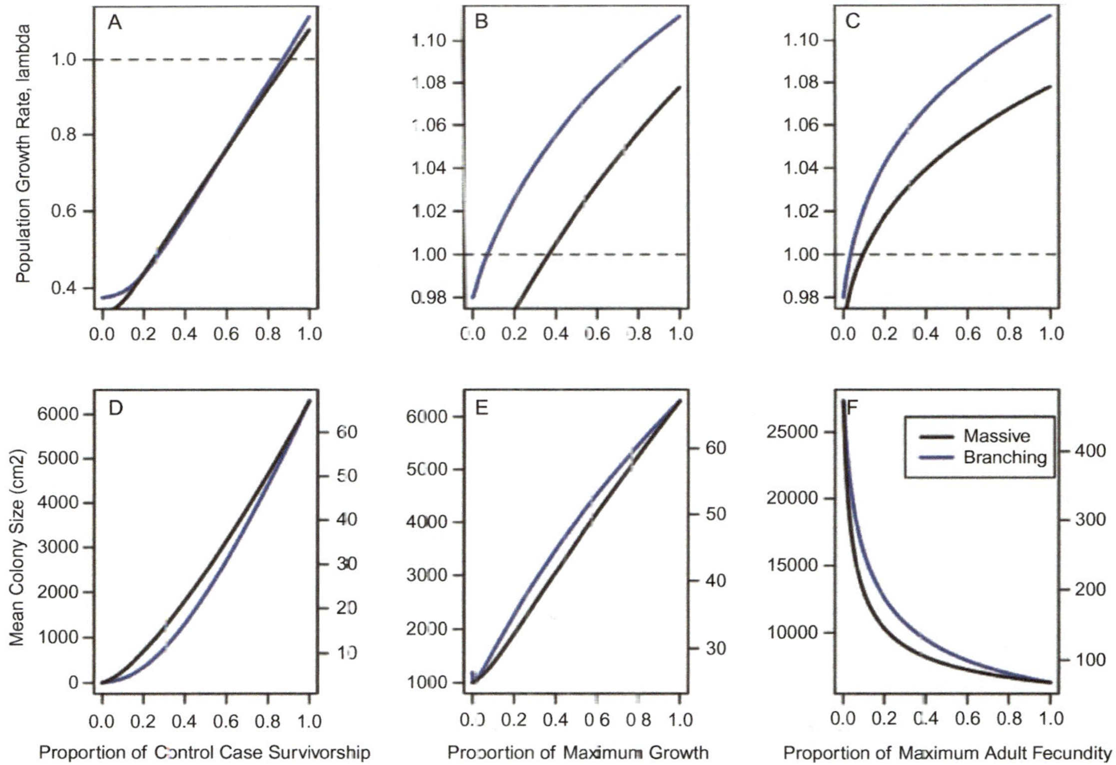

Fig. 3 Predicted coral population growth rates (top row A–C) and mean colony sizes (bottom row D–F) as vital rates are changed for both branching (blue) and massive (black) corals. Increases in mortality drive reductions in survivorship, reducing growth rate and mean size (left column). Decreases in growth rate drive similar patterns, though shifts are non-monotonic due to a loss of recruits that increases mean colony size at the lowest growth rates (middle column). While reductions in fecundity decrease the population growth rate, mean colony size actually increases due to the reduction in recruitment and, consequently, juvenile size classes. Trends are the same for both colony types, but note the larger mean sizes of branching corals. Dashed lines (top row) show the minimum population growth rate ($\lambda = 1$) for population persistence.

sustained. Predicting how population structure changes over time is more complex than evaluating whether the population increases in density. Mean colony size decreases with increasing mortality and decreasing colony growth rate (Figs. 3D and E, Figs. S1–2), though in the latter case this decrease is non-monotonic. In contrast, reductions in fecundity causes an increase in mean colony size, due to a reduction in recruitment that leads to a relative increase in abundance of the larger colonies (Figs. 3F, Fig. S3).

How long does it take a coral population to show a demographic response to a change in vital rates? Although multi-year (and possible decadal) datasets are necessary to empirically answer this question (particularly to establish baseline estimates of population size distributions and growth rates), our model reveals how changes in mortality, growth, and fecundity can cause shifts in size distributions within a few years (Fig. 4), though it may take several decades to reach the new stable stage distribution (Fig. 5). Variation in each vital rate produce distinctive shifts in the colony size distributions with differences between branching and massive morphologies: increased mortality decreases the probability of state transitions to larger size classes and reduces abundance especially of the largest size classes. Reduced growth preferentially removes individuals of intermediate size, initially bolstering the relative abundance of the smallest size-classes (either recruits or small tissue fragments) and, at steady state, a few large individuals (with higher abundance of larger corals for branching compared to massive corals). Reduced fecundity increases the relative abundance of the largest individuals, which survive but fail to support large propagule classes with increased abundance of larger colonies more pronounced for branching compared to massive corals (Fig. 4).

Our model simplifies reality in order to make conceptual advances, and in due course will require ecologically relevant treatment of a wider variety of demographic traits in order to achieve its potential in advancing coral reef science. For example, while there is strong evidence for some coral populations that they reach stable size distribution (Riegl and Purkis, 2015), this is unlikely to be the case in fully open populations, or those that reflect a combination of open and closed structuring that vary in relative importance over time (Caley et al., 1996). Thus, especially if strong connectivity among several populations is present, reductions in fecundity of the focal population may have a different, or entirely absent, expression in size distribution than our simplified model reveals. In the case of stabilization to favour abundant large colonies, such an outcome might reflect a diagnostic feature of the storage effect (Chesson, 1984) (e.g., large and old individuals

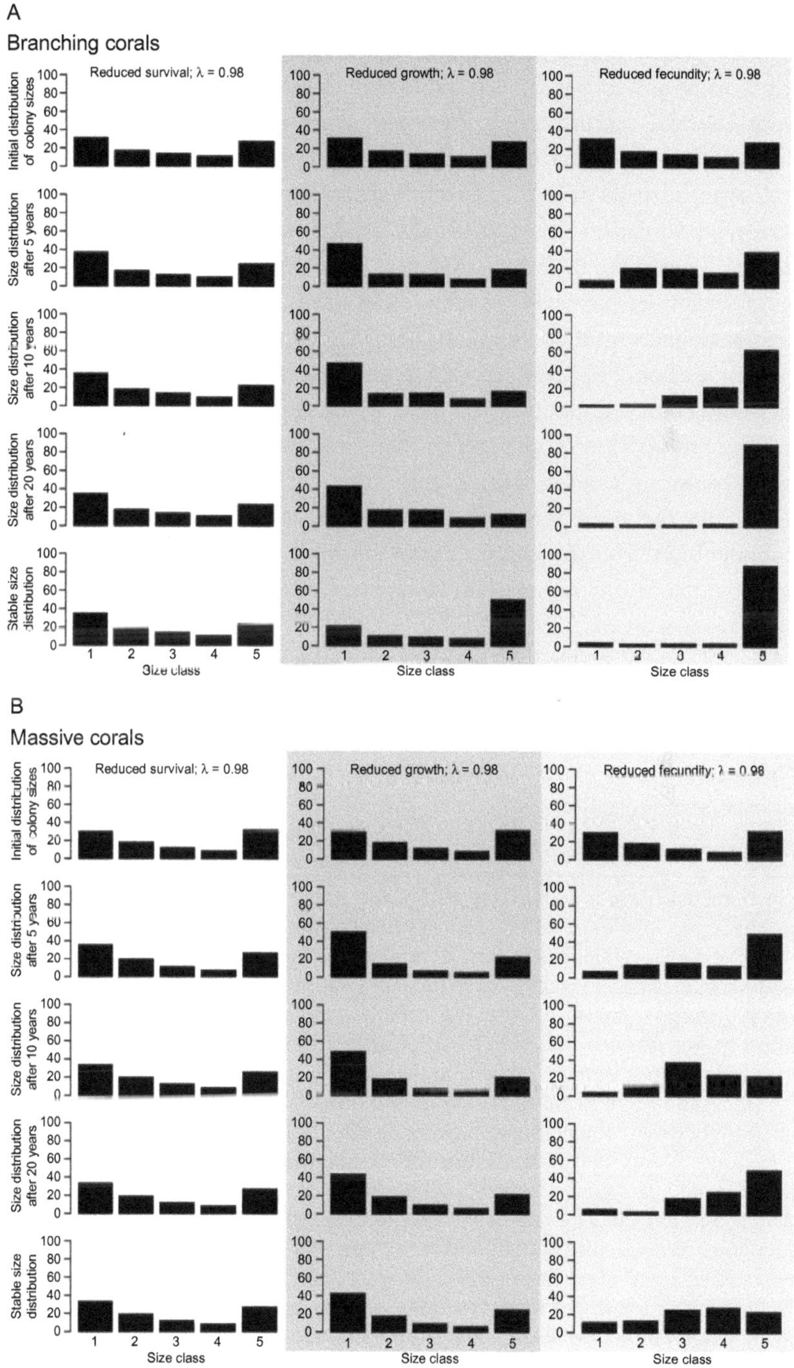

Fig. 4 See legend on next page.

"store" the demographic signal of historic recruitment until such point in the future when recruitment is/can be successful. Hence *Orbicella* in the Caribbean seems to persist as a stable population for decades without any effective sexual recruitment.). Also partial mortality can have important influences on size-structure, but it is highly context-specific with regards to the type of disturbance, environment, and species affected. In its effect on overall population size, it is usually subordinate to changes in fecundity and overall mortality. Similarly, fission is often an important process in population dynamics, however, its effect may also be subordinate. None of the aforementioned arguments suggests anything other than the overwhelming importance of obtaining colony size-dependent information in order to fully understand coral population dynamics. While real coral populations are subject to a combination of changes in their vital rates over time, our findings show how incorporating demography into coral monitoring can provide insights into the mechanisms driving changes in coral cover. Importantly, the model highlights the complex interaction of processes affecting size frequency distributions rather than document an absolute tendency toward reductions in colony size.

Overall, our model highlights two important results. First, the model formally represents the expectation that changes in coral vital rates (i.e., decreases in survivorship, linear extension, and fecundity) drive reductions in coral population growth. Importantly, the relationships between these parameters and population growth rate are non-linear (Fig. 3, top row).

Fig. 4 Coral populations shift in size distribution under different environmental conditions favouring dissimilar effects on key vital rates. For both (A) branching and (B) massive corals, reductions in survival and growth cause a shift towards smaller colony size classes, whereas reductions in fecundity cause an increase in mean colony size as small (young) colonies are lost from the population. For all cases, demographic parameters are set such that population growth rates are 0.98 (e.g., a 2% decrease in population size per year). While changes in colony size distribution are perceptible within a decade for reductions in survival and growth, it can take several decades for coral distributions to equilibrate to the new, larger, stable size distribution when fecundity is reduced. Top row: initial distribution of colony sizes, given by the stable stage distribution of the "control" matrix. Middle rows: stage distributions expected five and 10 years after the vital rate has shifted. Bottom row: stable stage distribution that will emerge from specific changes in one of three vital rates. Stable stage distributions are computed by dividing the juvenile and adult size classes (i.e., corals that would be detected when surveying a reef among five size classes defined by dividing the 90th percentile size range into five equal portions and adding larger colonies to size class 5 (as in Fig. 2).

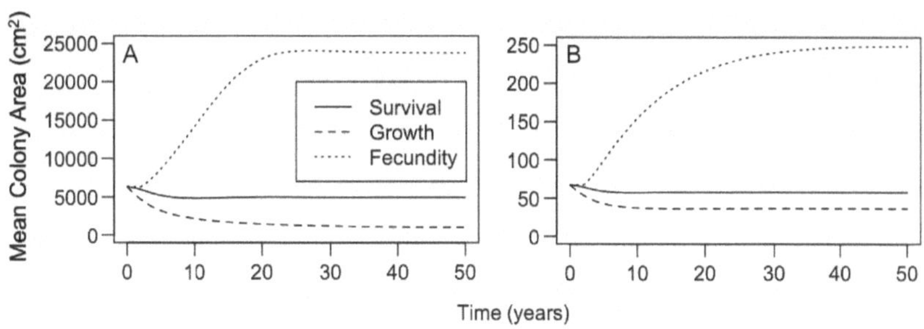

Fig. 5 Rate of change in mean coral colony size under different vital rates. For branching (A) and massive (B) corals, reductions in survival (solid line) and growth (dashed line) cause decreases in coral colony size. These declines are most precipitous when growth rates are affected and, in the latter case, may require decades (e.g., for branching corals) before the new mean size of corals is achieved. In contrast, reductions in fecundity (dotted lines) cause substantial increases in mean coral size due to the loss of the smallest (juvenile) size classes. For massive corals, which exhibit the slowest growth rates, this demographic shift takes the longest period of time, with the new stable size distribution not achieved for several decades.

Initial reductions in these rates result in relatively small shifts in population growth rates (λ), but as further declines in vital rates occur, the corresponding reductions in λ become more precipitous. Second, our model predicts that when each shift in vital rate acts in isolation, it produces a demographic "signature" in the colony size distribution that is expressed within years to decades (Fig. 5). These signatures highlight the intrinsic value of demographic data: knowledge of coral size distributions (and changes therein) may provide critical early warning signs for researchers and managers seeking to predict population trajectories, and qualitative shifts in size structure may reflect distinct demographic drivers (Fig. 2). However, care is needed not to over-interpret such patterns, since a variety of combinations of different demographic parameters can lead to similar size-distributions (Condit et al., 2017).

5. Tractable questions and the data with which they can be answered

We have placed our treatment of coral colony size structure in a context within which some of the key implications of declining size and abundance of coral colonies are highlighted. Even a cursory evaluation of these features reveals the broad extent to which coral colony size can affect the response of coral populations to environmental conditions and, therefore, the capacity for corals to persist in a world disturbed by anthropogenic effects. We present select evidence that coral colonies are becoming smaller and rarer on two reefs (Fig. 2), and through modelling, we reveal how variation in three vital rates can contribute to transitions of coral colonies into small size classes (Fig. 4). These effects are expressed through different demographic mechanisms creating distinctive signatures in the shape of the colony size distributions for the two most common morphological groups of reef-building corals (massive and branching taxa). By projecting a hypothetical coral population 10+ years into the future, our results suggest that rarity and small colony size does not necessarily lead to extinction ($\lambda \geq 1$) over decades. While decades are trivial in duration relative to the geological history of coral reefs, and the ecological history of many coral species (Babcock, 1991; Darke and Barnes, 1993; Devlin-Durante et al., 2016; Hughes and Jackson, 1980; Lough and Barnes, 1997), decades are relevant to the capacity of humans to develop tractable solutions to the coral reef crisis. Persistence of small and rare corals over multiple decades effectively buys time to develop solutions to the crisis, and staves off potential apathy and dismay at the prospect of preventing climate change from driving widespread extinctions of reef corals (Jackson, 2008).

Appeals for greater use of demographic tools to address the dynamics of coral populations have been made for more than 40 years (Connell, 1973; Hughes, 1984), and recently these appeals have been repeated (Edmunds and Riegl, 2020) as the global crisis affecting coral populations intensifies (Hughes et al., 2018). At a time when widespread coral extinctions are considered likely (Carpenter et al., 2008; Hoegh-Guldberg et al., 2007), the need for accurate and precise coral demography has never been more acute. As we highlight above, and has been described elsewhere (Edmunds and Riegl, 2020), there are unique needs for new data to which coral reef monitoring must respond in order to fully exploit the potential of demography to advance the discipline and answer critical questions addressing coral reef conservation, functionality, and management. Given the rapid advances in computer-aided image analysis, colony size distribution information probably will be easy to obtain from images in the near future.

A demographic approach can serve to identify early indicators of reef degradation and could provide insights into reef recovery potential. For instance, a distribution dominated by larger colonies may indicate that recruitment is currently limited, or that recent disturbances had a disproportionate impact on smaller colonies. Conversely a distribution where larger colonies are under-represented may indicate that recent disturbances caused selective loss of larger colonies and hence reproductive potential, which may reduce recovery capacity. It may however, be challenging to identify a particular distribution as "conducive to future coral population growth", thus a key need for coral reef monitoring is maintaining the capacity for repeated sampling of identical areas of reef. Some suggestions to support this outcome are detailed in Edmunds and Riegl (2020). Below we highlight four themes to which coral demography could be productively applied.

5.1 Risks of extinction

The possibility that reef corals will be locally extirpated or become extinct is the ultimate outcome of the coral reef crisis, and these possibilities are becoming more likely as the crisis intensifies (Hughes et al., 2018). Accurately characterizing these risks is a critical aspect of managing endangered species (Powles et al., 2000), and is a mandatory step in the development of effective conservation strategies (Jackson, 2008). The risks of premature estimates of extirpation and extinction are considerable, as they undermine efforts to develop solutions and challenge the credibility of scientific endeavour. Population viability analysis (PVA) is a well-established means to address risk of extinction for endangered species, and it offers a formal procedure to evaluate the probability that a population will persist for a specified time into the future (Boyce, 1992). To our knowledge, PVA has only been applied to one coral species in a single location (Muko et al., 2014), and thus consideration of risks of extinction for reef corals largely remain a matter of speculation (Carpenter et al., 2008). A wide variety of demographic tools, including those similar to the approach described above, offer great potential in addressing the critical questions of *which corals are threatened by extinction* and *how long can current populations persist with low densities of small colonies*?

5.2 Ecosystem functionality

The ecological value of coral reefs is to a large extent dependent on the framework-building capacity of reef corals and the maintenance of a positive carbonate budget. Only if reef structures can be maintained, will the habitat required for the reef's biodiversity continue to exist. Since the potential of corals to form interlocking frameworks of skeletons depends on their sizes, any prediction of future carbonate budgets and framework dynamics will be greatly assisted by adequate demographic information of the reef builders. Demographic approaches can evaluate the capacity of coral colonies to continue to contribute coral skeletons to future reef communities, and when combined with explicit models of community metabolism (such as MTE

(Brown et al., 2004)) and net accretion (Perry and Alvarez-Filip, 2019), can objectively evaluate the form in which the three-dimensional structure of coral reefs will persist.

5.2.1 Community and ecosystem resilience

Community and ecosystem resilience (Gunderson, 2000) are central concepts of modern coral reef ecology and their conservation (Hughes et al., 2010), yet these concepts continue to be addressed mostly with the same state variable used to quantify coral abundance (i.e., cover). Yet as dynamic and multifaceted concepts, coral cover offers an incomplete means to evaluate rates of future change. Coral habitats with similar coral cover may nonetheless have very different community structure (e.g., different size-frequency distribution of corals) which will have a plethora of implications (described above). Demographic tools are well designed to address these implications, and offer statistically robust tools to address the rate of future changes in coral community structure, and the extent to which future populations are functionally equivalent to (or different from) those of the past.

5.3 Necessary changes in coral reef monitoring

An implicit outcome of projected shifts in the size-structure of coral populations, where there are both fewer and smaller coral colonies, is that the needs for effective ecological sampling will differ from those designed and implemented in an era when corals thrived. These needs are also likely to differ from those of the recent past as photogrammetry (Burns et al., 2015; Ferrari et al., 2017) and computer-aided analyses (Beijbom et al., 2015; Pavoni et al., 2020; Williams et al., 2019) become economically and ecologically viable. Importantly, obtaining the necessary information from high quality images will likely require relative small changes in the next generation of software tools developed to support automated and semi-automated analysis of images. Without a time machine, the sampling conducted now, and the images that are archived, will remain the only means to quantify how reefs are changing, and thus there is an urgent need to build new capacity into ongoing sampling regimes for coral communities (Flower et al., 2017; Kramer et al., 2020). These should include sampling of larger numbers of replicate and permanently marked areas with high resolution images that preserve colour, shape, and size information for even the smallest of coral colonies.

Supplementary materials

An R markdown file containing the mathematical model and the simulations needed to reproduce the paper's results can be found at https://doi.org/10.5281/zenodo.3926845).

Acknowledgements

This paper originated from a workshop held at the Boone Center for Science and Environmental Leadership at the Wrigley Marine Science Center on Santa Catalina Island (13-16 September 2019), and was generously supported with funding from the ARC Centre of Excellence for Coral Reef Studies.

Author contributions

MSP motivated the questions that gave rise to the workshop in which all authors participated. MMcW crafted an outline of the paper that was fleshed out and written as a complete draft by CP, PJE, HVM, and BR. HVM drafted the demographic model, the products of which were fueled by input from CP, PJE, and BR. All authors contributed to revisions and edits. MSP largely funded the workshop.

References

Anthony, K.R., Hoogenboom, M.O., Maynard, J.A., Grottoli, A.G., Middlebrook, R., 2009. Energetics approach to predicting mortality risk from environmental stress: a case study of coral bleaching. Funct. Ecol. 23, 539–550.

Babcock, R.C., 1991. Comparative demography of three species of scleractinian corals using age-and size-dependent classifications. Ecol. Monogr. 61, 225–244.

Bak, R.P., Meesters, E.H., 1999. Population structure as a response of coral communities to global change. Am. Zool. 39, 56–65.

Barkley, H.C., Cohen, A.L., Mollica, N.R., Brainard, R.E., Rivera, H.E., DeCarlo, T.M., Lohmann, G.P., Drenkard, E.J., Alpert, A.E., Young, C.W., 2018. Repeat bleaching of a Central Pacific coral reef over the past six decades (1960–2016). Commun. Biol. 1, 177.

Beijbom, O., Edmunds, P.J., Roelfsema, C., Smith, J., Kline, D.I., Neal, B.P., Dunlap, M.J., Moriarty, V., Fan, T.-Y., Tan, C.-J., 2015. Towards automated annotation of benthic survey images: variability of human experts and operational modes of automation. PLoS One 10, e0130312.

Bell, J.D., Galzin, R., 1984. Influence of live coral cover on coral-reef fish communities. Mar. Ecol. Prog. Ser. 15, 265–274.

Boyce, M.S., 1992. Population viability analysis. Annu. Rev. Ecol. Syst. 23, 481–497.

Bramanti, L., Iannelli, M., Fan, T.-Y., Edmunds, P.J., 2015. Using demographic models to project the effects of climate change on scleractinian corals: *Pocillopora damicornis* as a case study. Coral Reefs 34, 505–515.

Brown, J.H., Gillooly, J.F., Allen, A.P., Savage, V.M., West, G.B., 2004. Toward a metabolic theory of ecology. Ecology 85, 1771–1789.

Bruno, J.F., Selig, E.R., 2007. Regional decline of coral cover in the Indo-Pacific: timing, extent, and subregional comparisons. PLoS One 2, e711.

Bruno, J.F., Côté, I.M., Toth, L.T., 2019. Climate change, coral loss, and the curious case of the parrotfish paradigm: why don't marine protected areas improve reef resilience? Ann. Rev. Mar. Sci. 11, 307–334.

Burgess, S.C., Ryan, W.H., Blackstone, N.W., Edmunds, P.J., Hoogenboom, M.O., Levitan, D.R., Wulff, J.L., 2017. Metabolic scaling in modular animals. Invertebr. Biol. 136, 456–472.

Burns, J.H.R., Delparte, D., Gates, R.D., Takabayashi, M., 2015. Utilizing underwater three-dimensional modeling to enhance ecological and biological studies of coral reefs. Int. Arch. Photogramm. Remote Sens. Spat. Inf. Sci. 40, 61.

Bythell, J.C., Gladfelter, E.H., Bythell, M., 1993. Chronic and catastrophic natural mortality of three common Caribbean reef corals. Coral Reefs 12, 143–152.

Caley, M.J., Carr, M.H., Hixon, M.A., Hughes, T.P., Jones, G.P., Menge, B.A., 1996. Recruitment and the local dynamics of open marine populations. Annu. Rev. Ecol. Syst. 27, 477–500.
Cantin, N.E., Cohen, A.L., Karnauskas, K.B., Tarrant, A.M., McCorkle, D.C., 2010. Ocean warming slows coral growth in the central Red Sea. Science 329, 322–325.
Carpenter, K.E., Abrar, M., Aeby, G., Aronson, R.B., Banks, S., Bruckner, A., Chiriboga, A., Cortés, J., Delbeek, J.C., DeVantier, L., 2008. One-third of reef-building corals face elevated extinction risk from climate change and local impacts. Science 321, 560–563.
Caswell, H., 2001. Matrix Population Models: Construction, Analysis, and Interpretation, second edn Sinauer Associates, Inc Sunderland MA.
Chesson, P.L., 1984. The storage effect in stochastic population models. In: Mathematical Ecology. Springer, pp. 76–89.
Condit, R., Pérez, R., Lao, S., Aguilar, S., Hubbell, S.P., 2017. Demographic trends and climate over 35 years in the Barro Colorado 50 ha plot. For. Ecosyst. 4, 17.
Connell, J.H., 1973. Population ecology of reef-building corals. Biol. Geol. Coral Reefs 2, 205–245.
Darke, W.M., Barnes, D.J., 1993. Growth trajectories of corallites and ages of polyps in massive colonies of reef-building corals of the genus *Porites*. Mar. Biol. 117, 321–326.
De'ath, G., Lough, J.M., Fabricius, K.E., 2009. Declining coral calcification on the Great Barrier Reef. Science 323, 116–119.
De'ath, G., Fabricius, K.E., Sweatman, H., Puotinen, M., 2012. The 27–year decline of coral cover on the Great Barrier Reef and its causes. Proc. Natl. Acad. Sci. 109, 17995–17999.
Devlin-Durante, M.K., Miller, M.W., Caribbean Acropora Research Group, Precht, W.F., Baums, I.B., Carne, L., Smith, T.B., Banaszak, A.T., Greer, L., Irwin, A., 2016. How old are you? Genet age estimates in a clonal animal. Mol. Ecol. 25, 5628–5646.
Doney, S.C., Ruckelshaus, M., Duffy, J.E., Barry, J.P., Chan, F., English, C.A., Galindo, H.M., Grebmeier, J.M., Hollowed, A.B., Knowlton, N., 2012. Climate change impacts on marine ecosystems. Annu. Rev. Mar. Sci. 4, 11–37.
Dornelas, M., Madin, J.S., Baird, A.H., Connolly, S.R., 2017. Allometric growth in reef-building corals. Proc. R. Soc. B Biol. Sci. 284, 20170053.
Dunstan, P.K., Johnson, C.R., 1998. Spatio-temporal variation in coral recruitment at different scales on Heron Reef, southern Great Barrier Reef. Coral Reefs 17, 71–81.
Eakin, C.M., Rauenzahn, J.L., Liu, G., Heron, S.F., Skirving, W.J., Geiger, E.F., Burgess, T.F., Strong, A.E., 2014. Will 2014–2015 be the next big El Niño? If so, what might it mean for coral reefs? Reef Encount. 29, 30–35.
Eakin, C.M., Liu, G., Gomez, A.M., De La Cour, J.L., Heron, S.F., Skirving, W.J., Geiger, E.F., Tirak, K.V., Strong, A.E., 2016. Global coral bleaching 2014–2017: status and an appeal for observations. Reef Encount. 31, 20–26.
Edmunds, P.J., 2000. Patterns in the distribution of juvenile corals and coral reef community structure in St. John, US Virgin Islands. Mar. Ecol. Prog. Ser. 202, 113–124.
Edmunds, P.J., 2007. Evidence for a decadal-scale decline in the growth rates of juvenile scleractinian corals. Mar. Ecol. Prog. Ser. 341, 1–13.
Edmunds, P.J., 2013. Decadal-scale changes in the community structure of coral reefs of St. John, US Virgin Islands. Mar. Ecol. Prog. Ser. 489, 107–123.
Edmunds, P.J., 2015. A quarter-century demographic analysis of the Caribbean coral, Orbicella annularis, and projections of population size over the next century. Limnol. Oceanogr. 60, 840–855.
Edmunds, P.J., Burgess, S.C., 2016. Size-dependent physiological responses of the branching coral Pocillopora verrucosa to elevated temperature and PCO2. J. Exp. Biol. 219, 3896–3906.

Edmunds, P.J., Elahi, R., 2007. The demographics of a 15-year decline in cover of the Caribbean reef coral Montastraea annularis. Ecol. Monogr. 77, 3–18.
Edmunds, P.J., Lasker, H.R., 2016. Cryptic regime shift in benthic community structure on shallow reefs in St. John, US Virgin Islands. Mar. Ecol. Prog. Ser. 559, 1–12.
Edmunds, P.J., Riegl, B., 2020. Urgent need for coral demography in a world where corals are disappearing. Mar. Ecol. Prog. Ser. 635, 233–242.
Ferrari, R., Figueira, W.F., Pratchett, M.S., Boube, T., Adam, A., Kobelkowsky-Vidrio, T., Doo, S.S., Atwood, T.B., Byrne, M., 2017. 3D photogrammetry quantifies growth and external erosion of individual coral colonies and skeletons. Sci. Rep. 7, 16737.
Flower, J., Ortiz, J.C., Chollett, I., Abdullah, S., Castro-Sanguino, C., Hock, K., Lam, V., Mumby, P.J., 2017. Interpreting coral reef monitoring data: a guide for improved management decisions. Ecol. Indic. 72, 848–869.
Fong, P., Glynn, P.W., 1998. A dynamic size-structured population model: does disturbance control size structure of a population of the massive coral Gardineroseris planulata in the Eastern Pacific? Mar. Biol. 130, 663–674.
Fong, P., Glynn, P.W., 2001. Population abundance and size-structure of an eastern tropical Pacific reef coral after the 1997–98 ENSO: a simulation model predicts field measures. Bull. Mar. Sci. 69, 187–202.
Gouezo, M., Golbuu, Y., Fabricius, K., Olsudong, D., Mereb, G., Nestor, V., Wolanski, E., Harrison, P., Doropoulos, C., 2019. Drivers of recovery and reassembly of coral reef communities. Proc. R. Soc. B 286, 20182908.
Graham, E.M., Baird, A.H., Connolly, S.R., 2008. Survival dynamics of scleractinian coral larvae and implications for dispersal. Coral Reefs 27, 529–539.
Graham, N.A., Jennings, S., MacNeil, M.A., Mouillot, D., Wilson, S.K., 2015. Predicting climate-driven regime shifts versus rebound potential in coral reefs. Nature 518, 94–97.
Gunderson, L.H., 2000. Ecological resilience—in theory and application. Annu. Rev. Ecol. Syst. 31, 425–439.
Hall, V.R., Hughes, T.P., 1996. Reproductive strategies of modular organisms: comparative studies of reef-building corals. Ecology 77, 950–963.
Harrison, P.L., Wallace, C.C., 1990. Reproduction, dispersal and recruitment of scleractinian corals. Ecosyst. World 25, 133–207.
Henry, L.-A., Hart, M., 2005. Regeneration from injury and resource allocation in sponges and corals–a review. Int. Rev. Hydrobiol. 90, 125–158.
Hernández-Pacheco, R., Hernández-Delgado, E.A., Sabat, A.M., 2011. Demographics of bleaching in a major Caribbean reef-building coral: *Montastraea annularis*. Ecosphere 2, 1–13.
Hoegh-Guldberg, O., Bruno, J.F., 2010. The impact of climate change on the world's marine ecosystems. Science 328, 1523–1528.
Hoegh-Guldberg, O., Mumby, P.J., Hooten, A.J., Steneck, R.S., Greenfield, P., Gomez, E., Harvell, C.D., Sale, P.F., Edwards, A.J., Caldeira, K., et al., 2007. Coral reefs under rapid climate change and ocean acidification. Science 318, 1737–1742.
Hughes, T.P., 1984. Population dynamics based on individual size rather than age: a general model with a reef coral example. Am. Nat. 123, 778–795.
Hughes, T.P., 1994. Catastrophes, phase shifts, and large-scale degradation of a Caribbean coral reef. Science 265, 1547–1551.
Hughes, R.N., 2005. Lessons in modularity: the evolutionary ecology of colonial invertebrates. Sci. Mar. 69, 169–179.
Hughes, T.P., Jackson, J.B.C., 1980. Do corals lie about their age? Some demographic consequences of partial mortality, fission, and fusion. Science 209, 713–715.
Hughes, T.P., Jackson, J.B.C., 1985. Population dynamics and life histories of foliaceous corals. Ecol. Monogr. 55, 141–166.

Hughes, T.P., Tanner, J.E., 2000. Recruitment failure, life histories, and long-term decline of Caribbean corals. Ecology 81, 2250–2263.
Hughes, T.P., Ayre, D., Connell, J.H., 1992. The evolutionary ecology of corals. Trends Ecol. Evol. 7, 292–295.
Hughes, T.P., Graham, N.A., Jackson, J.B., Mumby, P.J., Steneck, R.S., 2010. Rising to the challenge of sustaining coral reef resilience. Trends Ecol. Evol. 25, 633–642.
Hughes, T.P., Linares, C., Dakos, V., Van De Leemput, I.A., Van Nes, E.H., 2013. Living dangerously on borrowed time during slow, unrecognized regime shifts. Trends Ecol. Evol. 28, 149–155.
Hughes, T.P., Barnes, M.L., Bellwood, D.R., Cinner, J.E., Cumming, G.S., Jackson, J.B., Kleypas, J., van de Leemput, I.A., Lough, J.M., Morrison, T.H., et al., 2017a. Coral reefs in the Anthropocene. Nature 546, 82–90.
Hughes, T.P., Kerry, J.T., Álvarez-Noriega, M., Álvarez-Romero, J.G., Anderson, K.D., Baird, A.H., Babcock, R.C., Beger, M., Bellwood, D.R., Berkelmans, R., et al., 2017b. Global warming and recurrent mass bleaching of corals. Nature 543, 373–377.
Hughes, T.P., Kerry, J.T., Baird, A.H., Connolly, S.R., Dietzel, A., Eakin, C.M., Heron, S.F., Hoey, A.S., Hoogenboom, M.O., Liu, G., 2018. Global warming transforms coral reef assemblages. Nature 556 (7702), 492–496.
Hughes, T.P., Kerry, J.T., Baird, A.H., Connolly, S.R., Chase, T.J., Dietzel, A., Hill, T., Hoey, A.S., Hoogenboom, M.O., Jacobson, M., Kerswell, A., Madin, J.S., Mieog, A., Paley, A.S., Pratchett, M.S., Torda, G., Woods, R.M., 2019. Global warming impairs stock–recruitment dynamics of corals. Nature 568 (7752), 387–390. https://doi.org/10.1038/s41586-019-1081-y.
Jackson, J.B.C., 1977. Competition on marine hard substrata: the adaptive significance of solitary and colonial strategies. Am. Nat. 111, 743–767.
Jackson, J.B., 2008. Ecological extinction and evolution in the brave new ocean. Proc. Natl. Acad. Sci. 105, 11458–11465.
Jackson, J.B.C., Coates, A.G., 1986. Life cycles and evolution of clonal (modular) animals. Philos. Trans. R. Soc. Lond. B Biol. Sci. 313, 7–22.
Jackson, J.B.C., Donovan, M.K., Cramer, K.L., Lam, V.V., 2014. Status and Trends of Caribbean Coral Reefs. Global Coral Reef Monitoring Network, IUCN, Gland.
Jokiel, P.L., Morrissey, J.I., 1986. Influence of size on primary production in the reef coral *Pocillopora damicornis* and the macroalga *Acanthophora spicifera*. Mar. Biol. 91, 15–26.
Kayal, M., Lenihan, H.S., Brooks, A.J., Holbrook, S.J., Schmitt, R.J., Kendall, B.E., 2018. Predicting coral community recovery using multi-species population dynamics models. Ecol. Lett. 21, 1790–1799.
Kojis, B.L., Quinn, N.J., Claerboudt, M.R., 1985. Living coral reefs of northeast New-Guinea. In: Proc. Fifth Int. Coral Reef Congr. Tahiti, pp. 323–328.
Kramer, P.A., 2003. Synthesis of coral reef health indicators for the western Atlantic: results of the AGRRA program (1997–2000). Atoll Res. Bull. 496 (3), 1–57.
Lindenmayer, D.B., Likens, G.E., Krebs, C.J., Hobbs, R.J., 2010. Improved probability of detection of ecological "surprises." Proc. Natl. Acad. Sci. 107, 21957–21962.
Kramer, N, Tamir, R, Eyal, G, Loya, Y, 2020. Coral morphology portrays the spatial distribution and population size-structure along a 5–100 m depth gradient. Front. Mar. Sci.
Lindenmayer, D.B., Burns, E.L., Tennant, P., Dickman, C.R., Green, P.T., Keith, D.A., Metcalfe, D.J., Russell-Smith, J., Wardle, G.M., Williams, D., 2015. Contemplating the future: acting now on long-term monitoring to answer 2050's questions. Austral Ecol. 40, 213–224.
Little, A.F., Van Oppen, M.J., Willis, B.L., 2004. Flexibility in algal endosymbioses shapes growth in reef corals. Science 304, 1492–1494.
Lough, J.M., Barnes, D.J., 1997. Several centuries of variation in skeletal extension, density and calcification in massive Porites colonies from the great barrier reef: a proxy for

seawater temperature and a background of variability against which to identify unnatural change. J. Exp. Mar. Biol. Ecol. 211, 29–67.
MacNeil, M.A., Mellin, C., Matthews, S., Wolff, N.H., McClanahan, T.R., Devlin, M., Drovandi, C., Mengersen, K., Graham, N.A., 2019. Water quality mediates resilience on the great barrier reef. Nat. Ecol. Evol. 3, 620–627.
Madin, J.S., Baird, A.H., Dornelas, M., Connolly, S.R., 2014. Mechanical vulnerability explains size-dependent mortality of reef corals. Ecol. Lett. 17, 1008–1015.
Madin, J.S., Baird, A.H., Baskett, M.L., Connolly, S.R., Dornelas, M.A., 2020. Partitioning colony size variation into growth and partial mortality. Biol. Lett. 16, 20190727.
McCauley, D.J., Pinsky, M.L., Palumbi, S.R., Estes, J.A., Joyce, F.H., Warner, R.R., 2015. Marine defaunation: animal loss in the global ocean. Science 347, 1255641.
Meesters, E.H., Hilterman, M., Kardinaal, E., Keetman, M., De Vries, M., Bak, R.P.M., 2001. Colony size-frequency distributions of scleractinian coral populations: spatial and interspecific variation. Mar. Ecol. Prog. Ser. 209, 43–54.
Mellin, C., Aaron MacNeil, M., Cheal, A.J., Emslie, M.J., Julian Caley, M., 2016. Marine protected areas increase resilience among coral reef communities. Ecol. Lett. 19, 629–637.
Mellin, C., Matthews, S., Anthony, K.R., Brown, S.C., Caley, M.J., Johns, K.A., Osborne, K., Puotinen, M., Thompson, A., Wolff, N.H., 2019. Spatial resilience of the great barrier reef under cumulative disturbance impacts. Glob. Change Biol. 25, 2431–2445.
Mieog, J.C., Olsen, J.L., Berkelmans, R., Bleuler-Martinez, S.A., Willis, B.L., van Oppen, M.J., 2009. The roles and interactions of symbiont, host and environment in defining coral fitness. PLoS One 4, e6364.
Muko, S., Arakaki, S., Tamai, R., Sakai, K., 2014. An individual-based model for population viability analysis of the brooding coral *Seriatopora hystrix*. Ecol. Model. 277, 68–76.
Ortiz, J.-C., Wolff, N.H., Anthony, K.R., Devlin, M., Lewis, S., Mumby, P.J., 2018. Impaired recovery of the great barrier reef under cumulative stress. Sci. Adv. 4, eaar6127.
Osborne, K., Thompson, A.A., Cheal, A.J., Emslie, M.J., Johns, K.A., Jonker, M.J., Logan, M., Miller, I.R., Sweatman, H.P., 2017. Delayed coral recovery in a warming ocean. Glob. Change Biol. 23, 3869–3881.
Patterson, M.R., 1992a. A chemical engineering view of cnidarian symbioses. Am. Zool. 32, 566–582.
Patterson, M.R., 1992b. A mass transfer explanation of metabolic scaling relations in some aquatic invertebrates and algae. Science 255, 1421–1423.
Pavoni, G, Corsini, M, Cignoni, P, 2020. A state of the art technology in large scale underwater monitoring. ERCIM News.
Perry, C.T., Alvarez-Filip, L., 2019. Changing geo-ecological functions of coral reefs in the Anthropocene. Funct. Ecol. 33, 976–988.
Perry, C.T., Alvarez-Filip, L., Graham, N.A., Mumby, P.J., Wilson, S.K., Kench, P.S., Manzello, D.P., Morgan, K.M., Slangen, A.B., Thomson, D.P., 2018. Loss of coral reef growth capacity to track future increases in sea level. Nature 558, 396.
Pisapia, C., Anderson, K.D., Pratchett, M.S., 2016. Temporal consistency in background mortality of four dominant coral taxa along Australia's great barrier reef. Coral Reefs 35, 839–849.
Pisapia, C., Burn, D., Pratchett, M.S., 2019. Changes in the population and community structure of corals during recent disturbances (February 2016–October 2017) on Maldivian coral reefs. Sci. Rep. Nat. Publ. Group 9, 1–12.
Platt, J.R., 1964. Strong inference. Science 146, 347–353.
Powles, H., Bradford, M.J., Bradford, R.G., Doubleday, W.G., Innes, S., Levings, C.D., 2000. Assessing and protecting endangered marine species. ICES J. Mar. Sci. 57, 669–676.

Pratchett, M.S., Anderson, K.D., Hoogenboom, M.O., Widman, E., Baird, A.H., Pandolfi, J.M., Edmunds, P.J., Lough, J.M., 2015. Spatial, temporal and taxonomic variation in coral growth—implications for the structure and function of coral reef ecosystems. Oceanogr. Mar. Biol. Annu. Rev. 53, 215–295.
Pratchett, M.S., McWilliam, M.J., Riegl, B., 2020. Contrasting shifts in coral assemblages with increasing disturbances. Coral Reefs, 1–11.
Putnam, H.M., Barott, K.L., Ainsworth, T.D., Gates, R.D., 2017. The vulnerability and resilience of reef-building corals. Curr. Biol. 27, R528–R540.
Riegl, B., Purkis, S., 2015. Coral population dynamics across consecutive mass mortality events. Glob. Change Biol. 21, 3995–4005.
Riegl, B.M., Bruckner, A.W., Rowlands, G.P., Purkis, S.J., Renaud, P., 2012. Red Sea coral reef trajectories over 2 decades suggest increasing community homogenization and decline in coral size. PLoS One 7, e38396.
Riegl, B., Cavalcante, G., Bauman, A.G., Feary, D.A., Steiner, S., Purkis, S., 2017. Demographic mechanisms of reef coral species winnowing from communities under increased environmental stress. Front. Mar. Sci. 4, 344.
Roff, G., Bejarano, S., Bozec, Y.-M., Nugues, M., Steneck, R.S., Mumby, P.J., 2014. Porites and the Phoenix effect: unprecedented recovery after a mass coral bleaching event at Rangiroa atoll, French Polynesia. Mar. Biol. 161, 1385–1393.
Roy, K.J., Smith, S.V., 1971. Sedimentation and Coral Reef Development in Turbid Water: Fanning Lagoon.
Schlichting, H., 1979. Boundary layer theory. McGraw-Hill, New York, Fig. CAPTIONS Solid Curve Disp. Exact Solut. Differ. Exact Solut. Eighth QLM Iteration T Fig. Than, p. 10.
Schmidt-Nielsen, K., Knut, S.-N., 1984. Scaling: Why Is Animal Size So Important? Cambridge University Press.
Sebens, K.P., 1987. The ecology of indeterminate growth in animals. Annu. Rev. Ecol. Syst. 18, 371–407.
Stimson, J.S., 1978. Mode and timing of reproduction in some common hermatypic corals of Hawaii and Enewetak. Mar. Biol. 48, 173–184.
Stuart-Smith, R.D., Brown, C.J., Ceccarelli, D.M., Edgar, G.J., 2018. Ecosystem restructuring along the great barrier reef following mass coral bleaching. Nature 560, 92.
Szmant, A., Gassman, N.J., 1990. The effects of prolonged "bleaching" on the tissue biomass and reproduction of the reef coral Montastrea annularis. Coral Reefs 8, 217–224.
Szmant-Froelich, A., 1985. The effect of colony size on the reproductive ability of the Caribbean coral Montastrea annularis (Ellis and Solander). In: Proc 5th Int Coral Reef Symp, pp. 295–300.
Talbot, F.H., 1965. A description of the coral structure of Tutia reef (Panganyika territory, East Africa), and its fish funa. In: Proceedings of the Zoological Society of London. Wiley Online Library, pp. 431–470.
Wakeford, M., Done, T.J., Johnson, C.R., 2008. Decadal trends in a coral community and evidence of changed disturbance regime. Coral Reefs 27, 1–13.
Walther, G.-R., Post, E., Convey, P., Menzel, A., Parmesan, C., Beebee, T.J., Fromentin, J.-M., Hoegh-Guldberg, O., Bairlein, F., 2002. Ecological responses to recent climate change. Nature 416, 389.
Weil, E., 2004. Coral reef diseases in the wider Caribbean. In: Coral Health and Disease. Springer, pp. 35–68.
Williams, I.D., Couch, C., Beijbom, O., Oliver, T., Vargas-Angel, B., Schumacher, B., Brainard, R., 2019. Leveraging automated image analysis tools to transform our capacity to assess status and trends on coral reefs. Front. Mar. Sci. 6, 222.
Wismer, S., Tebbett, S.B., Streit, R.P., Bellwood, D.R., 2019. Spatial mismatch in fish and coral loss following 2016 mass coral bleaching. Sci. Total Environ. 650, 1487–1498.

CHAPTER THREE

Population dynamics of diseased corals: Effects of a Shut Down Reaction outbreak in Puerto Rican *Acropora cervicornis*

Alex E. Mercado-Molina[a,*], Alberto M. Sabat[b], and Edwin A. Hernández-Delgado[a]
[a]Sociedad Ambiente Marino, San Juan, Puerto Rico
[b]University of Puerto Rico, San Juan, Puerto Rico
[*]Corresponding author: e-mail address: alexmercado@sampr.org

Contents

Abstract

Chronic coral reef degradation has been characterized by a significant decline in the population abundance and live tissue cover of scleractinian corals across the wider Caribbean. *Acropora cervicornis* is among the species whose populations have suffered an unprecedented collapse throughout the region. This species, which once dominated the shallow-water reef communities, is susceptible to a wide range of stressors, resulting in a general lack of recovery following disturbances. *A. cervicornis* is a critical contributor to the structure, function, and resilience of Caribbean coral reefs. Therefore, it is essential to identify the factors that influence their demographic and population performance. Diseases are one of the factors that are compromising the recovery of coral populations. In this chapter, we use size-based population matrix models to evaluate the population-level effect of a Shut Down Reaction Disease (SDR) outbreak, one of the less-understood diseases affecting this coral. The model was parameterized by following the fate of 105 colonies for 2 years at Tamarindo reef in Culebra, Puerto Rico. SDR, which affected 78% of the population, led to a rapid decline in colony abundance. The estimated population

Advances in Marine Biology, Volume 87
ISSN 0065-2881
https://doi.org/10.1016/bs.amb.2020.08.001

growth rate (λ) for the diseased population was more than six times lower than would be expected for a population at equilibrium. It was found that colonies in the smaller size class (≤100 cm total linear length) were more likely to get infected and succumbing to the disease than larger colonies. Model simulations indicate that: (1) under the estimated λ, the population would reach extinction in 5 years; (2) an SDR outbreak as intense as the one observed in this study can lead to a notable decline in stochastic λs even when relatively rare (i.e. 10% probability of occurring); and (3) disease incidence as low as 5% can cause the population to lose its ecological functionality (e.g., reach a pseudo-extinction level of 10% of the initial population size) 33 years before disappearing. SDR and probably any other similarly virulent disease could thus be a major driver of local extinction events of *A. cervicornis*.

1. Introduction

With approximately 180 species (Veron, 2000), *Acropora* is one of the most common Scleractinian genera to inhabit tropical coral reefs. These corals are characterized by a branched/tabular morphology that often serves as a habitat and refuge for other vertebrate and invertebrate species, thus promoting the reef's ecological diversity. Acroporids are considered among the fastest-growing corals, with some species exceeding 20–30 cm linear extension per year (Crabbe and Smith, 2005; Weil et al., 2020; Yap and Gomez, 1984). Many of them also show high morphological plasticity in response to prevailing environmental conditions; an adaptation that broadens their range of habitats within distinct turbulence regimes. Under ideal conditions, their rapid growth and morphological plasticity provide acroporids with a competitive advantage in colonizing the reef substrate (Darling et al., 2012; Linares et al., 2011). Several species within this genus can monopolize large areas forming what it is known as coral thickets. Rapid growth makes acroporids a significant contributor to reef accretion and a vital functional group to maintain the functionality of coral reefs. Reefs dominated by acroporids have a coral community calcification potential of up to 62% higher than that of coral communities dominated by non-branching corals such as *Porites*, *Pavona*, and *Orbicella* (Cabral-Tena et al., 2018) and many acroporid-depleted reefs have negative $CaCO_3$ budgets with bioerosion exceeding $CaCO_3$ deposition (de Bakker et al., 2019; Perry et al., 2015).

According to the theory of "energetic trade-off dynamics" (Bazzaz et al., 1987), investing energy in accelerated growth may reduce the energy available to acroporids for reproduction, defence, maintenance, and immunity

(Palmer et al., 2010). This life strategy was effective in the past, and Caribbean acroporids covered about 90% of the substrate of flat reefs (Toth et al., 2019). Today, however, this life strategy may represent a disadvantage. The optimal development of acroporids occurs in oligotrophic warm waters that do not exceed 32 °C, and in areas free of aggressive competitive organisms. These environmental conditions have proven to be easily disrupted by multiple physical and biological factors, which have been exacerbated by a wide array of anthropogenic activities. Current change in climatic patterns are resulting in seawater temperatures exceeding the thermal optimum, can increase the frequency and intensity of hurricanes, and can provide a competitive advantage to non-coral benthic species (Kersting et al., 2015).

Numerous studies have demonstrated that the demographic performance of acroporids has been adversely affected by alterations to environmental conditions. Even in the absence of perceptible disturbances, survival and growth of acroporids (e.g. *Acropora cervicornis*) can be compromised (Mercado-Molina et al., 2015a). As ocean health continues to deteriorate, it could be argued that growing fast at the expense of a "poor" defence and maintenance system (immunology) and reproduction, puts the viability of acroporid populations at risk. It may be the case that under current conditions a life history that favours growth over maintenance and reproduction is not optimal, a contention supported by the inadequate capacity of many, but especially Caribbean, acroporids to recover after disturbances (Bramanti and Edmunds, 2016; Darling et al., 2012; Goergen et al., 2019). Indeed, approximately 74% of the acroporid species with enough demographic information to determine their population status can be categorized under the level of, at least, species of concern (IUCN, 2020) and about 50% as threatened (Carpenter et al., 2008).

In the Caribbean, the genus *Acropora* is represented by only two species, *A. cervicornis* and *A. palmata*; both are considered threatened and no other species can fill the ecological void that would result from the local and regional disappearance of these corals. Disease outbreaks have greatly reduced Caribbean *Acropora* and will likely continue to be a threat to their persistence, thus justifying a better understanding of disease-induced effects on the *Acropora* populations. The Caribbean has been regarded as a "coral disease research hot spot" as diseases emerge at a relatively fast pace and have high prevalence and virulence (Weil, 2004). In contrast to other physical (e.g. hurricanes), biological (e.g. predation) and anthropogenic (e.g. water pollution) stressors whose effects tend to be localized, diseases can spread rapidly across a sizeable area with the potential to develop into epizootic events if the conditions are favourable to speed up the rate of transmission (Goreau et al., 1998; Green and Bruckner, 2000; Hayes and Goreau, 1998). The potential of widespread distribution is exemplified by the development of the White Band Disease (WBD). When first recorded in the 1970s at Saint John, USVI (Gladfelter, 1982), WBD was also present at Curaçao, Sint Maarten, Antigua, Panamá, Nicaragua, and South Florida (Gladfelter, 1982; Peters et al., 1983). Currently, there are six diseases known to affect Caribbean acroporids (Croquer et al., 2006; Sutherland et al., 2004): 1—White band disease I (WBD-I, Gladfelter, 1982), 2—White band disease II (WBD-II, Ritchie, 1998), 3—Skeletal anomaly (SKA, Squires, 1965), 4—Shut Down Reaction (SDR, Antonius, 1981), 5—Black band disease (BBD, Sutherland et al., 2004) and 6—Caribbean ciliate infection (CCI, Croquer et al., 2006). Of those, WBD-I, SKA, SDR, and CCI infect both corals, BBD infects only *A. palmata* and WBD-II infects only *A. cervicornis*. All of the diseases have the potential to diminish the demographic performance of the two species.

Shut Down Reaction (SDR) is one of the less-understood diseases that can affect acroporids corals. It was described by Antonius (1977) as spontaneous disintegration of the tissue that can kill a coral fast. Injured colonies tend to be more susceptible and progression of the disease is faster on stressed corals suggesting that compromised colonies are less likely to overcome the condition (Antonius, 1981). There is no information about the potential pathogen causing SDR, but its onset has been associated with periods of suboptimal environmental conditions, such as increased sedimentation or high temperatures (Antonius, 1981). SDR (referred together with white band disease and bleaching as white syndromes) has also been correlated with an increase in corallivorous snails (Antonius and Riegl, 1997). But it remains unclear whether the snail spread the disease, or the disease made the corals more attractive for snail predation.

An SDR event affected a population of *A. cervicornis* in 2005 at the Island Municipality of Culebra, Northeastern Puerto Rico. Here we evaluate the population-level response of *A. cervicornis* to this disease because of the demonstrated detrimental impact on the demographics of acroporids (Aronson and Precht, 2001a; Gladfelter, 1982; Riegl et al., 2018). Little is known about population responses following infection, what life-cycle transition is most affected, what sizes are most susceptible to disease, and how fast is a population decline (if so) when affected by a disease. This information is helpful for developing targeted management, conservation, and restoration plans.

2. Material and methods

2.1 Study site

From 2003 to 2005, the demography of *A. cervicornis* was monitored at Tamarindo reef (TAM). TAM is located in the Island Municipality of Culebra, approximately 30 km east of Puerto Rico (Fig. 1). TAM is a shallow reef, 1–5 m in depth, with a consolidated bottom currently dominated by *Millepora* spp., *Pseudodiploria* spp., and *Porites* spp. Before the SDR outbreak that occurred during the summer of 2005, *A.cervicornis* used to be a common feature of the benthic sessile community structure, due, in part, to a restoration project launched by Sociedad Ambiente Marino (SAM) years before the onset of the disease. The *A. cervicornis* population at TAM was characterized by colonies scattered through the substrate rather than forming thickets.

2.2 Demographic data

The survival and growth of 105 colonies of *A. cervicornis* were tracked every month from December 2003 to December 2004 and then in June (when the diseased were first observed), September, October, and December of 2005. Each colony was identified with a numbered tag fixed with masonry nails to non-living substrate adjacent to the colony. Colonies were considered dead if no live tissue was distinguishable. Colony size at time t and time t + 1 was calculated as the sum of the lengths of all live portions of colony branches (Knowlton et al., 1990) and was defined as total linear length (TLL). All branches of a colony were measured in-situ to the nearest 0.5 cm using a measurement tape.

2.3 Population matrix models

We used the life history transition rates observed in the field to construct a size-based population matrix model (Caswell, 2001). The basic shape of such a model was:

$$\mathbf{n}_{t+1} = A^{*}\mathbf{n}_{t}.$$

where **n** is a column vector containing the number of colonies in each size category and time (t), and A is a square matrix of dimension **n** with the demographic transitions between size classes (survival, growth and shrinkage among size stages, as well as fertility, which is confined to the first row).

The population was divided in three size classes based on TLL (small: $\leq$100 cm; medium: 101–250 cm; large: $>$ 250 cm; following Mercado-Molina et al., 2015a; see Eq. 1). These three size classes were

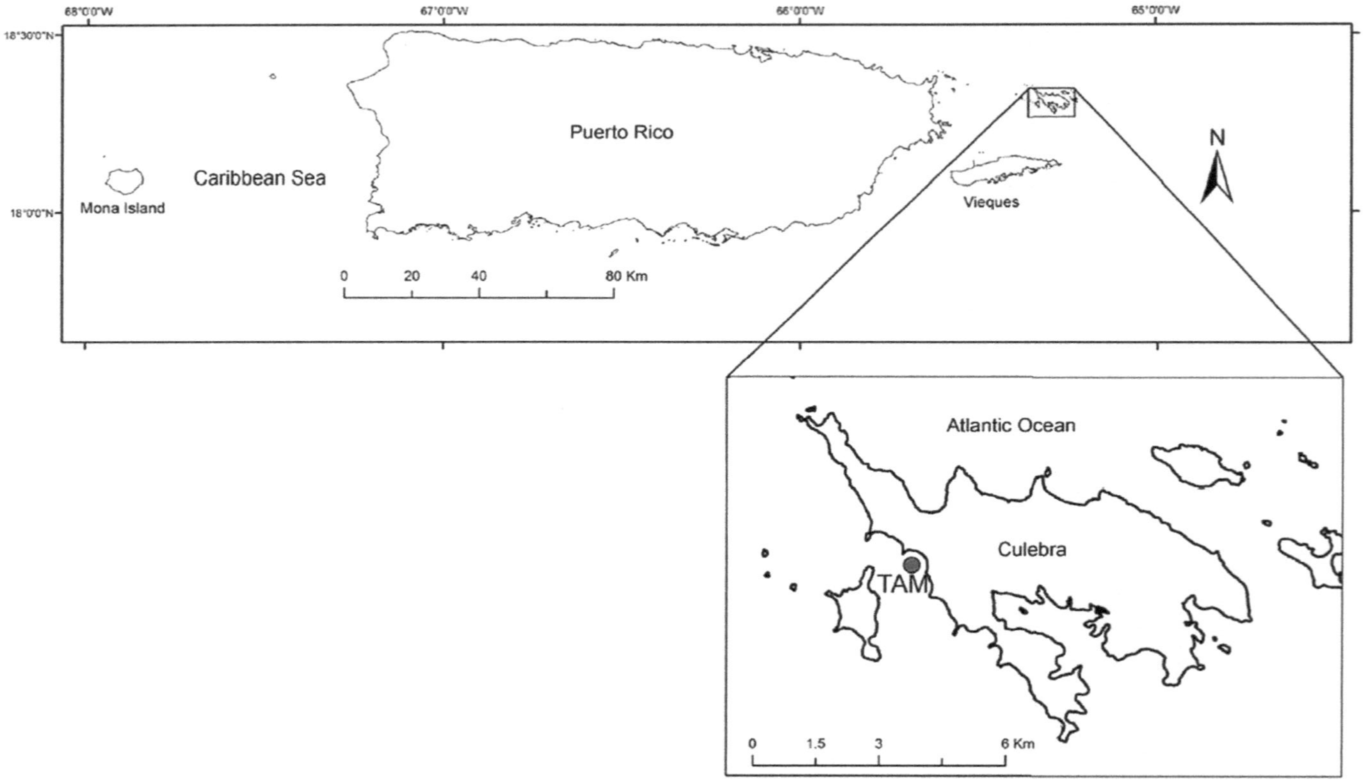

Fig. 1 Map showing the location where the Shut Down Reaction (SDR) outbreak occurred in 2005. TAM = Tamarindo.

chosen to incorporate size-specific patterns of vital transition rates while maintaining a sample size of a least 15 colonies for each size category (Mercado-Molina et al., 2015a). The diagonal elements in the matrix represent the probabilities of colonies surviving and remaining in their current size class (e.g. stasis, P_{ss}). The supra-diagonal (upper triangular) elements represent the contribution of the large-sized colonies classes to smaller ones by regressing in size due to partial mortality (e.g. retrogression or shrinkage, R_{sm}). The sub-diagonal (lower triangular) elements represent the probability of a colony to survive and advance to the next size class (e.g. growth, G_{ms}). Recruitment, either sexual or asexual, was not included in the model because no larval-derived recruits nor asexual fragments were observed during the study. Transition probabilities were estimated by constructing a transition frequency table (Caswell, 2001).

$$\begin{pmatrix} S \\ M \\ L \end{pmatrix}_{t+1} = \begin{pmatrix} P_{ss} & R_{sm} & R_{sl} \\ G_{ms} & P_{mm} & R_{ml} \\ G_{ls} & G_{lm} & P_{ll} \end{pmatrix} \times \begin{pmatrix} S \\ M \\ L \end{pmatrix}_{t} \quad (1)$$

The model was used to calculate the asymptotic rate of population growth (λ; the dominant right eigenvector of the matrix; Caswell, 2001) during the outbreak. The asymptotic analysis allows the evaluation of the long-term behaviour of the population (Caswell, 2001), which helps to determine possible population trajectories under different scenarios. The estimated λ and population trends for the diseased population were compared to those obtained from a model developed for a reference (healthy) population. This allowed determination of the effect of SDR on the population of *A. cervicornis*. The data for the reference population was obtained from Mercado-Molina et al. (2015a), who conducted a comprehensive study of the demography of *A. cervicornis* in Culebra. Transition matrices were subject to a bootstrapping re-sampling procedure (10,000 simulations) to obtain the 95% confidence intervals for λ (Caswell, 2001; Mercado-Molina et al., 2015a; Stubben and Milligan, 2007). Population trajectories were based on an initial vector of 1000 colonies with a colony size distribution characteristic of a healthy population (see Mercado-Molina et al., 2015a; Riegl et al., 2018). To determine the demographic transition (s) that accounted for most of the observed differences in growth rate between the healthy and diseased population, a retrospective analysis, Life Table Response Experiment (LTRE; Caswell, 2001; Riegl et al., 2017), was conducted.

We also modelled the effect of SDR on λ as a stochastic process by randomly selecting one of the two transition matrices constructed. A total of 1000 iterations (Stubben and Milligan, 2007) were performed in which the selected matrix was post-multiplied by the initial population vector to obtain the stochastic lambda (λs). During the iterations, transition matrices (healthy vs. diseased) had different probabilities of being selected. Probabilities varied between 0 and 1, in increments of 0.10, always assuring that the sum of the probabilities equalled 1. For instance, when the matrix representing the healthy population had a 0.90 probability of being selected, the matrix representing the diseased population had a 0.10 chance of being chosen.

A two-health state model (Eq. 2) was developed to determine how variation in disease incidence (i.e. the probability of infection of a healthy colony) would affect population growth rates. In the model, the population vector with three size classes was expanded to include two health states (healthy or diseased), which colonies of the three size classes could exhibit. The matrix was partitioned in four 3×3 submatrices each describing the transitional dynamics of this new population vector: (1) healthy colonies that remain healthy between t and t+1 (upper left submatrix); (2) healthy colonies that transition to a diseased state after infection (lower left submatrix); (3) diseased colonies that transition to a healthy state after healing (upper right submatrix); and (4) diseased colonies that remain diseased (lower right submatrix). As in Eq. (1), P, G, and R stand for stasis, growth, and retrogression, whereas h refers to healthy and d for diseased. The symbol α represents the probability of been infected, whereas β represents the probability of healing. The effect of SDR incidence rates on population trajectories was simulated by varying α from 0.00 to 0.50. On the other hand, β was set to 0 because no colony healed during the study. The trajectories resulting from the numerical simulations were used to determine both the time at which the population would disappear and the time to reach a quasi-extinction threshold of 10% of the initial population size. A Quasi-extinction threshold is a way of estimating the probability of extinction without knowing the critical level at which a given population becomes functionally extinct. Ten percent have been previously considered as a quasi-extinction threshold for *A. cervicornis* (Mercado-Molina et al., 2015a, 2018). Population trajectories were based on an initial vector of 1000 colonies with a colony size distribution characteristic of a healthy population (see Mercado-Molina et al., 2015a; Riegl et al., 2018).

All analyses were performed in R Core Team (2017) using the package Popbio (Stubben and Milligan, 2007).

$$
\begin{pmatrix} S_h \\ M_h \\ L_h \\ S_d \\ M_d \\ L_d \end{pmatrix}_{t+1}
=
\left(\begin{array}{ccc|ccc}
P_{sh,sh}(1-\alpha) & R_{sh,mh}(1-\alpha) & R_{sh,lh}(1-\alpha) & P_{sh,sd}\beta & R_{sh,md}\beta & R_{sh,ld}\beta \\
G_{mh,sh}(1-\alpha) & P_{mh,mh}(1-\alpha) & R_{mh,lh}(1-\alpha) & G_{mh,sd}\beta & P_{mh,md}\beta & R_{mh,ld}\beta \\
G_{lh,sh}(1-\alpha) & G_{lh,mh}(1-\alpha) & P_{lh,lh}(1-\alpha) & G_{lh,sd}\beta & G_{lh,md}\beta & P_{lh,ld}\beta \\
\hline
P_{sd,sh}\alpha & R_{sd,mh}\alpha & R_{sd,lh}\alpha & P_{sd,sd}(1-\beta) & R_{sd,md}(1-\beta) & R_{sd,ld}(1-\beta) \\
G_{md,sh}\alpha & P_{md,mh}\alpha & R_{md,lh}\alpha & G_{md,sd}(1-\beta) & P_{md,md}(1-\beta) & R_{md,ld}(1-\beta) \\
G_{ld,sh}\alpha & G_{ld,mh}\alpha & P_{ld,lh}\alpha & G_{ld,sd}(1-\beta) & G_{ld,md}(1-\beta) & P_{ld,ld}(1-\beta)
\end{array}\right)
\times
\begin{pmatrix} S_h \\ M_h \\ L_h \\ S_d \\ M_d \\ L_d \end{pmatrix}_{t}
\tag{2}
$$

3. Results

SDR affected 78% of the 105 *A. cervicornis* colonies that were evaluated from December 2003 to December 2005. The disease was very aggressive in the initial phases of the outbreak, first detected in June 2005. Seventy percent of the observed diseased colonies became infected within the first three months after the disease emerged. Overall survival rates declined from 100% before the outbreak, to about 30% at the end of the study (Fig. 2). The probability of being diseased ($\chi_{df,2}=11.20$; $P<0.004$) as well as the probability of survival after being infected ($\chi_{df,2}=9.85$; $P<0.007$) was related to colony size. For instance, the chance of small colonies (≤100 cm in TTL) getting the disease and succumbing to the infection was at least 79% and 105% higher than that of the larger colonies, respectively (Fig. 3).

SDR not only increased the probability of colony mortality but also reversed the dynamics of live tissue accumulation (Fig. 4). Before the outbreak, the cumulative live tissue increased linearly over time. However, the disease caused the population to lose living tissue at a rate of 26 linear cm per day. Small colonies lost live tissue at a more rapid pace (16 linear cm per day) than medium-sized colonies (7 linear cm per day) and large colonies (3 linear cm per day).

The outbreak of the SDR disease led to a considerable decline in the number of *A. cervicornis* colonies and the expectation of population stability. During the outbreak, the estimated λ was 0.16, a value 5.6 times lower than for a healthy population ($\lambda=0.90$) (Table 1) and over six times lower than would be expected for a population at equilibrium (i.e. $\lambda=1$). The threat of SDR to *A. cervicornis* population persistence was further demonstrated when

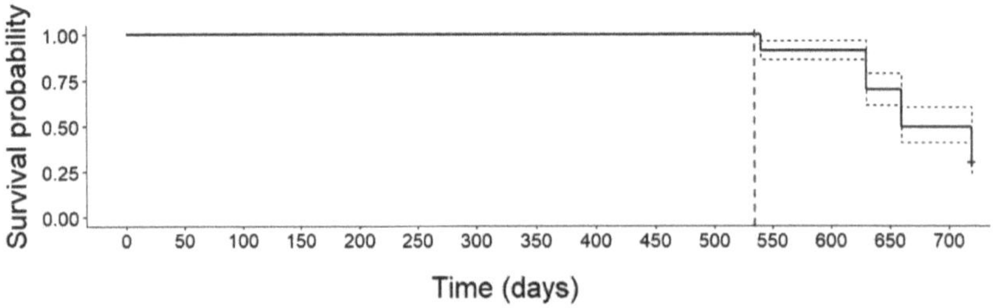

Fig. 2 Cumulative survival probability of *Acropora cervicornis* throughout the study period (2003–2005). The dotted vertical line represents the time at which the Shut Down Reaction (SDR) disease was first detected. Horizontal dotted lines represent the 95% confidence intervals.

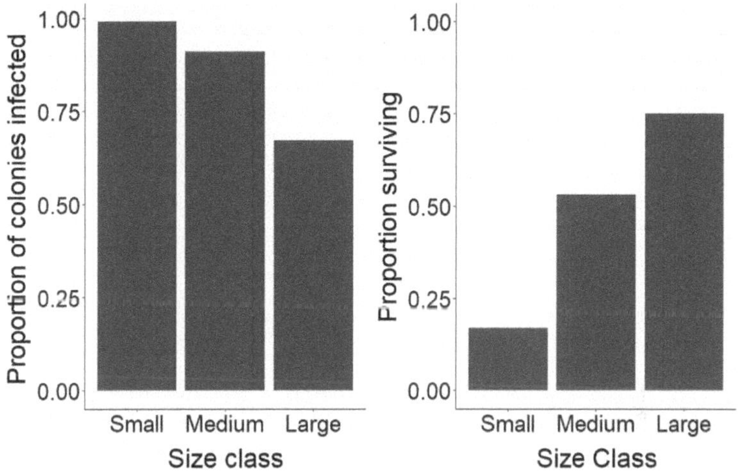

Fig. 3 Left: Size-specific susceptibility of *Acropora cervicornis* colonies of being infected by Shut Down Reaction disease (SDR). Right: Size-specific survival of *Acropora cervicornis* after being infected by SDR. Small ≤100cm in TLL; Medium = 101–250cm in TLL; Large ≥250cm in TLL. TLL = Total linear length.

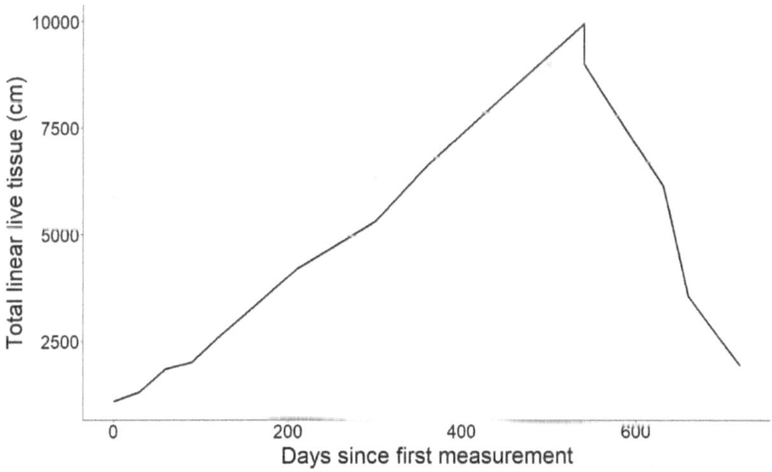

Fig. 4 Temporal change in the cumulative cover of live tissue (linear cm) of *Acropora cervicornis*.

projecting changes in colony abundances over time. A population characterized by the demographic transitions of diseased colonies (and based on an initial population size of 1000 colonies) can face extinction in about 5 years (Fig. 5). On the contrary, it would take a healthy population about 50 years to locally disappear (Fig. 5).

Table 1 Effect of increased probability of experiencing an Shut Down Reaction outbreak on *Acropora cervicornis* population growth rates.

Probability	λ	Lower limit	Upper limit	Percent decrease in λ
0	0.9079	0.9078	0.9100	–
0.1	0.8364	0.8349	0.8379	7.88
0.2	0.7544	0.7522	0.7566	9.80
0.3	0.6658	0.6632	0.6683	11.74
0.4	0.5750	0.5723	0.5777	13.64
0.5	0.4873	0.4847	0.4899	15.25
0.6	0.4074	0.4050	0.4097	16.40
0.7	0.3319	0.3300	0.3339	18.53
0.8	0.2638	0.2624	0.2652	20.52
0.9	0.2092	0.2083	0.2101	20.70
1	0.1600	0.1600	0.1600	23.52

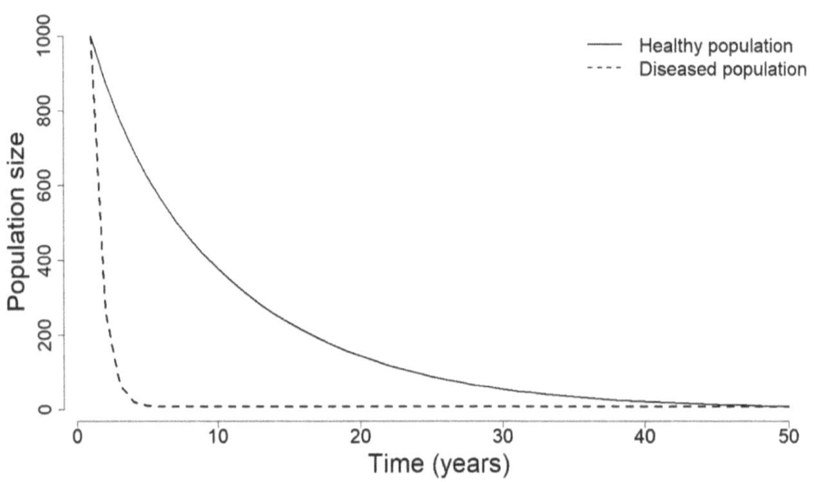

Fig. 5 Simulated population trajectories of *Acropora cervicornis* based on the size-class transition pattern observed for the diseased (dashed line) and healthy (straight line) populations.

Results of the LTRE analysis indicate that although large colonies are less likely to become infected, reduction in their probability to survive and remain in the same size class was the transition that contributed the most to the difference in λ between population states (healthy vs. diseased). There was also an important contribution from differences in the probability of a small colony to survive and grow to the next size class (Fig. 6).

Stochastic population growth rates decreased noticeably with an increased probability of a population suffering from SDR. In general, an increase of 10% in the chances of an outbreak event decreased λ by an average of 15.8% (range 8–24%, Table 1). Numerical simulations based on Eq. (2), indicate that the SDR can alter notably the dynamics of the population even under low incidence rates (Fig. 7). For instance, the population can reach a quasi-extinction threshold of 10% of the initial population size in less than 20 years under any disease incidence scenarios (Fig. 7). The trajectory of the population under a scenario of 5% disease incidence was similar to that of a healthy population in the sense that it would take approximately 50 years for the population to disappear. The time to extinction, however, is accelerated once disease incidence reaches 10%, with the population disappearing in ≤35 years (Fig. 7).

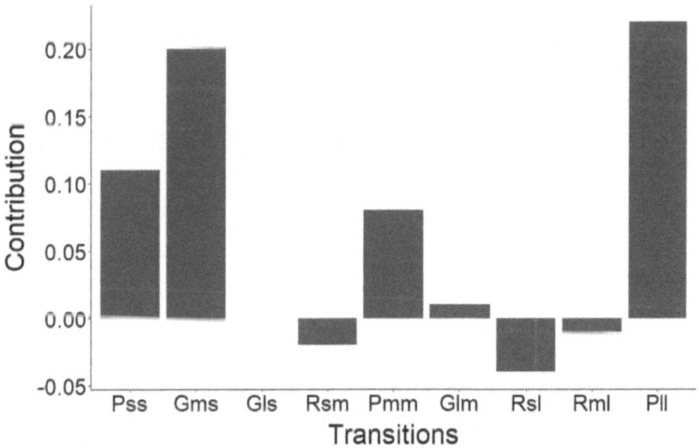

Fig. 6 Results of the life table response experiment analysis showing the contribution of each life cycle transition to differences in population growth rates of *Acropora cervicornis* between a healthy and a diseased population. P = stasis (a colony survived and remained in its original size class); G = growth (a colony survived but advanced to the next size class), R = retrogression (a colony survived but retrogressed to a smaller size class due to tissue loss), s = small-sized colonies, m = medium-sized colonies; l = large-sized colonies.

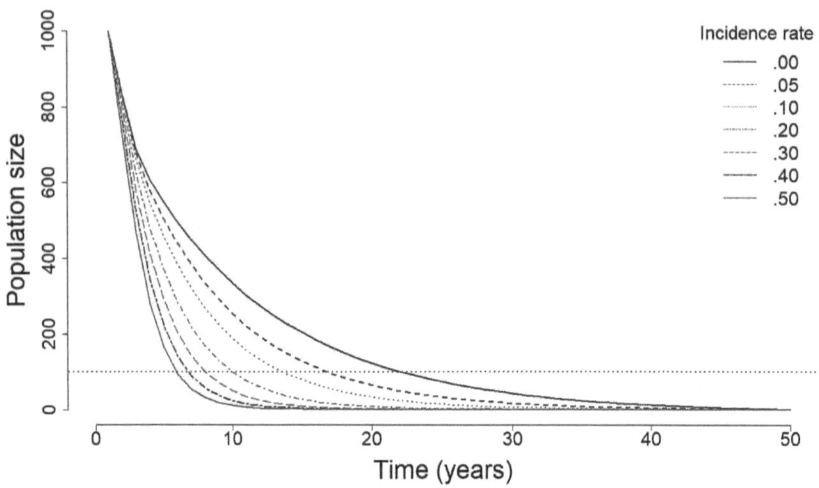

Fig. 7 Expected population trends as a function of disease incidence. The horizontal line represents the 10% quasi-extinction threshold.

4. Discussion

This study evaluated the effect of a Shut Down Reaction (SDR) disease outbreak on the population performance of the threatened coral *Acropora cervicornis*. It is not clearly understood what intrinsic or extrinsic factors trigger SDR emergence. The outbreak occurred during a time of high sediment discharge due to the combination of heavy rain and the deforestation of the coastline adjacent to the studied population. This observation concurs with Antonius (1981), who noted that sedimentation could be a causative factor leading to SDR. Antonius and Riegl (1997) found that SDR (referred as white syndrome) coincided with the presence of corallivores snails, which can be vectors for diseases (Williams and Miller, 2005). During the present study, however, no snail predators were feeding on coral colonies. Thus, it could be argued that changes in prevailing environmental conditions may be associated, at least in part, with the advent of the disease. High sediment loads could also explain the absence of sexual or asexual recruitment during the study. Sedimentation cannot only diminish the reproductive output of coral colonies by compromising fecundity (Kojis and Quinn, 1984) but can also preclude larval settlement and prevent fragment reattachment (Lirman, 2000; Mercado-Molina et al., 2014; Rogers, 1990; Speare et al., 2019).

SDR is highly contagious (Antonius, 1981), which may explain the high proportion (78%) of infected colonies in only 6 months. Little is known about how the disease is transmitted. The fact that the studied population was characterized by colonies scattered across the seedbed without being in physical contact suggests that the causative agent or pathogen may have been water-borne. For instance, it is possible that infection occurs when loose strands of contaminated tissue encounter healthy colonies (Antonius, 1981).

Small colonies were more likely to be infected. However, it is not clear whether the size-specific infection pattern was due to the higher number of small colonies within the population or because they were more susceptible to the disease. The fact that infected colonies in the smaller sized class ($\leq$100 cm in TLL) lost tissue more rapidly and were more likely to die than larger colonies point out to a possible trade-off in energy allocation among vital functions. Small coral colonies tend to invest more energy in growth to attain a "refuge" size in which they can cope better with stressful environments, i.e., higher competition (Hernández-Pacheco et al., 2011; Hughes, 1984). Diverting energetic resources to accelerate growth may come at the expense of maintenance (e.g. immunology), making small colonies more susceptible to infection, as proposed by the theory of "energetic trade-off dynamics" (Bazzaz et al., 1987).

As demonstrated by the model, SDR can endanger the persistence of *A. cervicornis* populations. Model simulations indicate that a disease event like the one observed in this study can lead to population disappearance in about 5 years. Such an outcome underlines three traits. First, many extant populations are smaller than the initial population vector used to forecast possible population trends (e.g. 1000 colonies). Second, most populations are characterized by small-sized colonies (Mercado-Molina et al., 2015a). Third, previous population studies have demonstrated that under current coral reef conditions (e.g. dominated by algae), recruitment rates are not adequate to promote population stability even under favourable environmental conditions (Mercado-Molina et al., 2015a). All factors combined can lead to population extinction faster than the model projected.

Disease outbreaks are highly dynamic, and their recurrence is not easily predicted. We modelled SDR as a stochastic process that could occur with a probability between 0 and 1 over 1000-time steps using a transition matrix developed for a healthy population as a reference point. Not surprisingly, there was a negative relationship between the probability of SDR occurrence and λ. Unexpectedly, however, SDR can lead to a notable decline in population growth rates even when relatively rare. For example, the estimated λ for a healthy population dropped from 0.90 to 0.84 in a scenario where SDR had only a 10% chance of occurring. The 7% (0.90 vs. 0.84) difference in λ may seem subtle, but it can result in the population reaching extinction approximately 10 years sooner than under a disease-free scenario. This suggests that even as an infrequent event, the effect of SDR on the population dynamics of *A. cervicornis* can be long-lasting; thereby, limiting the potential for population recovery. The result also supports the contention that this coral species is very susceptible to non-chronic disturbances.

The dynamics of coral epizootics depend on the prevailing environmental conditions (e.g. sediment loads) as well as in population-intrinsic factors (e.g. the spatial configuration of colonies, Sabat and Toledo-Hernández, 2015). Consequently, the proportion of diseased colonies within a population (e.g. prevalence) can vary considerably in space and time. One factor determining the prevalence of the disease is the rate at which new individuals are infected. In *A. cervicornis* populations, incidence, and thereby prevalence, can be lowered if colonies become resistant after being exposed to the disease (Libro and Vollmer, 2016; Vollmer and Kline, 2008) or if the colonies are dispersed in such a way as to limit the spread of infection. Because the population dynamics of *A. cervicornis* is inevitably linked to the demographic performance of diseased colonies, it could be expected that the population's behaviour would be conditional on the number of colonies infected. In this sense, we evaluated the effect of SDR as a function of disease incidence. Our analysis revealed that a 10% incidence rate would be enough to significantly alter population trajectories driving the population to extinction 15 years faster than what would be expected for a healthy population. These results highlight the susceptibility of *A. cervicornis* populations to even apparently low probabilities of being affected by a disease.

A population can become ecologically dysfunctional before disappearing. The size of a population at which it can become functionally extinct is known as a quasi-extinction threshold (Caswell, 2001; Ginzburg et al., 1982). Identifying such a level for *A. cervicornis* is essential to develop effective conservation and restoration strategies. Quasi-extinction thresholds, however, have not been established for any coral species (Hernández-Pacheco et al., 2011; but see Riegl et al., 2018). Following Mercado-Molina et al. (2015a,b), we considered 10% of the initial population size as a level in which the population could be considered as functionally extinct based on the poor reproductive output that characterize the species. Our simulation indicates that even in a scenario where the rate of infection is only

5%, a population can reach such a quasi-extinction level in 17 years; that is, 33 years before the population disappears. This period may be critical for conservation of the species, especially when taking into consideration that *A. cervicornis* is very susceptible to low/moderate variations in environmental conditions (Williams et al., 2006) and can suffer drastic declines in population growth rates very rapidly even in the absence of major disturbances (Mercado-Molina et al., 2015a). Similar dynamics of quasi-extinction was observed in Persian Gulf *A. downingi*, which suffer from disease mortality in the aftermath of heat events. Population response depended on the frequency of disturbance recurrence with populations entering a quasi-extinct state (persistence only at very low levels maintained by occasional recruitment) when crossing a threshold of seven disturbances in every 15-year period (Riegl et al., 2018).

The most evident non-lethal effect of diseases is the reduction in colony size due to tissue loss (Edmunds and Riegl, 2020; Mercado-Molina et al., 2018; Pisapia et al., 2020, in this volume; Riegl et al., 2017, 2018). Compared to large colonies, small-sized colonies show: 1) lower survival rates, 2) lower sexual reproductive performance (Soong and Lang, 1992), and 3) produce a lower number of branches, which is vital for the propagation of the Caribbean acroporids species (Mercado-Molina et al., 2016; Vardi et al., 2012). Moreover, the demographic performance of *A. cervicornis* is also associated with the amount of lost tissue (Mercado-Molina et al., 2018). Accordingly, small colonies' contribution to local population growth tends to be inferior to that of mid-to-large colonies (Mercado-Molina et al., 2015a). Hence, diseases can potentially reduce local population size by first rapidly killing colonies and indirectly by increasing the probability of larger colonies transitioning to lower-sized classes characterized by lower demographic performance. For example, the population growth rate (λ) of two healthy populations in Culebra comprised solely small colonies ($\leq$100 cm TLL) was estimated at 0.63 and 0.65 (Mercado Molina, unpublished data) compared to a λ of 0.90 estimated in this work.

The results of the LTRE further highlight the critical role of large colonies for the persistence of the population. The probability of a large colony to survive and remain in a large size-class (i.e. stasis), for instance, was the transition rate that best explained the difference in λ between population states (healthy vs. diseased). Similar findings were obtained for Persian Gulf *A. downingi* (Riegl et al., 2017). Such findings support previous studies demonstrating that large colonies contribute the most to the population growth of *A. cervicornis* in particular, and *Acropora* in general. Large colonies tend to survive better and produce more branches than small colonies as branching rates increase with colony size (Mercado-Molina et al., 2015b, 2016). This makes them a better source of asexual recruits. Therefore, conservation and restoration practices should be focused, whenever possible, on assuring that large coral colonies survive and remain large-sized. Populations composed of large colonies would potentially be better able to cope with a disease outbreak.

To the best of our knowledge, the present work is the first to evaluate the effect of the SDR in *A. cervicornis* at the population level. Projections from our modelling approach indicate that SDR could be a major driver of the population dynamic of the coral. Because *A. cervicornis* plays a vital role in promoting reef biodiversity, it could also be argued that SDR would have significant consequences for reef community dynamics. The expected rapid decline in population abundance, even under a low probability of occurring and low incidence, is undoubtedly worrisome. SDR could be as detrimental as other diseases like WBD (Aronson and Precht, 2001a,b; Gladfelter, 1982; Williams and Miller, 2012). Thus, understanding the aetiology and the dynamics of the disease is essential to determine the best approach to halt its spread when first observed and to prevent it from re-occurring.

The importance of SDR as a driver of population decline on *A. cervicornis* is particularly striking in light of its fossil history and importance in the pre-Anthropocene Caribbean. After arising in the Oligocene, *A. cervicornis* established dominance over Caribbean shallow-water coral reef communities (Budd et al., 1994; Johnson et al., 2008). Few subsequent changes (from the geological perspective) were observable during the Pleistocene and Holocene in the coral community assemblages (Aronson et al., 2002; Aronson and Precht, 1997; Budd and Johnson, 1997). This dominance of *A. cervicornis* throughout periods of high temperatures, variable salinity, hurricanes, and rapid sea-level rise (Greer et al., 2009) is remarkable because nowadays, this coral seems highly sensitive to stressful environmental conditions. That *A. cervicornis* was able to withstand previous climatic fluctuations (e.g. glacial periods) but have failed to cope with disease outbreaks emphasizes the danger of SDR.

The dynamics in modern reefs, like the fast disappearance of acroporid populations, seem to be without geological precedents (Aronson and Precht, 2001b; Greenstein et al., 1998; Greer et al., 2009; Wapnick et al., 2004). White band disease (WBD), possibly associated with unusually high seawater temperatures (Cramer et al., 2020a,b), may have been the catalyst leading to the demise of *A. cervicornis* populations (Aronson and Precht, 2001a,b; Gladfelter, 1982; Precht et al., 2020, in this volume). Nevertheless, reduction in region-wide abundances may have occurred decades before disease outbreaks (Cramer et al., 2012, Cramer et al., 2020a,b; but see Precht et al., 2020, in this volume). Pre-disease population declines have been associated with anthropogenic activities such as deforestation and the use of fertilizers (Cramer et al., 2012, 2020a,b). As human populations across zones near coral reefs increases, so does the pressure to local land-based resources such as forested areas. The clearance of forested lands coupled with expanded impervious areas would intensify sediment-laden runoff and nutrient discharge along coastal waters during strong precipitation events (especially during hurricanes; Otaño-Cruz et al., 2019) which could have devastating consequences for coral populations; especially if sedimentation is, indeed, the underlying factor prompting diseases like SDR.

Acknowledgements

Logistical and financial support was provided by Sociedad Ambiente Marino. The authors also thanks two anonymous reviewers as well as Bernhard Riegl for their useful comments that improved the earlier version of the manuscript.

References

Antonius, A., 1977. Coral mortality in reef: a problem for science and management. In: Proc. 3rd Int. Coral. Reef. Symp, 2, pp. 618–623.

Antonius, A., 1981. Coral reef pathology: a review. In: Proc. 4th Int. Coral. Reef. Symp, pp. 18–22.

Antonius, A., Riegl, B., 1997. A possible link between coral diseases and a corallivorous snail (*Drupella cornus*) outbreak in the Red Sea. Atoll Res. Bull. 447, 1–9.

Aronson, R.B., Precht, W.F., 1997. Stasis, biological disturbance, and community structure of a Holocene coral reef. Paleobiology 23 (3), 326–346.

Aronson, R.B., Precht, W.F., 2001a. White-band disease and the changing face of Caribbean coral reefs. Hydrobiologia 460, 25–38.

Aronson, R.B., Precht, W.F., 2001b. Evolutionary paleoecology of Caribbean coral reefs. In: Allmon, W.D., Bottjer, D.J. (Eds.), Evolutionary Paleoecology: The Ecological Context of Macroevolutionary Change. Columbia University Press, New York, pp. 171–233.

Aronson, R.B., Macintyre, I.G., Precht, W.F., Murdoch, T.J., Wapnick, C.M., 2002. The expanding scale of species turnover events on coral reefs in Belize. Ecol. Monogr 72 (2), 233–249.

Bazzaz, F.A., Chiariello, N.R., Coley, P.D., Pitelka, L.F., 1987. Allocating resources to reproduction and defense. Bioscience 37 (1), 58–67.

Bramanti, L., Edmunds, P.J., 2016. Density-associated recruitment mediates coral population dynamics on a coral reef. Coral Reefs 35 (2), 543–553.

Budd, A.F., Johnson, K.G., 1997. Coral reef community dynamics over 8 million years of evolutionary time; stasis and turnover. In: Proc of the 8th Int Coral Reef Symp. 8, p. 423.

Budd, A.N.N.F., Stemann, T.A., Johnson, K.G., 1994. Stratigraphic distributions of genera and species of Neogene to recent Caribbean reef corals. J. Paleo. 68 (5), 951–977.

Cabral-Tena, R.A., López-Pérez, A., Reyes-Bonilla, H., Calderon-Aguilera, L.E., Norzagaray-López, C.O., Rodríguez-Zaragoza, F.A., Cupul-Magaña, A., Rodríguez-Troncoso, A.P., Ayala-Bocos, A., 2018. Calcification of coral assemblages in the eastern Pacific: reshuffling calcification scenarios under climate change. Ecol. Ind. 95, 726–734.

Carpenter, K., Abrar, M., Aeby, G., Aronson, R.B., Banks, S., Bruckner, A., Chiriboga, A., Cortes, J., Delbeek, J.C., DeVantier, L., Graham, J.E., Edwards, A.J., Fenner, D., Guzma'n, H.M., Hoeksema, B.W., Hodgson, G., Johan, O., Licuanan, W.Y., Livingstone, S.R., Lovell, E.R., Moore, J.A., Obura, D.O., Ochavillo, D., Polidoro, B.A., Precht, W.F., Quibilan, M.C., Reboton, C., Richards, Z.T., Rogers, A.D., Sanciangco, J., Sheppard, A., Sheppard, C., Smith, J., Stuart, S., Turak, E., JEN, V., Wallace, C., Weil, E., Wood, E., 2008. One-third of reef-building corals face elevated extinction risk from climate change and local impacts. Science 321, 560–563.

Caswell, H., 2001. Matrix Population Models, second edn. John Wiley and Sons, Ltd.

Crabbe, M.J.C., Smith, D.J., 2005. Sediment impacts on growth rates of Acropora and Porites corals from fringing reefs of Sulawesi, Indonesia. Coral Reefs 24 (3), 437–441.

Cramer, K.L., Jackson, J.B., Angioletti, C.V., Leonard-Pingel, J., Guilderson, T.P., 2012. Anthropogenic mortality on coral reefs in Caribbean Panama predates coral disease and bleaching. Ecol. Lett. 15 (6), 561–567.

Cramer, K.L., Jackson, J.B.C., Donovan, M.K., Greenstein, B.J., Korpanty, C.A., Cook, G.M., Pandolfi, J.M., 2020a. Widespread loss of Caribbean acroporid corals was underway before coral bleaching and disease outbreaks. Sci. Adv 6 (17). p. eaax9395.

Cramer, K.L., O'Dea, A., Leonard-Pingel, J.S., Norris, R.D., 2020b. Millennial-scale change in the structure of a Caribbean reef ecosystem and the role of human and natural disturbance. Ecography 43 (2), 283–293.

Croquer, A., Bastidas, C., Lipscomp, D., Rodriguez-Martinez, R.E., Jordan-Dahlgren, E., Guzman, H.M., 2006. First report of folliculinid ciliates affecting Caribbean scleractinian corals. Coral Reefs 25 (3), 187–191.

Darling, E.S., Alvarez-Filip, L., Oliver, T.A., McClanahan, T.R., Côté, I.M., 2012. Evaluating life-history strategies of reef corals from species traits. Ecol. Lett. 15 (12), 1378–1386.

de Bakker, D.M., van Duyl, F.C., Perry, C.T., Meesters, E.H., 2019. Extreme spatial heterogeneity in carbonate accretion potential on a Caribbean fringing reef linked to local human disturbance gradients. Glob. Chang. Biol. 25 (12), 4092–4104.

Edmunds, P.J., Riegl, B., 2020. Urgent need for coral demography in a world where corals are disappearing. Mar. Ecol. Prog. Ser. 635, 233–242.

Ginzburg, L.R., Slobodkin, L.B., Johnson, K., Bindman, A.G., 1982. Quasiextinction probabilities as a measure of impact on population growth. Risk Anal. 2 (3), 171–181.

Gladfelter, W.B., 1982. White-band disease in *Acropora palmata*: implications for the structure and growth of shallow reef. Bull. Mar. Sci. 32 (2), 639–643.

Goergen, E.A., Moulding, A.L., Walker, B.K., Gilliam, D.S., 2019. Identifying causes of temporal changes in *Acropora cervicornis* populations and the potential for recovery. Front. Mar. Sci. 6 (FEB), 1–13.

Goreau, T.J., Cervino, J., Goreau, M., Hayes, R., Hayes, M., Richardson, L., et al., 1998. Rapid spread of diseases in Caribbean coral reefs. Rev. Biol. Trop. 46, 157–171.

Green, E.P., Bruckner, A.W., 2000. The significance of coral disease epizootiology for coral reef conservation. Biol. Conserv. 96 (3), 347–361.

Greenstein, B.J., Curran, H.A., Pandolfi, J.M., 1998. Shifting ecological baselines and the demise of *Acropora cervicornis* in the western North Atlantic and Caribbean Province: a Pleistocene perspective. Coral Reefs 17 (3), 249–261.

Greer, L., Jackson, J.E., Curran, H.A., Guilderson, T., Teneva, L., 2009. How vulnerable is *Acropora cervicornis* to environmental change? Lessons from the early to middle Holocene. Geology 37 (3), 263–266.

Hayes, R.L., Goreau, N.I., 1998. The significance of emerging diseases in the tropical coral reef ecosystem. Rev. Biol. Trop. 46, 173–185.

Hernández-Pacheco, R., Hernández-Delgado, E.A., Sabat, A.M., 2011. Demographics of bleaching in a major Caribbean reef-building coral: montastraea annularis. Ecosphere 2 (1), 1–13.

Hughes, T.P., 1984. Population dynamics based on individual size rather than age: a general model with a reef coral example. Am. Nat. 123 (6), 778–795.

IUCN, 2020. The IUCN Red List of Threatened Species. Version 2020-2. https://www.iucnredlist.org.

Johnson, K.G., Jackson, J.B., Budd, A.F., 2008. Caribbean reef development was independent of coral diversity over 28 million years. Science 319 (5869), 1521–1523.

Kersting, D.K., Cebrian, E., Casado, C., Teixidó, N., Garrabou, J., Linares, C., 2015. Experimental evidence of the synergistic effects of warming and invasive algae on a temperate reef-builder coral. Sci. Rep. 5 (18635), 1–8.

Knowlton, N., Lang, J.C., Keller, B.D., 1990. Case study of natural population collapse: post-hurricane predation on Jamaican staghorn corals. Smithson Contrib. Mar. Sci., 1–25.

Kojis, B.L., Quinn, N.J., 1984. Seasonal and depth variation in fecundity of *Acropora palifera* at two reefs in Papua New Guinea. Coral Reefs 3 (3), 165–172.

Libro, S., Vollmer, S.V., 2016. Genetic signature of resistance to white band disease in the Caribbean staghorn coral *Acropora cervicornis*. PloS One 11 (1), e0146636.

Linares, C., Pratchett, M.S., Coker, D.J., 2011. Recolonisation of *Acropora hyacinthus* following climate-induced coral bleaching on the Great Barrier Reef. Mar. Ecol. Prog. Ser. 438, 97–104.

Lirman, D., 2000. Fragmentation in the branching coral *Acropora palmata* (Lamarck): growth, survivorship, and reproduction of colonies and fragments. J. Exp. Mar. Biol. Ecol. 251 (1), 41–57.

Mercado-Molina, A.E., Ruiz-Diaz, C.P., Sabat, A.M., 2014. Survival, growth, and branch production of unattached fragments of the threatened hermatypic coral Acropora cervicornis. J. Exp. Mar. Biol. Ecol. 457, 215–219.

Mercado-Molina, A.E., Ruiz-Diaz, C.P., Pérez, M.E., Rodríguez-Barreras, R., Sabat, A.M., 2015a. Demography of the threatened coral *Acropora cervicornis*: implications for its management and conservation. Coral Reefs 34 (4), 1113–1124.

Mercado-Molina, A.E., Ruiz-Diaz, C.P., Sabat, A.M., 2015b. Demographics and dynamics of two restored populations of the threatened reef-building coral *Acropora cervicornis*. J. Nat. Conserv. 24, 17–23.

Mercado-Molina, A.E., Ruiz-Diaz, C.P., Sabat, A.M., 2016. Branching dynamics of transplanted colonies of the threatened coral *Acropora cervicornis*: morphogenesis, complexity, and modeling. J. Exp. Mar. Biol. Ecol. 482, 134–141.

Mercado-Molina, A.E., Ruiz-Diaz, C.P., Sabat, A.M., 2018. Tissue loss rather than colony size determines the demographic fate of the branching coral *Acropora cervicornis*. Mar. Ecol. Prog. Ser. 597, 147–159.

Otaño-Cruz, A., Montañez-Acuña, A.A., García-Rodríguez, N.M., Díaz-Morales, D.M., Benson, E., Cuevas, E., Ortiz-Zayas, J., Hernández-Delgado, E.A., 2019. Caribbean near-shore coral reef benthic community response to changes on sedimentation dynamics and environmental conditions. Front. Mar. Sci. 6, 1–19.

Palmer, C.V., Bythell, J.C., Willis, B.L., 2010. Levels of immunity parameters underpin bleaching and disease susceptibility of reef corals. FASEB J. 24, 1935–1946.

Perry, C.T., Steneck, R.S., Murphy, G.N., Kench, P.S., Edinger, E.N., Smithers, S.G., Mumby, P.J., 2015. Regional-scale dominance of non-framework building corals on Caribbean reefs affects carbonate production and future reef growth. Glob. Chang. Biol. 21 (3), 1153–1164.

Peters, E.C., Oprandy, J.J., Yevich, P.P., 1983. Possible causal agent of "white band disease" in Caribbean acroporid corals. J. Invertebr. Pathol. 41 (3), 394–396.

Pisapia, C., Edmunds, P.J., Moeller, H.V., Riegl, B., McWilliam, M., Wells, C.D., Pratchett, M.S., 2020. Projected shifts in coral size structure in the Anthropocene. In: Riegl, B. (Ed.), Population Dynamics of the Reef Crisis, vol. 87. Advances in Marine Biology. Elsevier, pp. 31–60 (in this volume). https://doi.org/10.1016/bs.amb.2020.07.003.

Precht, W.F., Aronson, R.B., Gardner, T.A., Gill, J.A., Hawkins, J.P., Hernández-Delgado, E.A., Jaap, W.C., Mcclanahan, T.R., Mcfield, M.D., Murdoch, T.J.T., Nugues, M.M.,

Roberts, C.M., Schelten, C.K., Watkinson, A.R., Côté, I.M., 2020. The timing and causality of ecological shifts on Caribbean reefs. In: Riegl, B. (Ed.), Population Dynamics of the Reef Crisis, vol. 87. Advances in Marine Biology. Elsevier, pp. 331–360 (in this volume). https://doi.org/10.1016/bs.amb.2020.08.008.

R Core Team, 2017. R: A Language and Environment for Statistical Computing. R Found. Stat. Comput, Vienna, Austria.

Riegl, B., Cavalcante, G., Bauman, A.G., Feary, D.A., Steiner, S., Purkis, S., 2017. Demographic mechanisms of reef coral species winnowing from communities under increased environmental stress. Front. Mar. Sci. 4, 1–14. doi.org/10.3389/fmars.2017.00344.

Riegl, B., Johnston, M., Purkis, S., Howells, E., Burt, J., Steiner, S.C.C., Sheppard, C.R.C., Bauman, A., 2018. Population collapse dynamics in Acropora downingi, an Arabian/ Persian Gulf ecosystem-engineering coral, linked to rising temperature. Glob. Chang. Biol. 24, 2447–2462.

Ritchie, K.B., 1998. Type II white-band disease. Rev. Biol. Trop., 199–203.

Rogers, C., 1990. Responses of coral reefs and reef organisms to sedimentation. Mar. Ecol. Prog. Ser. 62, 185–202.

Sabat, A.M., Toledo-Hernández, C., 2015. Viability of sea fan populations impacted by disease: recruitment versus incidence. J. Mar. Biol. 2015, 987060. doi.org/10.1155/2015/987060.

Soong, K., Lang, J.C., 1992. Reproductive integration in reef corals. Biol. Bull. 183 (3), 418–431.

Speare, K.E., Duran, A., Miller, M.W., Burkepile, D.E., 2019. Sediment associated with algal turfs inhibits the settlement of two endangered coral species. Mar. Pollut. Bull. 144, 189–195.

Squires, D.F., 1965. Neoplasia in a coral? Science 148 (3669), 503–505.

Stubben, C., Milligan, B.G., 2007. Estimating and analyzing demographic models using the popbio package in R. J. Stat. Softw. 22, 1–23.

Sutherland, K.P., Porter, J.W., Torres, C., 2004. Disease and immunity in Caribbean and indo-Pacific zooxanthellate corals. Mar. Ecol. Prog. Ser. 266, 273–302.

Toth, L.T., Stathakopoulos, A., Kuffner, I.B., Ruzicka, R.R., Colella, M.A., Shinn, E.A., 2019. The unprecedented loss of Florida's reef-building corals and the emergence of a novel coral-reef assemblage. Ecology 100 (9). https://doi.org/10.1002/ecy.2781.

Vardi, T., Williams, D.E., Sandin, S.A., 2012. Population dynamics of threatened Elkhorn coral in the northern Florida keys, USA, *Endanger.* Species. Res. 19 (2), 157–169.

Veron, J.E.N., 2000. Corals of the World (No. C/593.6 V4). Australian Institute of Marine Science and CRR Ald Pty Ltd.

Vollmer, S.V., Kline, D.I., 2008. Natural disease resistance in threatened staghorn corals. PLos One 3 (11), e3718.

Wapnick, C.M., Precht, W.F., Aronson, R.B., 2004. Millennial-scale dynamics of staghorn coral in Discovery Bay, Jamaica. Ecol. Lett. 7 (4), 354–361.

Weil, E., 2004. Coral reef diseases in the wider Caribbean. In: Coral Health and Disease. Springer, Berlin, Heidelberg, pp. 35–68.

Weil, E., Hammerman, N.M., Becicka, R.L., Cruz-Motta, J.J., 2020. Growth dynamics in *Acropora cervicornis* and *A. prolifera* in southwest Puerto Rico. PeerJ 8, e8435.

Williams, D.E., Miller, M.W., 2005. Coral disease outbreak: pattern, prevalence and transmission in *Acropora cervicornis*. Mar. Ecol. Prog. Ser. 301, 119–128.

Williams, D.E., Miller, M.W., 2012. Attributing mortality among drivers of population decline in *Acropora palmata* in the Florida keys (USA). Coral Reefs 31 (2), 369–382.

Williams, D.E., Miller, M.W., Kramer, K.L., 2006. Demographic Monitoring Protocols for Threatened Caribbean Acropora spp. Corals. NOAA Technical Memorandum NMFS-SEFSC-543. 91 pp.

Yap, H.T., Gomez, E.D., 1984. Growth of *Acropora pulchra*. Mar. Biol. 81 (2), 209–215.

CHAPTER FOUR

Spatial and temporal differences in *Acropora cervicornis* colony size and health

Elizabeth A. Goergen[a,*], Kathleen Semon Lunz[b], and David S. Gilliam[c]

[a]Department of Biological and Environmental Sciences, Qatar University, Doha, Qatar
[b]National Operations Center, Bureau of Land Management, Denver, CO, United States
[c]Halmos College of Arts and Sciences, Nova Southeastern University, Dania Beach, FL, United States
*Corresponding author: e-mail address: liz.goergen@gmail.com

Contents

Abstract

Little to no recovery in *Acropora cervicornis* populations has been documented since the 1970s and 1980s widespread disease events, and disease and predation appear to remain significant drivers of mortality. However, to date, demographic studies of *A. cervicornis* lack data temporally or spatially sufficient to quantify factors limiting recovery. *Acropora cervicornis* populations in three regions [Broward County (BWD), Middle Keys (MDK), and Dry Tortugas (DRTO)] of the Florida Reef Tract were surveyed up to three times per year from 2011 to 2015. Temporal and spatial differences were evaluated for colony size, live tissue volume, and prevalence and impact of disease and predation. Significantly larger colonies were reported in BWD, and at relatively deeper or more sheltered sites. At least 43% of colonies in each region were of reproductively capable size. Mean relative change in colony size between surveys (3–5 months) ranged from −20% to 19%. Disease and predation were consistently present in all regions, but levels varied significantly across space and time. Disease prevalence was the most variable condition (ranging from 0% to 28% per survey), increasing after periods of elevated temperatures and environmental disturbances, and caused significantly more partial

Advances in Marine Biology, Volume 87
ISSN 0065-2881
https://doi.org/10.1016/bs.amb.2020.08.004

mortality than fireworm (*Hermodice carunculata*) or snail (*Coralliophila* spp.) predation. Recovery potential and long-term persistence of this species may be limited due to the persistent presence of disease and predation, and reproductive limitations. However, there is still potential at sites of greater depth and/or more protection hosted larger and healthier colonies creating potential refugia for this species.

1. Introduction

Acropora cervicornis, staghorn coral, once dominated the fore reefs of many Caribbean, Florida, and Gulf of Mexico reefs (*Acropora* Biological Review Team, 2005; Bruckner, 2002; Davis, 1982). Its ability to reproduce through asexual fragmentation and its fast growth rate enables the species to quickly spread across surroundings, forming dense structures known as patches or thickets capable of covering hectares of reef habitat. These expansive areas played a significant role in the reef community by creating complex three-dimensional habitats utilized by a multitude of vertebrate and invertebrate species that is irreplaceable by any other Caribbean coral (Adey and Burke, 1977; Goreau, 1959; Neigell and Avise, 1983).

The health and survival of corals, particularly *A. cervicornis* in the Caribbean, and coral reefs throughout the world are affected by many environmental factors such as nutrient loading, pollution, increased hurricane frequency, and temperature stress (Hoegh-Guldberg et al., 2007; Hughes et al., 2017; Ortiz et al., 2014; van Hooidonk et al., 2013). Additionally, the persistence of *A. cervicornis* is affected by biological factors like disease, predation, and colony fragmentation (Gilmore and Hall, 1976; Highsmith et al., 1980; Knowlton et al., 1990; Shinn, 1976; Tunnicliffe, 1981), such as the unprecedented white band disease outbreak in the 1980s, leaving only a few remnant populations (Aronson and Precht, 2001; Davis, 1982). The populations that remained were spatially isolated and in relatively low abundance or cover (*Acropora* Biological Review Team, 2005; Miller et al., 2002), with only few documented large patches persisting into recent times (Busch et al., 2016; D'Antonio et al., 2016; Keck et al., 2005; Lirman et al., 2010; Vargas-Ángel et al., 2003; Walker et al., 2012). These remnant populations have not been able to facilitate large scale natural recovery, and decades later populations typically remain in low abundance and with few signs of range wide species recovery (Griffin et al., 2015; Lidz and Zawada, 2013; Lucas and Weil, 2016; Miller et al., 2008, 2016). By 2006, *A. cervicornis* was listed as threatened under the United States

Endangered Species Act (National Marine Fisheries Service, 2006) and in 2008 it was listed as critically endangered on the International Union for Conservation of Nature (IUCN) Red List (Aronson et al., 2008).

In response to these listings, numerous restoration programs have been established across the species' range to facilitate recovery (Lirman and Schopmeyer, 2016; Young et al., 2012). However, understanding of species dynamics such as identifying the current drivers of both population recovery and decline is at an early stage (Goergen et al., 2019; Mercado-Molina et al., 2015a). Biological threats, such as white band disease (Antonius, 1981; Gladfelter, 1982; Peters et al., 1983) and rapid tissue loss (Goergen et al., 2019; Williams and Miller, 2005), and predation by the bearded fireworm, *Hermodice carunculata* (Antonius, 1977; Goergen et al., 2019; Knowlton et al., 1990; Marsden, 1962; Miller et al., 2014b), the corallivorous gastropod, *Coralliophila abbreviata* (Hayes, 1990; Knowlton et al., 1990), and the three-spot damselfish, *Stegastes planifrons* (Brawley and Adey, 1977; Kaufman, 1977, 1981; Knowlton et al., 1990), have all been reported as drivers of mortality. The most eminent of these threats to its recovery, documented in one region of the Florida Reef Tract (FRT), are environmental disturbances and chronic disease and predation (Goergen et al., 2019), however the spatial characteristics of the impact of these threats are not fully understood.

Species recovery and conservation plans require long-term, range wide, studies documenting temporal and spatial changes in population condition. The main objectives of this multiyear study were to compare colony size and live tissue volume, and describe the prevalence and impact of major conditions affecting *A. cervicornis* among three regions on the FRT. Results of this research will improve our knowledge of the species dynamics and should influence the planning of species recovery, conservation, and restoration efforts such as determining priority habitats for restoration or identification of activities that could improve recovery (e.g., predator removal).

2. Materials and methods

Wild *A. cervicornis* populations were surveyed between 2011 and 2015 in three regions spanning the species range on the Florida Reef Tract: (1) Broward County (BWD), (2) Middle Keys (MDK), and (3) Dry Tortugas (DRTO; Fig. 1). The regional distinctions were developed based on management entity boundaries (State of Florida waters, and federal

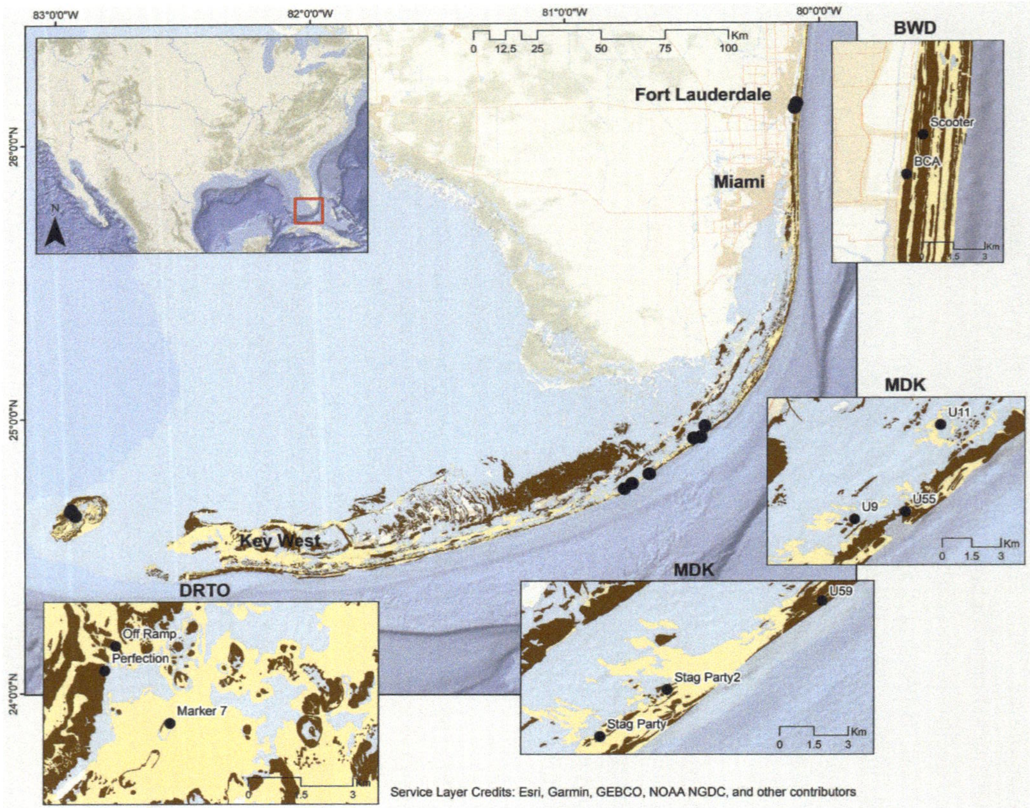

Fig. 1 Map showing locations of study sites in Broward (BWD), Middle Keys (MDK), and Dry Tortugas (DRTO). Coral reef or hard bottom habitat are shown in brown.

waters managed by either the National Park Service or the National Marine Fisheries Service) and regional boundaries commonly employed by Florida reef monitoring programs (Ruzicka et al., 2013). Within each region, multiple sites were established with permanent monitoring plots (Table 1). A centre pin and identification tag marked each plot (3.5 m radius). Temporary transect lines, 7 m in length, laid perpendicular across the centre of each plot defined the survey area during each monitoring event.

The original goal of this study was to compare similar sites (density and cover) across all three regions; however, during reconnaissance and site selection, high cover patch sites could not be found in the MDK and DRTO regions that were comparable to the patches in BWD (Goergen et al., 2019; Vargas-Ángel et al., 2003; Walker et al., 2012). While site level characteristics such as geomorphology, stony coral and octocoral density, and overall *A. cervicornis* distribution (except for BWD—see Walker et al., 2012), were not specifically measured as part of this project, differences were apparent through anecdotal observations and images (Fig. 2). *Acropora cervicornis* cover in the BWD region sites ranged from nearly 100% to sporadic colonies in low densities towards the periphery of the site, whereas typical *A. cervicornis* colony distribution in both the MDK and DRTO sites was sporadic clusters of low to moderate densities separated by 10s of metres. BWD sites were flat with low complexity, whereas the MDK and DRTO sites had variability in complexity (higher complexity—see G, H, J, and K in Fig. 2). Density of octocorals was also variable (high density—see C, D, E, and I in Fig. 2). Many sites in the MDK region were on isolated patch reefs surrounded by rubble or sand and seagrass, whereas the sites in BWD and DRTO were part of larger reefs. In the BWD region, plots included in this study were not established in high cover areas, where >50% of the plot had living *A. cervicornis*, in order to maintain a level of similarity between regions. As a minimum requirement, 10 individual *A. cervicornis* colonies had to be present in at least two non-overlapping radial plots at a site to be included.

BWD and MDK sites were monitored three times per year: winter/spring (February–April; WS), summer (June–August; SM), and fall (October–November; F); however, due to travel logistics, DRTO was visited during each summer (June) and fall (September) and only during one winter/spring monitoring event in 2012. In the MDK region, site SP2 was added in Fall 2012 and U59 was not monitored during the final event, SM15 (Table 1).

Table 1 *Acropora cervicornis* colony monitoring schedule, site depth, number of plots, and initial number of colonies per site.

Region	Site name	Depth (m)	Protection	No. of plots	Total colonies	SM11	F11	WS12	SM12	F12	WS13	SM13	F13	WS14	SM14	F14	WS15	SM15
Broward County	BCA	6.3	S	10	735	X	X	X	X	X	X	X	X	X	X	X	X	X
	Scooter	4.8	S	10	1174	X	X	X	X	X	X	X	X	X	X	X	X	X
Middle Keys	Stag Party 2 (SP2)	4.6	E	4	305					X	X	X	X	X	X	X		X
	Stag Party(SP)	6.1	S	2	212	X	X	X	X	X	X	X	X	X	X	X		X
	U11	4.3	E	4	463	X	X	X	X	X	X	X	X	X	X	X		X
	U55	4	E	10	1200	X	X	X	X	X	X	X	X	X	X	X		X
	U59	12.5	S	3	251	X	X	X	X	X	X	X	X	X	X	X		
	U9	6.1	S	4	370	X	X	X	X	X	X	X	X	X	X	X		X
Dry Tortugas National Park	Marker 7 (M7)	3.7	S	3	265	X	X	X	X	X		X	X		X			X
	Off Ramp (OR)	4.6	E	3	270	X	X	X	X	X		X	X		X			X
	Perfection (FF)	4.6	E	3	270	X	X	X	X	X		X	X		X			X

Protection ranking, sheltered (S) or exposed (E), was determined by the authors based on site characteristics, depth, knowledge of the site, and geographic location.

Fig. 2 Representative site images depicting typical characteristics of the site within each region. Broward County sites: BCA (A) and Scooter (B), Middle Keys sites: Staghorn Party 2 (C), Staghorn Party (D), U11 (E), U55 (F), U59 (G), and U9 (H), and Dry Tortugas sites: Marker 7 (I), Off Ramp (J), and Perfection (K).

Up to 10 colonies were surveyed in each plot during every monitoring event by starting due north from the plot centre pin and working clockwise around the plot. A colony was defined as a secure individual with living tissue, continuous skeletal structure, and distinct separation from other colonies; colony measurements included both living tissue and dead skeleton as long as it was continuous. Colony dimensions (length, width, and height), estimation of per cent recent and old mortality, cause of recent mortality, and a photograph were documented during each event.

Our demographic approach to monitoring *A. cervicornis* individuals focuses on surveying colonies within a designated area with the likelihood that some colonies are repeated across monitoring periods, but are not targeted. Although colony fragmentation, dislodgement, and reattachment are a natural part of the life history of this species, it makes tracking of individual colonies impractical and uninformative over the long-term (Bruckner et al., 2009; Garrison and Ward, 2012; Smith et al., 2005; Tunnicliffe, 1981; Williams et al., 2006). By using this approach, we obtained a consistent sample size to describe temporal and spatial patterns in colony size, distribution, and condition. The purpose of this survey technique was not to obtain individual colony growth measurements or rates of colony survival. These metrics can be determined using colony fate tracking over a short term, although Goergen and Gilliam (n.d.) describe how this method herein can also be used to describe *A. cervicornis* colony retention and residence rates.

Recent mortality, defined as stark white skeleton on which turf algae had not colonized, was estimated and attributed to one of the following conditions: rapid tissue loss, white band disease, fireworm predation (*Hermodice carunculata*), snail predation (*Coralliophila* spp.), and "other." Because the distinction between rapid tissue loss and white band disease is unclear, they were grouped as "white disease" for analyses. The category of "other" included conditions not frequently encountered, or causes of mortality that could not confidently be identified. Damselfish predation (*Stegastes planifrons*), and bleaching (including partial bleaching) were only noted as present or absent.

Temperature loggers, HOBO© Pendant Temperature and Light Data Loggers, were deployed at each site recording temperature every two hours. These data were used to determine daily averages by region. Loggers were deployed in BWD June 2011–July 2015, MDK July 2011–November 2014, and DRTO April 2012–June 2015.

2.1 Data analysis

2.1.1 Colony diameter and volumetric index

Two colony metrics were evaluated, maximum colony diameter (d) and colony volumetric index (CVI), an index corresponding to the per cent live tissue (PL) of the colony. CVI was calculated using the shape of an ellipsoid $[((4/3)\pi \times d/2 \times w/2 \times h/2) \times \text{PL}]$, where w is the width of the colony perpendicular to d, and h is the maximum height of the colony measured through the axis of growth (Huntington and Miller, 2013). Both metrics were evaluated at the plot level, yielding an average colony maximum diameter and CVI by site and region. One-way analysis of variance (ANOVA) were used for intra and inter regional comparisons of colony maximum diameter and CVI. Colony diameter and CVI were log($x+1$) transformed to meet normality (Kolmogorov-Smirnov) and homogeneity assumptions (Levene's Test). Post hoc comparisons between groups were performed using Tukey's HSD tests. Relative change in colony maximum diameter and mean CVI per plot were calculated for each monitoring period and for the overall project (SM11–SM15). To evaluate the sexual reproductive potential of each of region, colonies were organized by size class (Soong and Lang, 1992): less than 10 cm diameter (recruit), 11–30 cm diameter (non-reproductive size), 31–89 cm diameter (reproductive size), or over 90 cm diameter (large reproductive colonies). Sites SP2 and U59 were not included in temporal analysis of colony diameter or CVI because these sites were not surveyed during all monitoring events.

2.1.2 Colony condition

The impact of white disease, fireworm predation, or snail predation was evaluated through per cent recent partial mortality and was averaged by site or region. Only those colonies affected were included in the condition average, representing the mortality caused by each condition when present. Kruskal-Wallis tests by rank were used to explore per cent recent mortality caused by each condition across monitoring event, year, season, and sites within each region to identify patterns in tissue loss. When significant, a Dunn's test with a Bonferroni correction (p values were multiplied by number of comparisons) was performed post hoc to determine significance between factor levels. Prevalence of a condition, defined as the number of colonies identified as having a condition divided by the total number of colonies assessed during that event per plot, was analysed for spatial (site and region) and temporal differences (year, season, and monitoring event). Colonies with more than one condition were included in the prevalence for each condition. Binomial generalized linear models were used to describe the temporal variation in prevalence of each condition, using monitoring event, year, season, and site as factors. Region was also used as an interaction term with monitoring event to evaluate if temporal changes were similar between regions. When the model identified significant factors, post hoc multiple comparisons with a Bonferroni correction were employed to define specific contrasts of factor levels. All analyses were performed in R Studio 3.3.1 (RStudio Team, 2016) using the multicomp (Hothorn et al., 2008) and dunn.test packages (Dinno, 2017).

3. Results

A total of 5515 colonies were surveyed seasonally across 5 years at 11 sites within three regions of the Florida Reef Tract (Fig. 1; Table 1). Colonies were between 2 and 240 cm in diameter. Overall, all data pooled, mean colony maximum diameter was 43.6 ± 0.37 cm with 72.3 ± 0.35% of the colony having live tissue, resulting in a mean CVI of 22,615 ± 631.2 cm^3. Recent mortality affected 17 ± 0.5% of all colonies surveyed, causing 8.4 ± 0.37% tissue loss per colony effected. The most prevalent condition was fireworm predation, affecting 8.3 ± 0.37% of colonies; however, white disease caused the largest amount of tissue loss per colony affected 13.0 ± 0.89%.

3.1 Colony diameter and volumetric index

3.1.1 Regional

Mean colony maximum diameter decreased regionally from North to South, varying significantly between regions ($F_{2, 5511} = 63.745$, $P < 0.001$; Fig. 3), with the largest colonies in the BWD region (northern most region)

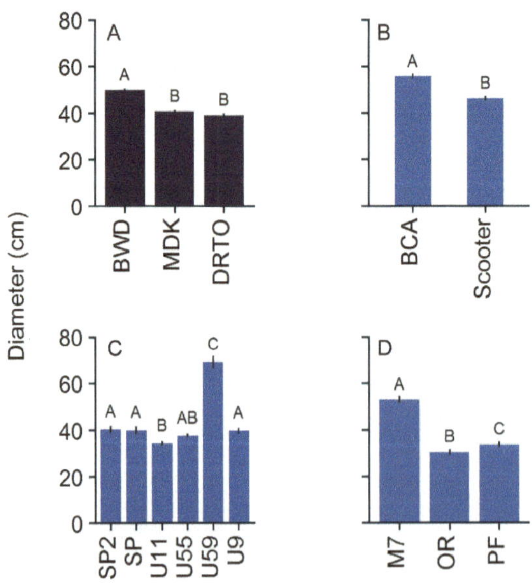

Fig. 3 Mean colony maximum diameter by region all events pooled (A) and site within region: Broward County (B), Middle Keys (C), and Dry Tortugas (D). Letters over bars indicate significant differences within groups, Tukey HSD $P < 0.05$. Error bars indicate ±1 SE.

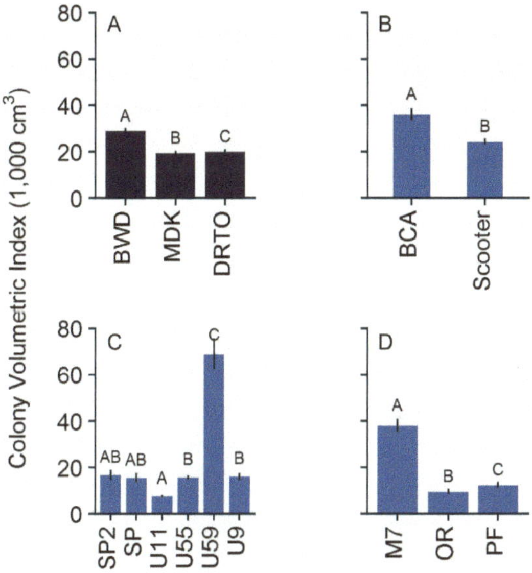

Fig. 4 Mean colony volumetric index by region all events pooled (A) and site within region: Broward County (B), Middle Keys (C), and Dry Tortugas (D). Letters over bars indicate significant differences within groups, Tukey HSD $P < 0.05$. Error bars indicate ±1 SE.

at 49.8±0.70 cm (Tukey HSD, $P < 0.001$). A similar pattern was observed for colony volumetric index (CVI), which was also significantly different between regions ($F_{2,\ 5511} = 37.58$, $P < 0.001$; Fig. 4). Even though DRTO had smaller colonies, the CVI for this region was significantly greater than the MDK region due to colonies having more live tissue per colony (Tukey HSD, $P < 0.001$).

All size classes were recorded in all regions (Fig. 5). The BWD region had substantially more large colonies (>90 cm diameter; dark red Fig. 5) than MDK or DRTO, indicating a higher likelihood of potential reproduction. A majority (60–80%) of colonies measured during each survey in BWD were of reproductive size (>30 cm diameter; Soong and Lang 1992). The proportion of mature colonies per survey in the MDK region was 43–60%, and slightly higher for the DRTO at 45–65% (shades of red—Fig. 5). When per cent mortality and the volume of the colony were included in this estimate (CVI), which is likely a more accurate estimate for reproduction, reproductive potential decreased for all regions (Fig. 5D–F). The mean proportion of mature colonies based on CVI was 60% in BWD, 43% in MDK, and 51% in DRTO.

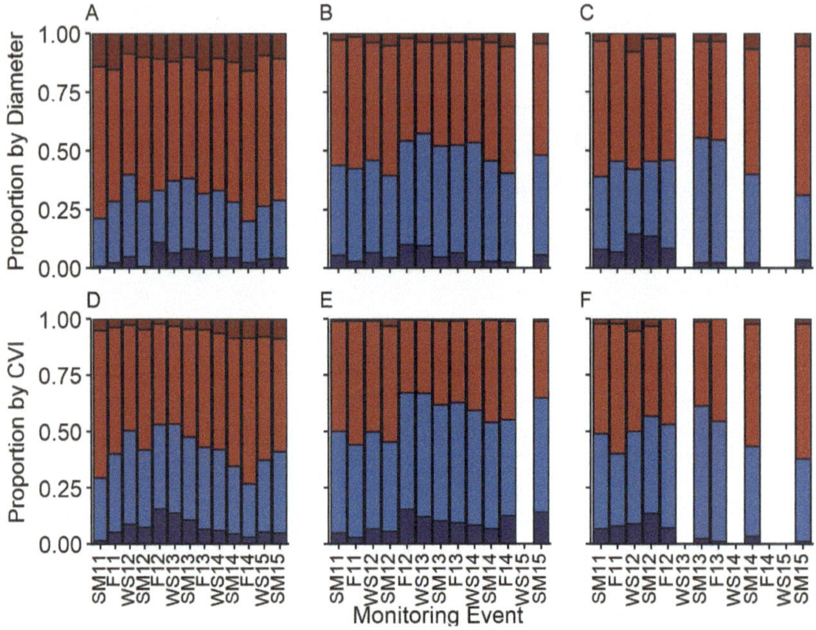

Fig. 5 Distribution of colony sizes based on mean maximum diameter (A–C) and colony volumetric index (D–F) within four classes by region: Broward County (A and D), Middle Keys (B and E), and Dry Tortugas (C and F). Dark blue bars represent recruits (<10 cm diameter/1–291 cm^3), light blue are non-reproductive size colonies (11–30 cm/292–6271 cm^3), light red are reproductive size colonies (31–89 cm/6272–125,308 cm^3), and dark red are large reproductive colonies (>90 cm/>125,308 cm^3).

3.1.2 Local

Local level variation also occurred in both colony diameter and CVI. Mean colony maximum diameter and CVI were significantly different between sites within all regions (Tukey HSD, $P<0.05$ for all comparisons; Figs. 3 and 4). The deeper sites (BCA and U59) had significantly larger colonies than the shallower sites in their regions (Tukey HSD, $P<0.05$ for all comparisons), whereas Marker 7, the shallowest, but possibly more sheltered site in DRTO, had significantly larger colonies than the other two sites. Over the length of the project, change in colony diameter by site ranged from a decrease of 32% (Stag Party) to an increase of 43% (Off Ramp), DRTO was the only region where colony diameter increased for all sites. The overall change in CVI ranged from a loss of 73% (U9) to an increase of 271% (BCA). Colony diameter and CVI exhibited similar patterns within each region between sites.

3.1.3 Temporal

Mean colony maximum diameter increased from the initial to final events for the MDK and DRTO regions by, 9 ± 10% and 22 ± 12%, respectively; in contrast, mean colony maximum diameter for BWD decreased by 1 ± 11%. Relative change in colony diameter was in constant flux, the largest range of change (−69% to 100%) occurred in BWD. On average, colony diameter changed between −20% and 19% between surveys (Fig. 6). The winter/spring to summer time period (yellow in Fig. 6) resulted in a positive change in mean colony diameter in all regions for all years besides 2015 in MDK and 2012 in DRTO. Colony diameter generally decreased between the fall and winter/spring time period (blue in Fig. 6), whereas change in colony diameter from the summer to the fall was variable across years and region (red in Fig. 6). Colony diameter in 2012 for all regions decreased substantially compared to other years.

Mean CVI per site varied strongly between sites, ranging between 2773 and 141,229 cm^3 for any given monitoring event. In addition, the relative change in CVI by plot was highly variable per monitoring event (Fig. 7). Overall mean CVI within BWD and DRTO increased towards the end

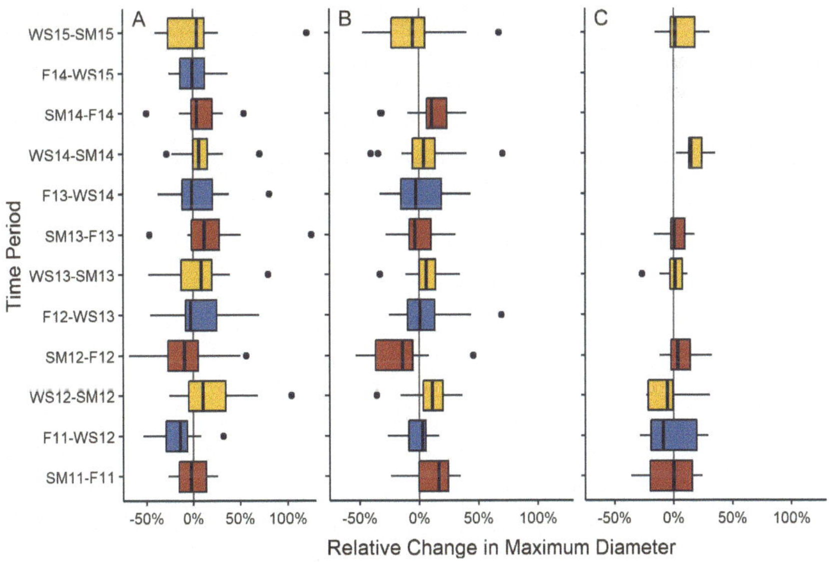

Fig. 6 Relative change in mean maximum colony diameter between monitoring events by region: Broward County (A), Middle Keys (B), and Dry Tortugas (C). SM11-F11 represents the time period between monitoring events Summer 2011 and Fall 2011. Box and whisker plots represent upper and lower quartile and minimum and maximum, respectively. Line across box represents the median and dots are outliers.

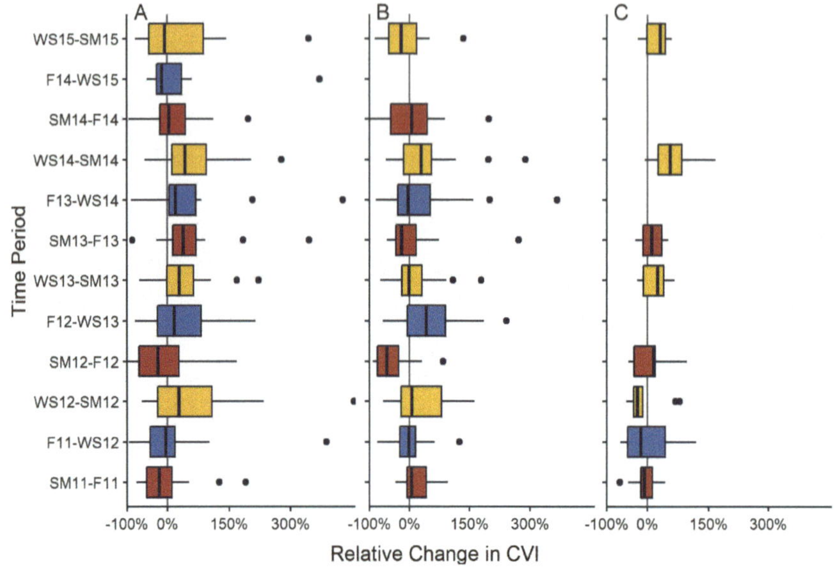

Fig. 7 Relative change in mean colony volumetric index between monitoring events by region: Broward County (A), Middle Keys (B), and Dry Tortugas (C). SM11-F11 represents the time period between monitoring events Summer 2011 and Fall 2011. Box and whisker plots represent upper and lower quartile and minimum and maximum, respectively. Line across box represents the median and dots are outliers. Two extreme values were removed from the BWD figure to improve figure clarity-3100 WS12-SM12 and 1400 SM13-F3.

of the project (118 ± 70% and 117 ± 82%, respectively) whereas a decrease (−13 ± 41%) was observed in the MDK. Even though CVI from the initial to final survey may have increased or decreased, the fluctuations between surveys caused dramatic changes in the volume of tissue per plot that were not always maintained or regained. For example, while the BWD region showed an overall increase in mean CVI, the net tissue change was negative—net loss of 411,453 cm^3 of tissue (a 50 × 40 × 30 cm colony is approximately 34,000 cm^3). In other words, the tissue lost between surveys was more than the tissue gained from the initial survey to the end. In the DRTO, fluctuations (range: −68% to 167%) were not as great as BWD (range: −97% to 3100%; Fig. 7) resulting in a net increase in tissue volume of 271,715 cm^3. Plots in the MDK region had a net loss of 1,038,334 cm^3 of tissue. Seasonal changes in CVI were similar in pattern to change in colony diameter, but with a larger range.

Change in colony diameter was influenced by the mean diameter of the colony during the prior survey (colony diameter SM11 vs change in colony diameter SM11-F11, etc.; Fig. 8). As colonies became larger within their

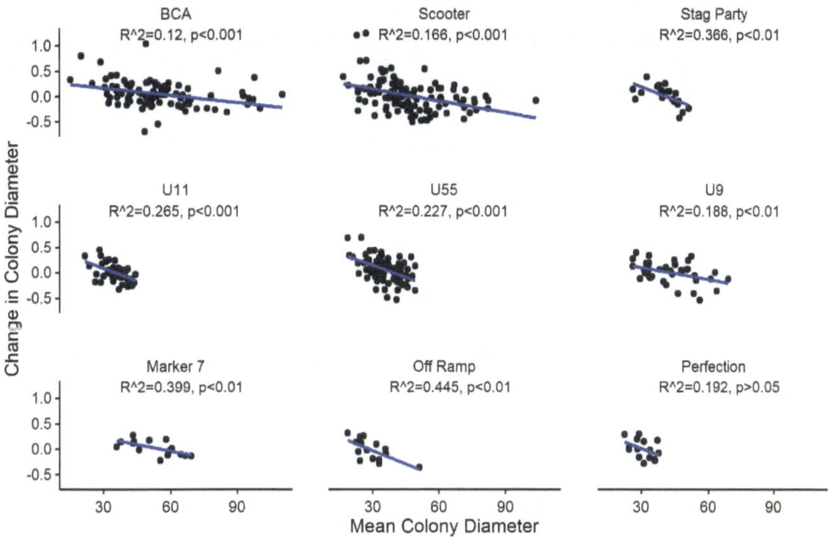

Fig. 8 Linear regressions of mean colony diameter and change in colony diameter by site.

region, the likelihood of a negative change in diameter over the next time period increased; sites with smaller colonies would maintain size or increase.

3.2 Colony condition

3.2.1 Regional

The MDK region had significantly more colonies (19±0.7%) affected by recent mortality than BWD (16±0.8%) and DRTO (11±1%; Table 2). The mean amount of recent mortality (per cent tissue recently lost) per colony was greatest in the MDK region (8.6±11%), followed by DRTO (8.5±11.8%) then BWD (7.9±11.5%).

Disease and fireworm, snail, and damselfish predation were present in all regions; however, region significantly influenced the rate at which each caused mortality (Fig. 9). The prevalence of fireworm and snail predation in the MDK region was twice that of the other two regions and damselfish predation prevalence was four times higher in the DTRO. White disease was the leading cause of colony partial mortality for all regions (Fig. 10). Snail and fireworm predation caused similar levels of colony partial mortality (4.9±0.25% and 4.6±0.35%, respectively; $P > 0.05$) and were significantly less than white disease in all regions ($P < 0.05$). Mortality caused by fireworms was significantly lower in BWD (Fig. 10; $P < 0.05$), ranging from 1% to 7.4% per monitoring event, than in MDK (3.3–7.4%) and DRTO (1–16.8%).

Table 2 Results of generalized linear models by sub-region and condition using Year, Season, and Site as factors.

			White disease				Fireworm				Snail				Damselfish				Recent mortality			
Region	**Factor**	**df**	***N***	***D***	**RD**	***P***	***N***	***D***	**RD**	***P***	***N***	***D***	**RD**	***P***	***N***	***D***	**RD**	***P***	***N***	***D***	**RD**	***P***
BWD (1908)	Year	4	140	39.77	961.1	0.000	123	25.40	886.93	0.000	31	18.66	338.52	0.001	17	5.91	188.44	0.206	308	41.74	1645	0.000
	Season	2		83.31	877.79	0.000		1.27	885.67	0.531		8.41	330.1	0.015		2.36	186.07	0.307		35.53	1609.5	0.000
	Site	1		3.05	874.74	0.081		2.22	883.45	0.136		0.00	330.1	0.997		1.57	184.5	0.210		2.34	1607.2	0.126
MDK (2802)	Year	4	139	39.71	1066.3	0.000	303	6.96	1913	0.138	141	5.29	1130	0.259	69	12.77	634.67	0.012	541	5.01	2744.6	0.286
	Season	2		0.89	1065.4	0.641		2.45	1910.5	0.294		1.39	128.7	0.500		3.26	631.41	0.196		0.03	2744.6	0.988
	Site	5		28.69	1036.7	0.000		19.91	1890.6	0.001		44.10	1084.6	0.000		319.50	311.91	0.000		28.27	2716.3	0.000
DRTO (805)	Year	4	36	11.56	282.53	0.021	33	17.92	257.53	0.001	5	3.67	57.11	0.452	99	4.66	595.59	0.324	88	14.56	541.03	0.006
	Season	2		6.44	276.09	0.040		1.42	256.11	0.491		5.08	52.032	0.079		0.62	594.97	0.733		3.56	537.47	0.169
	Site	2		0.33	275.77	0.848		6.18	249.93	0.046		1.46	50.57	0.481		101.20	493.77	0.000		0.41	537.07	0.816
Overall	Region	2		14.17	2401.00	0.001		53.27	3107.70	0.000		70.21	1495.50	0.000		177.80	1442.00	0.000		34.70	4992.00	0.000

N = number of colonies affected by each condition, D = model deviance, RD = residual deviance, and P = factor significance $\alpha = 0.05$.
Total number of colonies surveyed in parentheses.

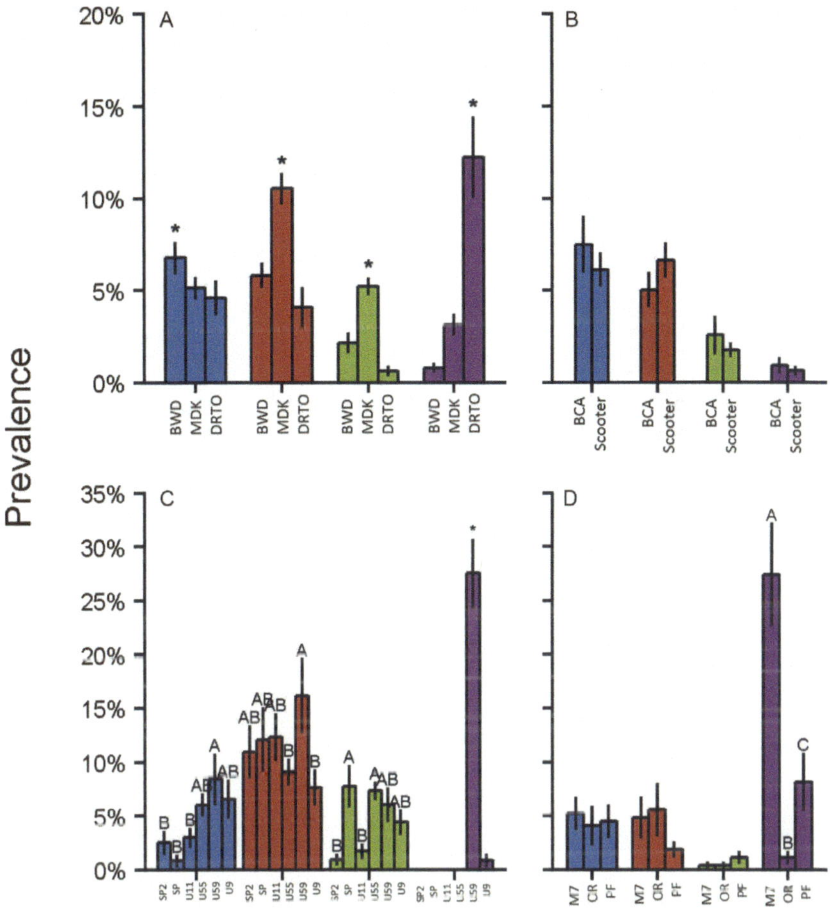

Fig. 9 Mean prevalence of white disease (blue), fireworm predation (red), snail predation (green), and damselfish predation (purple) by region (A) and site within region: Broward County (B), Middle Keys (C), and Dry Tortugas (D). Letters or asterisk over bars indicate significant differences within groups, Tukey HSD test $P < 0.05$. For clarity, letters were removed from groups where there were no significant differences. Error bars indicate ±1 SE.

3.2.2 Local

Recent mortality was observed at all sites, but not all conditions were present at all sites (Fig. 9). Site prevalence of recent mortality ranged from 10% to 32%, but only MDK sites were significantly different: U59 (32 ± 3%) had significantly higher prevalence than all other sites in the MDK ($P < 0.05$). Prevalence of all conditions in the MDK region were variable across the sites (Fig. 9). There was significantly more damselfish predation at MDK and DRTO sites which had larger colonies (Fig. 9; $P < 0.05$). However,

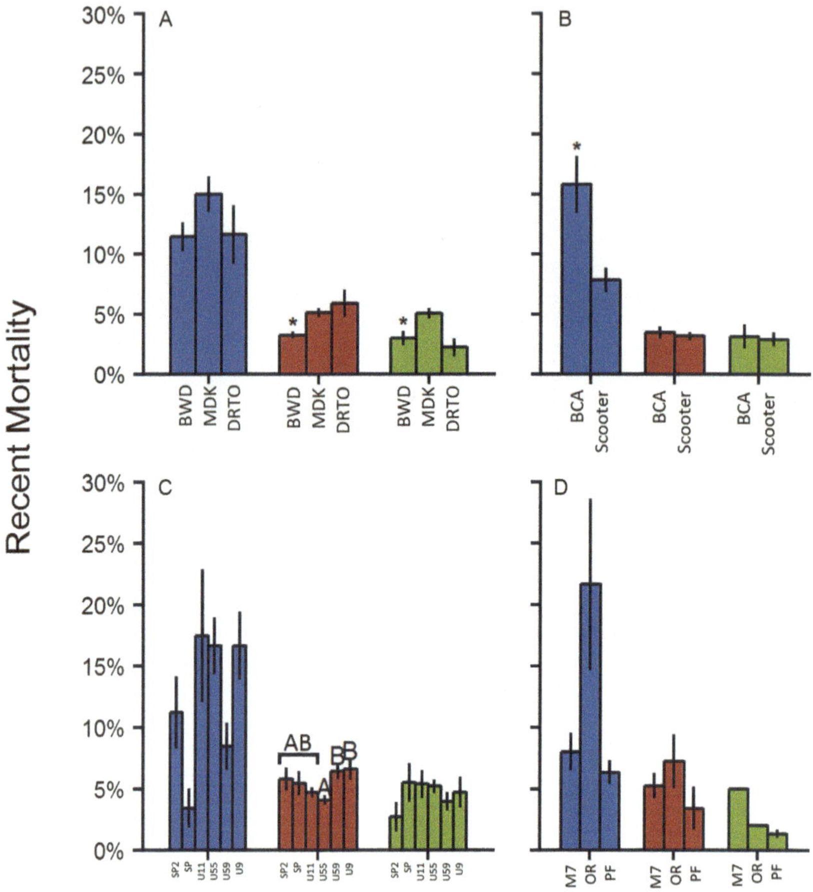

Fig. 10 Mean per cent recent mortality caused by white disease (blue), fireworm predation (red), and snail predation (green) by region (A) and site within region, Broward County (B), Middle Keys (C), and Dry Tortugas (D). Letters or asterisk over bars indicate significant differences within groups, Dunns test $P < 0.05$. For clarity, letters were removed from groups where there were no significant differences. Error bars indicate ± 1 SE.

the only site level differences for the total amount of recent tissue lost was within BWD: Scooter had significantly less recent mortality than BCA ($P < 0.05$). Due to the low abundance of colonies affected by each condition, site level differences in amount of tissue lost per condition were minimal (Fig. 10; Table 2).

3.2.3 Temporal

The influence of year, season, and site were variable among conditions and region (Table 2). Year then season were the most significant factors in explaining the prevalence of disease and predation in BWD (whereas site drove differences in the MDK), and in DRTO, year or site influenced prevalence, depending on condition (Table 2). Models indicate significant influence of year in all regions, season in BWD and DRTO, and site in MDK on white disease (Table 2). Significant seasonal changes in recent mortality frequency were only observed in the BWD region (Table 2), where colonies exhibited more recent mortality in the summer and fall than the winter/spring ($P < 0.05$). 2015 was the worst year for recent mortality in BWD (Tukey, $P < 0.05$). Recent mortality was more prevalent on colonies during 2014 than 2013 in all regions, but was only significant for BWD and DRTO ($P < 0.05$). Per cent recent mortality was similar among seasons ($P > 0.05$), but 2011 was significantly higher than 2015 in MDK and 2013 and 2015 in BWD ($P < 0.05$; Fig. 11).

Temporal patterns in the prevalence of all conditions were apparent and varied by region (Fig. 12). In BWD, summer events had significantly more disease than the fall or winter/spring events ($P < 0.05$). 2013 was a significantly low year for white disease prevalence in all regions and 2015 was a significantly high year in BWD and MDK ($P < 0.05$). Prevalence of fireworm predation in the MDK region was consistently higher than BWD and DRTO until the last two years of the project when prevalence in BWD increased ($P < 0.05$; Fig. 12). Damselfish predation remained constant across the project for all regions ($P > 0.05$). Snail predation had consistently low prevalence across all regions, but was showing signs of increase in BWD during the last 2 years of the project. Bleaching was rarely observed, but was significantly higher in F11 than all other events for the DRTO region.

The amount of tissue lost to snail predation did not fluctuate by season or year, but was significantly higher in the MDK region (Table 2; $P < 0.05$). There was no seasonality to the amount of tissue lost to fireworm predation, but year was a significant factor in BWD and MDK (Table 2; $P < 0.05$). The average amount of tissue loss per colony from white disease was highly variable between monitoring events, ranging between 1% and 24% in

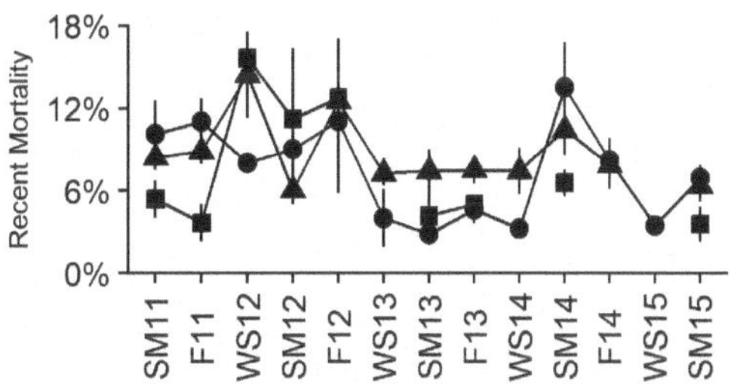

Fig. 11 Temporal pattern in mean recent mortality by region, Broward County (circles), Middle Keys (triangles), and Dry Tortugas (squares). Error bars indicate ±1 SE.

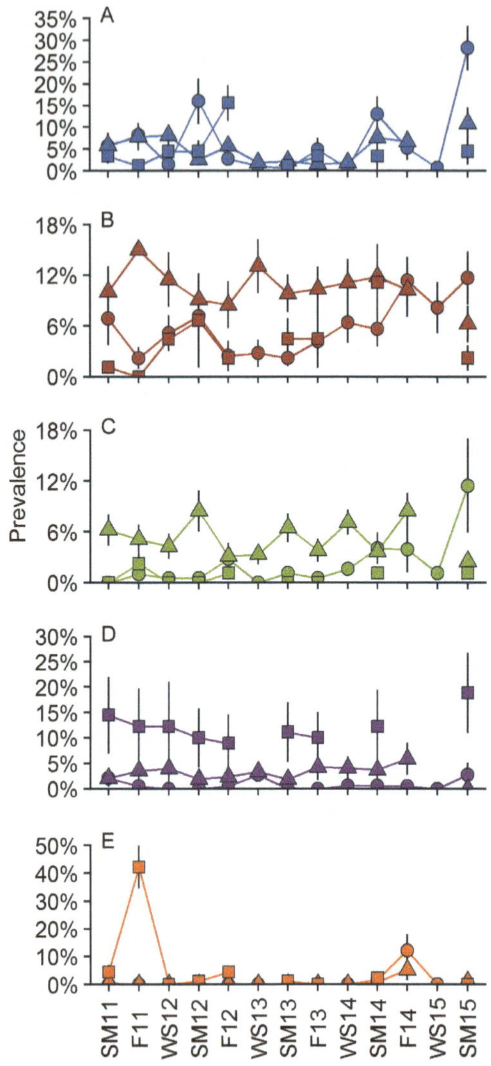

Fig. 12 Temporal patterns in mean prevalence of white disease (A), fireworm predation (B), snail predation (C), damselfish predation (D), and bleaching by region (E). Broward County (circles), Middle Keys (triangles), and Dry Tortugas (squares). Error bars indicate ±1 SE.

BWD, 7–29% in MDK, and 5–17% in DRTO. Kruskal-Wallis tests indicate year influenced per cent tissue loss by white disease in both BWD ($\chi^2=9.8182$, $P=0.04$) and MDK ($\chi^2=25.508$, $P=0$); however, post hoc analysis only indicate 2015 being significantly less than 2012 in the MDK region ($P<0.05$).

3.3 Temperature

Temperature variation within each region indicates that disease prevalence appears to be influenced by temperature. Lower mean maximum temperatures for all regions in 2013 were associated with low disease prevalence and higher mean maximum temperatures in 2014 were associated with elevated disease prevalence and mean colony partial mortality (Figs. 12 and 13). Bleaching was mostly recorded in low prevalence, apart from Fall 2011 when 42±7.6% of the colonies were bleached in the DRTO region;

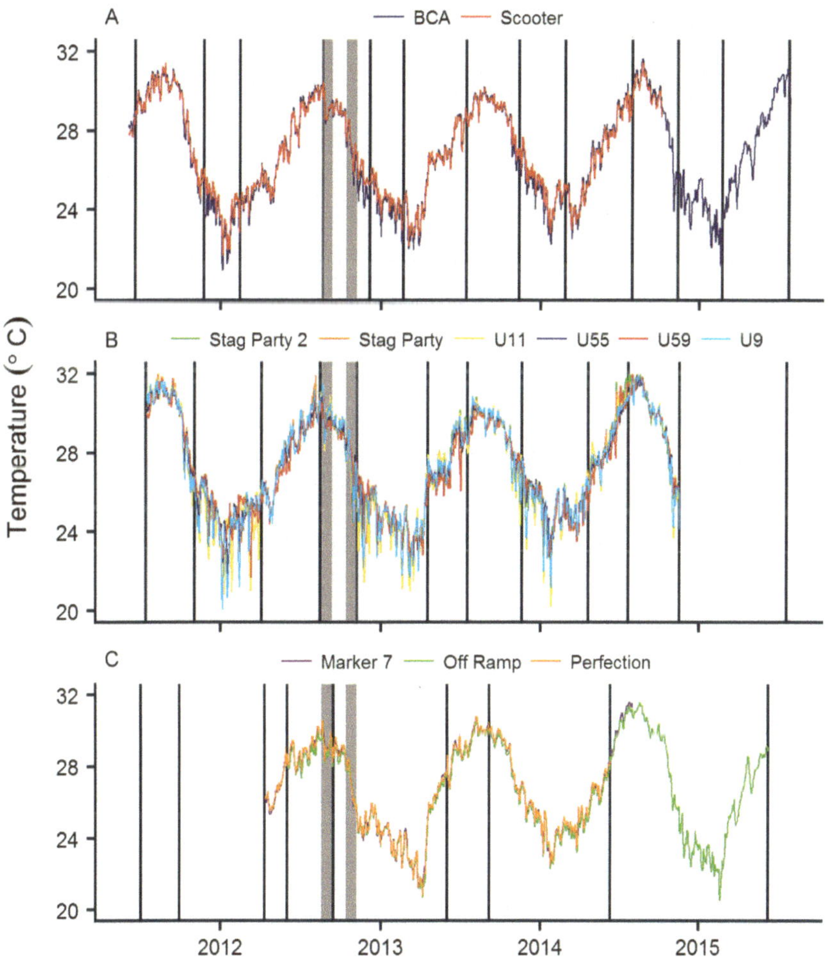

Fig. 13 Daily mean in situ water temperature by region, Broward County (A), Middle Keys (B), and Dry Tortugas (C). Black lines represent monitoring events and grey bars represent Tropical Storm Isaac and Hurricane Sandy from left to right, respectively.

however, temperature loggers had not yet been deployed at these sites. Bleaching was more commonly recorded during the Fall monitoring events, following peak temperatures in each region. The warmest year, based on number of days where daily mean temperatures were above 29.5 °C, was 2011 in BWD ($n=89$ days) and 2014 for MDK ($n=108$ days) and DRTO ($n=113$ days), but data were not recorded for 2011 in DRTO and MDK began in mid-2011.

4. Discussion

In this study, we documented the spatial and temporal variability of *A. cervicornis* in terms of colony size and volume, and predation and disease prevalence across the Florida Reef Tract (FRT). Environmental characteristics (depth, protection from wave energy) of each region and site influenced colony diameter and volume, resulting in larger colonies at deeper sites. The MDK and DRTO regions were composed of mainly small immature colonies, with little indication of change in size class structure across 5 years. This relatively small change in colony size class structure may be surprising for a species with such a fast growth rate. Limited change in these regions is likely due to this species' ability to constantly change through fragmentation, dislodgment, and reattachment (asexual recruitment). As we show in the data, larger colonies are more likely to have a negative change in diameter among surveys (although individual colonies were not fate tracked); it is likely that these colonies are generating new colonies through fragmentation. Unless a colony was completely lost through fragmentation, it will maintain its size due to fast growth rates and therefore maintaining the size class structure. In addition, disease and predation were constantly present (18% of colonies were affected each monitoring event), but varied widely in terms of prevalence and resulting partial mortality on *A. cervicornis* populations, further affecting the ability of colonies to grow. These continuous background levels of colony shrinkage with intermittent high rates of disease could be devastating for the long-term survival and recovery of this species, because colonies must constantly battle adverse health conditions that stunt their growth and reduce their reproductive capabilities. Similar temporal patterns in disease prevalence were observed in all regions, with increases during the Summer and Fall events, whereas seasonal trends were not observed in predation prevalence. As the frequency of thermal anomalies and disturbance events increase

(Hoegh-Guldberg et al., 2007; Hughes et al., 2017; Ortiz et al., 2014; van Hooidonk et al., 2013), disease is likely to become more widespread, and potential recovery periods with lower prevalence will become more infrequent.

Colony diameters measured at all regions were larger than what has previously been reported for these regions (Dustan and Halas, 1987; Huntington and Miller, 2013; Huntington et al., 2017; Lidz and Zawada, 2013; Vargas-Ángel et al., 2003; Williams and Miller, 2006), but smaller than colonies in the Dominican Republic (Lirman et al., 2010) and Venezuela (Agudo-Adriani et al., 2016). Still, the ambiguity of defining a "colony" (entire skeletal unit (Dustan and Halas, 1987) to live area units on a larger skeletal unit (Huntington and Miller, 2013; Miller et al., 2008)) and the variability in reporting methods of colony size (maximum diameter, live tissue area, length of live tissue, total linear extension, and volume) can be problematic when comparing studies. This highlights the need to have standardized metrics for monitoring. For a demographic study, it is important that the entire skeletal structure of a colony, both living and dead be measured indicating the percentage of that colony that is alive. These data can then provide more informative analyses such as size structure, partial mortality, volume (structural complexity), and colony abundance, which would not be possible if only the live tissue were measured.

Acropora cervicornis is prone to frequent fragmentation due to its fragile skeleton. Fragmentation of colonies (evident through a decrease in colony size) was observed following tropical disturbances or elevated sea states, which for the FRT commonly occurs during the fall. Increases in colony diameter aligned with summer, a time where calmer sea states are more frequently observed and have been documented as the growth and branching period for this species (Shinn, 1976). Furthermore, the frequency and intensity of change in colony size may be due to colony morphology and skeletal density, which are likely adapted to each regional hydrodynamic regime (Bottjer, 1980; Chamberlain Jr and Graus, 1975; Gladfelter, 1984; Kuffner et al., 2017; Schumacher and Plewka, 1981). Although colony morphology was not measured, casual observations indicate differences with more compact, densely branched colonies in DRTO than in BWD. Morphological differences were also apparent at a few of the MDK sites. A more compact growth form and possibly denser skeleton may be less prone to colony fragmentation, whereas a less dense skeleton (likely in more sheltered areas) would be more prone to fragmentation during high energy events. Less dense skeletons in BWD colonies, in addition to larger colonies being more prone to fragmentation (Fig. 8), may explain the high variability (75–125%) in colony diameter change. The range of change in CVI can be similarly explained, but because it is a metric of colony health, patterns also emerged with the prevalence of disease. This was most evident in the BWD region, where the highest prevalence resulted in negative changes in CVI.

The structural complexity of an *A. cervicornis* colony is irreplaceable on Atlantic and Caribbean reefs in terms of the shelter it provides for small fishes and invertebrates. Colony size and morphological differences, such as branch frequency and length, will inherently affect complexity, habitat creation (space available for protection), and influence the size structure of the fish community inhabiting a colony (Huntington et al., 2017; Wilson et al., 2010). Therefore, the role that *A. cervicornis* colonies play may vary among regions and sites with theoretically higher fish abundance and diversity surrounding larger colonies.

The observed colony size class composition among region indicates reduced sexual reproductive potential for MDK and DRTO, since under half the colonies were of reproductive size (Soong and Lang, 1992). This potential is further reduced when using the CVI, a more accurate estimate as it accounts for colony partial mortality (i.e., removing from the size estimate the dead portion of the colony that cannot spawn). The higher abundance of large colonies (>90 cm) in BWD suggests a higher sexual reproductive potential, but contribution to population growth may be limited due the region's northern (thus, generally downstream) location on the FRT.

Overall disease prevalence (<6%) was similar to what others have reported for the FRT (Miller et al., 2014b; Vargas-Ángel et al., 2003) and the Dominican Republic (Lirman et al., 2010). Disease prevalence was relatively low with increases during the summer months (up to 28% of colonies) influenced by water temperatures and disturbance events. Tropical Storm Isaac passed west of the DRTO region on 26 August 2012 (Berg, 2013) just prior to the Fall 2012 DRTO monitoring event, when the highest disease prevalence for that region was observed. Miller et al. (2014b) also reported temporal and spatial variability in disease prevalence with increases surrounding a disturbance event, but these increases were not associated with elevated temperatures. Disease was more prevalent at sites with larger colonies, and although they are more capable of overcoming adverse conditions such as disease (Forsman et al., 2006;

Loya, 1976; Sato, 1985), it is still of concern for reproductive potential as disease was most commonly seen in colony centres where colonies are most fecund (Soong and Lang, 1992).

Colony partial mortality (amount of tissue lost) and prevalence (how widespread) were both good indicators of disease impact on a population. However, describing the impact using just one of these indicators is inadequate since variation in these metrics does not always follow the same pattern, and prevalence rates can vary widely over short periods of time (Goergen et al., 2019; Miller et al., 2014a, b). The instances in which rates did not correlate may be due to spatial or temporal characteristics such that the data collection period may be at the beginning or end of an event (high prevalence/low mortality), or that the condition is not widespread and is heavily impacting only a few colonies. A good example of this anomaly occurred during Summer 2015, when extremely high prevalence rates were observed for almost all conditions (especially in the BWD region), but the amount of recent mortality per colony was on the lower end of the spectrum. During this same period, decreases in relative change in CVI for both the BWD and MDK regions were evident, which indicate that while recent mortality was no longer extensive, mortality had indeed occurred since the last monitoring event. Furthermore, annual temperature fluctuations likely influenced the health and subsequent growth of *A. cervicornis*; in all regions, a reduction in recent mortality and an increase in colony CVI were observed during 2013, which was during a 2 year period where mean high temperatures were relatively low.

Chronic predation by fireworms (ranging from 0% to 52% of colonies per site per monitoring event) poses a major threat to *A. cervicornis* growth and production. Previously reported fireworm predation prevalence in the BWD region of 4.8–65% of survey quadrats indicate that the presence of predation was already widespread eight years prior to this study (Vargas-Ángel et al., 2003). Similarly, in the MDK region, Miller et al. (2014b) reported a wide range in prevalence (4–43% of colonies) with significant changes between years and sites. Snail predation was difficult to quantify because snails typically reside and predate in cryptic locations on the colony (e.g., the underside of branches, around the bases or in branching junctures (Johnston and Miller, 2014)). When predation was observed, per cent of tissue loss was similar to fireworm predation, typically less than 5% and no more that 30%, which is substantially lower than a manipulative feeding behaviour study where 70% of transplanted colonies were completely consumed by snails within 23 weeks (Johnston and Miller, 2014). While the direct mortality caused by predation was minimal, snail and fireworm impacts have been associated with increased disease prevalence, disturbance events, and as vectors of disease (Bright et al., 2016; Knowlton et al., 1990; Miller et al., 2014b). Furthermore, predation reduces a colony's ability to contribute to population growth, reproduction, and recovery due to the typical location of predation on a colony, apical ends (fireworm) and central portions of the colony (snail). Therefore, management of predators within a site could decrease the prevalence or impact of coral disease (Miller, 2001; Miller et al., 2014b).

Large colony diameters in the BWD region were likely due to local environmental conditions (depth, location on the reef, hydrodynamics) supportive of faster growth (Bliss, 2015), broad distribution (D'Antonio et al., 2016), and long-term colony persistence (Vargas-Ángel et al., 2003; Walker et al., 2012). Given the location of these populations at the northernmost extent of the species' range, and the dominant Florida Current running north, the likelihood of these populations contributing to population growth in other regions through sexual reproduction is limited, and would depend on a countercurrent (Lee, 1975; Soloviev et al., 2017).

There is also evidence of the impact of environmental conditions at the site level within each region. Sites in the BWD region may be thought of as more sheltered because of their depth and location leeward of the nearshore ridge complex, a shallow habitat that attenuates wave energy, whereas most of the MDK and DRTO sites (except for U59 and Marker 7) could be considered less sheltered. Sites with more protection (BCA, Scooter, U59, and Marker 7) had larger colonies and higher disease prevalence, but per cent mortality caused by disease was highest at the sites with the smallest colonies. This pattern is concerning, especially for the MDK and DRTO where most of the colonies are of smaller sizes. Site depth also contributed to differences in colony diameter; the deepest site U59 (12.5 m) had significantly larger colonies than any other site, which were similar to colony sizes Lidz and Zawada (2013) reported at depths greater than 7 m in the DRTO region, although protection from wave energy at shallow sites lead to increases in colony size (Marker 7). Meanwhile, shallower depths and lack of protection from hydrodynamic forces may cause more frequent fragmentation limiting colony growth (Bright et al., 2016; Hughes, 1994; Hughes et al., 2017; Hughes and Connell, 1999; Meesters et al., 1994), although this was not observed for BWD where more sheltered sites had a large variability in change of colony diameter (possibly fragmentation). This could be caused by the data collection methods used herein. Because colonies were not individually fate tracked over time, there is the possibility that the change in colony size is due to colonies either being included or excluded from the dataset during each time period. However, we also show that larger colonies are more likely to have a negative change in size between surveys than smaller colonies regardless of the site characteristics. Larger colonies feel more drag and as branches extend further from their central attachment point, the force required for fragmentation will decrease. Therefore, large colonies at a sheltered healthy site should benefit from good conditions to grow to a larger size, naturally fragment, and reattach to the surrounding reef.

As restoration practices continue to expand, these data will be valuable for gauging success through comparison of what could be expected of a restoration program in terms of disease and predation presence and the impact that each may have. The results of this study could also provide insight into outplant site selection for locations that offer protection (deeper or sheltered from strong wave energy), low predator abundances, or have an active management plan for predators (in particular for *H. carunculata* and *C. abbreviata*). Additionally, as results suggested herein, restoration programs that outplant larger colonies may have a greater chance of success (Goergen and Gilliam, 2018; Lirman et al., 2014; Mercado-Molina et al., 2015b).

This study provides a multiyear look at the size, growth, and health of numerous *A. cervicornis* populations across the FRT. Our data show that change can be experienced within a population (site or region) over a short period of time and that these changes do not always occur similarly across the entire reef tract, supporting the need for site or region specific conservation and restoration strategies as also suggested by Mercado-Molina et al. (2015a). The impact of environmental conditions, likely influenced by the site location, played an integral part in colony size, growth, and health. Site depth or protection allowed for larger colony growth. Sites with larger colonies had higher prevalence of disease and predation, but lower colony partial mortality. Conditions were not unique to one region. The emergent pattern of consistent disease and predation presence, coupled with the documented relationship with larger colonies/densities across the entire range of this study raises concern for the future persistence of this species. Whether or not these results suggest a cyclic population it is worthy of further investigation.

Acknowledgements

This study was made possible through the support of National Coral Reef Institute (NCRI), Florida Fish and Wildlife Conservation Commission (FWC), and Nova Southeastern University (NSU). This study was partially funded by, the National Oceanic and Atmospheric Administration Coastal Ocean Program award (NA03NOS4260046) to NSU for the NCRI, as part of an award from the National Marine Fisheries Service (NA1ONMF4720029) under contract #10275 with the FWC, and by a NSU Presidents Faculty Research and Development Grant. Sampling was completed under permits FKNMS-2010-126-A2, FKNMS-2013-104, DRTO-2010-SCI-0025, DRTO-2011-SCI-0016, DRTO-2012-SCI-0018, and DRTO-2013-SCI-0016. This work was submitted as a portion of Elizabeth A. Goergen's Dissertation at Nova Southeastern University, Dania Beach, FL, USA. We thank all the students of Nova Southeastern University's Coral Reef Restoration, Assessment, and Monitoring Lab, Meaghan Johnson, Cory Walter, and Caitlin Lustic for assisting in data collection. Cindy Lewis and the staff of the Keys Marine Laboratory and Tracy Ziegler of the National Park Service who assisted with project logistics. Dale Goergen wrote R code for figure creation.

References

Acropora Biological Review Team, 2005. Atlantic *Acropora* status review. Report to National Marine Fisheries Service, Southeast Regional Office, p. 152 p+ App.

Adey, WH, Burke, RB, 1977. Holocene bioherms of Lesser Antilles—geologic control of development. In: Frost, SH, Weiss, MP, Saunders, JB (Eds.), Reef and Related Carbonates—Ecology and Sedimentology. American Association of Petroleum Geologists, Tulsa, OK, pp. 67–81. https://doi.org/10.1306/St4393C6.

Agudo-Adriani, E.A., Cappelletto, J., Cavada-Blanco, F., Croquer, A., 2016. Colony geometry and structural complexity of the endangered species *Acropora cervicornis* partly explains the structure of their associated fish assemblage. PeerJ 4, e1861.

Antonius, A., 1977. Coral Mortality in reefs: a problem for science and management. Proc. 3rd Int. Coral Reef Symp. 2, 617–623.

Antonius, A., 1981. The "band" diseases in coral reefs. Proc. 4th Int. Coral Reef Symp. 2, 7–14.

Aronson, R.B., Bruckner, A.W., Moore, J., Precht, W.F., Weil, E., 2008. IUCN Red List of Threatened Species: *Acropora cervicornis*. International Union for Conservation of Nature and Natural Resources, Cambridge, UK.

Aronson, R.B., Precht, W.F., 2001. White-band disease and the changing face of Caribbean coral reefs. Hydrobiologia 460, 25–38.

Berg, R.J., 2013. Tropical cyclone report Hurricane Isaac (AL092012). National Nurricane Center, p. 78.

Bliss, B.C., 2015. Translocation of *Acropora cervicornis* across geographic regions: investigating species recovery and restoration, Oceanographic Center. Nova Southeastern University, p. 102.

Bottjer, D.J., 1980. Branching morphology of the reef coral *Acropora cervicornis* in different hydraulic regimes. J. Paleo. 54, 1102–1107.

Brawley, S.H., Adey, W.H., 1977. Territorial behavior of threespot damselfish (*Eupomacentrus planifrons*) increases reef algal biomass and productivity. Environ. Biol. Fishes 2, 45–51.

Bright, A.J., Rogers, C.S., Brandt, M.E., Muller, E., Smith, T.B., 2016. Disease prevalence and snail predation associated with swell-generated damage on the threatened coral, *Acropora palmata* (Lamarck). Front. Mar. Sci. 3, 1–13. https://doi.org/10.3389/fmars.2016.00077.

Bruckner, A., Bruckner, R., Hill, R., 2009. Improving restoration approaches for *Acropora palmata*: lessons from the Fortuna Reefer grounding in Puerto Rico. Proc. 11th Int. Coral Reef Symp. 1, 1199–1203.

Bruckner, A.W., 2002. Proceedings of the Caribbean *Acropora* workshop: potential application of the U.S. Endangered Species Act as a conservation strategy. NOAA Technical Memorandum NMFS-OPR-24, Silver Spring, MD. 199 pp.

Busch, J., Greer, L., Harbor, D., Wirth, K., Lescinsky, H., Curran, H.A., de Beurs, K., 2016. Quantifying exceptionally large populations of *Acropora* spp. corals off Belize using submeter satellite imagery classification. Bull. Mar. Sci. 92, 265–283.

Chamberlain Jr., J.A., Graus, R.R., 1975. Water flow and hydromechanical adaptations of branched reef corals. Bull. Mar. Sci. 25, 112–125.

D'Antonio, N.L., Gilliam, D.S., Walker, B.K., 2016. Investigating the spatial distribution and effects of nearshore topography on *Acropora cervicornis* abundance in Southeast Florida. PeerJ 4, e2473.

Davis, G.E., 1982. A century of natural change in coral distribution at the Dry Tortugas: a comparison of reef maps from 1881-1976. Bull. Mar. Sci. 32, 608–623.

Dinno, A., 2017. dunn.test: Dunn's Test of multiple comparisons using rank sums. R package version 1.3.5.

Dustan, P., Halas, J.C., 1987. Changes in the reef-coral community of Carysfort reef, Key Largo, Florida: 1974 to 1982. Coral Reefs 6, 91–106.

Forsman, Z.H., Rinkevich, B., Hunter, C.L., 2006. Investigating fragment size for culturing reef-building corals (*Porites lobata* and *P. compressa*) in ex situ nurseries. Aquaculture 261, 89–97.

Garrison, V.H., Ward, G., 2012. Transplantation of storm-generated coral fragments to enhance Caribbean coral reefs: a successful method but not a solution. Rev. Biol. Trop. 60, 59–70.

Gilmore, M.D., Hall, E.R., 1976. Life history, growth habitats and constructional roles of *Acropora cervicornis* in the patch reef environment. J. Sediment. Petrol. 46, 519–522.

Gladfelter, E.H., 1984. Skeletal development in *Acropora cervicornis* III. A comparison of monthly rates of linear extension and calcium carbonate accretion measured over a year. Coral Reefs 3, 51–57.

Gladfelter, W., 1982. White-band disease in *Acropora palmata*: implications for the structure and growth of shallow reefs. Bull. Mar. Sci. 32, 639–643.

Goergen, E.A., Gilliam, D.S., 2018. Outplanting technique, host genotype, and site affect the initial success of outplanted *Acropora cervicornis*. PeerJ 6, e4433.

Goergen, E. A., Gilliam, D. S., n.d. *Acropora cervicornis* colony residence time and retention rates implications for long-term monitoring. Ecol. Indic. In preparation.

Goergen, E.A., Moulding, A.L., Walker, B.K., Gilliam, D.S., 2019. Identifying causes of temporal changes in *Acropora cervicornis* populations and the potential for recovery. Front. Mar. Sci. 6, 13.

Goreau, T.F., 1959. The ecology of Jamaican coral reefs I. Species composition and zonation. Ecology 40, 67–90.

Griffin, S., Nemeth, M., Moore, T., Gintert, B., 2015. Restoration using *Acropora cervicornis* at the T/V MARGARA grounding site. Coral Reefs 34, 885.

Hayes, J.A., 1990. Distribution, movement and impact of the coralivorous gastropod *Coralliophila abbreviata* (Lamarck) on a Panamanian patch reef. J. Exp. Mar. Biol. Ecol. 142, 25–42.

Highsmith, R.C., Riggs, A.C., D'Antonio, C.M., 1980. Survival of hurricane-generated coral fragments and a disturbance model of reef calcification/growth rates. Oecologia 46, 322–329.

Hoegh-Guldberg, O., Mumby, P.J., Hooten, A.J., Steneck, R.S., Greenfield, P., Gomez, E., Harvell, C.D., Sale, P.F., Edwards, A.J., Caldeira, K., Knowlton, N., Eakin, C.M.,

Iglesias-Prieto, R., Muthiga, N., Bradbury, R.H., Dubi, A., Hatziolos, M.E., 2007. Coral reefs under rapid climate change and ocean acidification. Science 318, 1737–1742.
Hothorn, T., Bretz, F., Westfall, P., 2008. Simultaneous inference in general parametric models. Biom. J. 50, 346–363.
Hughes, T.P., 1994. Catastrophes, phase shifts and large-scale degredation of a caribbean coral reef. Science 265, 1547–1550.
Hughes, T.P., Barnes, M.L., Bellwood, D.R., Cinner, J.E., Cumming, G.S., Jackson, J.B.C., Kleypas, J., van de Leemput, I.A., Lough, J.M., Morrison, T.H., Palumbi, S.R., van Nes, E.H., Scheffer, M., 2017. Coral reefs in the Anthropocene. Nature 546, 82–90.
Hughes, T.P., Connell, J.H., 1999. Multiple stressors on coral reefs: a long-term perspective. Limnol. Oceanogr. 44, 932–940.
Huntington, B.E., Miller, M.W., 2013. Location-specific metrics for rapidly estimating the abundance and condition of the threatened Coral *Acropora cervicornis*. Restor. Ecol. 22, 299–303.
Huntington, B.E., Miller, M.W., Pausch, R., Richter, L., 2017. Facilitation in Caribbean coral reefs: high densities of staghorn coral foster greater coral condition and reef fish composition. Oecologia 184, 247–257.
Johnston, L., Miller, M.W., 2014. Negative indirect effects of neighbors on imperiled scleractinian corals. Coral Reefs 33, 1047–1056.
Kaufman, L., 1977. The threespot damselfish: effects on benthic biota of Caribbean coral reefs. Proc. 3rd Int. Coral Reef Symp. 1, 559–564.
Kaufman, L., 1981. There was biological disturbance on Pleistocene coral reefs. Paleobiology 7, 527–532.
Keck, J., Houston, R.S., Purkis, S., Riegl, B.M., 2005. Unexpectedly high cover of *Acropora cervicornis* on offshore reefs in Roatán (Honduras). Coral Reefs 24, 509.
Knowlton, N., Lang, J.C., Keller, B.D., 1990. Case study of natural population collapse: post-hurricane predation on Jamaican Staghorn coral. Smithsonian Contrib. Mar. Sci. 31, 36.
Kuffner, I.B., Bartels, E., Stathakopoulos, A., Enochs, I.C., Kolodziej, G., Toth, L.T., Manzello, D.P., 2017. Plasticity in skeletal characteristics of nursery-raised staghorn coral, *Acropora cervicornis*. Coral Reefs 36, 679–684.
Lee, T.N., 1975. Florida current spin-off eddies. Deep-Sea Res. Oceanogr. Abstr. 22, 753–765.
Lidz, B.H., Zawada, D.G., 2013. Possible Return of *Acropora cervicornis* at Pulaski Shoal, Dry Tortugas National Park, Florida. J. Coast. Res. 287, 256–271.
Lirman, D., Bowden-Kerby, A., Schopmeyer, S., Huntington, B., Thyberg, T., Gough, M., Gough, T., Gough, R., Gough, Y., 2010. A window to the past: documenting the status of one of the last remaining 'megapopulations' of the threatened staghorn coral *Acropora cervicornis* in the Dominican Republic. Aquat. Conserv. 20, 773–781.
Lirman, D., Schopmeyer, S., 2016. Ecological solutions to reef degradation: optimizing coral reef restoration in the Caribbean and Western Atlantic. PeerJ 4, e2597.
Lirman, D., Schopmeyer, S., Galvan, V., Drury, C., Baker, A.C., Baums, I.B., 2014. Growth dynamics of the threatened Caribbean staghorn coral *Acropora cervicornis*: influence of host genotype, symbiont identity, colony size, and environmental setting. PLoS One 9, e107253.
Loya, Y., 1976. Skeletal regeneration in a Red Sea scleractinian coral population. Nature 261, 490–491.
Lucas, M.Q., Weil, E., 2016. Recent recovery in *Acropora cervicornis* and abundance of *A. prolifera* off La Parguera, Puerto Rico. Mar. Biodivers. 46, 531–532.
Marsden, J.R., 1962. A coral-eating polychaete. Nature 193, 598.

Meesters, H., Noordeloos, M., Bak, R., 1994. Damage and regeneration links to coral growth in the reef-building coral *Montastrea annularis*. Mar. Ecol. Prog. Ser. 112, 119–128.
Mercado-Molina, A.E., Ruiz-Diaz, C.P., Pérez, M.E., Rodríguez-Barreras, R., Sabat, A.M., 2015. Demography of the threatened coral *Acropora cervicornis*: implications for its management and conservation. Coral Reefs 34 (4), 1113–1124.
Mercado-Molina, A.E., Ruiz-Diaz, C.P., Sabat, A.M., 2015b. Demographics and dynamics of two restored populations of the threatened reef-building coral *Acropora cervicornis*. J. Nat. Conserv. 24, 17–23.
Miller, M.W., 2001. Corallivorous snail removal: evaluation of impact on *Acropora palmata*. Coral Reefs 19, 293–295.
Miller, M.W., Bourque, J., Bohnsack, J., 2002. An analysis of the loss of Acroporid corals at Looe Key, Florida, USA 1983-2000. Coral Reefs 21, 179–182.
Miller, M.W., Kerr, K., Williams, D.E., 2016. Reef-scale trends in Florida *Acropora* spp. abundance and the effects of population enhancement. PeerJ 4, e2523.
Miller, M.W., Lohr, K.E., Cameron, C.M., Williams, D.E., Peters, E.C., 2014a. Disease dynamics and potential mitigation among restored and wild staghorn coral, *Acropora cervicornis*. PeerJ 2, e541.
Miller, M.W., Marmet, C., Cameron, C.M., Williams, D.E., 2014b. Prevalence, consequences, and mitigation of fireworm predation on endangered staghorn coral. Mar. Ecol. Prog. Ser. 516, 187–194.
Miller, S.L., Chiappone, M., Rutten, L.M., Swanson, D.W., 2008. Population status of *Acropora* coral in the Florida keys. Proc. 11th Int. Coral Reef Symp. 2, 781–785.
National Marine Fisheries Service, 2006. Endangered and threatened species: final listing determinations for elkhorn coral and staghorn coral. Fed. Regist. 71, 26852–26872.
Neigell, J.E., Avise, J.C., 1983. Clonal diversity and population structure in a reef-building coral, *Acropora cervicornis*: self recognition analysis and demographic interpretation. Evolution 37, 437–453.
Ortiz, J.C., Bozec, Y.-M., Wolff, N.H., Doropoulos, C., Mumby, P.J., 2014. Global disparity in the ecological benefits of reducing carbon emissions for coral reefs. Nat. Climate Change 4, 1090–1094.
Peters, E.C., Oprandy, J.J., Yevich, P.P., 1983. Possible causal agent of "white band disease" in Caribbean Acroporid corals. J. Invertebr. Pathol. 41, 394–396.
RStudio Team, 2016. RStudio: Integrated Development for R. RStudio, Inc., Boston, MA.
Ruzicka, R., Colella, M., Porter, J., Morrison, J., Kidney, J., Brinkhuis, V., Lunz, K., Macaulay, K., Bartlett, L., Meyers, M., 2013. Temporal changes in benthic assemblages on Florida Keys reefs 11 years after the 1997/1998 El Niño. Mar. Ecol. Prog. Ser. 489, 125–141.
Sato, M., 1985. Mortality and growth of juvenile coral *Pocillopora damicornis* (Linnaeus). Coral Reefs 4, 27–33.
Schumacher, H., Plewka, M., 1981. The adaptive significance of mechanical properties versus morphological adjustments in skeletons of *Acropora palmata* and *Acropora cervicornis* (Cnidaria, Scleractinia). Proc. 4th Int. Coral Reef Symp. 2, 121–128.
Shinn, E.A., 1976. Coral reef recovery in Florida and the Persian Gulf. Environ. Geol. 1, 241–254.
Smith, L.D., Devlin, M., Haynes, D., Gilmour, J.P., 2005. A demographic approach to monitoring the health of coral reefs. Mar. Pollut. Bull. 51, 399–407.
Soloviev, A.V., Hirons, A., Maingot, C., Dean, C.W., Dodge, R.E., Yankovsky, A.E., Wood, J., Weisberg, R.H., Luther, M.E., McCreary, J.P., 2017. Southward flow on the western flank of the Florida current. Deep-Sea Res. I Oceanogr. Res. Pap. 125, 94–105.

Soong, K., Lang, J.C., 1992. Reproductive integration in reef corals. Biol. Bull. 183, 418–431.

Tunnicliffe, V., 1981. Breakage and propagation of the stony coral *Acropora cervicornis*. Proc. Natl. Acad. Sci. U. S. A. 78, 2427–2431.

van Hooidonk, R., Maynard, J.A., Planes, S., 2013. Temporary refugia for coral reefs in a warming world. Nat. Climate Change 3, 508–511.

Vargas-Ángel, B., Thomas, J.D., Hoke, S.M., 2003. High-latitude *Acropora cervicornis* thickets off Fort Lauderdale, Florida, USA. Coral Reefs 22, 465–473.

Walker, B.K., Larson, E.A., Moulding, A.L., Gilliam, D.S., 2012. Small-scale mapping of indeterminate arborescent acroporid coral (*Acropora cervicornis*) patches. Coral Reefs 31, 885–894.

Williams, D.E., Miller, M.W., 2005. Coral disease outbreak: pattern, prevalence and transmission in *Acropora cervicornis*. Mar. Ecol. Prog. Ser. 301, 119–128.

Williams, D.E., Miller, M.W., 2006. Importance of disease and predation to the growth and survivorship of juvenile *Acropora palmata* and *A. cervicornis*: a demographic approach. Proc. 10th Int. Coral Reef Symp. 1, 1096–1104.

Williams, D.E., Miller, M.W., Kramer, K.L., 2006. Demographic monitoring protocols for threatened Caribbean *Acropora* spp. corals. NOAA Technical Memorandum NMFS-SEFSC-543, 91 pp.

Wilson, S., Fisher, R., Pratchett, M., Graham, N., Dulvy, N., Turner, R., Cakacaka, A., Polunin, N., 2010. Habitat degradation and fishing effects on the size structure of coral reef fish communities. Ecol. Appl. 20, 442–451.

Young, C., Schopmeyer, S., Lirman, D., 2012. A review of reef restoration and coral propagation using the threatened genus *Acropora* in the Caribbean and Western Atlantic. Bull. Mar. Sci. 88, 1075–1098.

CHAPTER FIVE

Population dynamics and growth rates of free-living mushroom corals (Scleractinia: Fungiidae) in the sediment-stressed reefs of Singapore

Rian Prasetia[a,*], Zi Wei Lim[a], Aaron Teo[a], Tom Shlesinger[b], Yossi Loya[c], and Peter A. Todd[a]

[a]Experimental Marine Ecology Laboratory, Department of Biological Sciences, National University of Singapore, Singapore, Singapore
[b]Institute for Global Ecology, Florida Institute of Technology, Melbourne, FL, United States
[c]School of Zoology, George S. Wise Faculty of Life Sciences, Tel-Aviv University, Tel-Aviv, Israel
*Corresponding author: e-mail address: dbsrian@nus.edu.sg

Contents

Abstract

The reefs of Singapore provide an excellent opportunity to study the population dynamics and growth rates of free-living mushroom corals (Fungiidae) under sediment-stressed conditions. Transect surveys at four study sites revealed a total of 11 free-living mushroom coral species—the same 11 species as those found by local studies since the 1980s. The abundance of the four most common species ranged from 1.0 to 68.3 Ind. per 100 m^2, while their population size-structure showed a common pattern of a higher proportion of small-sized corals than large-sized ones (i.e. positively

Advances in Marine Biology, Volume 87
ISSN 0065-2881
https://doi.org/10.1016/bs.amb.2020.08.002

skewed size-structure), although very few individuals of the smallest-size classes were recorded for any of the four species. A more positively skewed size-structure for each of the four most common species was observed at the reef slope (5–6 m depth) than at the reef crest (2–3 m depth), possibly due to a slower growth rate caused by light reduction with depth. All the mushroom corals studied exhibited a decline in growth rate with increasing size and weight, indicating determinate growth. Growth rate of each of the four most common species was similar among the study sites, despite variation in environmental conditions. Our results demonstrate species richness stability over the past three decades, suggesting that these free-living mushroom coral assemblages comprise species that are well-adapted to the chronic high sedimentation characteristic of Singapore's reefs. However, if the paucity of individuals of the smallest-size classes reflects poor recruitment and/or early mortality, there may be some cause for concern. Our robust baseline data can contribute to a long-term monitoring strategy for determination of changes in mushroom coral population dynamics.

1. Introduction

Despite their ecological and economic importance, coral reefs are globally threatened (Hughes et al., 2017; Sully et al., 2019). Local anthropogenic stressors, such as overfishing and coastal development (Burke et al., 2011; Jackson et al., 2001), are being compounded by increasing seawater temperatures and ocean acidification (Hoegh-Guldberg et al., 2007; Hoegh-Guldberg and Bruno, 2010). These impacts are often associated with various changes in coral-reef communities (e.g. Guest et al., 2016b; Loya et al., 2001; Mellin et al., 2019; Stuart-Smith et al., 2018; Zinke et al., 2018). Although some coral reefs have recovered from disturbances (Gilmour et al., 2013; Graham et al., 2015; Shlesinger and Loya, 2016) or were less affected by certain stressors (Fine et al., 2013; Riegl and Piller, 2003; Sully and van Woesik, 2020), many are expected to experience further decline (Hughes et al., 2017).

Population dynamics of corals offer an indicator of the health status of coral reefs (Bak and Meesters, 1998; Hughes and Jackson, 1980; Shlesinger and Loya, 2019; Smith et al., 2005), i.e. the size and frequency of individuals at various life-history stages, as well as the spatio-temporal changes that can occur as a result of birth and death within a population (Meesters et al., 2001). For example, in Curaçao, a size-structure assessment of coral populations revealed relatively few colonies in the smaller size classes and relatively more larger colonies (increased negative skewness), probably reflecting a response to the degraded environmental conditions there (Meesters et al., 2001). In contrast, four locally-abundant corals in the

Persian Gulf displayed a significantly more positively skewed size-structure, with smaller-sized colonies at sites with higher sea surface temperatures, salinity, and bleaching, compared to sites with relatively stable environmental conditions (Bauman et al., 2013). Also, size-structure of coral populations changed across several subsequent disturbances. Dynamics of changed size-structure was taxon specific and resulted in cover loss due to reduced coral sizes (Riegl et al., 2018; Riegl and Purkis, 2015). Investigating the size-structure of coral populations can provide valuable insights into how they are affected by local environmental conditions and how they respond to disturbances (Bak and Meesters, 1998; Bauman et al., 2013; Riegl et al., 2017).

Mushroom corals (family: Fungiidae) represent a potential model taxon for the study of coral population dynamics, since most of them are solitary individuals that grow in distinct forms, with life-history stages that can be easily identified (Hoeksema, 1989, 1991a). The majority of mushroom coral species are free-living and capable of movement, which enables them to colonize new areas, initiate reef growth, and create new patch reefs (Chadwick-Furman and Loya, 1992; Littler et al., 1997). The typical life cycle of mushroom corals begins when a sexually-produced larva in the planktonic stage seeks a suitable substrate for settlement (Abe, 1937). The settled larva then develops into the sessile stage, in which the individual coral polyp becomes attached to the substratum via a stalk termed the anthocaulus. At the end of the sessile stage, the coral develops a mushroom-like appearance before detaching itself and transitioning to the free living (or anthocyathus) stage (Hoeksema and Yeemin, 2011). While new polyps may develop asexually from the anthocaulus, which remains attached to the substratum (Hoeksema and Yeemin, 2011; Vizel et al., 2009), the free-living coral will be able to initiate sexual reproduction and produce gametes upon maturity (Hoeksema, 1989). In some mushroom coral species, the free-living coral can achieve asexual reproduction through the formation of new polyps that bud off from an injured or dead parent (Gilmour, 2004a; Hoeksema, 1989, 2004). In *Cycloseris* spp., asexual reproduction can also take place through natural autotomy, in which individuals repetitively split into wedge-shaped segments (Hoeksema et al., 2018; Hoeksema and Waheed, 2011; Yamashiro and Nishihira, 1994, 1998).

Several previous studies have examined mushroom corals population dynamics and size-structures (Chadwick-furman et al., 2000; Feliciano et al., 2018; Gilmour, 2004b; Goffredo and Chadwick-Furman, 2003; Hoeksema, 1991b; Knittweis et al., 2009). For example, the size-structure of the *Fungia fungites* population at a site exposed to frequent cyclones and high sedimentation rates underwent a dramatic decline of individuals at all life-history stages compared to a nearby sheltered site in the Dampier Archipelago, north-western Australia (Gilmour, 2004b). The size-structure of the *Heliofungia actiniformis* population in the Spermonde Archipelago, Indonesia, reflected higher mortality rates among the young attached polyps compared to the free-living individuals (Knittweis et al., 2009). To date, no extensive study on the population dynamics of mushroom corals has been conducted within a heavily urbanized reef system. The coral reefs in Singapore offer an excellent opportunity to investigate coral demographics in such an environment, where >80% of the coastline has been modified due to reclamation and shoreline development (Hilton and Manning, 1995; Lai et al., 2015). Levels of sedimentation, turbidity and eutrophication around the coral reefs of Singapore are higher than what is considered 'optimal' for tropical reefs (Browne et al., 2015; Dikou and van Woesik, 2006; Kleypas et al., 1999). In addition, bleaching events occurred there in 1998, 2010, 2013, and 2016 (Chou et al., 2016; Guest et al., 2012, 2016a; Ng et al., 2020). Quantifying demographic parameters will provide new insights into how Singapore's coral reefs are coping with these stressors, as the common metric of assessing the percentage of coral cover alone can overlook critical aspects regarding the state of coral communities (Edmunds and Riegl, 2020; Shlesinger and Loya, 2019; Smith et al., 2005).

Most research on free-living mushroom corals in Singapore has focused on their biology and taxonomy (Hoeksema and Koh, 2009; Huang et al., 2009; Koh, 1987). A total of 15 free-living mushroom coral species have been recorded from museum collections (since the 1860s) and fieldwork observations (since the 1970s), with four species reported to have become locally extinct (Hoeksema and Koh, 2009). In the current study, we investigated the distribution and population dynamics and structure of free-living mushroom corals in the heavily sedimented reefs of Singapore. We also analysed the sizes and growth rates of the four most common species.

2. Materials and methods

2.1 Study sites

Singapore is located just over 1° north of the equator, at the southernmost tip of the Malay Peninsula. Densely populated and highly urbanized, Singapore's coastal waters are impacted by numerous stressors, especially sediments originating from coastal development and dredging (Heery et al., 2018; Junjie et al., 2014; Todd et al., 2010). Little coral cover exists

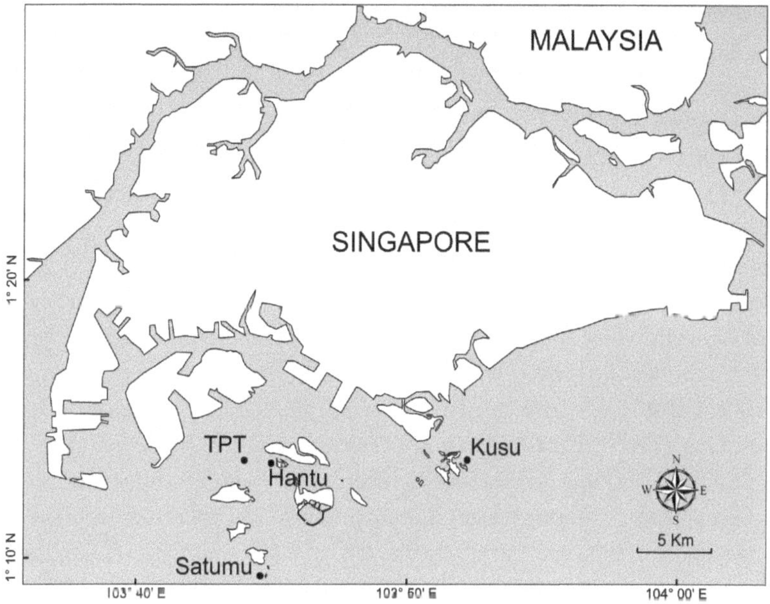

Fig. 1 The four study sites within the Southern Islands of Singapore.

below 6 m, due to turbidity-driven 'reef compression' (i.e. a reduction in bathymetric range with increasing turbidity and decreasing water clarity) (Chow et al., 2019; Guest et al., 2016b; Heery et al., 2018). The present study was conducted at four fringing reefs adjacent to four of Singapore's Southern Islands (Fig. 1), in May and November–December 2018. The four selected sites represented a range of distances (5–15 km) from the main island of Singapore: Kusu (1°13′32″N, 103°51′33″), Hantu (1°13′39″N, 103° 44′48″E), Satumu (1°09′37″N, 103°44′26″E), and Terumbu Pempang Tengah (hereafter TPT; 1°13′38″N, 103°43′41″E).

2.2 Mushroom coral distribution and size-structure

At the Kusu and Hantu study sites, eight belt transects ($50 \times 2\,m^2$, $n=4$ at each depth) were haphazardly deployed at the reef crest (2–3 m depth) and the reef slope (5–6 m depth) by SCUBA diving. Due to the large number of mushroom corals at the Satumu and TPT sites, the length of the eight belt transects at each of these two sites was shortened to $30 \times 2\,m^2$. To account for different transect lengths, data were presented as the number of individuals divided by area of observation (see below). All the free-living mushroom corals within the belt transects were counted and their life-history stages were classified as follows: free-living corals that were not attached to the substrata; sexual recruits that were attached to the substrata via a stalk; and asexual recruits that were attached to a parent polyp (following Gilmour, 2004b). Each of the mushroom corals was then photographed with a digital camera (Sony RX100) next to a ruler for scale. All individuals of the four most common fungiid species (i.e. *Ctenactis echinata*, *Herpolitha limax*, *Danafungia scruposa* and *Fungia fungites*) within the transects were collected for growth measurements (see *Growth rates* below) and were then photographed on the support boat. Close-up images of the oral and aboral surfaces (only for corals at the free-living stage) were taken for each coral, since identification of free-living mushroom corals requires a close examination of both sides (Hoeksema, 1989). Species identification was conducted based on the taxonomy and morphological characteristics described by Hoeksema (1989), Koh and Chou (1989), and Gittenberger et al. (2011). An additional 15 min swim was conducted at each study site outside of the belt transects (depth range: 1–6 m) in order to increase both the area of observation for creating a species list and the number of samples for the size-structure analysis.

To examine the size-structure of the free-living mushroom corals, individual sizes of the observed four above-noted most common species were measured (with more than 100 specimens per species recorded across all sites). Coral size was determined by measurement of maximum length or maximum diameter, depending on whether their morphology was elongate or disc-shaped, respectively (Gilmour, 2004b; Goffredo and Chadwick-Furman, 2003), using ImageJ (Schneider et al., 2012).

2.3 Growth rates

During the surveys conducted in May 2018, all the collected specimens of the four most common species were transferred to large containers (frequently replenished with seawater) on the support boat. These were labelled individually with a numeric plastic or aluminium tag attached by nylon fishing cord, following Loya and Sakai (2008), before being photographed together with a ruler for scale. To weigh the corals, the boat was docked and a weighing station using digital scales (±0.1 g) was established on a jetty for stability. Individuals were patted dry with a towel before weighing (following Loya and Sakai, 2008). All specimens were returned to the reef environment on the same day of sampling. Similar measurements were conducted just over a year later, in June and July 2019. From a total of 281 corals that were tagged in 2018, 13 and 8 corals could not be retrieved for the size and weight measurements, respectively, and hence only 268 and

273 corals were used for these analyses. Growth rates were expressed as annual linear growth (Lg) and weight increase (Wg) rates. Linear growth (Lg) rate was estimated by subtracting maximum polyp length (mm) of the first (L_0) from the last (L_t) measurements ($Lg = L_t - L_0$). Weight increase rate (Wg) was estimated following Eyal-Shaham et al. (2019): $= \frac{W_t - W_0}{W_0}$, where W_t = weight (g) at the last measurement and W_0 = weight (g) at first measurement. Both growth rates (Lg and Wg) were standardized to 12 mo.

2.4 Data analysis

The abundance of each mushroom coral species from the belt transects was used to compare the free-living mushroom coral assemblages among sites and reef zones (i.e. four transects as replicates per site and per reef zone). For descriptive statistics, because of the difference in the transect size ($50 \times 2\,m^2$ in Kusu and Hantu; $30 \times 2\,m^2$ in Satumu and TPT), the abundance was standardized to number of individuals per $100\,m^2$.

All statistical analyses were conducted using R software v3.6.1 (R Core Team, 2015). Multivariate generalized linear models with negative binomial distribution were performed using the *mvabund* package (Wang et al., 2012) to test for differences in the mushroom coral assemblage composition among reef sites and reef zones. In this analysis, the individual counts were offset by their respective survey area (Warton et al., 2012) to account for the differences in transect length between sites.

Size-structure of the four species was evaluated for each site and each reef zone. Since there were two different types of mushroom coral shape (i.e. elongated: *C. echinata*, *H. limax*; and disc-shaped: *D. scruposa*, *F. fungites*) and their size ranges differed, the size-structure of the elongated corals was grouped to the nearest 5 cm (0–5, 5–10, etc., up to 70–75 cm), while disc-shape corals were grouped to the nearest 2 cm (0–2, 2–4, etc., up to 26–28 cm). The calculated size-structure parameters included the means, standard deviations, minima, maxima, skewness (g_1), and kurtosis (g_1) of mushroom coral length or diameter distribution.

Differences in mean coral size (length or diameter) among sites and among reef zones for each species were compared using two-way ANOVA followed by Tukey HSD post hoc for pairwise comparison. Prior to analyses, the statistical assumptions (i.e. normality and homogeneity) were verified through graphical analyses of residuals. Mean coral size of *H. limax* and *D. scruposa* were square root transformed to meet statistical assumptions. Skewness, kurtosis and standard deviations of the size-structure were compared among species, among sites and between reef zones using one-way ANOVA following Meesters et al. (2001). Kolmogorov-Smirnov (KS) two-sample comparison test was used for pairwise comparison of the size-structure among sites and reef zones within species.

ANCOVAs using individual samples as replicates were used to test for differences in linear growth rate for each species among sites, using initial coral size (length or diameter) as the covariate. The same tests were used to identify differences in weight increase for each species and among sites, using initial coral weight as the covariate. Weight increase and initial coral weight were log transformed to meet statistical assumptions. The statistical assumptions were tested through graphical analyses of residuals and the post hoc analyses were conducted using the *emmeans* package (Lenth et al., 2018).

3. Results

3.1 Mushroom coral distribution

A total of 776 free-living mushroom corals were recorded at the four study sites from both the belt transects (680 individuals) and the timed swims (96 individuals), with a total of 11 species recorded (Table S1 in Supplementary Material 1 in the online version at https://doi.org/10.1016/bs.amb.2020.08.002). Out of the 776 corals, 17 were sexual recruits (one *D. scruposa*, one *F. fungites* and 15 unidentified) and seven were asexual recruits (all unidentified). Out of the 680 individuals at the belt transects, most of the recruits were observed at TPT and Satumu both on the reef slope, with mean abundance (mean ± SD) 4.6 ± 5.7 and 3.8 ± 4.8 Ind. per 100 m^2, respectively (Fig. 2). Among the study sites, TPT had the highest mean abundance of adult corals (mean ± SD; reef crest: 61.7 ± 44.7, reef slope: 72.9 ± 45.1 Ind. per 100 m^2, Fig. 2) and highest species richness (10 species, Table S1 in Supplementary Material 1 in the online version at https://doi.org/10.1016/bs.amb.2020.08.002). Kusu had the lowest mean abundance of adult corals (mean ± SD; reef crest: 2.3 ± 3.3, reef slope: 2.3 ± 1.7 Ind. per 100 m^2, Fig. 2) and lowest species richness (seven species, Table S1 in Supplementary Material 1 in the online version at https://doi.org/10.1016/bs.amb.2020.08.002). The free-living mushroom coral assemblage (both for recruits and adult corals) at the four study sites combined was dominated by *F. fungites* (23.1% of the total), followed by *H. limax* (21.3%), *D. scruposa* (18.8%), and *C. ehinata* (16.0%). The abundance of these four most common species ranged from 1.0 to 68.3 Ind. per 100 m^2 at each site and each reef zone (Fig. 2). Within each site, these four species dominated Hantu, Satumu and TPT (ranged from 10.8 to

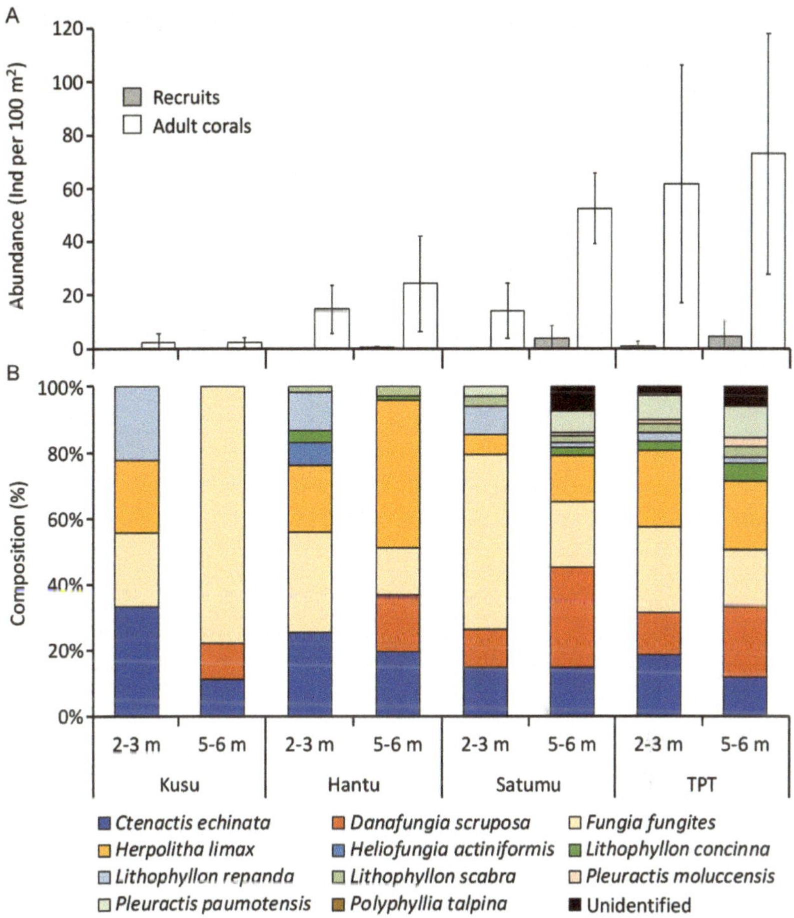

Fig. 2 Abundance (A) and percentage composition (B) of mushroom coral species per $100\,m^2$ on the reef crest (2–3 m depth) and reef slope (5–6 m depth) at the four study sites in Singapore.

35.7%), while at Kusu *Lithophyllon repanda*, together with these four species, dominated (ranged from 11.1% to 50.0%; Fig. 2).

The mushroom coral assemblages varied significantly among sites (GLM; $P=0.001$) and between reef zones (GLM; $P=0.004$) (Table 1). All pairwise comparisons between sites (except between Satumu and TPT and between Satumu and Hantu) showed significant differences in species composition (Table 1), driven mostly by variations in the abundances of *C. echinata*, *D. scruposa*, *F. fungites*, *H. limax*, *L. concinna*, *L. scabra*, *Pleuractis moluccensis*, and *P. paumotensis* (Table 1). Significant differences between reef zones were driven mostly by variations in the abundance of *D. scruposa* (Table 1).

Table 1 Multivariate generalized linear models (GLMs) results comparing free-living mushroom coral assemblage composition across the two reef zones (reef crest: 2–3 m depth and reef slope: 5–6 m depth) and across four study sites (Kusu, Hantu, Satumu, and TPT) in Singapore.

Sources	Res. d.f.	d.f.	Dev.	*P*
Site	28	3	163.80	**0.001**
Reef zone	27	1	36.95	**0.004**
Pairwise test for 'Site'				***P***
Hantu-Kusu				**0.030**
Hantu-Satumu				0.306
Hantu-TPT				**0.024**
Kusu-Satumu				**0.012**
Kusu-TPT				**0.006**
Satumu-TPT				0.438
			P	
Within taxa			Site	Reef zone
Ctenactis echinata			**0.006**	0.771
Danafungia scruposa			**0.004**	**0.002**
Fungia fungites			**0.003**	0.771
Heliofungia actiniformis			0.720	0.578
Herpolitha limax			**0.004**	0.436
Lithophyllon concinna			**0.022**	0.578
Lithophyllon repanda			0.720	0.521
Lithophyllon scabra			**0.022**	0.578
Pleuractis moluccensis			**0.022**	0.578
Pleuractis paumotensis			**0.003**	0.560
Polyphyllia talpina			0.766	0.632

Significant differences are shown in bold.

3.2 Size-structure of the four most common species

A total of 129 individuals of *C. echinata*, 159 *H. limax*, 139 *D. scruposa*, and 197 *F. fungites* were recorded at the four study sites. All corals were observed in the free-living stage except one *D. scruposa* and one *F. fungites*, which were still attached to the substrate via stalks as sexual recruits. The mean length (±SD) of *C. echinata* was 29.8 (±9.9) cm with the smallest and largest individuals were observed in Kusu and Hantu both at the reef slopes, respectively; while the mean length (±SD) of *H. limax* was 32.3 (±12.3) cm with the smallest and largest individuals were observed in TPT and Hantu both at the reef slopes, respectively (Table 2). The mean diameter (±SD) of *D. scruposa* and *F. fungites* were 14.0 (±5.5) and 13.2 (±4.4) cm, respectively. The smallest and largest individuals of *D. scruposa* were observed both at the reef crests in TPT, while for *F. fungites* were observed at the reef crest in Hantu and at the reef slope in Kusu, respectively (Table 2). For each species there was no significant difference in mean length and diameter among sites except for *C. echinata* at TPT, which was significantly longer compared to *C. echinata* at Satumu (Table S2 in Supplementary Material 1 in the online version at https://doi.org/10.1016/bs.amb.2020.08.002). Between reef zones,

Table 2 Size-structure distribution parameters for *Ctenactis echinata, Herpolitha limax, Danafungia scruposa,* and *Fungia fungites* at each of the four study sites in Singapore.

	Kusu		Hantu		Satumu		TPT	
	2–3m	5–6m	2–3m	5–6m	2–3m	5–6m	2–3m	5–6m
Ctenactis echinata								
N	6	6	15	19	5	29	28	21
Mean	30.5	25.5	28.9	32.7	20.9	26.8	32.2	32.0
SD	9.5	12.3	5.7	12.0	10.1	8.9	9.0	10.8
Min	17.5	8.1	17.2	18.5	8.8	9.0	15.6	10.2
Max	40.1	40.1	36.5	63.4	36.9	45.2	52.6	48.3
g1	−0.3	−0.2	−0.5	1.1	0.9	0.1	−0.1	−0.3
g2	−2.0	−1.4	−0.6	0.8	2.4	0.0	−0.3	−0.6
Herpolitha limax								
N	2	1	12	44	1	29	32	38
Mean	29.7	26.5	38.1	30.4	29.4	31.2	36.6	30.1
SD	20.3	NA	8.6	12.1	NA	14.2	10.4	12.6
Min	15.3	26.5	25.6	12.4	29.4	11.0	21.4	10.4
Max	44.0	26.5	51.6	70.8	29.4	62.2	59.9	59.2
g1	NA	NA	−0.2	1.5	NA	0.7	0.3	0.6
g2	NA	NA	−0.9	3.2	NA	−0.4	−0.9	−0.2

Continued

Table 2 Size-structure distribution parameters for *Ctenactis echinata, Herpolitha limax, Danafungia scruposa,* and *Fungia fungites* at each of the four study sites in Singapore.—cont'd

	Kusu		Hantu		Satumu		TPT	
	2–3 m	5–6 m	2–3 m	5–6 m	2–3 m	5–6 m	2–3 m	5–6 m
Danafungia scruposa								
N	NA	11	NA	17	4	48	19	40
Mean	NA	11.6	NA	12.6	12.1	15.2	13.6	14.1
SD	NA	4.4	NA	5.6	4.3	5.1	7.3	5.4
Min	NA	6.7	NA	3.9	7.2	5.4	3.0	4.8
Max	NA	19.8	NA	20.1	17.7	26.4	26.6	24.6
g1	NA	0.9	NA	0.0	0.5	0.0	0.1	−0.1
g2	NA	−0.3	NA	−1.7	1.5	−0.6	−1.3	−1.0
Fungia fungites								
N	3	40	18	14	18	33	39	32
Mean	20.1	13.4	14.7	10.4	11.3	14.4	13.8	11.8
SD	3.8	4.4	5.8	2.4	4.0	3.9	4.0	4.3
Min	16.3	4.5	3.7	6.1	5.0	7.0	5.1	4.3
Max	23.9	27.4	25.4	16.7	17.4	22.4	19.1	19.2
g1	−0.1	0.4	0.1	1.0	−0.1	0.2	−0.9	0.2
g2	NA	1.6	0.1	3.1	−1.3	−0.7	−0.3	−0.7

N: total number of individuals measured; Mean, SD, Min, Max: mean, standard deviation, minimum, and maximum length or diameter of individuals (cm), respectively; g_1: skewness; g_2: kurtosis.

H. limax and *F. fungites* were significantly larger at the reef crest compared to their counterparts at the reef slope (Table S2 in Supplementary Material 1 in the online version at https://doi.org/10.1016/bs.amb.2020.08.002).

In total, for all of the species studied, there was more positive ($+g_1$; 15 size-structures) than negative ($-g_1$; 10 size-structures) skewness among sites and reef zones (Table 2; Fig. 3). Within each species, there was more negative skewness (i.e. more larger corals) for *C. echinata* than the positive skewness (i.e. more smaller corals) among sites and reef zones, while for *H. limax*, *D. scruposa*, and *F. fungites* there was more positive than negative skewness (Table 2; Fig. 3). Within each site, Kusu had more negative than positive skewness for all the studied species, while Hantu and Satumu had

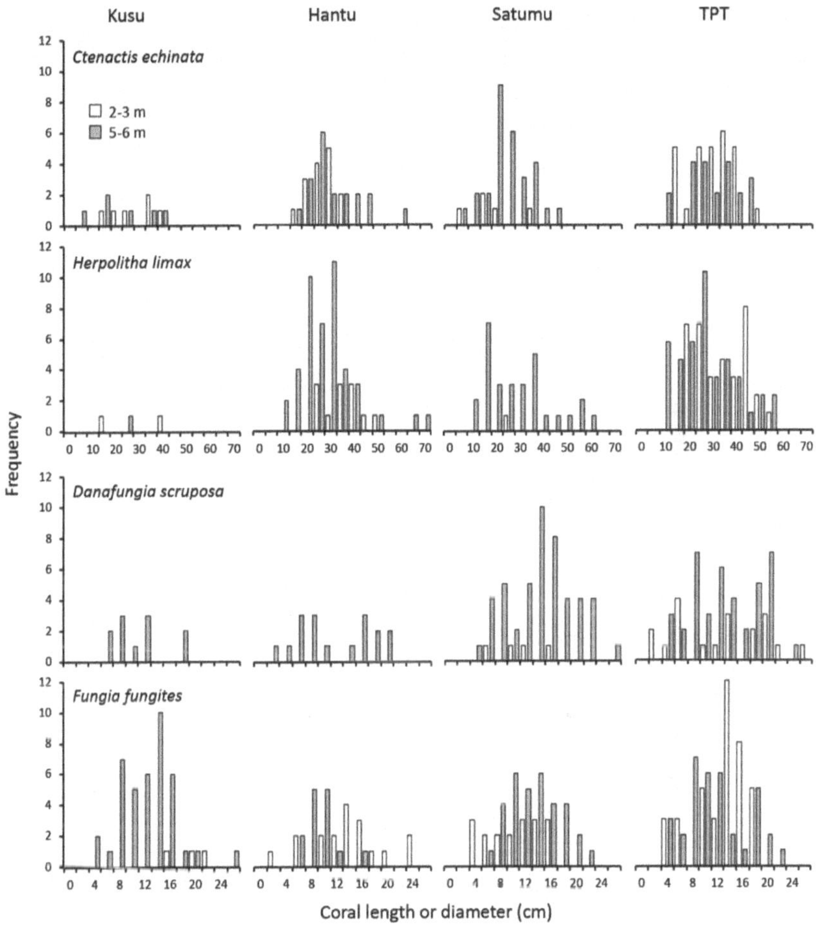

Fig. 3 Size-structure of *Ctenactis echinata* ($n = 129$), *Herpolitha limax* ($n = 159$), *Danafungia scruposa* ($n = 139$), and *Fungia fungites* ($n = 197$) at each of the four study sites in Singapore.

more positive than negative skewness, and TPT had the same skewness (Table 2; Fig. 3). Within each reef zone, the reef crest had more negative than positive skewness, while the reef slope had more positive than negative skewness. A significant difference in skewness was found between the reef crest and the reef slope (means $\pm$ SD $= -0.03 \pm 0.47$ and 0.41 ± 0.54, respectively), while no significant difference in skewness was found among species or among sites (Table S3 in Supplementary Material 1 in the online version at https://doi.org/10.1016/bs.amb.2020.08.002).

More platykurtic population structures (18 size-structures; $-g_2$; flatter population distribution) were observed than leptokurtic population structures (6; $+g_2$; peaky population distribution) for all the studied species among sites and reef zones (Table 2; Fig. 3). All species at TPT had platykurtic population structures, while variable kurtosis was observed at the other sites with platykurtic distributions being the more common (Table 2; Fig. 3). Similarly, more platykurtic distributions were more common in three of the studied species (*C. echinata*, *H. limax* and *D. scruposa*) and in all reef zones (Table 2; Fig. 3). However, no significant difference was found when comparing kurtosis among species, among sites and between reef zones (Table S3 in Supplementary Material 1 in the online version at https://doi.org/10.1016/bs.amb.2020.08.002).

Standard deviations of *C. echinata* and *H. limax* length were significantly higher than those of *D. scruposa* and *F. fungites* diameter (Table S3 in Supplementary Material 1 in the online version at https://doi.org/10.1016/bs.amb.2020.08.002). No significant difference was found comparing standard deviations among sites or between reef zones (Table S3 in Supplementary Material 1 in the online version at https://doi.org/10.1016/bs.amb.2020.08.002). The general shape of the size-structure distributions was compared among sites and reef zones within each species (Table S4 in Supplementary Material 1 in the online version at https://doi.org/10.1016/bs.amb.2020.08.002) with several pairs of site and reef zone comparisons presenting a significant difference of size-structure distributions (Table S4 in Supplementary Material 1 in the online version at https://doi.org/10.1016/bs.amb.2020.08.002). Distributions were more significantly different in *F. fungites* (i.e. pairwise comparisons) than in the other three species (Table S4 in Supplementary Material 1 in the online version at https://doi.org/10.1016/bs.amb.2020.08.002).

3.3 Growth rates of the four most common species

Growth rates were measured in 281 corals of the four most common species (*C. echinata*, $N=60$; *H. limax*, $N=102$; *D. scruposa*, $N=34$; and *F. fungites*, $N=85$), and were found to range from 80 to 490 mm (*C. echinata*), 100 to 700 mm (*H. limax*), 40 to 260 (*D. scruposa*), and 40 to 270 mm (*F. fungites*). While the linear growth rate of the four coral species declined with increasing length or diameter (Fig. 4), this decline was statistically significant for only three out of the four species (*H. limax*, *D. scruposa*, *F. fungites*; Table S5 in Supplementary Material 1 in the online version at https://doi.org/10.1016/bs.amb.2020.08.002). The 10 smallest

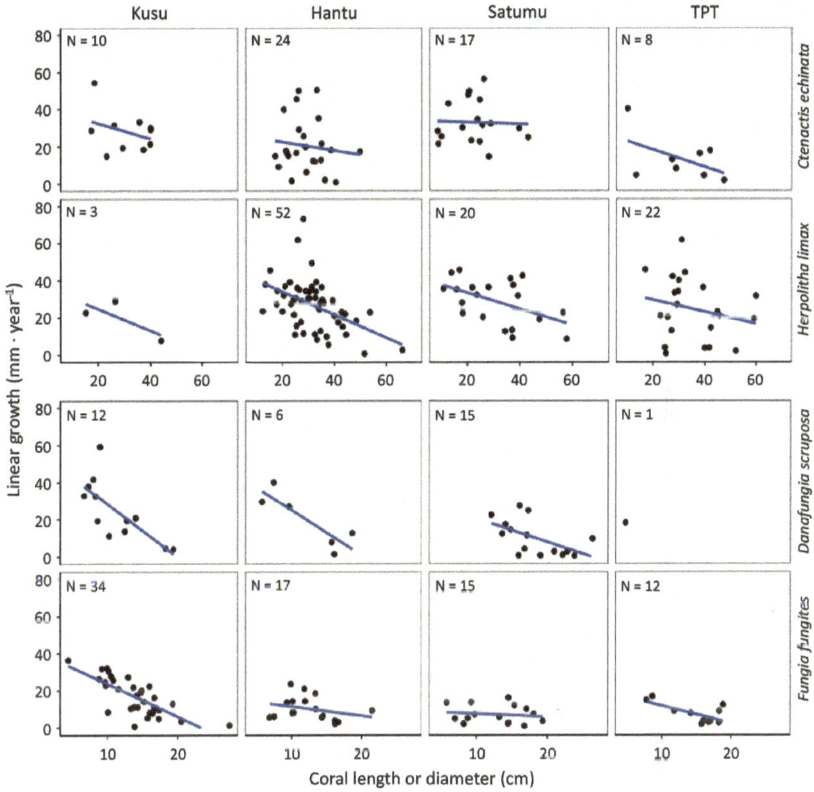

Fig. 4 The relationship between annual linear growth rate (cm) and initial coral length for *Ctenactis echinata* and *Herpolitha limax*; and between coral diameter for *Danafungia scruposa* and *Fungia fungites* at each of the four study sites in Singapore.

corals of *C. echinata* and *H. limax* had mean linear growth rate of 24.5 ± 12.4 and 36.3 ± 9.1 mm yr^{-1} (mean ± SD), respectively, while their 10 largest corals had mean linear growth rate of 17.6 ± 11.5 and 14.7 ± 10.6 mm yr^{-1} (mean ± SD), respectively. For *D. scruposa* and *F. fungites*, the 10 smallest corals had mean linear growth rate of 33.9 ± 12.0 and 13.2 ± 11.0 mm yr^{-1} (mean ± SD), respectively, while their 10 largest corals had mean linear growth rate of 6.4 ± 7.6 and 7.2 ± 4.1 mm yr^{-1} (mean ± SD), respectively. Within each species, site had no significant effect on linear growth rate for *H. limax* and *D. scruposa*, while there was an effect of site for *C. echinata* and *F. fungites*. For *C. echinata*, corals at TPT had a significantly lower linear growth rate than those at Satumu; while for *F. fungites*, linear growth rate was highest at Kusu (Table S5 in Supplementary Material 1 in the online version at https://doi.org/10.1016/bs.amb.2020.08.002).

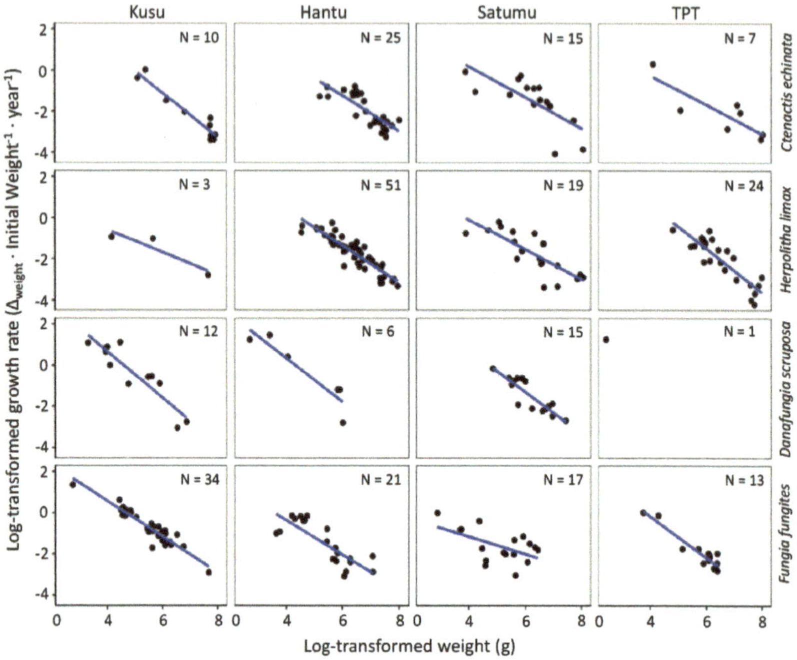

Fig. 5 The relationship between annual rate of weight increase (g) per year and initial weight for *Ctenactis echinata*, *Herpolitha limax*, *Danafungia scruposa*, and *Fungia fungites* at each of the four study sites in Singapore.

Weight increase in all four studied coral species significantly declined with increasing weight (Fig. 5; Table S5 in Supplementary Material 1 in the online version at https://doi.org/10.1016/bs.amb.2020.08.002). The 10 lightest corals of *C. echinata*, *H. limax*, *D. scruposa* and *F. fungites* had mean weight increase of 0.4 ± 0.2, 0.3 ± 0.2, 1.3 ± 0.5 and $0.7 \pm 0.8\,\text{yr}^{-1}$ (mean ± SD), respectively; while the 10 heaviest corals had the same mean weight increase of $0.1 \pm 0.1\,\text{yr}^{-1}$ (mean ± SD) for all the species. Within each species, site had no significant effect on weight increase for *C. echinata*, *H. limax* and *D. scruposa*, while for *F. fungites*, weight increase was highest at Kusu (Table S5 in Supplementary Material 1 in the online version at https://doi.org/10.1016/bs.amb.2020.08.002).

4. Discussion

Among the 11 species of mushroom corals in Singaporean waters, *C. echinata*, *H. limax*, *F. fungites*, and *D. scruposa* were the most frequently found at all four studied reef sites. The size distributions of these dominant taxa varied among sites and between reef zones, although positively skewed (i.e. having a higher proportion of small-size classes) distributions were more common. However, despite more representation of small-size classes, very few individuals of the smallest-size classes were recorded for any species. Interestingly, corals at the reef slope (5–6 m depth) presented significantly more positively skewed distributions than their reef crest (2–3 m depth) counterparts. Determinate growth was identified for all four species with no major differences in growth rate observed among sites.

The 11 species (Table S1 in Supplementary Material 1 in the online version at https://doi.org/10.1016/bs.amb.2020.08.002) found in the present study were the same as found by Koh (1987) and Hoeksema and Koh (2009), suggesting no change in species richness during the past three decades (i.e. from 1986). Historically, a total of 15 free-living mushroom coral species has been recorded in Singapore (Hoeksema and Koh, 2009), but four of these, i.e. *Ctenactis albitentaculata*, *Ctenactis crassa*, *Danafungia horrida*, and *Pleuractis granulosa*, have been considered locally extinct since surveys performed in 2006 (Hoeksema and Koh, 2009; Huang et al., 2009). As with other coral extirpations in Singapore (e.g. Poquita-Du et al., 2019) these extinctions are thought to be a result of chronic sediment and turbidity related to land-use change and urbanization (Hoeksema and Koh, 2009). The remaining 11 free-living mushroom coral species would appear to be able withstand the various stressors associated with high levels of sediments in the water column. All of these 11 species have also been found in the nearshore reef zones off Jakarta and Makassar, Indonesia, where the reefs are similarly characterized by low visibility and high sedimentation (Hoeksema, 2012; Hoeksema et al., 2019). Indeed, the coral fauna in Singapore is generally composed of species that are tolerant to low light and turbid waters (Dikou and van Woesik, 2006).

Species composition and abundance of the free-living mushroom corals varied among reef sites and between reef zones, potentially a result of local biotic and abiotic conditions. For example, most of the free-living mushroom corals were found on the reef slopes (depth of 5–6 m depth) irrespective of taxa. This is in contrast to what is usually observed in Singapore, where coral cover (all taxa) is higher on the reef crests than on the slopes (e.g. Guest et al., 2016b). Free-living mushroom corals are known to detach from the anthocaulus at a relatively small size, and to avoid competition by migrating down the reef crest to colonize sandy habitats that are otherwise inaccessible or unsuitable to other coral taxa (Chadwick-Furman and Loya, 1992; Hoeksema, 1988). Previous studies have documented the active movement of free-living mushroom corals due to competition with other organisms (Elahi, 2008; Hoeksema et al., 2014; Hoeksema and de Voogd, 2012). They can also be passively transported down-slope by wave action (Hoeksema, 1988). In Singapore, wave energy is relatively low (Sin et al., 2016), but large and frequent ship wakes (Browne et al., 2015) may generate enough energy to shift the free-living mushroom corals from the exposed reef crest down to the reef slope.

Among the four reef sites, Kusu generally has higher sedimentation levels and greater exposure to frequent ship wakes and strong tidal currents (Low and Chou, 1994; Ng et al., 2019). Although mushroom corals can usually remove sediments through polyp inflation, mucus entanglement, and ciliary action (Bongaerts et al., 2012; Schuhmacher, 1977), chronic or acute sedimentation can cause injury and mortality, particularly in smaller-sized individuals (Gilmour, 2002). This may explain why they were the least abundant at Kusu. In contrast, due to its more offshore position, Satumu is considered the least impacted among the Southern Islands reefs (Dikou and van Woesik, 2006; Ng et al., 2019), which probably explains the higher mushroom coral species richness and abundance recorded there. A similar trend of greater richness and abundance at more offshore sites has also been observed in Makassar and Jakarta (Hoeksema, 2012; Hoeksema et al., 2019). However, out of the four reefs studied here, TPT supported the most individuals, in agreement with Koh (1987), who reported that one of the highest free-living mushroom coral abundances ever observed was at this site. This suggests a suitable habitat for free-living mushroom corals but, given that TPT is much less studied than the other reefs in Singapore, exactly what biotic and/or abiotic conditions at this site are promoting these high numbers remains unknown.

Positively skewed size-structures of the four most common free-living mushroom coral indicate a greater representation of smaller-sized corals than larger-sized ones (Babcock, 1991; Gilmour, 2004b; Hughes and Jackson, 1980). Size-structure was generally bell-shaped (Fig. 3), and with low numbers of the smallest corals. In relatively non-degraded scleractinian coral populations, the size-structure is typically dominated by small-sized individuals, with a subsequent exponential decrease in number of the large-sized ones (Chadwick-furman et al., 2000; Gilmour, 2004b; Goffredo and Chadwick-Furman, 2003; Knittweis et al., 2009) or with a peak of mid-sized individuals (Riegl et al., 2018). In the present study, the low number of very young individuals might indicate early high mortality. Very young corals are sensitive to anthropogenic disturbances such as high sedimentation (Fabricius, 2005) and smaller-sized mushroom corals are known to be more prone to sediment-induced injury and mortality (Gilmour, 2002).

Interestingly, a more positively skewed size-structure was observed on the reef slope (5–6 m depth) not on the reef crest (2–3 m depth). This trend was paralleled by marked differences in the mean size of *H. limax* and *F. fungites,* where individuals on the reef slope were significantly smaller than their counterparts on the reef crest. One possible explanation for this is that light availability on the reef slope affected growth rate of the free-living mushroom corals, as light attenuates rapidly in the highly sedimented marine environment of Singapore (Todd et al., 2004). Alternatively, the relatively smaller-sized detached corals might actively migrate down to the less occupied reef slope in order to avoid competition with other organisms (Elahi, 2008; Hoeksema et al., 2014; Hoeksema and de Voogd, 2012) on the reef crest, or be passively transported downwards by wave action (Hoeksema, 1988).

For all the four species we analysed, we found a decline in growth rate with increasing size and weight (i.e. determinate growth) despite lack of representation of the smallest size of individuals. Determinate growth has also been observed for other coral species, such as the branching *Pocillopora meandrina* (Grigg and Maragos, 1974) and the massive *Goniastrea aspera* (Sakai, 1998), as well as for most species of free-living mushroom corals, including *Pleuractis granulosa* (Chadwick-furman et al., 2000), *Ctenactis echinata, Lobactis scutaria, Fungia fungites,* and *Danafungia* spp. (Goffredo and Chadwick-Furman, 2003), *Heliofungia actiniformis* (Knittweis et al., 2009), and *Herpolitha limax* (Eyal-Shaham et al., 2019). Determinate growth is a genetically controlled feature in animals and can be limited by energetic costs (Sebens, 1987). In corals, as the individual gets older (i.e. increases in size/weight), the excess energy conserved by reducing growth can be allocated to other processes, such as reproduction and maintenance (Elahi and Edmunds, 2007; Hall and Hughes, 1996).

For each species, we found no major differences in growth rate among the four sites except for *F. fungites* at Kusu. This latter species demonstrated the highest linear growth and weight increase at Kusu despite this site supporting the lowest total species abundance (all species combined). Given that the population size-structure of *F. fungites* and its abundance at Kusu were comparable to the other sites (Fig. 3), the high growth rates of *F. fungites* may be indicative of selection due to size-dependent mortality factors. For instance, at Kusu, following detachment from its stalk, a coral that rapidly grows to a larger size would be less likely to be dislodged and swept down the reef slope by Kusu's stronger wave action and currents (Low and Chou, 1994; Ng et al., 2019).

Our results show that the number of species of Singapore's mushroom corals has remained unchanged over the past three decades, suggesting that these species are well-adapted to highly sedimented reefs. Despite a positively skewed size-structure being prevalent for the four most common mushroom corals studies, the paucity of the smallest-size corals suggest high mortality during the early life stage that could threaten their long-term persistence. However, the low number of very small individuals found during the field surveys may be an underestimation due to their cryptic nature (Chadwick-Furman and Loya, 1992), and a more intensive survey dedicated to finding recruits is recommended. The findings presented here provide new insights into free-living mushroom coral population dynamics in sediment-stressed reefs, and provide a comprehensive baseline for monitoring mushroom coral populations in Singapore's challenging urban marine environment.

Acknowledgements

We thank Daisuke Taira, Jovena Seah, Ambert Ang, Samantha Lai, Amanda Hsiung Rouwen, Jenny Fong, Andrew Bauman, Dillen Ng, Zack Chen and Hui En from the Experimental Marine Ecology Laboratory, National University of Singapore for field assistance in collecting the mushroom corals. Danwei Huang and Bert Hoeksema provided invaluable assistance in species identification. Thanks also go to Bernhard Riegl and Bert Hoeksema for their insightful comments on this manuscript. Sample collection was carried out under permit issued by Singapore National Parks Board (Permit No. NP/RP18-042). This research was funded by the National Research Foundation, Prime Minister's Office, Singapore (NRF) and the Israel Science Foundation (ISF) Joint Research Program No. NRF-ISF-2654/17.

References

Abe, N., 1937. Post-larval development of the coral *Fungia actiniformis* var. palawensis Doederlein. Palao Trop. Biol. Stat. Stud. 1, 73–93.

Babcock, R.C., 1991. Comparative demography of three species of scleractinian corals using age- and size-dependent classifications. Ecol. Monogr. 61, 225–244. https://doi.org/10.2307/2937107.

Bak, R.P.M., Meesters, E.H., 1998. Coral population structure: the hidden information of colony size-frequency distributions. Mar. Ecol. Prog. Ser. 162, 301–306. https://doi.org/10.3354/meps162301.

Bauman, A.G., Pratchett, M.S., Baird, A.H., Riegl, B., Heron, S.F., Feary, D.A., 2013. Variation in the size structure of corals is related to environmental extremes in the Persian Gulf. Mar. Environ. Res. 84, 43–50. https://doi.org/10.1016/j.marenvres.2012.11.007.

Bongaerts, P., Hoeksema, B.W., Hay, K.B., Hoegh-Guldberg, O., 2012. Mushroom corals overcome live burial through pulsed inflation. Coral Reefs 31, 399. https://doi.org/10.1007/s00338-011-0862-z.

Browne, N.K., Tay, J.K.L., Low, J., Larson, O., Todd, P.A., 2015. Fluctuations in coral health of four common inshore reef corals in response to seasonal and anthropogenic changes in water quality. Mar. Environ. Res. 105, 39–52. https://doi.org/10.1016/j.marenvres.2015.02.002.

Burke, L., Reytar, K., Spalding, M., Perry, A., 2011. Reefs At Risk Revisited. World Resources Institute, Washington, DC, p. 124.

Chadwick-Furman, N.E., Loya, Y., 1992. Migration, habitat use, and competition among mobile corals (Scleractinia: Fungiidae) in the Gulf of Eilat, Red Sea. Mar. Biol. 114, 617–623. https://doi.org/10.1007/BF00357258.

Chadwick-furman, N.E., Goffredo, S., Loya, Y., 2000. Growth and population dynamic model of the reef coral *Fungia granulosa* Klunzinger, 1879 at Eilat, northern Red Sea. J. Exp. Mar. Biol. Ecol. 249, 199–218. https://doi.org/10.1016/S0022-0981(00)00204-5.

Chou, L.M., Toh, T.C., Toh, K.B., Soon, C., Ng, L., Cabaitan, P., Tun, K., Goh, E., Afiq-rosli, L., Taira, D., Poquita-du, R.C., Loke, H.X., Khalis, A., Li, J., Song, T., 2016. Differential response of coral assemblages to thermal stress underscores the complexity in predicting bleaching susceptibility. PLoS One 11, e0159755. https://doi.org/10.1371/journal.pone.0159755.

Chow, G.S.E., Chan, Y.K.S., Jain, S.S., Huang, D., 2019. Light limitation selects for depth generalists in urbanised reef coral communities. Mar. Environ. Res. 147, 101–112. https://doi.org/10.1016/j.marenvres.2019.04.010.

Dikou, A., van Woesik, R., 2006. Survival under chronic stress from sediment load: spatial patterns of hard coral communities in the southern islands of Singapore. Mar. Pollut. Bull. 52, 1340–1354. https://doi.org/10.1016/j.marpolbul.2005.07.021.

Edmunds, P.J., Riegl, B., 2020. Urgent need for coral demography in a world where corals are disappearing. Mar. Ecol. Prog. Ser. 635, 233–242. https://doi.org/10.3354/meps13205.

Elahi, R., 2008. Effects of aggregation and species identity on the growth and behavior of mushroom corals. Coral Reefs 27, 881–885. https://doi.org/10.1007/s00338-008-0403-6.

Elahi, R., Edmunds, P., 2007. Determinate growth and the scaling of photosynthetic energy intake in the solitary coral *Fungia concinna* (Verril). J. Exp. Mar. Biol. Ecol. 349, 183–193. https://doi.org/10.1016/j.jembe.2007.05.007.

Eyal-Shaham, L., Eyal, G., Sakai, K., Nozawa, Y., Harii, S., Sinniger, F., Bronstein, O., Ben-zvi, O., Shlesinger, T., Loya, Y., 2019. Repetitive sex change in the stony coral *Herpolitha limax* across a wide geographic range. Sci. Rep. 9, 2936. https://doi.org/10.1038/s41598-018-37619-y.

Fabricius, K.E., 2005. Effects of terrestrial runoff on the ecology of corals and coral reefs: review and synthesis. Mar. Pollut. Bull. 50, 125–146. https://doi.org/10.1016/j.marpolbul.2004.11.028.

Feliciano, G.N.R., Gagalac, L.M.F., Maria, A., Nava, C.G., Licuanan, W.Y., 2018. Size structure and possible bleaching susceptibility of mushroom corals in the Philippines following a bleaching event. Reg. Stud. Mar. Sci. 24, 288–295. https://doi.org/10.1016/j.rsma.2018.09.003.

Fine, M., Gildor, H., Genin, A., 2013. A coral reef refuge in the Red Sea. Glob. Chang. Biol. 19, 3640–3647. https://doi.org/10.1111/gcb.12356.

Gilmour, J.P., 2002. Acute sedimentation causes size-specific mortality and asexual budding in the mushroom coral, *Fungia fungites*. Mar. Freshw. Res. 53, 805–812. https://doi.org/10.1071/MF01219.

Gilmour, J.P., 2004a. Asexual budding in Fungiid corals. Coral Reefs 23, 595. https://doi.org/10.1007/s00338-004-0426-6.

Gilmour, J.P., 2004b. Size-structures of populations of the mushroom coral *Fungia fungites*: the role of disturbance. Coral Reefs 23, 493–504. https://doi.org/10.1007/s00338-004-0427-5.

Gilmour, J.P., Smith, L.D., Heyward, A.J., Baird, A.H., Pratchett, M.S., 2013. Recovery of an isolated coral reef system following severe disturbance. Science 340, 69–71. https://doi.org/10.1126/science.1232310.

Gittenberger, A., Reijnen, B.T., Hoeksema, B.W., 2011. A molecularly based phylogeny reconstruction of mushroom corals (Scleractinia: Fungiidae) with taxonomic consequences and evolutionary implications for life history traits. Contrib. Zool. 80, 107–132. https://doi.org/10.1163/18759866-08002002.

Goffredo, S., Chadwick-Furman, N.E., 2003. Comparative demography of mushroom corals (Scleractinia: Fungiidae) at Eilat, northern Red Sea. Mar. Biol. 142, 411–418. https://doi.org/10.1007/s00227-002-0980-9.

Graham, N.A.J., Jennings, S., Macneil, M.A., Mouillot, D., Wilson, S.K., 2015. Predicting climate-driven regime shifts versus rebound potential in coral reefs. Nature 518, 94–97. https://doi.org/10.1038/nature14140.

Grigg, R.W., Maragos, J.E., 1974. Recolonization of hermatypic corals on submerged lava flows in Hawaii. Ecology 55, 387–395. https://doi.org/10.2307/1935226.

Guest, J.R., Baird, A.H., Maynard, J.A., Muttaqin, E., Edwards, A.J., Campbell, S.J., Yewdall, K., Affendi, Y.A., Chou, L.M., 2012. Contrasting patterns of coral bleaching susceptibility in 2010 suggest an adaptive response to thermal stress. PLoS One 7, e33353. https://doi.org/10.1371/journal.pone.0033353.

Guest, J.R., Low, J., Tun, K., Wilson, B., Ng, C., Raingeard, D., Ulstrup, K.E., Tanzil, J.T.I., Todd, P.A., Toh, T.C., McDougald, D., Chou, L.M., Steinberg, P.D., 2016a. Coral community response to bleaching on a highly disturbed reef. Sci. Rep. 6, 20717. https://doi.org/10.1038/srep20717.

Guest, J.R., Tun, K., Low, J., Vergés, A., Marzinelli, E.M., Campbell, A.H., Bauman, A.G., Feary, D.A., Chou, L.M., Steinberg, P.D., 2016b. 27 years of benthic and coral community dynamics on turbid, highly urbanised reefs off Singapore. Sci. Rep. 6, 36260. https://doi.org/10.1038/srep36260.

Hall, V.R., Hughes, T.P., 1996. Reproductive strategies of modular organisms: comparative studies of reef-building corals. Ecology 77, 950–963. https://doi.org/10.2307/2265514.

Heery, E.C., Hoeksema, B.W., Browne, N.K., Reimer, J.D., Ang, P.O., Huang, D., Friess, D.A., Chou, L.M., Loke, L.H.L., Saksena-Taylor, P., Alsagoff, N., Yeemin, T., Sutthacheep, M., Vo, S.T., Bos, A.R., Gumanao, G.S., Hussein, M.A.S., Waheed, Z., Lane, D.J.W., Johan, O., Kunzmann, A., Jompa, J., Suharsono, T.D., Bauman, A.G., Todd, P.A., 2018. Urban coral reefs: degradation and resilience of hard coral assemblages in coastal cities of East and Southeast Asia. Mar. Pollut. Bull. 135, 654–681. https://doi.org/10.1016/j.marpolbul.2018.07.041.

Hilton, M.J., Manning, S.S., 1995. Conversion of coastal habitats in Singapore: Indications of unsustainable development. Environ. Conserv. 22, 307–322. https://doi.org/10.2307/44519144.

Hoegh-Guldberg, O., Bruno, J.F., 2010. The impact of climate change on the world's marine ecosystem. Science 328, 1523–1528. https://doi.org/10.1126/science.1189930.

Hoegh-Guldberg, O., Mumby, P.J., Hooten, A.J., Steneck, R.S., Greenfield, P., Gomez, E., Harvell, C.D., Sale, P.F., Edwards, A.J., Caldeira, K., Knowlton, N., Eakin, C.M., Iglesias-Prieto, R., Muthiga, N., Bradbury, R.H., Dubi, A., Hatziolos, M.E., 2007. Coral reefs under rapid climate change and ocean acidification. Science 318, 1737–1742. https://doi.org/10.1126/science.1152509.

Hoeksema, B.W., 1988. Mobility of free-living fungiid corals (Scleractinia), a dispersion mechanism and survival strategy in dynamic reef habitats. Proc. 6th Int. Coral Reef Symp. 2, 715–720.

Hoeksema, B.W., 1989. Taxonomy, phylogeny and biogeography of mushroom corals (Scleractinia: Fungiidae). Zool. Verh. 254, 1–295.

Hoeksema, B.W., 1991a. Evolution of body size in mushroom corals (Scleractinia: Fungiidae) and its ecomorphological consequences. Netherlands J. Zool. 41, 112–129. https://doi.org/10.1163/156854291X00072.

Hoeksema, B.W., 1991b. Control of bleaching in mushroom coral populations (Scleractinia: Fungiidae) in the Java Sea: stress tolerance and interference by life history strategy. Mar. Ecol. Prog. Ser. 74, 225–237. https://doi.org/10.3354/meps074225.

Hoeksema, B.W., 2004. Impact of budding on free-living corals at East Kalimantan, Indonesia. Coral Reefs 23, 492. https://doi.org/10.1007/s00338-004-0402-1.

Hoeksema, B.W., 2012. Distribution patterns of mushroom corals (Scleractinia: Fungiidae) across the Spermonde Shelf, South Sulawesi. Raffles Bull. Zool. 60, 183–212.

Hoeksema, B.W., de Voogd, N.J., 2012. On the run: free-living mushroom corals avoiding interaction with sponges. Coral Reefs 31, 455–459. https://doi.org/10.1007/s00338-011-0856-x.

Hoeksema, B.W., Giyanto, Suharsono, 2019. The role of maximum shelf depth versus distance from shore in explaining a diversity gradient of mushroom corals (Fungiidae) off Jakarta. Diversity 11, 46. https://doi.org/10.3390/d11030046.

Hoeksema, B.W., Koh, E.G.L., 2009. Depauperation of the mushroom coral fauna (Fungiidae) of Singapore (1860s-2006) in changing reef conditions. Raffles Bull. Zool. 22, 91–101. https://doi.org/10.1007/s00338-011-0784-9.

Hoeksema, B.W., Waheed, Z., 2011. Initial phase of autotomy in fragmenting *Cycloseris* corals at Semporna, eastern Sabah, Malaysia. Coral Reefs 30, 1087. https://doi.org/10.1007/s00338-011-0807-6.

Hoeksema, B.W., Yeemin, T., 2011. Late detachment conceals serial budding by the free-living coral *Fungia fungites* in the Inner Gulf of Thailand. Coral Reefs 30, 975. https://doi.org/10.1007/s00338-011-0784-9.

Hoeksema, B.W., Dekker, F., de Voogd, N.J., 2014. Free-living mushroom corals strike back by overtopping a coral-killing sponge. Mar. Biodivers. 44, 3–4. https://doi.org/10.1007/s12526-013-0188-x.

Hoeksema, B.W., Bouwmeester, J., Range, P., Ben-Hamadou, R., 2018. A large aggregation of self-fragmenting mushroom corals in the Arabian/Persian Gulf. Ecology 99, 1236–1238. https://doi.org/10.1002/ecy.2139.

Huang, D., Tun, K.P.P., Chou, L.M., Todd, P.A., 2009. An inventory of zooxanthellate scleractinian corals in Singapore, including 33 new records. Raffles Bull. Zool. 22, 69–80.

Hughes, T.P., Jackson, J.B.C., 1980. Do corals lie about their age? Some demographic consequences of partial mortality, fission, and fusion. Science 209, 713–715. https://doi.org/10.1126/science.209.4457.713.

Hughes, T.P., Barnes, M.L., Bellwood, D.R., Cinner, J.E., Cumming, G.S., Jackson, J.B.C., Kleypas, J., Van De Leemput, I.A., Lough, J.M., Morrison, T.H., Palumbi, S.R., Van Nes, E.H., Scheffer, M., 2017. Coral reefs in the Anthropocene. Nature 546, 82–90. https://doi.org/10.1038/nature22901.

Jackson, J.B., Kirby, M.X., Berger, W.H., Bjorndal, K.A., Botsford, L.W., Bourque, B.J., Bradbury, R.H., Cooke, R., Erlandson, J., Estes, J.A., Hughes, T.P., Kidwell, S., Lange, C.B., Lenihan, H.S., Pandolfi, J.M., Peterson, C.H., Steneck, R.S., Tegner, M.J., Warner, R.R., 2001. Historical overfishing and the recent collapse of coastal ecosystems. Science 293, 629–638. https://doi.org/10.1126/science.1059199.

Junjie, R.K., Browne, N.K., Erftemeijer, P.L., Todd, P.A., 2014. Impacts of sediments on coral energetics: partitioning the effects of turbidity and settling particles. PLoS One 9 (9). https://doi.org/10.1371/journal.pone.0107195.

Kleypas, J.A., McManus, J.W., Menez, L.A.B., 1999. Environmental limits to coral reef development: where do we draw the line? Am. Zool. 39, 146–159. https://doi.org/10.1093/icb/39.1.146.

Knittweis, L., Jompa, J., Richter, C., Wolff, M., 2009. Population dynamics of the mushroom coral *Heliofungia actiniformis* in the Spermonde Archipelago, South Sulawesi, Indonesia. Coral Reefs 28, 793–804. https://doi.org/10.1007/s00338-009-0513-9.

Koh, E.G.L., 1987. Systematics and ecology of mushroom corals (Scleractinia: Fungiidae) in Singapore reefs. BSc thesis, Department of Zoology, National University of Singapore, Singapore. 113 pp.

Koh, E.G.L., Chou, L.M., 1989. The Mushroom Corals of Singapore. Department of Zoology, National University of Singapore, Singapore. 45 pp.

Lai, S., Loke, L.H.L., Hilton, M.J., Bouma, T.J., Todd, P.A., 2015. The effects of urbanisation on coastal habitats and the potential for ecological engineering: a Singapore case study. Ocean Coast. Manag. 103, 78–85. https://doi.org/10.1016/j.ocecoaman.2014.11.006.

Lenth, R., Singmann, H., Love, J., Buerkner, P., Herve, M., 2018. *emmeans*: Estimated Marginal Means, aka Least-Squares Means. https://doi.org/https://CRAN.R-project.org/package=emmeans.

Littler, M.M., Littler, D.S., Brooks, B.L., Koven, J.F., 1997. A unique coral reef formation discovered on the Great Astrolabe Reef, Fiji. Coral Reefs 16, 51–54. https://doi.org/10.1007/s003380050059.

Low, J.K.Y., Chou, L.M., 1994. Sedimentation rates in Singapore waters. In: Sudara, S., Wilkinson, C.R., Chou, L.M. (Eds.), Third ASEAN-Australia Symposium on Living Coastal Resources, pp. 697–701.

Loya, Y., Sakai, K., 2008. Bidirectional sex change in mushroom stony corals. Proc. Biol. Sci. 275, 2335–2343. https://doi.org/10.1098/rspb.2008.0675.

Loya, Y., Sakai, K., Yamazato, K., Nakano, Y., Sambali, H., van Woesik, R., 2001. Coral bleaching: the winners and the losers. Ecol. Lett. 4, 122–131. https://doi.org/10.1046/j.1461-0248.2001.00203.x.

Meesters, E.H., Hilterman, M., Kardinaal, E., Keetman, M., de Vries, M., Bak, R.P.M., 2001. Colony size-frequency distributions of scleractinian coral populations: spatial and interspecific variation. Mar. Ecol. Prog. Ser. 209, 43–54. https://doi.org/10.3354/meps209043.

Mellin, C., Thompson, A., Jonker, M.J., Emslie, M.J., 2019. Cross-shelf variation in coral community response to disturbance on the Great Barrier Reef. Diversity 11, 38. https://doi.org/10.3390/d11030038.

Ng, C.S.L., Lim, J.X., Sam, S.Q., Kikuzawa, Y.P., Toh, T.C., Wee, T.W., Sim, W.T., Ng, N.K., Huang, D., Chou, L.M., 2019. Variability in skeletal bulk densities of common hard corals in Southeast Asia. Coral Reefs 38, 1133–1143. https://doi.org/10.1007/s00338-019-01852-2.

Ng, C.S.L., Huang, D., Toh, K.B., Sam, S.Q., Kikuzawa, Y.P., Toh, T.C., Taira, D., Chan, Y.K.S., Hung, L.Z.T., Sim, W.T., Rashid, A.R., Afiq-Rosli, L., Ng, N.K., Chou, L.M., 2020. Responses of urban reef corals during the 2016 mass bleaching event. Mar. Pollut. Bull. 154, 111111. https://doi.org/10.1016/j.marpolbul.2020.111111.

Poquita-Du, R.C., Quek, Z.B.R., Jain, S.S., Schmidt-Roach, S., Tun, K., Heery, E.C., Chou, L.M., Todd, P.A., Huang, D., 2019. Last species standing: loss of Pocilloporidae corals associated with coastal urbanization in a tropical city state. Mar. Biodiversity 49 (4), 1727–1741. https://doi.org/10.1007/s12526-019-00939-x.

R Core Team, 2015. R: A Language and Environment for Statistical Computing. http://www.R-project.org.

Riegl, B., Piller, W.E., 2003. Possible refugia for reefs in times of environmental stress. Int. J. Earth Sci. 92, 520–531. https://doi.org/10.1007/s00531-003-0328-9.

Riegl, B., Purkis, S., 2015. Coral population dynamics across consecutive mass mortality events. Glob. Chang. Biol. 21, 3995–4005. https://doi.org/10.1111/gcb.13014.

Riegl, B., Cavalcante, G., Bauman, A.G., Feary, D.A., Steiner, S., Purkis, S., 2017. Demographic mechanisms of reef coral species winnowing from communities under increased environmental stress. Front. Mar. Sci. 4, 344. doi.org/10.3389/fmars.2017.00344.

Riegl, B., Johnston, M., Purkis, S., Howells, E., Burt, J., Steiner, S.C.C., Sheppard, C.R.C., Bauman, A., 2018. Population collapse dynamics in *Acropora downingi*, an Arabian/Persian Gulf ecosystem-engineering coral, linked to rising temperature. Glob. Chang. Biol. 24, 2447–2462. https://doi.org/10.1111/gcb.14114.

Sakai, K., 1998. Delayed maturation in the colonial coral *Goniastrea aspera* (Scleractinia): Whole-colony mortality, colony growth and polyp egg production. Res. Popul. Ecol. 40, 287–292. https://doi.org/10.1007/BF02763460.

Schneider, C.A., Rasband, W.S., Eliceiri, K.W., 2012. NIH image to ImageJ: 25 years of image analysis. Nat. Methods 9, 671–675. https://doi.org/10.1038/nmeth.2089.

Schuhmacher, H., 1977. Ability in fungiid corals to overcome sedimentation. Proc. 3rd Int. Coral Reef Symp. 1, 503–509.

Sebens, K.P., 1987. The ecology of indeterminate growth in animals. Annu. Rev. Ecol. Syst. 18, 371–407. https://doi.org/10.1146/annurev.ecolsys.18.1.371.

Shlesinger, T., Loya, Y., 2016. Recruitment, mortality, and resilience potential of scleractinian corals at Eilat, Red Sea. Coral Reefs 35, 1357–1368. https://doi.org/10.1007/s00338-016-1468-2.

Shlesinger, T., Loya, Y., 2019. Breakdown in spawning synchrony: a silent threat to coral persistence. Science 365, 1002–1007. https://doi.org/10.1126/science.aax0110.

Sin, T.M., Ang, H.P., Buurman, J., Lee, A.C., Leong, Y.L., Ooi, S.K., Steinberg, P., Teo, S.L., 2016. The urban marine environment of Singapore. Reg. Stud. Mar. Sci. 8, 331–339. https://doi.org/10.1016/j.rsma.2016.01.011.

Smith, L.D., Devlin, M., Haynes, D., Gilmour, J.P., 2005. A demographic approach to monitoring the health of coral reefs. Mar. Pollut. Bull. 51, 399–407. https://doi.org/10.1016/j.marpolbul.2004.11.021.

Stuart-Smith, R.D., Brown, C.J., Ceccarelli, D.M., Edgar, G.J., 2018. Ecosystem restructuring along the Great Barrier Reef following mass coral bleaching. Nature 560, 92–96. https://doi.org/10.1038/s41586-018-0359-9.

Sully, S., van Woesik, R., 2020. Turbid reefs moderate coral bleaching under climate-related temperature stress. Glob. Chang. Biol. 26, 1367–1373. https://doi.org/10.1111/gcb.14948.

Sully, S., Burkepile, D.E., Donovan, M.K., Hodgson, G., van Woesik, R., 2019. A global analysis of coral bleaching over the past two decades. Nat. Commun. 10, 1264. https://doi.org/10.1038/s41467-019-09238-2.

Todd, P.A., Ladle, R.J., Lewin-Koh, N.J.I., Chou, L.M., 2004. Genotype x environment interactions in transplanted clones of the massive corals *Favia speciosa* and *Diploastrea heliopora*. Mar. Ecol. Prog. Ser. 271, 167–182. https://doi.org/10.3354/meps271167.

Todd, P.A., Ong, X., Chou, L.M., 2010. Impacts of pollution on marine life in Southeast Asia. Biodivers. Conserv. 19, 1063–1082. https://doi.org/10.1007/s10531-010-9778-0.

Vizel, M., Kramarsky-Winter, E., Loya, Y., 2009. Mushroom coral regeneration from a detached stalk. Coral Reefs 28, 939. https://doi.org/10.1007/s00338-009-0496-6.

Wang, Y., Naumann, U., Wright, S.T., Warton, D.I., 2012. mvabund-an R package for model-based analysis of multivariate abundance data. Methods Ecol. Evolution 3, 471–474. https://doi.org/10.1111/j.2041-210X.2012.00190.x.

Warton, D.I., Wright, S.T., Wang, Y., 2012. Distance-based multivariate analyses confound location and dispersion effects. Methods Ecol. Evol. 3, 89–101. https://doi.org/10.1111/j.2041-210X.2011.00127.x.

Yamashiro, H., Nishihira, M., 1994. Radial skeletal dissolution to promote vegetative reproduction in a solitary coral *Diaseris distorta*. Experientia 50, 497–498. https://doi.org/10.1007/BF01920756.

Yamashiro, H., Nishihira, M., 1998. Experimental study of growth and asexual reproduction in *Diaseris distorta* (Michelin, 1843), a free-living fungiid coral. J. Exp. Mar. Biol. Ecol. 225, 253–267. https://doi.org/10.1016/S0022-0981(97)00229-3.

Zinke, J., Gilmour, J.P., Fisher, R., Puotinen, M., Maina, J., Darling, E., Stat, M., Richards, Z.T., Mcclanahan, T.R., Beger, M., Moore, C., Graham, N.A.J., Feng, M., Hobbs, J.-P.A., Evans, S.N., Field, S., Shedrawi, G., Babcock, R.C., Wilson, S.K., 2018. Gradients of disturbance and environmental conditions shape coral community structure for south-eastern Indian Ocean reefs. Divers. Distrib. 24, 605–620. https://doi.org/10.1111/ddi.12714.

CHAPTER SIX

Population fluctuations of the fungiid coral *Cycloseris curvata*, Galápagos Islands, Ecuador

Joshua S. Feingold* and Brandon Brulé
Department of Marine and Environmental Sciences, Halmos College of Arts and Sciences, Nova Southeastern University, Dania Beach, FL, United States
*Corresponding author: e-mail address: joshua@nova.edu

This paper is dedicated to the memory of my father, Edgar Lee Feingold (1927–2020), whose unflagging faith in my abilities and support of my education and research made this possible.

Contents

Abstract

Fungiid corals (Cnidaria: Anthozoa: Scleractinia) occur at isolated locations scattered throughout the eastern tropical Pacific. They can be reef-associated but are often found on sand and rubble substrata distant from reef coral habitat. *Cycloseris curvata* is known in this region from the southern Gulf of California, through Mexico, Costa Rica, and Panamá, and with the southern-most populations occurring in the Galápagos Islands, Ecuador. During Archipelago-wide surveys (1988–2019), living individuals of *Cycloseris curvata* were observed at only two locations, Devil's Crown (near Floreana Island) and Xarifa Island (near Española Island). The Devil's Crown population was observed from 1988 to 2017, whereas living individuals in the Xarifa population were observed from 2005 to 2009. In 2012 a death assemblage (dead skeletons) was discovered at Darwin Island, at the northern-most extent of the Archipelago. At Devil's Crown, visual surveys were performed annually or biennially from 1990 to 2012, with two more

Advances in Marine Biology, Volume 87
ISSN 0065-2881
https://doi.org/10.1016/bs.amb.2020.08.005

surveys in 2017 and 2019. The living *Cycloseris curvata* population consisted of 15 individuals in 1990 that gradually increased to 78 individuals by 1995. Over 200 individuals were observed in 1996, and high numbers persisted through 1998 with 335 individuals. Live tissue surface area per polyp ranged from 0.5 to 95.0 cm^2. The population decreased to 112 individuals in 1999 (following warming associated with the 1997–98 El Niño), with further declines to 20 in 2009 (following cooling associated with the 2007 La Niña) and a rebound to 91 in 2012. After a 5y break in data collection, only one individual (28.3 cm^2) was observed in 2017, and in 2019 none were observed. Although undetected living *Cycloseris curvata* populations may exist, and renewed recruitment provides some hope for population reestablishment, it is possible that this fungiid coral species is now extirpated from the Galápagos Archipelago.

1. Introduction

Fungiid corals possess endosymbiotic dinoflagellates (zooxanthellae), however, they are distinct from their zooxanthellate reef-building relatives because they are non-reef-forming, solitary (vs colonial) and free-living (vs attached). Fungiids are widespread throughout the Indo-Pacific, although those in the genus *Cycloseris* occur in three disjunct regions (western Indian Ocean and Red Sea, western Pacific, and the eastern tropical Pacific (ETP)) (Veron, 2000). Three fungiid species occur in the ETP, all in the genus *Cycloseris* (Glynn et al., 2017a; Hoeksema, 1989), *Cycloseris curvata* (Hoeksema, 1989), *Cycloseris* (*Diaseris*) *distorta* (Michelin 1842), and (provisionally) *Cycloseris vaughani* (Boschma 1923). Two species (*C. curvata* and *C.* (*D.*) *distorta*) range from the southern Gulf of California, Mexico (Reyes-Bonilla et al., 1997) through Central America (Cortés, 1997) to the Galápagos Islands, Ecuador (Wellington, 1975). *Cycloseris* (provisionally *C. vaughani*) is reported in the ETP only from Rapa Nui Island, Chile (Hoeksema et al., 2019). Typical habitats for fungiids include sand and rubble substrata either adjacent to or distant from reef coral accumulations (Veron, 2000). *Cycloseris curvata* is known in the ETP from collections of dead skeletons that were dredged (Durham, 1966) or polyps observed alive (Glynn and Wellington, 1983; Feingold, 1995, 1996; Reyes-Bonilla et al., 1997). This study focuses on the population dynamics of this species in the Galápagos Islands over a 29y period (1990–2019) and reports its possible extirpation from the Archipelago.

The best known Galápagos Islands population of living *C. curvata* individuals was observed adjacent to the nearly submerged volcanic cone Devil's Crown (Onslow Island, Corona del Diablo) off the north coast of Floreana

Island (Glynn and Wellington, 1983; Wellington, 1975). Wellington (1975) first reported this species (= *Cycloseris elegans*) as a minor constituent of a free-living coral community composed of large numbers of the related fungiid *Cycloseris* (*Diaseris*) *distorta* (= *Cycloseris mexicana*) and the siderastreid coral *Psammocora stellata*. These species contributed to a living stratum over an accumulation of coarse calcareous sand and rubble (Feingold, 1995, 1996). Although no *C. curvata* was recovered from sediment cores performed in this area, likely due to its relative rarity compared to the other two coral species, the *C.* (*D.*) *distorta*/*P. stellata* community has persisted in this location for the past 7700y based on radiometric dating (Feingold et al., 2020). Holocene coral reef growth initiated in the ETP around 3000–7000 yBP (Macintyre et al., 1992; Toth et al., 2017), making this coral community at Devil's Crown the oldest known coral assemblage in this biogeographical province.

The oceanographic setting of the Galápagos Archipelago places it at the confluence of several major current systems. The cool Peru Oceanic Current, Peru Coastal Current, and South Equatorial Current transport water in a westerly direction and promote mixed-layer shoaling along the South American coast. The warm Panamá Current transports water seasonally from Central America to the southwest creating generally warmer conditions in the northern islands (e.g. Darwin) compared to the southern islands (e.g. Floreana and Española) (see Glynn et al., 2018, Fig. 2 for an overview). These general patterns are affected by El Niño—Southern Oscillation events that both augment (warm-phase El Niño) and diminish (cool-phase La Niña) seawater temperatures (Glynn et al., 2018). Widespread bleaching and mortality associated with the 1982–83 El Niño dramatically affected (~95% mortality) reef-building coral in the Galapagos Archipelago (Glynn, 1998), although corals in the deeper habitat at Devil's Crown bleached but did not die (Robinson, 1985).

Isolated populations, particularly those with low numbers of individuals, are at risk of local extinction (extirpation) from their geographic outposts (Carpenter et al., 2008). Since *C. curvata* conforms to this categorization, the primary goal of this paper is to describe population fluctuations of *C. curvata* at Devil's Crown, with particular reference to seawater temperature variations, to provide context for future extirpations and possible Archipelago-wide losses in other species. Secondary objectives are to document the short-lived living population at Xarifa Island, a death assemblage also at Xarifa, a new death assemblage discovered at Darwin Island, and report on interspecific interactions, and basic reproductive capabilities.

2. Methods

Archipelago-wide visual surveys for fungiid corals were performed by surface or SCUBA diving over 31 years (1988–2019). Sites for more focused surveys were chosen based on reports by Durham and Barnard (1952) and Glynn and Wellington (1983), and interviews with local fishermen, naturalists, and scientists. Individuals (living polyps and/or dead skeletons) of *Cycloseris curvata* were counted and measured at the three sites where they were encountered, Devil's Crown (~15 m depth) on the north coast of Floreana Island, Xarifa Island (~12 m depth), on the northeast coast of Española Island, and Darwin Island (~25 m depth) (Fig. 1A–D). Survey areas

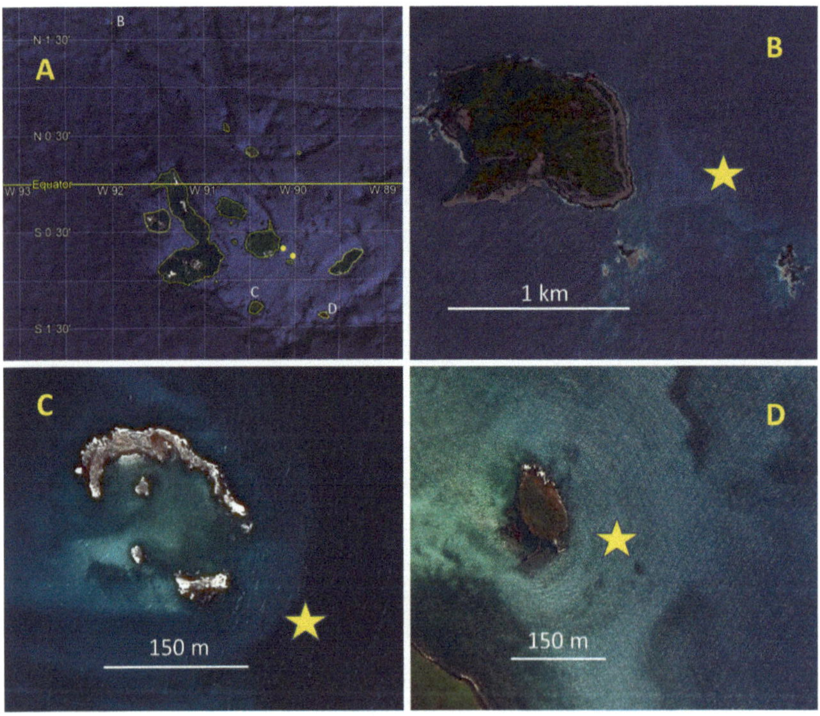

Fig. 1 Satellite imagery of the Galápagos Archipelago (A), and study sites at Wellington Reef (B), Darwin Island (N01.676226° W091.992738°), Devil's Crown (C), Floreana Island (S01.217184° W090.421960°), and Xarifa Island (D), Española Island (S01.358059° W089.642605°). White letters in panel (A) denote the locations of study sites at three islands (B–D). Stars ★ indicate the location of the study site at each island. Observations of a single individual at Santa Fé Island and a skeleton at Punta Nuñez, Santa Cruz Island are indicated with yellow dots • (A). *Images from Google Earth 2020*

were ~750 m^2 at Devil's Crown and Xarifa in an area approximately 15 m × 50 m at each site. The survey area at Darwin (~500 m^2) was a belt approximately 5 m wide and 100 m long at the base of Wellington Reef where the steeply sloping reef structure transitions into a sand and rubble plain. Living *Cycloseris curvata* was distinguished in the field from its congener *Cycloseris* (*Diaseris*) *distorta* by its generally larger size, strongly convex shape, denser skeleton, a slit-shaped mouth, and the presence of four white exsert septa on the oral surface (Fig. 2A). Skeletons of *C. curvata* were distinguished from *C.* (*D.*) *distorta* skeletons by their generally larger size, strongly convex shape, greater density, more strongly developed septa, and more regular

Fig. 2 *Cycloseris curvata*, Galápagos Islands, Ecuador. (A) *C. curvata* polyp adjacent to ~17 *Cycloseris* (*Diaseris*) *distorta* individuals, Devil's Crown (14 m depth) 9 June 2011. (B) Profile view of large *C. curvata*, Devil's Crown (13 m depth) 27 April 2005. (C) Dead skeletons of *C. curvata*, base of Wellington Reef, Darwin Island (26 m depth) 8 June 2012. (D) Asexually derived anthocauli (buds) on aboral surface of *C. curvata* skeleton, Devil's Crown (14 m depth) collected May 1995. Polyps in natural positions in (A) and (B). Skeletons arranged in image (C) from nearby (<2 m radius) accumulations. Skeleton in (D) was dead when collected.

shape (circular or elliptical) without evidence of prior autochthonous fragmentation that produces wedge-shaped skeletons in *C.* (*D.*) *distorta*.

The primary study site at Devil's Crown was first visited on 16 June 1988, the distributional limit of the *C. curvata* population determined during exploratory dives in 1989, and populations were quantified beginning in 1990. The low-relief bottom topography was sketched, and a permanent anchor deployed that was later used with the addition of a surface marker buoy to obtain GPS coordinates. The site was then demarcated by an array of iron reinforcing bar stakes that were used as a frame of reference during subsequent dives. When shifts in the location of corals was noticed, searches were extended to explore the periphery of known distribution. Visual SCUBA diving surveys were performed annually or biennially from 1990 until 2012 over the entire known *C. curvata* habitat ($\sim$750 m^2). After a 5y hiatus, two more surveys were performed, one on 20 May 2017 and another on 24 May 2019, resulting in quantitative data spanning a 29y period. The 2017 and 2019 searches were performed across all locations at the study site where this species was previously observed and focused on the area that supported the highest densities of individuals. However, survey times were limited to a single 30-min period in each of these years as opposed to multiple dives and >1 h total time allocated to searching in prior visits. Since one individual was observed in 2017 and none were observed in 2019, relatively more bottom time was devoted to searching for live specimens than in measuring discovered polyps compared to previous visits, so it is unlikely that any were overlooked.

Surveys at Xarifa were initiated in 1991 and were conducted annually or biennially until the last was performed in 2011. Living individuals were discovered on 22 May 2005 and were counted but not measured. During the next visit in 2007 they were all counted and measured and were last observed alive there on 2 June 2009. Sufficient time was allocated to cover the entire habitat ($\sim$750 m^2) so that all individuals that were present would be observed.

Although there were prior visits to Wellington Reef at Darwin Island by JSF and other coral scientists, *C. curvata* skeletons were first discovered there on 6 June 2012. Six more dives ($\sim$90 min total survey time) were performed during that trip to count and measure all skeletons in the $\sim$500 m^2 habitat. No living *C. curvata* were observed.

Due to their scattered presence at Devil's Crown, a roving diver protocol was followed encompassing all areas of the *C. curvata* known habitat ($\sim$750 m^2). All encountered living individuals were counted and measured except for the surveys in 1996, 1997 and 1998. In 1996 and 1997 there was insufficient time to count and measure all corals, so these were recorded as 200+ and 300+ count values, and of these, 139 and 223 polyps were measured, respectively. All corals were counted in 1998 though not all of these were measured (203 of 334 polyps). This means that the entire population of observed living corals were measured on each survey date, except for 1996–98 when over half of the population was measured. When shifts in the location of corals were observed, especially when the population decreased, additional wider-ranging searches were performed around the known habitat to ensure that all individuals were included in the survey. Measurements were performed with plastic rulers or Vernier callipers with major and minor axes of the individual polyps recorded to the nearest 0.5 cm (1990–2011). When polyps displayed inflated coenosarcs, the size of the skeleton was used for the measurements, not the margin of living tissue. The 2-dimensional area was calculated using the formula for an ellipse. Digital photographic images were obtained in 2012 and 2017 with a Nikon D5100 DSLR camera in an Ikelite submersible housing, including a knife (20.9 cm length) in the frame that was used for scale. Each image was taken as close to parallel to the coronal plane of the coral as possible. These images were processed with the programme CPCe to determine their respective 2-dimensional area (Kohler and Gill, 2006). Both methods of determining live tissue surface area per polyp (LTPP) are underestimates (2-d ellipses are approximately one half the surface area of 3-d ellipsoid hemispheres of the coral) but are nevertheless useful for comparisons of size trends over time and among populations. Due to rapid cover of skeletons by benthic algae upon death, dead individuals were difficult to detect at the Devil's Crown study site, and since a reliable count and measurement of skeletons was not possible, these data were not used in this study.

At Xarifa Island (2007), all encountered living individuals were counted and measured using plastic rulers or Vernier callipers. Also, a haphazard measurement of 189 skeletons was performed as there was insufficient bottom time to measure all of them (estimated in the thousands). Surface areas were determined as described above.

At Darwin Island (2012), all encountered skeletons ($n=166$) were photographed (e.g. Fig. 2C) and surface areas determined using CPCe (Kohler and Gill, 2006).

A mixed effects model Analysis of Variance (ANOVA) with Kenward-Roger degrees of freedom adjustments was performed on the time-series population data from Devil's Crown with survey year as a random effect, followed by a pair-wise Tukey-Kramer *a posteriori* test, to determine if coral polyp mean sizes and trends were different among three periods of interest. These were (A) prior to the 1997–98 El Niño, (B) post-El Niño recovery until the 2007 La Niña, and (C) post-La Niña recovery. A second type of analysis (Welch two sample *t*-Test with *p*-values adjusted for multiple testing using the Bonferroni–Holm method) compared differences in live tissue surface area per polyp (LTPP) between the population at Devil's Crown and the one at Xarifa, both in 2007. This test was also used to compare mean dead skeleton surface areas per polyp between Xarifa (2007) and Darwin (2012).

The rationale for making these comparisons were: (1) To determine if there were significant differences in live polyp size at Devil's Crown among years leading up to, between and following ENSO disturbance events (warm-phase El Niño in 1997–98 and cool-phase La Niña in 2007). This allows interpretations of changes in polyp size as responses to environmental stress. (2) To determine if there were differences in average polyp size at the two study sites with living corals—Devil's Crown and Xarifa Island—in the same year (2007). This allowed comparisons between sites to infer environmental and survivorship differences. (3) To determine if there were differences in average skeleton size between the two death assemblages at Xarifa Island and Darwin Island. This allowed inferences concerning environmental differences on coral growth and survivorship in past populations of *C. curvata* from two locations, each at latitudinal extremes of the Archipelago.

Two *C. curvata* polyps were collected for reproductive analysis, one on 20 April 1989 and the other on 1 June 1996. Each polyp was approximately 5 cm in diameter (19.6 cm^2). These samples were histologically prepared and assessed for gonadal presence and state following methods described in Colley et al. (2002).

Sea water temperatures were measured *in situ* within the area populated by *C. curvata* at the Devil's Crown study site (15 m depth) with submersible temperature loggers. Measurements were made at either 2-h or hourly intervals. Four models of loggers were used, Ryan TempMentor (accuracy $\pm 0.1\,°C$), and Onset Instruments Hobo, Stowaway and Water Temp Pro (accuracy $\pm 0.2\,°C$). Data were obtained for most of the survey period (1990–2019), except for when loggers were lost or flooded. Two periods of interest were May 1996–May 1999 (2 h intervals), before, during and after the 1997–98 El Niño, and December 2005–December 2008 (hourly intervals), before, during and after the 2007 La Niña.

3. Results

Two populations of living *C. curvata* were observed in the Galápagos Islands (Fig. 1), one at Devil's Crown (~15 m depth) on the north coast of Floreana Island and the second at Xarifa Island (~12 m depth) near Gardner Bay, Española Island. A death assemblage (skeletons) was observed at Xarifa from early surveys (1991) and a second one discovered in 2012 at the base of Wellington Reef, Darwin Island (~25 m depth). Dead individuals were observed at Devil's Crown, but they were often covered with macroalgae and rapidly became difficult to locate compared to the relatively bare and easily observed skeletons at Xarifa and Darwin. The living Xarifa population was present from 2005 to 2009, whereas the Devil's Crown population persisted throughout nearly all the observation period (1988–2019) with robust numbers (200+ individuals) from 1996 to 1998. Following declines beginning in 1999, *C. curvata* was last observed alive at Devil's Crown in 2017. Peter Glynn (pers comm, STRI coral collection) observed one living *C. curvata* polyp in the channel between Xarifa and Española Island in the late 1980s. Fernando Rivera (pers comm) reported *C. curvata* at Santa Fé Island, and André De Roy collected a skeleton on the beach at Punta Nuñez, Santa Cruz Island (invertebrate collection at the Charles Darwin Research Station). None (living or dead) were observed at either location during focused searches via SCUBA diving, nor were any seen during more generalized searches at Santa Cruz Island (Coamaño Island, Canal de Itabaca near Punta Carrion), east coast of Mosquera Island, southeast coast of North Seymour, Marchena Island (Punta Espejo), Genovesa Island (Darwin Bay), and Wolf Island. Thus, the presence of *C. curvata* is documented from five locations within the Archipelago from the northern-most, south-central, and southern-most islands, with living populations only observed in the south.

At Devil's Crown, 22 surveys were performed over 29y (Fig. 3). A total of 2119 polyps were counted, and of these 1888 were measured. In 1990 the observed living *C. curvata* population was 15 individuals that gradually increased to 78 individuals by 1995. Over 200 individuals were observed in 1996, and high population numbers persisted through 1998 when 334 individuals were counted. The population decreased to 112 in 1999, with further declines to 20 in 2009 and a rebound to 91 in 2012. Following a 5y hiatus in data collection, only one individual (28 cm^2) was observed in 2017, and in 2019 none were observed. Discounting 2017 when

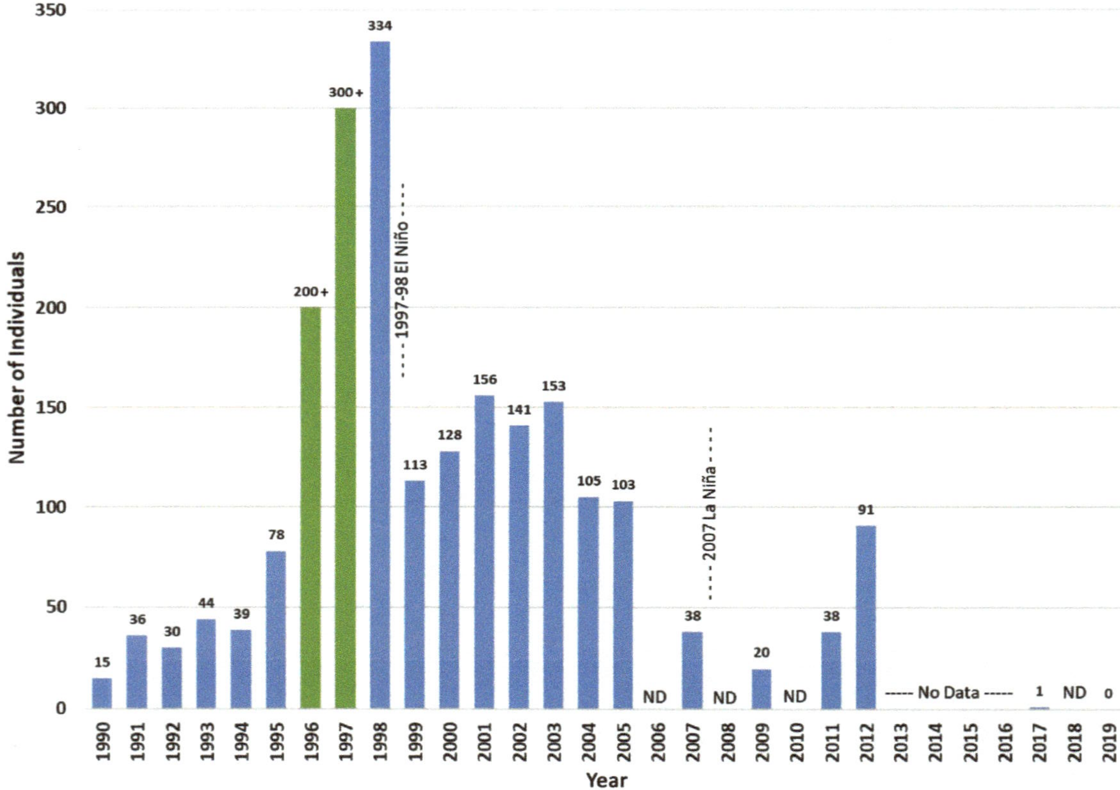

Fig. 3 Numbers of living *Cycloseris curvata* individuals at Devil's Crown (13–15 m depth), 1990–2019. Counts of 200+ and 300+ denote minimum abundances due to insufficient bottom time to count all individuals (green-coloured bars).

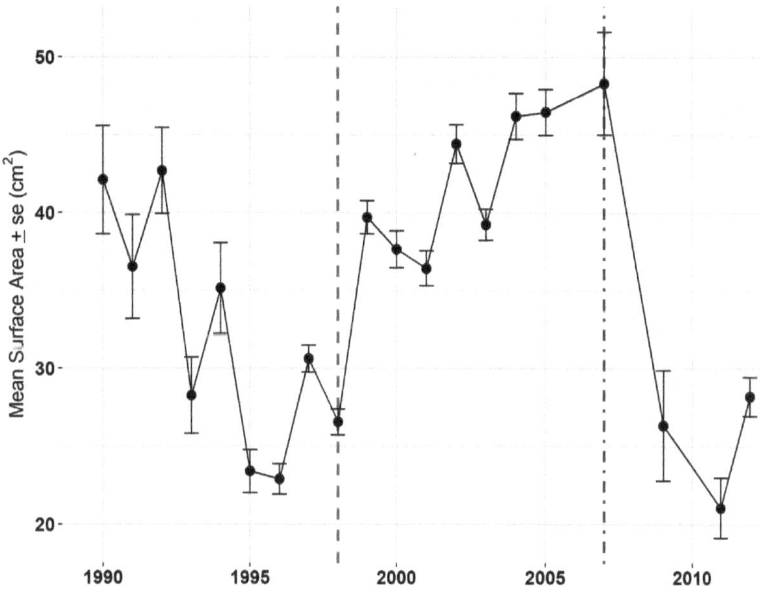

Fig. 4 Mean (±se) live tissue surface area per polyp (LTPP) of living individuals of *Cycloseris curvata* at Devil's Crown (13–15 m depth), 1990–2019. Only one individual was observed in 2017, and none was observed in 2019, and these are not plotted. Mean LTPP (±se) ranged from 21.1 ± 1.9 to $48.3 \pm 3.3\,cm^2$ (2011 and 2007 respectively). The vertical dashed lines indicate the 1997–98 El Niño (red), and 2007 La Niña (blue).

only one individual was observed, densities of *C. curvata* ranged from 0.02 individuals m^{-2} in 1990 to 0.45 individuals m^{-2} in 1998.

Coral live tissue surface area per polyp (LTPP) at Devil's Crown ranged from 0.5 to $95\,cm^2$ in all measured individuals, with the smallest value observed in 1993 and the largest in 1997. Average LTPP area (±se) ranged from 21.1 ± 1.9 to $48.3 \pm 3.3\,cm^2$ (2011 and 2007 respectively) (Fig. 4). Analysis by mixed model ANOVA documented significant differences in trends of live tissue surface area per polyp (LTPP) in two of three survey periods (Fig. 5). Mean LTPP decreased significantly ($p = 0.004$) prior to the 1997–98 El Niño and increased significantly ($p = 0.037$) during recovery from El Niño. There was no significant change in mean LTPP during post-La Niña recovery ($p = 0.86$). Percentages of the smaller size class (0–20 cm^2) of living polyps increased prior to the 1997–98 El Niño and decreased during El Niño recovery (Fig. 6). Conversely, polyps in the larger size class (20.0 + cm^2) decreased leading up to the El Niño and increased afterwards. These patterns repeated after the 2007 La Niña. Larger individuals,

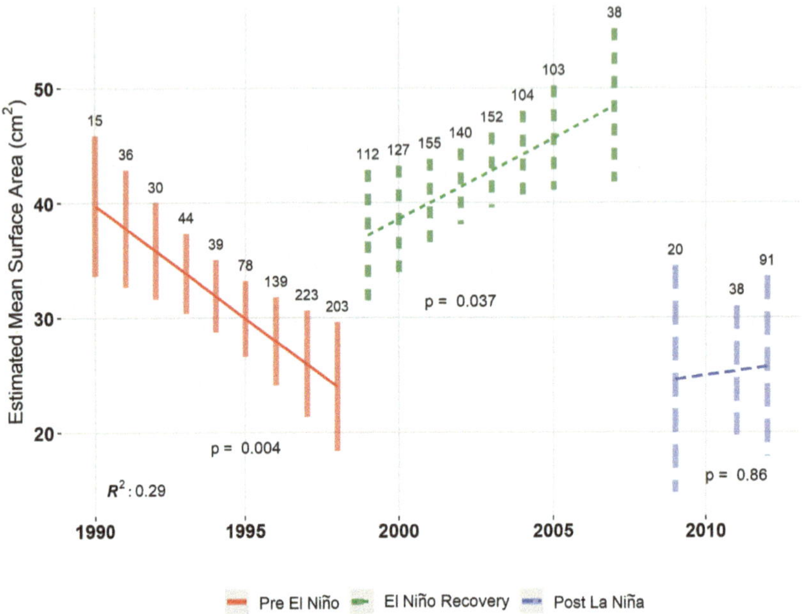

Fig. 5 Differences in response to environmental perturbations on terms of mean live tissue area per polyp (LTPP) from three periods, pre-1997–98 El Niño, El Niño recovery, Post-La Niña recovery. Sample size (n) is shown above each error bar (95% C.I.). There are significant decreases (Mixed effects model ANOVA, $p = 0.004$) in mean surface area prior to the 1997–98 El Niño, significant increases ($p = 0.037$) in LTPP during El Niño recovery, and no changes ($p = 0.86$) in LTPP post-La Niña 2007.

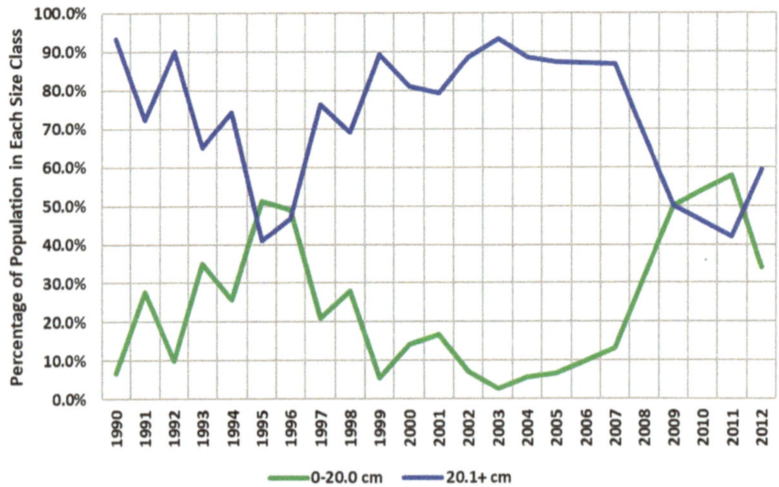

Fig. 6 Percentage of *Cycloseris curvata* polyps in small (0–20.0 cm^2) and large (>20.0 cm^2) size-classes over time at Devil's Crown. Note that the percentages derived in 1996, 1997, and 1998 were from the population of measured polyps and not the total number counted as reported in Fig. 3.

generally >7 cm major axis diameter (>38.5 cm^2), develop strongly folded outer margins of the skeleton and a highly arched overall shape (Fig. 2B).

Over the 22 surveys at Devil's Crown and 1888 measured individuals, anthocauli (asexual buds) were observed on the aboral surface of only 23 living individuals and three additional dead specimens (e.g. Fig. 2D), however, it is estimated that less than half of the measured corals were handled in such a way to observe the aboral surface where anthocauli occur. In May 1993, a tiny (0.5 cm^2) live individual was observed adjacent to a senescent, much larger (38.5 cm^2) individual, and on 30 May 2002 a live 7.0 cm^2 attached bud was observed on the aboral surface of an otherwise dead *C. curvata* individual (33.0 cm^2).

There was often direct contact between *C. curvata* individuals and the other two species of corals that share the same habitat at Devil's Crown, *C. (D.) distorta* and *Psammocora stellata* (Fig. 2A). Other animal associates were rarely observed directly interacting with *C. curvata*. Small invertebrates such as the ubiquitous bioeroding sea urchin *Eucidaris galapagensis* and the arrow crab *Stenorhynchus debilis* were observed on several occasions sheltering in the concavity on the aboral surface of upright *C. curvata* polyps. Also, small brachyuran crabs (~1.0 cm carapace width) were observed on only one survey date, 3 October 1991, similarly sheltering under 16 of 22 inspected polyps. Crabs were in pairs in 7 of the 16 observations. On two survey dates (30 May 2002 and 21 May 2004), snails (*Babelomurex hindsii*) (Fig. 7) were observed foraging on *C. curvata* (14 snails on one individual in 2002 and 5, 8, and 15 snails on each of 3 individuals in 2004). The corallivorous pufferfish *Arothron meleagris* was rarely observed at the study site and was never observed foraging on corals. Defecated material from *A. meleagris* held overnight in buckets consisted of *Psammocora stellata* fragments that could have originated from the colonies living in the study site, or from another community of this species in the shallow channel within Devil's Crown (Glynn et al., 2018, Fig. 9C). Other instances of corallivory were not observed. See Feingold (1995) for a more detailed description of benthic associates in this community.

Dead skeletons of *C. curvata*, and *Cycloseris* (*Diaseris*) *distorta*, were first observed at Xarifa Island (~12 m depth) during Archipelago-wide coral surveys in 1991. This site was visited yearly or biennially, and only dead skeletons were observed there until 2005 when a population of 60 small living individuals was discovered. The site was visited again in 2007 and 50 living individuals were observed and measured. One living individual

Fig. 7 Gastropod mollusc (*Babelomurex hindsii*) corallivory on *Cycloseris curvata*, Devil's Crown (14 m depth) 21 May 2004. Individual snails ($n = 15$) are outlined in yellow.

was observed during the next visit in 2009, living polyps were not observed in 2011, and no subsequent visits were performed. Densities of *C. curvata* were 0.08 individuals m^{-2} in 2005 and 0.07 individuals m^{-2} in 2007. These corals were small (mean ± se, 13.0 ± 0.8 cm^2) compared to living individuals at Devil's Crown (48.3 ± 3.3 cm^2) also in 2007, and there was a significant difference in mean surface area of living tissue per polyp (LTPP) between sites (Welch 2 sample *t*-Test, $p < 0.001$) (Fig. 8).

A death assemblage of *C. curvata* was discovered on 6 June 2012 at the base of Wellington Reef, Darwin Island (~25 m depth). Six additional surveys along this deep habitat (~90 min total survey time) were performed at this time, and all encountered skeletons were measured, but no living individuals were found. Analysis of skeletal taphonomic degradation was not performed, but all skeletons possessed sufficient detail to determine species identity (Fig. 2C). There was no significant difference in mean size between skeletons at Darwin in 2012 (18.4 ± 0.9 cm^2) and those measured at the Xarifa survey site in 2007 (20.2 ± 0.7 cm^2) (Welch 2 sample *t*-Test, $p = 0.10$) (Fig. 8).

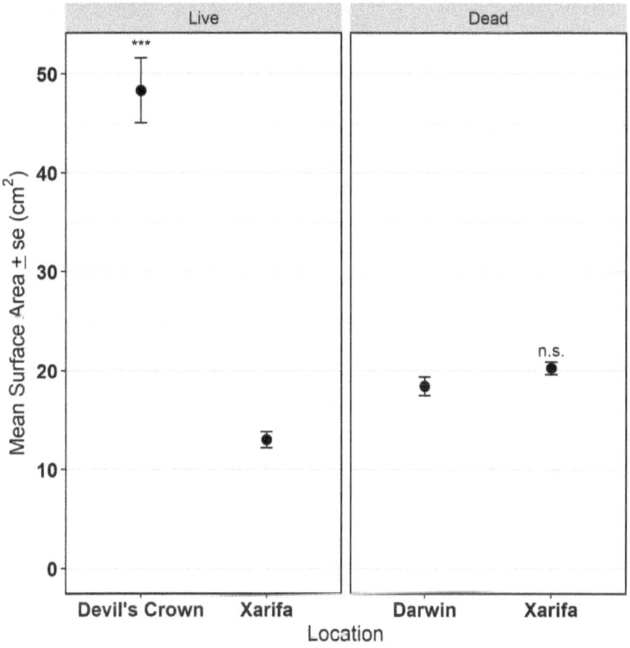

Fig. 8 Comparisons of *Cycloseris curvata* mean surface area (± se) per polyp/skeleton. Live mean tissue surface area per polyp (LTPP) was significantly greater at Devil's Crown ($48.3 \pm 3.3\,cm^2$, $n=38$) compared to Xarifa ($13.0 \pm 0.8\,cm^2$, $n=50$) (Welch 2-sample *t*-Test, $p<0.001$). There was no significant difference in mean surface area per skeleton between Darwin ($18.4 \pm 0.9\,cm^2$, $n=166$) and Xarifa ($20.2 \pm 0.7\,cm^2$, $n=189$) (Welch 2-sample *t*-Test, $p=0.10$). All data is from 2007 except for the Darwin data that is from 2012.

Both polyps that were histologically examined displayed stage IV ovaries with some areas showing lacunae indicative of spawned eggs. No spermaries were observed.

Sea water temperatures at the Devil's Crown study site ranged from 14.0 °C (21 November 2007) to 30.2 °C (2 March and 6 April 1998) (Fig. 9) for all data obtained during the 29y survey period. These minimum and maximum values respectively corresponded with cooling associated with the 2007 La Niña and warming associated with the 1997–98 El Niño. More typically, temperatures during non-ENSO conditions ranged from 16.0 to 28.0 °C. On an annual basis, coldest temperatures occurred in November and warmest temperatures occurred in March–April.

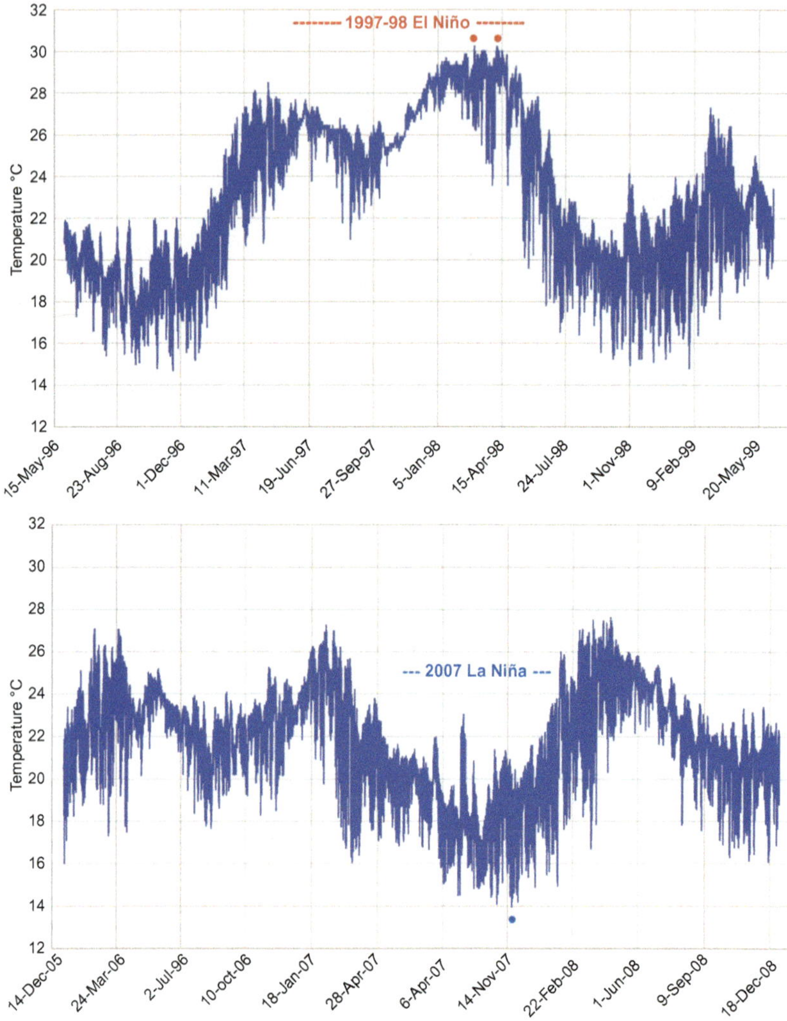

Fig. 9 Seawater temperatures at Devil's Crown (15 m depth) before, during and after (A) the 1997–98 El Niño, and (B) the 2007 La Niña. The most extreme temperatures measured in 29y of seawater temperature monitoring at this location occurred in each of these periods, 30.2 °C on 2 March and 6 April 1998 (red dots), and 14.0 °C on 21 November 2007 (blue dot).

4. Discussion

Although our surveys covered many sites in the Galápagos Archipelago, focusing on locations with known or suspected fungiid coral populations, many areas remain unexplored. Since fungiid corals are less common in the ETP than other scleractinian species (Glynn and Ault, 2000), and occur on substrata often distant from reef-constructing coral habitat, there may be additional locations in the Islands where *Cycloseris curvata* can be found (living and/or dead). Since all five known Galápagos locations are on the upstream margins of islands on coastlines with regular current flow, this type of habitat should be explored in the future. Currents in the Archipelago are typically westward, suggesting east-southeast coastlines as possible habitat, particularly where coarse calcareous sand and rubble accumulates.

Durham and Barnard (1952) first reported *C. curvata* (= *Cycloseris elegans*) in the Galápagos Islands from dredged collections made at Xarifa Island, though it is not clear if the specimens were living or dead. Early reports of living *C. curvata* in the Archipelago noted its presence and gave qualitative assessments of numbers. In an Archipelago-wide summary of coral distribution, Wellington (1975) reported there were 'several individuals' of *C. curvata* at Devil's Crown. Also, at that location, Glynn and Wellington (1983) reported densities of approximately one *C. curvata* individual per 500 *Cycloseris* (*Diaseris*) *distorta* (= *Cycloseris mexicana*) individuals, similar to densities that we observed. The maximum density of *C. curvata* at Devil's Crown was 0.45 individuals m^{-2} in 1998. Densities of *C.* (*D.*) *distorta* ranged from ~20 to >2000 individuals m^{-2} in this area yielding a ratio of one *C. curvata* individual per 44 to 4444 *C.* (*D.*) *distorta* individuals. However, *C. curvata* was restricted to an approximately 15 m-wide band on the western-most, shallowest (13–15 m depth) margin of the coral community where the calcareous sand and rubble substrata grade upward towards Devil's Crown, whereas *C.* (*D.*) *distorta* was observed up to 500 m to the east of this margin in water up to 34 m depth. Robinson (1985) did not mention *C. curvata* in his summary of ENSO 1982–83 bleaching impacts on the *C.* (*D.*) *distorta* (= *C. mexicana*) population at Devil's Crown.

The environmental conditions of the Galápagos Archipelago are considered marginal for coral persistence. Physiological stress from generally cool conditions is exacerbated by aperiodic, anomalous warming and cooling from warm-phase (El Niño) and cool-phase (La Niña) components of El-Niño-Southern Oscillation (ENSO) (Glynn, 1998; Glynn et al., 2018). The oceanographic setting of the Galápagos Archipelago places it at the confluence of several major current systems. The cool Peru Oceanic Current, Peru Coastal Current, and South Equatorial Current move water in a generally westerly direction across the Archipelago and drive upwelling along the South American coast. The warm Panamá Current delivers water seasonally from the Panamá bight and Gulf of Panamá to the southwest creating generally warmer conditions in the northern Islands (e.g. Darwin) compared to the south (e.g. Floreana and Española) (see Glynn et al., 2018, Fig. 2 for an overview). However, mixed-layer shoaling was reported from all three study sites; Darwin (Riegl et al., 2019), Devil's Crown (Feingold, 1995, 1996), and Xarifa (J.S. Feingold, unpublished data). For example, during 1997–98 El Niño warming, temperatures in excess of 30.0 °C were recorded at Devil's Crown, however, temperature reductions to 24 °C were also measured in that same period (Fig. 9). Incursion of cooler water mitigates warming and provides these deeper dwelling corals with a more moderate thermal refuge at depth (Witman et al., 2010) allowing survival of warm-phase El Niño-Southern Oscillation conditions. High temperatures associated with the 1982–83 El Niño resulted in mass mortalities of shallow (<10 m depth) Galápagos reef-building corals in 1982–83 (Glynn, 1998), whereas periodic cooling contributed to survival of deeper (>13 m) dwelling corals at Devil's Crown (Feingold, 1996). At Devil's Crown (15 m depth) harmonics variance analysis of *in situ* seawater temperature data documented a predominantly lunar semidiurnal periodicity to temperature variances (Feingold, 1995), suggesting internal tides as the driving mechanism of these thermal shifts.

Increased variability of ENSO from ~4000 to 2000 yBP was implicated in millennial-scale negative impacts on coral reef development in Pacific Panamá (Toth et al., 2012). Similar processes may be currently underway as sea water temperatures continue to increase. At the study site in March 1998 during warming (28.0–30.0 °C) associated with the severe 1997–98 ENSO event (Fig. 9A) *C. curvata* polyps were mostly bleached with stark white upper surfaces and some noticeable pigment on their flanks. In June 1999 the recovering population (112 individuals vs 334 in May 1998) had normal pigmentation (Feingold, 2001). Even though the congener *C.* (*D.*) *distorta* in Galápagos was reported to survive extended (>90d) periods of bleaching without suffering mortality (Feingold, 1995, 1996), it is possible that *C. curvata* is more sensitive to these conditions, and/or the stresses associated with zooxanthellae loss and increased metabolic rates. *Cycloseris curvata* at the Devil's Crown study site suffered a 66.5% population reduction (Fig. 3) following sea water warming associated with the 1997–98 ENSO event (Fig. 9A), reductions that were not detected in their more numerous congener *C.* (*D.*) *distorta*. Seawater cooling often below

16.0 °C associated with the 2007 La Niña (Fig. 9B) was measured in August–December 2007, following the May 2007 survey. In the subsequent (May 2009) survey there were 48.0% fewer individuals (Fig. 3) and 46% smaller polyps (Figs. 4 and 5). This suggests that both warm and cool-phase ENSO events negatively affected the *C. curvata* population at Devil's Crown. However, Hoeksema et al. (2019) reported a mesophotic *Cycloseris* community near Rapa Nui at 79–85 m depth, presumably exposed to cooler conditions than this. Also, according to Romero-Torres et al. (2020), Eastern Pacific corals demonstrate great resilience to aperiodic exposure to thermal stress. This summary concerned more typical reef-building species, underscoring the need to more detailed investigations into responses of ahermatypic coral species such as *C. curvata* to environmental stress.

Fungiid corals display both sexual and asexual modes of reproduction. Although sexual reproduction (spawning) in the Galápagos population of *C. curvata* was not observed, even during several night dives performed for this purpose, there is indirect evidence of spawning. Out of an abundance of caution to minimize impact to this small population, only two specimens of *C. curvata* were collected for reproductive analysis. One was collected in April (1989), in the middle of the warm season, and the other was collected at the end of the warm season in early June (1996). Both possessed stage IV (mature) ovaries and the one collected in June had lacunae indicative of spawned ova. This is consistent with other fungiid species that spawn during warm months (Colley et al., 2002). No spermaries were observed, suggesting that this species may be gonochoric, although the sample size ($n=2$) is very small. Histological examination of the co-occurring gonochoric congener *C.* (*D.*) *distorta* documented the presence of male and female gametes in separate individuals, although no spawning was observed in this species either (Colley et al., 2002; Glynn et al., 2017b). Furthermore, the appearance of living individuals at Xarifa Island in 2005 was likely via larval dispersal since no live specimens were observed there during multiple prior visits over 14y. This also suggests a possible larval source from an unknown population upstream of Xarifa. Several focused searches were performed up to 4 km up-current (SE) of Xarifa along the coast of Española Island, but neither living polyps nor dead skeletons of *C. curvata* were observed. Furthermore, in the more distant past when the death assemblage specimens were alive, the Xarifa population may have provided a larval source for the population at Devil's Crown, located to the west-northwest, ~88 km down-current. On a broader scale, a model of coral larval connectivity within the Archipelago suggested that the southern Islands could be sources of coral larvae for the northern Islands more so than vice-versa. Connectivity between the Islands and mainland was weak, but the Panamá current, especially when strongest in the southern hemisphere warm season (January–April), could intermittently transport larvae to the Archipelago from Central and South America (Riegl et al., 2019). This could explain the sporadic appearance of *C. curvata* at various locations within the Galápagos Islands.

Asexual reproduction in the Devil's Crown population of *C. curvata* via anthocauli (buds) was observed on the aboral surface of 23 living polyps during surveys and three dead specimens (e.g. Fig. 2D). According to Kramarsky-Winter and Loya (1996) exposure to stressful conditions and associated tissue degradation can induce this type of budding, perhaps serving as a mechanism of survival during inimical conditions. On one occasion a relatively large live bud (7.0 cm^2) was observed still attached to an otherwise dead (33.0 cm^2) individual, suggestive of this type of survival strategy. In addition to both high and low thermal stress, tissue damage from corallivorous snails may have also triggered bud formation. Live polyps generally displayed clumped distributions, further suggesting asexual reproduction as an important mode of population maintenance and increasing susceptibility to corallivory. Alternatively, production of competent, large planula larvae that settle near to the adults could result in clumped distribution. In May 1993, a tiny (0.5 cm^2) live individual was observed adjacent to a senescent, much larger (38.5 cm^2) individual, likely the result of either a pinched off anthocauli bud or a recently settled sexual recruit. Fragmentation through the self-directed dissolution of skeletal material (Yamashiro et al., 1989) was reported by Wells (1983) for the Devil's Crown population of *C. curvata*, however, it was never observed in the 1888 polyps examined during this project. This form of fragmentation was observed in the co-occurring congener *C.* (*D.*) *distorta* (Colley et al., 2002) at Devil's Crown, and in *Cycloseris* (provisionally *C. vaughani*) at Rapa Nui (Hoeksema et al., 2019). Asexual reproduction is important in the maintenance of populations once they become established (presumably by larval dispersal), and facilitated fragmentation explains the abundance of *C.* (*D.*) *distorta* at Devil's Crown, estimated to be ~500,000 individuals with densities up to ~2000 individuals m^{-2} in some areas (Feingold, 1995; Glynn et al., 2018) compared to the relatively small numbers (maximum of 334 individuals and densities of 0.45 individuals m^{-2} in 1998) of *C. curvata* observed there. The observation of widely scattered, isolated populations of *C. curvata* within the Archipelago and more broadly across the ETP (Cortés, 1997) suggests sporadic larval dispersal, followed by periods of population stability, decline, and then extirpation. Since these corals live in marginal habitats, aperiodic anomalous warming (El Niño) and cooling (La Niña) events combined with a broader trend of warming creates variable environmental conditions challenging the persistence of these corals.

Relationships among benthic organisms were not a focus of this research, however, several interspecific interactions of interest were observed. Direct contact between *C. curvata* and the other two coral species present at Devil's Crown, *C. (D.) distorta* and *Psammocora stellata*, was often observed (Fig. 2A). Chadwick-Furman and Loya (1992) reported compatible interspecific interactions among fungiid coral species in the Gulf of Eilat, Red Sea, but damage was inflicted by fungiids to other non-fungiid scleractinians. At Devil's Crown contact between *C. curvata* and the siderastreid *Psammocora stellata* was commonly observed with no evidence of tissue damage to either species. The pencil sea urchin *Eucidaris galapagensis* and the arrow crab *Stenorhynchus debilis* were observed on multiple occasions throughout the surveys sheltering in the concavity on the aboral surface of upright *C. curvata* polyps. An unidentified brachyuran crab similarly sheltered under 16 of 22 polyps on one sample date (3 October 1991), but was not seen on any other survey. The corals provided shelter for these invertebrates and these interactions did not appear to harm the coral. Corallivory was observed on two survey dates (30 May 2002 and 21 May 2004) when snails (*Babelomurex hindsii*) (Fig. 7) were observed foraging on *C. curvata*. The cryptic nature of these snails (Enochs and Glynn, 2017), and the infrequency of visitations to the site (generally several dives during one time of year) means snail predation on *C. curvata* may be more meaningful than can be determined in this study.

Analysis by mixed model ANOVA documented significant differences in trends of live tissue surface area per polyp (LTPP) in two of three survey periods (Fig. 5). Mean LTPP decreased significantly ($p=0.004$) prior to the 1997–98 El Niño as populations were growing and increased significantly ($p=0.037$) during recovery from El Niño when populations were relatively stable (Fig. 3). There was no significant change in mean LTPP during post-La Niña recovery ($p=0.86$). These analyses are consistent with the initial increasing percentages of smaller individuals in 1990–96 (Fig. 6) and subsequent increase in percentages of larger size-class polyps in 1996–98. Overall, the Devil's Crown population increased from 15 individuals in 1990 to 334 individuals in 1998. This population declined to 113 individuals in 1999, and coral polyp size increased in post-El Niño recovery (Figs. 4 and 5) even though the population did not increase. This further supports the interpretation that increased numbers of smaller polyps drove the initial population increases leading up to the 1997–98 El Niño, larger polyps (through growth) lead to peak numbers in 1998, and decreased numbers with increasing polyp size occurred following this event. Impacts to the overall population of ETP reef-building coral populations are similar (Glynn, 1998; Glynn et al., 2001).

In a comparison of living coral size between study sites (Devil's Crown vs Xarifa) in 2007 (the only year that live corals were measured at both sites), mean polyp size was significantly larger at Devil's Crown ($48.3 \pm 3.3\,cm^2$, $n=38$) than at Xarifa ($13.0 \pm 0.8\,cm^2$, $n=50$) (Welch 2-sample t-Test, $p<0.001$) (Fig. 8), which suggests polyps were likely alive longer and grew larger at Devil's Crown. The longer persistence of living polyps at Devil's Crown compared to Xarifa (27 vs 6y) may also suggest more favourable environmental conditions for this species at Devil's Crown during the survey period.

There was no significant difference in average skeleton size between the two dead populations of *C. curvata* (Darwin vs Xarifa, Welch two sample t-Test, $p=0.10$) (Fig. 8). Even though the living polyps measured at Xarifa were small (mean $\pm$ se $= 13.0 \pm 0.8\,cm^2$), this documents that in the past *C. curvata* grew larger, leaving a death assemblage composed of skeletons with a mean size $\pm$ se of $20.2 \pm 0.7\,cm^2$. Furthermore, it is interesting to note that *C. curvata* has the capacity to grow to the same size at the two latitudinal extremes of the archipelago. Darwin is a relatively warm, more tropical environment compared to the rest of the archipelago (Glynn et al., 2018), bathed by the warm, seasonally variable Panamá Current, whereas Devil's Crown and Xarifa are located in the cooler southern zone, primarily affected by the oceanic extension of the Peru Currents. However, since the Darwin site is notably deeper than the Xarifa site (~25 vs ~12 m), shoaling of cool water from the sub-surface mixed layer at Darwin (Riegl et al., 2019) may cause similar *in situ* seawater temperatures to Xarifa.

This research documents that *C. curvata* populations have occurred from the southern-most to northern-most limits of the Galápagos Archipelago, a distance of 425 km. These widely separated, small populations suggest that *C. curvata* maintains a tenuous hold on survival in the marginal environmental conditions in the Galápagos Islands, located near the southern limit of coral distribution in the Eastern Pacific (Glynn, 2003). Anomalously warm conditions associated with 1997–98 El Niño preceded population reductions (−66.5%) observed between the 1998 and 1999 survey dates, and cool conditions associated with the 2007 La Niña preceded population reductions (−48.0%) observed in 2009. Additionally, periodic flushing by shoaling mixed-layer water (Feingold, 1996) may mitigate more severe effects of anomalous warming and allow survival in deeper, cooler refugia (Riegl and Piller, 2003).

Despite their rarity, small populations, and atypical habitat, *C. curvata* nevertheless contribute to habitat complexity and ecosystem diversity. Sea urchins and crabs were observed sheltering in the concavity of the aboral surface of this coral. Several instances of corallivory by snails were observed (Fig. 9) (Enochs and Glynn, 2017), documenting at least one potential biological control on numbers of *C. curvata*. Continued monitoring of this intriguing species is critical for assessing its abundance and developing a better understanding of its life history and ecological function. Furthermore, these isolated populations of *C. curvata* serve as exemplars of coral resilience and capacity for biogeographic dispersion and demonstrate the ability to persist in atypical coral habitats for at least several decades.

As of May 2019, no living *C. curvata* individuals were observed at Devil's Crown, the location that previously supported the largest known population in the Galápagos Islands. Only dead skeletons were observed at Xarifa Island in 2011 and Darwin Island in 2012 (the last time these sites were surveyed). Thus, living *C. curvata* was not observed at the three study sites the last time each site was surveyed. However, *Cycloseris* (provisionally *C. vaughani*) was observed at mesophotic depths (79–85 m) at Rapa Nui, Chile (Hoeksema et al., 2019). Perhaps undetected *C. curvata* populations occur in these deeper and less accessible habitats in the Galápagos Islands as well, or other populations remain to be discovered in shallower coastal waters. If none currently exist, sources of recruitment propagules from other locations in the ETP provide some hope for population reestablishment; however, it is possible that *C. curvata* is now extirpated from the Galápagos Archipelago.

Acknowledgements

Peter Glynn continues to be an invaluable friend, mentor, and source of inspiration for this work. We are grateful to The Galápagos National Park and Charles Darwin Research Station for permissions and logistical support of this research for the duration of the project. We thank Lenin Cruz, captain of M/V Pirata, for providing a reliable base of operations and camaraderie on our field trips. Travel and shipboard support in 2012 were provided by the Khaled bin Sultan Living Ocean Foundation. Intercultural Outreach Initiative provided logistical support in 2017 and 2019. Godfrey Merlen, René Espinosa, Fernando Rivera, David Day, David Obura, Christianne Huerkamp, Susan Thornton, Elizabeth Glynn, Edward Major, Katy Brown, Iliana Baums, Dana Wingate, Joselyn Bassantes, Sarah Meltzoff, and Tito Mejía Sánchez provided field assistance. Kelly Larkin provided assistance with image processing and data compilation. Kyle Dettloff and J. Matthew Hoch aided data analysis and interpretation. Ian Enochs and Peter Glynn identified the molluscan corallivore *Babelomurex hindsii*. The manuscript was improved by comments and suggestions from Bernhard Riegl, Pedro Alexandro Garcia Range and Peter Glynn. Galápagos field research was supported by National Science Foundation grant OCE-9018392 and OCE-9218197 to Peter Glynn, a Grant-in-Aid of Research from Sigma Xi, and President's Faculty Research Development Grants from Nova Southeastern University to Joshua Feingold.

References

Carpenter, K.E., Abrar, M., Aeby, G., Aronson, R.B., Banks, S., et al., 2008. One third of reef-building corals face elevated extinction risk from climate change and local impacts. Science 321, 560–563.

Chadwick-Furman, N., Loya, Y., 1992. Migration, habitat use, and competition among mobile corals (Scleractinia: Fungiidae) in the Gulf of Eilat, Red Sea. Mar. Biol. 114, 617–623.

Colley, S., Feingold, J.S., Peña, J., Glynn, P.W., 2002. Reproductive ecology of *Diaseris distorta* (Michelin) (Fungiidae) in the Galápagos Islands, Ecuador. In: Proceedings 9th International Coral Reef Symposium, Bali, Indonesia. vol. 1, pp. 373–379.

Cortés, J., 1997. Biology and geology of Eastern Pacific coral reefs. Coral Reefs 16, 39–46.

Durham, J.W., 1966. Coelenterates, especially corals, from the Galapagos and Cocos Islands. In: Bowman, R.I. (Ed.), The Galapagos: Proceedings of the Symposia of the Galapagos International Scientific Project. University of California Press, Berkeley, pp. 123–135.

Durham, J.W., Barnard, J.L., 1952. Stony Corals of the Eastern Pacific Collected by the Velero III and Velero IV. vol. 16. Allan Hancock Pacific Expeditions, pp. 1–110. no. 1.

Enochs, I.C., Glynn, P.W., 2017. Corallivory in the Eastern Pacific. In: Glynn, P.W., Manzello, D.P., Enochs, I.C. (Eds.), Coral Reefs of the Eastern Tropical Pacific. Springer, Dordrecht, pp. 315–335.

Feingold, J.S., 1995. Effects of Elevated Water Temperature on Coral Bleaching and Survival during El Niño Disturbance Events (dissertation). Univ. of Miami, Coral Gables, FL.

Feingold, J.S., 1996. Coral survivors of the 1982-83 El Niño-Southern Oscillation, Galápagos Islands. Ecuador. Coral Reefs 15, 108.

Feingold, J.S., 2001. Responses of three coral communities to the 1997-98 El Niño-Southern Oscillation: Galápagos Islands, Ecuador. Bull. Mar. Sci. 69, 61–77.

Feingold, J.S., Riegl, B., Hendrickson, K., Toth, L.T., Cheng, H., et al., 2020. 7700-year persistence of an isolated, free-living coral assemblage in the Galápagos Islands: a model for coral refugia? Coral Reefs 39, 639–647. https://doi.org/10.1007/s00338-020-01935-5.

Glynn, P.W., 1998. El Niño-Southern Oscillation 1982–1983: nearshore population, community, and ecosystem responses. Annu. Rev. Ecol. Syst. 19, 309–345.

Glynn, P.W., 2003. Coral communities and coral reefs of Ecuador. In: Cortés, J. (Ed.), Latin American Coral Reefs. Elsevier, Amsterdam, pp. 449–472.

Glynn, P.W., Ault, J.S., 2000. A biogeographic analysis and review of the far eastern Pacific coral reef region. Coral Reefs 19, 1–23.

Glynn, P.W., Wellington, G.M., 1983. Corals and Coral Reefs of the Galápagos Islands. Univ. California Press, Berkeley.

Glynn, P.W., Maté, J.L., Baker, A.C., Calderón, M.O., 2001. Coral bleaching and mortality in Panamá and Ecuador during the 1997-1998 El Niño-Southern Oscillation event: spatial/temporal patterns and comparisons with the 1982-1983 event. Bull. Mar. Sci. 69, 79–109.

Glynn, P.W., Alvarado, J.J., Banks, S., Cortés, J., Feingold, J.S., et al., 2017a. Eastern Pacific coral reef provinces, coral community structure and composition: an overview. In: Glynn, P.W., Manzello, D.P., Enochs, I.C. (Eds.), Coral Reefs of the Eastern Tropical Pacific. Springer, Dordrecht, pp. 107–176.

Glynn, P.W., Colley, S.B., Carpizo-Ituarte, E., Richmond, R., 2017b. Coral reproduction in the Eastern Pacific. In: Glynn, P.W., Manzello, D.P., Enochs, I.C. (Eds.), Coral Reefs of the Eastern Tropical Pacific. Springer, Dordrecht, pp. 435–476.

Glynn, P.W., Feingold, J.S., Baker, A., Banks, S., Baums, I., et al., 2018. Status of corals and coral reefs of the Galapagos Islands (Ecuador): past, present, and future. Mar. Pollut. Bull. 133, 717–733.

Hoeksema, B.W., 1989. Taxonomy, phylogeny, and biogeography of mushroom corals (Scleractinia: Fungiidae). Zool. Verh. 254, 1–295.

Hoeksema, B.W., Sellanes, J., Easton, E.E., 2019. A high-latitude, mesophotic *Cycloseris* field at 85m depth off Rapa Nui (Easter Island). Bull. Mar. Sci. 95, 101–102.

Kohler, K.E., Gill, S.M., 2006. Coral Point Count with Excel extensions (CPCe): a Visual Basic program for the determination of coral and substrate coverage using random point count methodology. Comput. Geosci. 32 (9), 1259–1269. https://doi.org/10.1016/j.cageo.2005.11.009.

Kramarsky-Winter, E., Loya, Y., 1996. Regeneration versus budding in fungiid corals: a trade-off. Mar. Ecol. Prog. Ser. 134, 179–185.

Macintyre, I.M., Glynn, P.W., Cortés, J., 1992. Holocene reef history in the Eastern Pacific: mainland Costa Rica, Caño Island, Cocos Island, and Galápagos Islands. In: Proceedings of 7th International Coral Reef Symposium. vol. 2, pp. 1174–1184.

Reyes-Bonilla, H., Riosmena-Rodriguez, R., Foster, M.S., 1997. Hermatypic corals associated with rhodolith beds in the Gulf of California. México. Pac. Sci. 51, 328–337.

Riegl, B., Piller, W.E., 2003. Possible refugia for reefs in times of environmental stress. Int. J. Earth Sci. 92, 520–531.

Riegl, B., Johnston, M., Glynn, P.W., Keith, I., Rivera, F., et al., 2019. Some environmental and biological determinants of coral richness, resilience and reef building in Galapagos (Ecuador). Sci. Rep. 9, 10322. www.nature.com/articles/s41598-019-46607-9.

Robinson, G., 1985. Influence of the 1982-83 El Niño on Galapagos marine life. In: Robinson, G., del Pino, E.M. (Eds.), El Niño en las Islas Galápagos: El Evento de 1982–1983. Fundación Charles Darwin para las Islas Galápagos, Quito, Ecuador, pp. 153–190.

Romero-Torres, M., Acosta, A., Palacio-Castro, A.M., Treml, E.A., Zapata, F.A., et al., 2020. Coral reef resilience to thermal stress in the Eastern Tropical Pacific. Glob. Chang. Biol. 26, 3880–3890. https://doi.org/10.1111/gcb.15126.

Toth, L.T., Aronson, R.B., Vollmer, S.V., Hobbs, J.W., Urrego, D.H., et al., 2012. ENSO drove 2500-year collapse of eastern Pacific coral reefs. Science 337, 81–84.

Toth, L.T., Macintyre, I.G., Aronson, R.B., 2017. Holocene reef development in the eastern tropical Pacific. In: Glynn, P.W., Manzello, D.P., Enochs, I.C. (Eds.), Coral Reefs in the Eastern Tropical Pacific. Springer, Dordrecht, pp. 177–203.

Veron, J.E.N., 2000. Corals of the World. Australian Institute of Marine Science, Townsville, Australia.

Wellington, G.M., 1975. The Galápagos Coastal Marine Environments: A Resource Report to the Department of National Parks and Wildlife. Ministry of Agriculture and Livestock (MAG), Quito, Ecuador.

Wells, J.W., 1983. Annotated list of the scleractinian corals of the Galápagos. In: Glynn, P.W., Wellington, G.M. (Eds.), Corals and Coral Reefs of the Galápagos Islands. University of California Press, Berkeley, pp. 213–217.

Witman, J.D., Brandt, M., Smith, F., 2010. Coupling between subtidal prey and consumers along a mesoscale upwelling gradient in the Galapagos Islands. Ecol. Monogr. 80, 153–177.

Yamashiro, H., Hidaka, M., Nishihira, M., Poung-In, S., 1989. Morphological studies on skeletons of *Diaseris fragilis*, a free-living coral which reproduces asexually by natural autotomy. Galaxea 8, 283–294.

CHAPTER SEVEN

Considering the rates of growth in two taxa of coral across Pacific islands

Stuart A. Sandin[a,*,†], Clinton B. Edwards[a,†], Nicole E. Pedersen[a,†], Vid Petrovic[b], Gaia Pavoni[c], Esmeralda Alcantar[a], Kendall S. Chancellor[a], Michael D. Fox[d], Brenna Stallings[e], Christopher J. Sullivan[a], Randi D. Rotjan[e], Federico Ponchio[c], and Brian J. Zgliczynski[a]

[a]Scripps Institution of Oceanography, UC San Diego, La Jolla, CA, United States
[b]Department of Computer Science and Engineering, UC San Diego, La Jolla, CA, United States
[c]Visual Computing Lab, Istituto di Scienza e Tecnologie dell'Informazione "A. Faedo", Consiglio Nazionale delle Ricerche, Pisa, Italy
[d]Woods Hole Oceanographic Institution, Woods Hole, MA, United States
[e]Department of Biology, Boston University, Boston, MA, United States
*Corresponding author: e-mail address: ssandin@ucsd.edu

Contents

Abstract

Reef-building coral taxa demonstrate considerable flexibility and diversity in reproduction and growth mechanisms. Corals take advantage of this flexibility to increase or decrease size through clonal expansion and loss of live tissue area (i.e. via reproduction and mortality of constituent polyps). The biological lability of reef-building corals may be expected to map onto varying patterns of demography across environmental contexts which can contribute to geographic variation in population dynamics. Here we

† SAS, CBE, and NEP contributed equally to the manuscript.

Advances in Marine Biology, Volume 87
ISSN 0065-2881
https://doi.org/10.1016/bs.amb.2020.08.006

explore the patterns of growth of two common coral taxa, corymbose *Pocillopora* and massive *Porites*, across seven islands in the central and south Pacific. The islands span a natural gradient of environmental conditions, including a range of pelagic primary production, a metric linked to the relative availability of inorganic nutrients and heterotrophic resources for mixotrophic corals, and sea surface temperature and thermal histories. Over a multi-year sampling interval, most coral colonies experienced positive growth (greater planar area of live tissue in second relative to first time point), though the distributions of growth varied across islands. Island-level median growth did not relate simply to estimated pelagic primary productivity or temperature. However, at locations that experienced an extreme warm-water event during the sampling interval, most *Porites* colonies experienced net losses of live tissue and nearly all *Pocillopora* colonies experienced complete mortality. While descriptive statistics of demographics offer valuable insights into trends and variability in colony change through time, simplified models predicting growth patterns based on summarized oceanographic metrics appear inadequate for robust demographic prediction. We propose that the complexity of life history strategies among colonial reef-building corals introduces unique demographic flexibility for colonies to respond to a wide breadth of environmental conditions.

1. Introduction

The colonial growth form has evolved in parallel across multiple metazoan phyla (bryozoans, chordates (e.g. tunicates), and cnidarians (e.g. coral)) allowing for a number of unique ecological and demographic advantages not available to solitary animals. Of particular interest is the manner in which these species occupy space by production of clonal units and their ability to reproduce via sexually or asexually produced gametes (Harper et al., 1986; Jackson, 1986). This flexibility, sometimes termed indeterminate growth, results in less constraint on growth rates and final adult sizes compared to taxa whose demographic patterns, while still influenced by environmental controls, are largely determinate (Sebens, 1987). Further, colonial organisms often are characterized by extensive regenerative capabilities (Mackie, 1986) and thus can avoid entire colony mortality during stressful events via partial mortality of individual clones (Hughes et al., 1992), broadening demographic potential.

Observations of tradeoffs in demographic traits and associated vital rates, and the mathematical techniques to study population dynamics through time, form the basis of a substantial body of empirical and theoretical considerations of life history strategies (Grime, 1977; Pianka, 1970; Tilman, 1994; Werner and Caswell, 1977). Early studies of coral community dynamics adapted such frameworks to the study of coral reefs, considering explicitly the exceptional demographic flexibility of corals (Highsmith, 1982; Hughes and Jackson, 1980; Jackson and Coates, 1986). However, following the catastrophic decline of Caribbean corals (Hughes, 1994) and the first reported worldwide coral bleaching event in 1997/98 (Lough, 2000), research has focused increasingly on the role of recruitment in community dynamics (Bellwood et al., 2004; Connell et al., 2004; Doropoulos et al., 2015; Hughes and Tanner, 2000; Hughes et al., 2019; Szmant, 1986). It has been estimated that returns to pre-disturbance coral cover or community structure under scenarios involving principally recruitment may take decades (Gilmour et al., 2013; Roff et al., 2014), as coral larvae must successfully settle and survive the post-settlement demographic bottleneck (Vermeij and Sandin, 2008). On the other hand, under a regrowth scenario, surviving colonies may recover lost tissue (e.g. resheeting) and expand to occupy new benthic habitat, more rapidly restoring elements of benthic composition. Based on knowledge of the biology of scleractinian corals as well as anecdotal observations, the pathway of colony-level growth and regrowth can lead to recovery of populations (and coral cover) within a decade (Diaz-Pulido et al., 2009; Gouezo et al., 2019).

Patterns of change in response to disturbance, however, can vary significantly across coral taxa. As a result, several studies have documented taxonomic shifts in coral community composition in response to each continuous and episodic environmental forcing (Adjeroud et al., 2009; Aronson et al., 2004; Loya et al., 2001). It has been proposed that intense or persistent disturbances result in taxonomic shifts favouring dominance by so-called 'weedy' species (Green et al., 2008; Knowlton, 2001). For example, one coral taxon commonly referred to as weedy, the corymbose *Pocillopora*, has experienced rapid increases in abundance following disturbances in some locations (Bramanti and Edmunds, 2016; Holbrook et al., 2018). On the other hand, massive species of the genus *Porites*, generally referred to as stress tolerant, have shown resistance to many forms of disturbance, resulting in systematic increases in their proportional representation among stony coral communities following disturbances (Golbuu et al., 2007; Heery et al., 2018; Loya et al., 2001; Smith et al., 2008). Recently, researchers interested in fundamental questions of coral biology have applied trait-based approaches to increase the understanding of coral community dynamics (Darling et al., 2012; Madin et al., 2016), providing an empirically derived framework for the understanding of life history strategies in coral communities. While the degree to which physiological traits map onto demographic patterns at the community level remains to be tested

(Madin et al., 2016), the accelerated frequency and intensity of global bleaching events and other stressors makes it increasingly important to understand the capacity of coral taxa from across the spectrum of life history strategies to change and persist.

In addition to intrinsic drivers of variation in coral vital rates among taxa, there is an increasing body of literature showing demographic flexibility of corals in response to local environmental gradients. The acute effects of extreme thermal events on the survivorship of corals are well known (Baker et al., 2008; Hughes et al., 2018; Loya et al., 2001). Differences in bleaching responses have been observed both among and within taxa (e.g. via adaptive bleaching) (Buddemeier and Fautin, 1993; Loya et al., 2001), suggesting an ensemble of realized responses to thermal stress. Further, it has been shown that the effects of stress events can be long-lasting, with quantifiable impacts on coral growth rates (Cantin and Lough, 2014; Goreau and Macfarlane, 1990; Razak et al., 2020). There is also emerging evidence indicating that oceanic primary productivity impacts the functioning of coral reef ecosystems through increased supply of inorganic nutrients and particulate food resources (Fox et al., 2018; Yeakel et al., 2015). Corals growing in more productive regions can access relatively more allochthonous resources through heterotrophy, which may confer physiological benefits to coral growth observed as enhanced ecosystem calcification (Fox et al., 2018; Kealoha et al., 2019; Yeakel et al., 2015). In one example from the Caribbean, corals in areas with more frequent internal waves, a primary mechanism of food supply at intra-island scales, experienced higher rates of skeletal growth (Leichter and Genovese, 2006); thus, it may be expected that coral colonies in areas with higher oceanographic productivity could have capacity for higher realized growth rates.

In order to test hypotheses regarding the variability and putative drivers of demographic rates, it is critical to assemble appropriate demographic information. The description of colonial demography—rates of recruitment, survivorship, growth, and tissue loss (i.e. 'shrinkage')—requires data collected at the proper temporal, spatial, and ecological scales. In order to determine accurately context-dependent demographic rates, the fate of many individual coral colonies should be tracked though time to extract accurate and precise measurements of coral size and fate through time (Babcock, 1991; Bak and Nieuwland, 1995; Connell et al., 2004; Doropoulos et al., 2015; Edmunds, 2019; Fong and Glynn, 1998; Hughes and Tanner, 2000; Kayal et al., 2018; Riegl and Purkis, 2015; Vardi et al., 2012). While much has been learned from site-based and comparative studies of demographics, an opportunity exists to conduct comparative demographics across natural experimental contexts spanning large geographies. Generating reliable and repeated measurements of coral fates and sizes stands as a significant bottleneck for expanded geographic analyses. As a result, there is still much to be learned regarding the geographic and temporal variability of community-level coral demography.

For over a decade, marine scientists interested in questions at scales relevant to community ecology have taken advantage of the proliferation of digital photography, developing software platforms that blend thousands of individual images together to create detailed two-dimensional orthorectified imagery (Gracias and Santos-Victor, 2000; Lirman et al., 2007) and 3-dimensional reconstructions of benthic habitats using platforms such as Structure from Motion (SfM) software (Burns et al., 2015; Ferrari et al., 2016; Pizarro et al., 2009). More recently, the robust software platforms and advanced computational infrastructure necessary to generate these products for large areas has become more accessible. As a result, ecological applications of these techniques in marine habitats have increased, allowing researchers to extract information such as coral sizes, percent cover, or metrics of complexity and rugosity at scale (Burns et al., 2015; Edwards et al., 2017; Ferrari et al., 2016; González-Rivero et al., 2017; Pedersen et al., 2019). Such approaches offer a complement to *in situ* methodologies where search times are limited by the constraints of sampling in subtidal habitats, allowing for nearly limitless 'bottom-time', facilitating searches of large areas of the benthos in a relatively rapid manner.

In this study, we explore patterns and variability of coral life history across a broad geographic range using a large area imaging approach. For two widely distributed coral taxa, *Porites* and *Pocillopora*, we test whether rates of growth vary across environmental contexts, and, if so, whether oceanographic correlates (e.g. oceanic primary production, thermal history) help to interpret this variability. We focus our study on coral reef communities that are exposed to relatively low levels of local anthropogenic disturbance (e.g. pollution and overfishing), thus offering insights into the natural demographic potential of these coral taxa.

2. Methods

2.1 Study locations

The study was completed on the coral reefs of seven islands of the central tropical Pacific that span approximately 78 degrees of longitude (Table 1; Fig. 1). From west to east, the islands are: (i) Helen reef (Republic of Palau),

Table 1 Location and sampling metadata for each island included in the analysis.

Island	Latitude	Longitude	Sampling month, t_0	Sampling month, t_1	Sampling interval (months)
Mo'orea	17.54° S	149.83° W	Mar 2017	Feb 2019	23
Pohnpei	6.85° N	158.26° E	Aug 2016	Aug 2018	24
Millennium	9.96° S	150.21° W	Oct 2013	Aug 2017	46
Nikumaroro	4.68° S	174.52° W	Jun 2012	Sep 2015	39
Palmyra	5.88° N	162.08° W	Sep 2016	Oct 2018	25
Helen	2.90° N	131.78° E	Dec 2014	Jul 2017	31
Enderbury	3.13° S	171.08° W	Jun 2012	Sep 2015	39

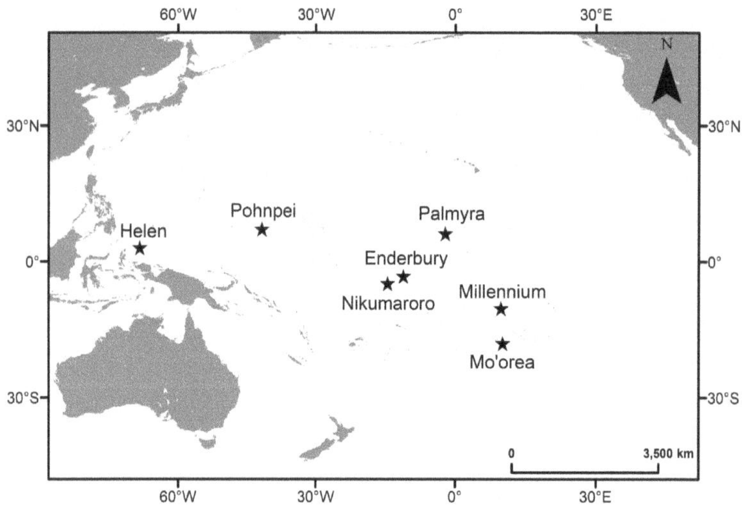

Fig. 1 Map of the seven sampling locations, chosen to span gradients of physical oceanographic parameters and human habitation. The uninhabited islands of Millennium Atoll, Enderbury and Nikumaroro are administered by the Republic of Kiribati, the latter two of which are part of the Phoenix Islands Protected Area. Palmyra Atoll is part of the US Pacific Remote Islands Marine National Monument and supports a research station maintained by The Nature Conservancy; Helen Reef is a no-take marine protected area in the Hatohobei State, Republic of Palau. Both Palmyra and Helen are home to temporary research, enforcement, and monitoring staff that maintain a year-round presence. Mo'orea (French Polynesia) and Pohnpei (Federated States of Micronesia) support populations of 16,000 and 36,000 permanent residents, respectively. At each island imaging was conducted within 100-m^2 plots positioned along the leeward 8–12 m fore reef isobath.

(ii) Pohnpei (Federated States of Micronesia), (iii) Nikumaroro and (iv) Enderbury (Republic of Kiribati), (v) Palmyra (USA), (vi) Millennium (Republic of Kiribati), and (vii) Mo′orea (French Polynesia). The central Pacific is oceanographically diverse, and these islands span a range of climatologies, linked in large part to equatorial upwelling and patterns of recent thermal history. All data were collected from similar habitats on each island, specifically the leeward fore reef habitats at 8–12m depth.

2.2 Study taxa

Two regionally common coral taxa were selected—massive *Porites* and corymbose *Pocillopora*. The taxa are among the most ubiquitous corals in the central Pacific, with distributional ranges extending across most of the tropical Indo-Pacific (Veron, 2000). Massive *Porites* is a common target for paleoecological studies in the tropical Pacific, providing robust records linking patterns of skeletal extension to thermal (Barkley and Cohen, 2016; Barkley et al., 2018) and oceanographic conditions (Carilli et al., 2017). Further, a 2018 report on the status and trends of coral reef communities in the Pacific identified evidence of stasis or decline in mean coral cover over the past 40 years; the only coral taxon to deviate from this regional pattern was massive *Porites* that showed evidence of net increase in total cover during this time window (Moritz et al., 2018). Corymbose *Pocillopora* has been shown to be a primary taxon in the rapid loss (Adjeroud et al., 2009) and recovery (Bramanti and Edmunds, 2016) of coral cover on Mo′orea in French Polynesia. Further, a synoptic study of functional traits of corals classified these taxa into distinct life history strategies, with *Pocillopora* referred to as weedy or ruderal and *Porites* as stress tolerant (Darling et al., 2012). These taxa define much of the change in total coral cover across the Indo-Pacific and contribute to our understanding of coral demography and dominant trends in coral community change in many locations.

Given that our goal here is to study a wide geographic range of environmental characteristics and that coral species are known to exhibit variable population structure within species across geographic ranges, we used taxonomic classifications broader than the species level to focus upon common, pooled functional characteristics. A functional trait-based rather than species-specific approach focused on the fundamental niche thus offers value in exploring demographic variability across environmental gradients (McGill et al., 2006). The massive *Porites* across this study area represent a tight taxonomic cluster dominated by three species, *P. arnaudi*, *P. lobata*, and *P. lutea*, congeners that are closely related (Forsman et al., 2009) with constrained reported growth rates (0.3–2.3 cm y^{-1} and 0.4–2.4 cm y^{-1} for *P. lobata* and *P. lutea*, respectively), tissue thickness (4–8 mm and 3–8 mm for *P. lobata* and *P. lutea*, respectively), morphology (massive), reproductive mode (spawner), and life history strategy (stress-tolerant) (Darling et al., 2012). The corymbose *Pocillopora* represent a species complex composed of two species, *P. meandrina* and *P. verrucosa*, which are monophyletic and show high similarity with functional traits (Pinzón and LaJeunesse, 2011) including growth rates (1.5–4.5 cm y^{-1} and 1.7–3.8 cm y^{-1} for *P. meandrina* and *P. verrucosa*, respectively), tissue thickness (0.4 mm), morphology (corymbose), reproductive mode (spawner), and life history strategy (competitive) (Darling et al., 2012). Given the similarities within defined groups, demographic descriptions of each aggregated taxon offer robust functional perspective of how realized demographic rates vary as a function of environmental context.

2.3 Survey design

Growth and survivorship were estimated across populations by tracking the fates of individual coral colonies through time. A target minimum sample size of 60 colonies of each taxon per island were sought for identification and tracking of live tissue through time. Colonies were identified from 3 to 5 sites on each island, with sites haphazardly selected across the leeward fore reef habitat. The time interval over which growth and fate were estimated differed based upon field access, with the target being integer multiples of years—2 years (Mo′orea, Palmyra, and Pohnpei), 3 years (Enderbury, Helen, Nikumaroro), and 4 years (Millennium). Note that reported sampling interval was rounded to nearest year and was within 3 months of multiple of 12-month interval with the exception of Helen (2 years and 7 months) (Table 1).

Data were collected using large-area imagery, in which 2D orthoprojections derived from photogrammetrically constructed 3D models of 10 m × 10 m reef areas were generated for the first and second annual time point (t_0 and t_1, respectively) for each study site. The workflow of ecological data extraction from composite, large-area imagery has been described elsewhere (Fox et al., 2019; Kodera et al., 2020; McNamara et al., 2019; Pedersen et al., 2019); an overview of the approach follows.

2.3.1 *Image acquisition*

Large-area imagery was collected within defined 100-m^2 plots. All plots were established with two geo-referenced steel pins marked by GPS (defining the 10-m isobathic centreline of the plot), which enable straightforward relocation of plots in subsequent surveys and are used to coregister models through time. At each sampling interval, six reference markers are deployed to identify the boundaries of the plot (at the four corners of the 10×10 m plot, and adjacent to the two permanent steel pins) and depths are collected at each to define the plane of projection (Kodera et al., 2020). Additionally, four 50-cm scale bars are deployed inside the plot and captured in image data to provide scale.

Images were collected using two Nikon D7000 16.2 megapixel DSLR still cameras mounted to a custom frame. The camera used to generate models is equipped with a wide-angle 18-mm focal length lens to ensure high overlap (>80%) among adjacent images. The second camera is equipped with a 55-mm focal length lens to capture images with ≤ 1 mm resolution to aid in taxonomic identification and definition of colony borders (Pedersen et al., 2019). To obtain continuous coverage of the reef floor within a plot, the diver operating the camera system swims a gridded pattern approximately 1.5 m above the benthos at a speed sufficient to maintain maximum overlap between adjacent images (5–7 $m\,min^{-1}$). Images are captured every second from each still camera using the built-in intervalometers, yielding approximately 2500 individual images per camera per plot.

2.3.2 *Large-area model construction and coregistration*

Images were processed using commercially available Structure from Motion (SfM) software, Agisoft *Metashape* (previously called *Photoscan*; Agisoft LLC., St. Petersburg, Russia) to create 3D point clouds. The 3D point clouds were manipulated and coregistered in *Viscore*, a custom integrated visualization and analytical platform (Petrovic et al., 2014). 3D models from multiple time points of a single plot were input into *Viscore* as layers and manually coregistered to ensure that derived 2D orthoprojections were captured at the same orientation (e.g. projection) across time. Scale was assigned within *Viscore* by inputting the known distance between the targets on 50-cm scale bars. Scaling of imagery has shown model errors relative to the fixed length scale bars are unbiased with 95% confidence intervals of <0.5% (+/−0.24 cm for 50 cm distance). The reference plane used for orthorectification is the plane orthogonal to gravity (or the average plane of the sea surface) and is defined in *Viscore* using depth measurements taken in the field at the 6 plot boundary reference markers and which are visible in the model. To account for errors from *in situ* depth measurements (e.g. due to wave action and instrument precision), a nonlinear least squares optimization procedure was used that produces a plane of best fit across depth measurements.

Distance and angle-based distortion bias in single-image and composite photomosaic imagery can be corrected through orthoprojection (Habib et al., 2007; Nicosevici et al., 2009). However, most approaches for generating orthorectified imagery suffer from geometric deformations and problems such as ghosting that result from re-projection errors involved in mapping multiple dense 2D images onto an approximate 3D model. *Viscore* implements a 3D-to-2D approach, which enables the generation of orthoprojections directly from the dense point cloud, circumventing the need to work with original 2D imagery minimizes distortion and enables the creation of geometrically accurate 2D orthoprojections with direct access to the underlying high-resolution imagery (Naughton et al., 2015; Petrovic et al., 2014).

2.3.3 Ecological data extraction

All colony annotations and planar area measurements on 2D orthoprojections were made using the software platform *TagLab* (Visual Computing Lab, CNR-ISTI) available on Github (https://github.com/cnr-isti-vclab/TagLab). *TagLab* allows for direct comparisons of annotated, co-registered orthoprojections which facilitates matching of colony segmentations and measured areas across time. Corals were identified within imagery in a stratified random design. For each taxon, a point was dropped randomly within each 2 m × 2 m quadrat of the 100-m^2 plot from t_0, and the coral colony with its centroid closest to the point was selected. Given the clustered distribution of many corals, some quadrats did not contain colonies, and additional randomized points were dropped until the target sample size was met (in few cases, there were fewer than the target number available across sites).

Colonies were defined as contiguous patches of live tissue, and their size was determined by measuring the planar area of live tissue within the scaled and orthorectified orthoprojection. The same colonies were identified in co-registered orthoprojections from t_1 and again sized based on the planar area of live tissue. For colonies experiencing fission during the study period, the planar area measured in t_1 was taken as the summed planar area of all daughter colonies in t_1 associated with a single contiguous patch of live tissue during t_0 (defined here as a colony), with the reverse being true for colonies experiencing fusion. In some study locations where *Porites* often grew in an overlapping plated structure, a colony was delineated as only area of the top tier, as it could not be determined reliably if there were tissue connections between plate tiers. Colonies that were dislodged in t_1 were not included in the study, given that the demographic fate of these colonies could not be assessed; colonies that lost all live tissue but with skeleton still in place were included in this study. Proportional growth was calculated as the colony-specific ratio of planar area of live tissue in t_1 to that in t_0, standardized to units of annual change by taking the n-th root of the ratio, where n is the number of years between sampling intervals.

2.3.4 Oceanographic context

For each island, levels of oceanographic primary production were estimated as concentrations of chlorophyll-*a* in nearshore surface waters from 8-day, 0.0417° (4 km) Moderate Resolution Imaging Spectroradiometer (MODIS) data (Gove et al., 2013). To avoid data quality concerns associated with shallow water bottom reflectance (Boss and Zaneveld, 2003), we excluded data from within less than 3.27 km from the 30 m isobath or visible edge of the reef crest in aerial imagery when bathymetric data are not available (Fox et al., 2018; Gove et al., 2013). To explore the influence of primary production on coral growth we quantified island-scale differences in oceanographic conditions at two time scales. First, we established climatological chl-*a* baselines for each island over a fourteen-year period (2005–18). For each island, we also estimated the mean chl-*a* concentrations during the interval between the first and second sampling event, thus describing the productivity context for organisms during the temporal window over which coral colonies were observed (Table 1).

Similarly, we estimated thermal conditions at each island. Sea surface temperature (SST) records were retrieved for each island using data from the NOAA Coral Reef Watch operational daily near-real-time global 5-km satellite coral bleaching monitoring products (https://pae-paha.pacioos.hawaii.edu/erddap/griddap/dhw_5km.html), using data pixels nearest to our sampling sites. We calculated the climatological SST over the 25-year period specific to each island (25 years ending at second sampling date) as well as the mean SST during the sampling interval for each island (Tables 1 and 2). To quantify the recent history of thermal stress, we

Table 2 Oceanographic metadata for each island included in the analysis.

Island	Chl *a* (long-term mean; mg m^{-3})	Chl *a* (interval mean; mg m^{-3})	SST (long-term mean; °C)	SST (interval mean; °C)	Months since >8 DHW
Mo'orea	0.066	0.062	27.8	28.0	>100
Pohnpei	0.070	0.069	29.0	29.6	0
Millennium	0.106	0.094	28.6	29.0	0
Nikumaroro	0.125	0.130	29.1	29.1	29
Palmyra	0.126	0.113	28.2	28.6	12
Helen	0.145	0.173	29.2	29.3	>100
Enderbury	0.148	0.149	28.6	29.0	0

Islands are ordered from lowest to highest long-term mean concentration of chlorophyll-*a* (Chl *a*) calculated as the 8-day moving window average from 2005 to 2018 using remote-sensed estimates. Long-term sea surface temperature (SST) estimates are presented as long-term means across the 25-year window extending historically from the second sampling month for each island (Table 1). Complementary estimates of Chl *a* concentrations and SST are reported for the sampling interval for each island. Time since bleaching was calculated for each location as the number of months between the first sampling month and the most recent thermal event above 8 degree heating weeks (DHW), using NOAA Coral Reef Watch data (see Methods); if an event >8 DHW happened during the survey interval, the island was coded as 0 months.

determined the most recent extreme thermal stress events (>8 degree heating weeks; DHW). We used the climatological SST data, complemented by regionally tailored reports of thermal history (Mollica et al., 2019), to define recent events. We report only events within 100 months of the first sampling interval, and we report 0 months if an event of >8 DHW occurred during the sampling interval.

2.4 Statistical analysis

To determine whether the proportional change in planar area of live tissue of colonies varied across islands, we performed a non-parametric Kruskal-Wallis rank sum test. Dunn's post hoc test for multiple comparisons was used to investigate significant difference between island pairs as data violated assumptions of normality. Given evidence of significant difference among islands, we investigated the relationship between coral growth rate and oceanographic context. We independently tested the association between median proportional change in area and estimates of climatological mean chl-*a* and SST values across the seven islands using Spearman's rank correlation test for *Porites* and *Pocillopora*, independently. Comparable correlation tests were performed with interval-specific estimates of mean chl-*a* and SST. All analyses were conducted in R, version 3.5.1 (http://www.r-project.org/).

3. Results

A total of 946 coral colonies were included in this study, 435 *Pocillopora* and 511 *Porites* colonies. Across all islands, 130 *Pocillopora* and 29 *Porites* colonies suffered complete mortality of visible live tissue during the sampling interval (Table 3). *Pocillopora* mortality was most common at Millennium (72 of 74 colonies) and Pohnpei (20 of 20 colonies), less common at Palmyra (22 of 96 colonies), and least common at the remaining four islands (<10% at each) (Table 3). *Porites* mortality exceeded 10% only at Mo′orea (12 of 97 colonies) (Table 3). Given that mortality is cumulative with time, we may expect that the proportion of colonies suffering whole-colony mortality would increase with sampling interval. Notably, the islands experiencing the highest overall mortality were principally those with the shortest sampling interval (2 years between sampling for Mo′orea, Palmyra, and Pohnpei), with the exception of Millennium (4 years). Inclusive of colonies with complete tissue mortality, the summed area of live tissue across sampled colonies revealed an increase in four of seven islands for *Pocillopora* and six of seven islands for *Porites* (Table 3).

Colony-level patterns of change in tissue area were variable, including evidence of both tissue growth (larger colony size in t_1 relative to t_0) and shrinkage (smaller colony size in t_1 relative to t_0) (Fig. 2). Across all islands, more colonies grew than shrank or suffered total tissue loss; 54.5% of *Pocillopora* colonies and 66.1% of *Porites* colonies were observed with a greater area of live tissue in t_1 relative to t_0. However, the proportion of colonies that grew differed among islands. For *Pocillopora*, no colonies were observed to grow on Millennium or Pohnpei, while over 50% of colonies grew on the remaining islands, ranging from 57.3% on Palmyra to 90% or higher on Enderbury and Nikumaroro (Fig. 2A, C and E). For *Porites*, only on Millennium did less than 50% of colonies experience growth (38.5%), ranging from 61.9% on Mo′orea to 96.8% on Enderbury (Fig. 2B, D and F).

The median of coral colony area distributions generally increased between sampling intervals. Inclusive of colonies with complete tissue mortality, the median area of live tissue was larger in t_0 than in t_1 (indicating a loss of live coral tissue) for focal *Pocillopora* colonies only on Millennium and Pohnpei, the islands that experienced extreme mortality during the time interval, and for focal *Porites* colonies only on Millennium (Fig. 3, Tables 2 and 3).

Table 3 Summary of sampling statistics from each island for *Pocillopora* and *Porites*.

Island	# of colonies, t_0	Total area, t_0 (cm^2)	Median area, t_0 (cm^2)	# of surviving colonies, t_1	Total area, t_1 (cm^2)	Median area, t_1 (cm^2)	Median proportional change in area (yr^{-1})
Pocillopora							
Mo'orea	96	15,639.5	135.6	89	19,097.8	178.4	1.16[b]
Pohnpei	20	650.7	27.6	0	0.0	0.0	0.00[a]
Millennium	74	19,579.6	225.7	2	335.1	0.0	0.00[a]
Nikumaroro	43	7156.0	122.2	43	15,031.8	293.5	1.35[c]
Palmyra	96	18,167.6	169.0	74	18,556.7	198.5	1.12[c]
Helen	86	24,458.7	247.1	78	23,491.4	285.4	1.06[b]
Enderbury	20	3028.2	110.4	19	6219.1	320.3	1.34[c]
Porites							
Mo'orea	97	10,565.2	51.2	85	12,324.8	67.3	1.14[b,c]
Pohnpei	102	83,275.8	113.6	98	87,219.6	189.2	1.10[b,c]
Millennium	78	108,950.8	827.4	72	78,290.0	524.6	0.94[a]
Nikumaroro	55	47,972.1	611.7	55	60,977.6	948.6	1.14[c]
Palmyra	92	41,764.9	291.5	85	45,880.4	324.4	1.09[b]
Helen	56	46,133.2	276.4	56	53,271.6	326.5	1.07[b,c]
Enderbury	31	19,446.6	564.2	31	28,315.2	800.9	1.14[b,c]

Islands are ordered from lowest to highest mean concentration of chlorophyll-*a*.
A minimum sample size of 60 colonies was targeted from 3 to 5 sites at each island, though, at some locations colonies of both *Pocillopora* and *Porites* were rare and we were unable to locate the target 60 colonies. The median annual proportional change was calculated as the *n*-th root of the ratio of area at t_1 divided by the area at t_0, where n is the rounded number of years between surveys. Letters in median proportional change indicate significant groupings from the Dunn's multiple comparison post hoc test.

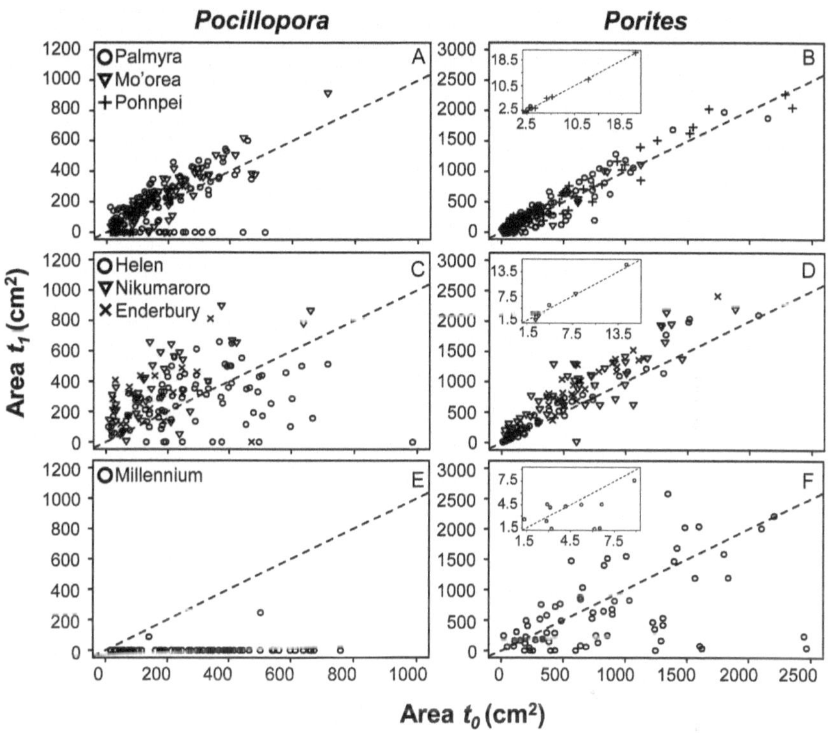

Fig. 2 Colony-specific changes in live tissue area across each of the islands separated by length of sampling interval (2-year interval (A and B); 3-year interval (C and D); 4-year interval (E and F)). Axes for insets (B, D and F) are scaled 1×10^3, indicating the regular presence of exceptionally large colonies of *Porites* at all islands. The black line represents the 1:1 line of area at t_0 to area at t_1, indicating no change over the course of the sampling interval. In general, complete mortality was more common in *Pocillopora*, while a greater proportion of *Porites* colonies avoided total mortality (A–F). *Pocillopora* colonies at both Pohnpei (A) and Millennium experienced total mortality (except for two colonies at Millennium, E), likely reflecting the impact of the warm-water events which occurred at both locations during the sampling interval.

The annual median of proportional change in colony-specific area showed a similar pattern as the distributional changes. The median proportional change for *Pocillopora* differed significantly by island (Kruskal-Wallis test, $X^2 = 202.4$, $p < 0.001$), ranging from $0.00\,\mathrm{yr}^{-1}$ on Millennium and Pohnpei (where most colonies suffered complete mortality) to over $1.3\,\mathrm{yr}^{-1}$ on Nikumaroro and Enderbury (Fig. 4). The median proportional change for *Porites* differed significantly by island (Kruskal-Wallis test, $X^2 = 48.5$, $p < 0.001$), spanning a range from $0.94\,\mathrm{yr}^{-1}$ on Millennium to approximately

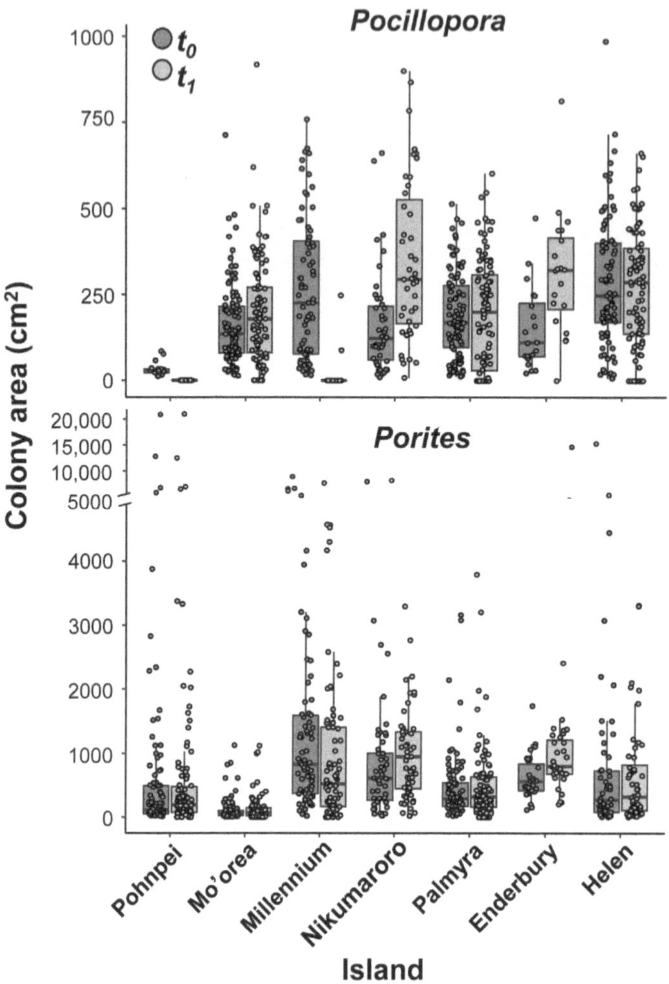

Fig. 3 Area of all *Pocillopora* (top) and *Porites* colonies at each location across the time series. Horizontal lines represent median proportional change of all colonies at each island, upper and lower limits of boxes represent the 75th and 25th percentiles, respectively; vertical lines delineate values 1.5 times greater or lower than the upper and lower ranges, respectively. Except for Pohnpei and Millennium where nearly all colonies died during the study period, the median size of *Pocillopora* increased at all islands. Though size distributions of *Porites* are dominated by colonies smaller than 3000 cm^2 at all islands, across the time series exceptionally large colonies of *Porites* were present at all islands (see axis break) were relatively common.

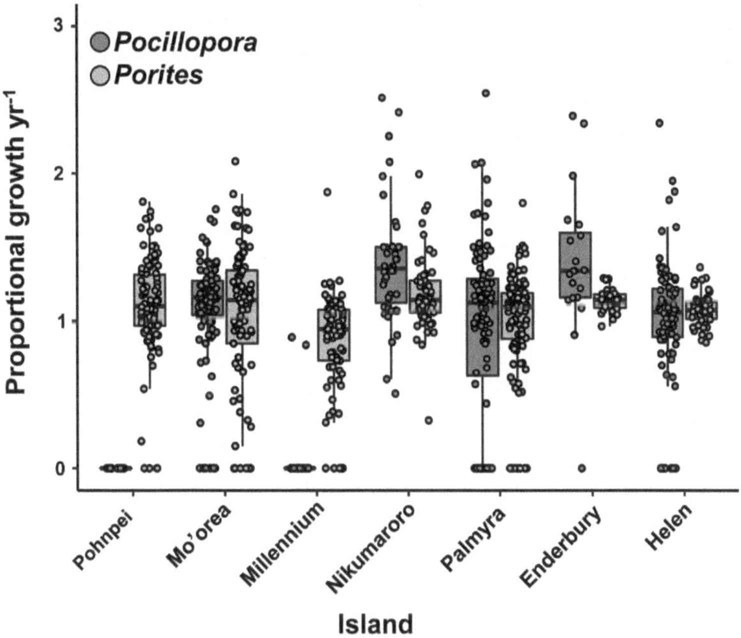

Fig. 4 Yearly proportional change in planar area of live tissue for individual *Pocillopora* (pink (grey in the print version)) and *Porites* (blue (light grey in the print version)) colonies at each island, calculated as n-th root of the ratio of area at t_1 divided by the area at t_0, where n is the rounded number of years between surveys. Points represent individual colonies with random horizontal offset for display. Horizontal lines represent median proportional change of all colonies at each island, upper and lower limits of boxes represent the 75th and 25th percentiles, respectively; vertical lines delineate values 1.5 times greater or lower than the upper and lower ranges, respectively. The range in median proportional change for both *Pocillopora* and *Porites* was remarkably consistent across islands, however notable differences exist in the overall variability of change experienced by individual colonies. Growth of *Pocillopora* colonies at Palmyra was highly variable, including a single colony (not shown here) which experienced yearly proportional growth of 3.65 yr^{-1} over the course of the study period.

1.14 yr^{-1} on Mo′orea, Nikumaroro, and Enderbury (Fig. 4). Annual proportional change across the two taxa was consistently ranked for three islands—lowest on Millennium and highest on Nikumaroro and Enderbury (Table 3). The remaining four islands had inconsistent relative ranking for the two taxa.

The annual median proportional change in area was not significantly correlated to estimates of climatological mean oceanographic primary productivity (chl-*a*) across islands (Spearman correlation; $r_s = 0.31$, $p > 0.1$ for each *Pocillopora* and *Porites*). Constraining the estimate of primary productivity to the inter-survey time window changes only some rankings of relative chl-*a* concentrations across islands (Table 2) and did not affect correlation results. The annual median proportional change in area was not significantly correlated to estimates of climatological mean SST across islands (Spearman correlation; $r_s=0.00$, $p>0.1$ for each *Pocillopora* and *Porites*). The relative values of mean SST across islands changed only slightly when considering the sampling interval; correlation of these metrics with median proportional change comparably were not significant (Spearman correlation; $r_s=-0.25$, $p>0.1$ for each *Pocillopora* and *Porites*).

4. Discussion

In order to understand changes in coral communities through time, especially colony-specific life, death, growth, and shrinkage, it is essential to quantify demographic dynamics explicitly. Given the complex nature of benthic interactions as well as geographic and temporal variability of oceanographic conditions, there is a need for information to be collected across sufficiently heterogeneous ecological scales to reflect the linkage between realized demographic rates and environmental context. A primary challenge to the study of regional patterns of and correlates to variability in coral demography is the challenge associated with collection of large sets of life historical data. *In situ* tagging, tracking, and measurement of individual colonies is sufficiently burdensome to limit the commonness of such data collection efforts (notable examples include Connell et al., 2004; Edmunds, 2019; Hughes and Tanner, 2000). Predictive modelling of benthic community structure builds from robust demographic data (Hughes and Tanner, 2000; Kayal et al., 2018; Riegl and Purkis, 2015; Tanner et al., 2009); prerequisite to the expansion of such modelling efforts to regional scales is the expanded collection of reliable and extensive life history data.

In this study, we investigated the patterns and variability of coral growth across a broad geographic range. Here, we tracked 435 *Pocillopora* and 511 *Porites* colonies at seven islands spanning a gradient of oceanographic conditions. We found evidence for significant differences in growth rates, reflecting differences in the realized growth distributions at the various islands (Table 3). We leveraged the use of digital imagery and tools of photogrammetric analysis, supporting a repeatable and archivable workflow. The application of image-based products enabled a survey design that was less constrained by limits of diving time and minimized challenges of underwater search and recovery for focal features (i.e. coral colonies observed in sequential time points). We tested whether patterns of growth and survival varied across geographic units, and if so, whether this variability can be partly explained by the environmental context of each island, quantified here as the climatological mean of nearshore chl-*a* concentration and thermal history.

We found significant variability in coral growth rates across islands but no evidence that the variability was affected strongly by spatial patterns of oceanographic primary production or seawater temperature. Several alternative hypotheses can be considered that could lead to inter-island differences in distributions of growth rates. Our analyses focused upon description of growth patterns of representative colonies from each island, having been selected randomly from the population of the two taxa within the 10-m depth fore reef habitat. One hypothesis is that systematic differences in size distributions of colonies selected during t_0 would result in systematic differences in growth rates, as estimated here. In a classic study of 405 colonies tracked in a reef crest habitat on the Great Barrier Reef, Hughes and Connell found significant size-dependence in observed growth rate, with smaller colonies revealing higher annual relative growth (Hughes and Connell, 1987). Similarly, Kodera and colleagues found size-dependency in the probability of survival and growth with size for corymbose *Pocillopora* on a central Pacific fore reef habitat (Kodera et al., 2020). Although size distributions at t_0 for each taxon showed some variability (Fig. 3), there were no consistent patterns relating median annual growth to median size by island for either taxon (Table 3). For example, the islands revealing the lowest median proportional change for *Pocillopora* spanned from lowest to second-highest median starting size (Pohnpei and Millennium, respectively; Table 3). An opportunity exists to explore individual-based models of coral growth, parameterizing with size effects, environmental effects, benthic context, and other putative influences on realized colony-level changes; such statistical efforts will become more viable as colony-specific growth data become more abundant to support quantitatively rigorous modelling exercises.

Another hypothesis suggests that recent thermal history may have a consistent influence on the realized demographic rates of colonies. Two of the three islands that experienced an extreme warm-water event during the interval between surveys revealed clear evidence of coral stress. *Pocillopora* on Millennium and Pohnpei suffered over 95% mortality after experiencing a thermal anomaly of at least 8 degree heating-weeks (DHW; NOAA Bleaching Alert Level 2) during the survey interval

(Tables 2 and 3); *Porites* on Millennium alone revealed evidence of demographic 'costs' associated with the thermal anomaly, as the only island-level median growth rate less than 1.0 (Table 3; Fig. 4). While the outlying demographic data (high mortality of *Pocillopora*, low growth of *Porites*) were linked with an extreme temperature anomaly, the demographic signal of thermal data was not consistent. *Porites* on Pohnpei, for example, showed growth rates comparable to islands that did not experience recent thermal stress (Tables 2 and 3; Fig. 4). Further, despite experiencing an extreme DHW event during the survey interval, the corals of both taxa on Enderbury showed among the highest median growth (Fig. 4). Perhaps the use of classical metrics of thermal stress are less functional for equatorial islands (Donner, 2011; Mollica et al., 2019), or the recent history of repeated thermal stress in the Phoenix Islands has led to systematic shifts in the population structure and realized demography of corals in the region (Mangubhai et al., 2019).

Demographic studies of stony corals confirm the broad flexibility in survivorship and growth potential within and among taxa. Given the breadth of biological potential for these taxa, including sexual reproduction, asexual reproduction (e.g. fragmentation), and flexible within-colony 'demography' (e.g. potential for each loss and addition of live polyps resulting in shrinkage and growth through time, fission/fusion dynamics), it is not surprising that population-level descriptions of demography vary across environmental contexts. Local environmental conditions, defined by specifics of resource availability, oceanographic context, thermal history, and benthic competitive landscape, among myriad others, will be met with diverse survival and growth realizations from individual corals. We have shown that realized growth distributions of two Pacific coral taxa vary appreciably across seven islands representing seven distinct sets of environmental conditions. The results suggest that simple relations to oceanographic summaries may not capture the complexity of coral demographic dynamics. More detailed and comprehensive studies of population dynamics across environmental contexts hold promise to reveal more about the factors influencing such demographic differences.

Acknowledgements

This research was made possible through efforts of the 100 Island Challenge, and the various organizations that have supported this project, including the Scripps family and the Moore Family Foundation. We would like to acknowledge the efforts of Lindsay Bonito, Angelica Dimas, and Mary Liesegang for their assistance in the field and in the laboratory. Special thanks to the people of the Republic of Kiribati, colleagues from the New England Aquarium, Prince Albert of Monaco II, and Sangeeta Mangubhai for coordinating and supporting access to the Phoenix Islands. The authors declare no conflicts of interest.

References

Adjeroud, M., Michonneau, F., Edmunds, P., Chancerelle, Y., De Loma, T.L., Penin, L., Thibaut, L., Vidal-Dupiol, J., Salvat, B., Galzin, R., 2009. Recurrent disturbances, recovery trajectories, and resilience of coral assemblages on a South Central Pacific reef. Coral Reefs 28, 775–780.

Aronson, R.B., Macintyre, I.G., Wapnick, C.M., O'Neill, M.W., 2004. Phase shifts, alternative states, and the unprecedented convergence of two reef systems. Ecology 85, 1876–1891.

Babcock, R.C., 1991. Comparative demography of three species of scleractinian corals using age-and size-dependent classifications. Ecol. Monogr. 61, 225–244.

Bak, R.P.M., Nieuwland, G., 1995. Long-term change in coral communities along depth gradients over leeward reefs in the Netherlands Antilles. Bull. Mar. Sci. 56, 609–619.

Baker, A.C., Glynn, P.W., Riegl, B., 2008. Climate change and coral reef bleaching: an ecological assessment of long-term impacts, recovery trends and future outlook. Estuar. Coast. Shelf Sci. 80, 435–471.

Barkley, H.C., Cohen, A.L., 2016. Skeletal records of community-level bleaching in *Porites* corals from Palau. Coral Reefs 35, 1407–1417.

Barkley, H.C., Cohen, A.L., Mollica, N.R., Brainard, R.E., Rivera, H.E., DeCarlo, T.M., Lohmann, G.P., Drenkard, E.J., Alpert, A.E., Young, C.W., 2018. Repeat bleaching of a central Pacific coral reef over the past six decades (1960–2016). Commun. Biol. 1, 1–10.

Bellwood, D.R., Hughes, T.P., Folke, C., Nystrom, M., 2004. Confronting the coral reef crisis. Nature 429, 827–833.

Boss, E., Zaneveld, J.R.V., 2003. The effect of bottom substrate on inherent optical properties: evidence of biogeochemical processes. Limnol. Oceanogr. 48, 346–354.

Bramanti, L., Edmunds, P.J., 2016. Density-associated recruitment mediates coral population dynamics on a coral reef. Coral Reefs 35, 543–553.

Buddemeier, R.W., Fautin, D.G., 1993. Coral bleaching as an adaptive mechanism. Bioscience 43, 320–326.

Burns, J., Delparte, D., Gates, R., Takabayashi, M., 2015. Integrating structure-from-motion photogrammetry with geospatial software as a novel technique for quantifying 3D ecological characteristics of coral reefs. PeerJ 3, e1077.

Cantin, N.E., Lough, J.M., 2014. Surviving coral bleaching events: Porites growth anomalies on the Great Barrier Reef. PLoS One 9, e88720.

Carilli, J.E., Hartmann, A.C., Heron, S.F., Pandolfi, J.M., Cobb, K., Sayani, H., Dunbar, R., Sandin, S.A., 2017. *Porites* coral response to an oceanographic and human impact gradient in the Line Islands. Limnol. Oceanogr. 62, 2850–2863.

Connell, J.H., Hughes, T.P., Wallace, C.C., Tanner, J.E., Harms, K.E., Kerr, A.M., 2004. A long-term study of competition and diversity of corals. Ecol. Monogr. 74, 179–210.

Darling, E.S., Alvarez-Filip, L., Oliver, T.A., McClanahan, T.R., Côté, I.M., 2012. Evaluating life-history strategies of reef corals from species traits. Ecol. Lett. 15, 1378–1386.

Diaz-Pulido, G., McCook, L.J., Dove, S., Berkelmans, R., Roff, G., Kline, D.I., Weeks, S., Evans, R.D., Williamson, D.H., Hoegh-Guldberg, O., 2009. Doom and boom on a resilient reef: climate change, algal overgrowth and coral recovery. PLoS One 4, e5239.

Donner, S.D., 2011. An evaluation of the effect of recent temperature variability on the prediction of coral bleaching events. Ecol. Appl. 21, 1718–1730.

Doropoulos, C., Ward, S., Roff, G., González-Rivero, M., Mumby, P.J., 2015. Linking demographic processes of juvenile corals to benthic recovery trajectories in two common reef habitats. PLoS One 10, e0128535.

Edmunds, P.J., 2019. Three decades of degradation lead to diminished impacts of severe hurricanes on Caribbean reefs. Ecology 100, e02587.

Edwards, C.B., Eynaud, Y., Williams, G.J., Pedersen, N.E., Zgliczynski, B.J., Gleason, A.C.R., Smith, J.E., Sandin, S.A., 2017. Large-area imaging reveals biologically driven non-random spatial patterns of corals at a remote reef. Coral Reefs 36, 1291–1305.

Ferrari, R., Bryson, M., Bridge, T., Hustache, J., Williams, S.B., Byrne, M., Figueira, W., 2016. Quantifying the response of structural complexity and community composition to environmental change in marine communities. Glob. Chang. Biol. 22, 1965–1975.

Fong, P., Glynn, P., 1998. A dynamic size-structured population model: does disturbance control size structure of a population of the massive coral *Gardineroseris planulata* in the Eastern Pacific? Mar. Biol. 130, 663–674.

Forsman, Z.H., Barshis, D.J., Hunter, C.L., Toonen, R.J., 2009. Shape-shifting corals: molecular markers show morphology is evolutionarily plastic in *Porites*. BMC Evol. Biol. 9, 45.

Fox, M.D., Williams, G.J., Johnson, M.D., Radice, V.Z., Zgliczynski, B.J., Kelly, E.L., Rohwer, F.L., Sandin, S.A., Smith, J.E., 2018. Gradients in primary production predict trophic strategies of mixotrophic corals across spatial scales. Curr. Biol. 28, 3355–3363. e3354.

Fox, M.D., Carter, A.L., Edwards, C.B., Takeshita, Y., Johnson, M.D., Petrovic, V., Amir, C.G., Sala, E., Sandin, S.A., Smith, J.E., 2019. Limited coral mortality following acute thermal stress and widespread bleaching on Palmyra Atoll, central Pacific. Coral Reefs 38, 701–712.

Gilmour, J.P., Smith, L.D., Heyward, A.J., Baird, A.H., Pratchett, M.S., 2013. Recovery of an isolated coral reef system following severe disturbance. Science 340, 69–71.

Golbuu, Y., Victor, S., Penland, L., Idip, D., Emaurois, C., Okaji, K., Yukihira, H., Iwase, A., Van Woesik, R., 2007. Palau's coral reefs show differential habitat recovery following the 1998-bleaching event. Coral Reefs 26, 319–332.

González-Rivero, M., Harborne, A.R., Herrera-Reveles, A., Bozec, Y.-M., Rogers, A., Friedman, A., Ganase, A., Hoegh-Guldberg, O., 2017. Linking fishes to multiple metrics of coral reef structural complexity using three-dimensional technology. Sci. Rep. 7, 13965.

Goreau, T., Macfarlane, A., 1990. Reduced growth rate of Montastrea annularis following the 1987–1988 coral-bleaching event. Coral Reefs 8, 211–215.

Gouezo, M., Golbuu, Y., Fabricius, K., Olsudong, D., Mereb, G., Nestor, V., Wolanski, E., Harrison, P., Doropoulos, C., 2019. Drivers of recovery and reassembly of coral reef communities. Proc. R. Soc. B 286, 20182908.

Gove, J.M., Williams, G.J., McManus, M.A., Heron, S.F., Sandin, S.A., Vetter, O.J., Foley, D.G., 2013. Quantifying climatological ranges and anomalies for Pacific coral reef ecosystems. PLoS One 8, e61974.

Gracias, N., Santos-Victor, J., 2000. Underwater video mosaics as visual navigation maps. Comput. Vis. Image Underst. 79, 66–91.

Green, D.H., Edmunds, P.J., Carpenter, R.C., 2008. Increasing relative abundance of *Porites astreoides* on Caribbean reefs mediated by an overall decline in coral cover. Mar. Ecol. Prog. Ser. 359, 1–10.

Grime, J., 1977. Evidence for the existence of three primary strategies in plants and its relevance to ecological and evolutionary theory. Am. Nat. 111, 1169–1194.

Habib, A.F., Kim, E.-M., Kim, C.-J., 2007. New methodologies for true orthophoto generation. Photogramm. Eng. Remote. Sens. 73, 25–36.

Harper, J.L., Rosen, B.R., White, J., 1986. The Growth and Form of Modular Organisms. Royal Society, London.

Heery, E.C., Hoeksema, B.W., Browne, N.K., Reimer, J.D., Ang, P.O., Huang, D., Friess, D.A., Chou, L.M., Loke, L.H., Saksena-Taylor, P., 2018. Urban coral reefs: degradation and resilience of hard coral assemblages in coastal cities of East and Southeast Asia. Mar. Pollut. Bull. 135, 654–681.

Highsmith, R.C., 1982. Reproduction by Fragmentation in Corals. Marine Ecology Progress Series, vol. 7 Oldendorf, pp. 207–226.

Holbrook, S.J., Adam, T.C., Edmunds, P.J., Schmitt, R.J., Carpenter, R.C., Brooks, A.J., Lenihan, H.S., Briggs, C.J., 2018. Recruitment drives spatial variation in recovery rates of resilient coral reefs. Sci. Rep. 8, 1–11.

Hughes, T.P., 1994. Catastrophes, phase shifts, and large scale degradation of a Caribbean coral reef. Science 265, 1547–1551.

Hughes, T.P., Connell, J.H., 1987. Population dynamics based on size or age? A reef-coral analysis. Am. Nat. 129, 818–829.

Hughes, T., Jackson, J., 1980. Do corals lie about their age? Some demographic consequences of partial mortality, fission, and fusion. Science 209, 713–715.

Hughes, T.P., Tanner, J.E., 2000. Recruitment failure, life histories, and long-term decline in Caribbean corals. Ecology 81, 2250–2263.

Hughes, T.P., Ayre, D., Connell, J.H., 1992. The evolutionary ecology of corals. Trends Ecol. Evol. 7, 292–295.

Hughes, T.P., Kerry, J.T., Baird, A.H., Connolly, S.R., Dietzel, A., Eakin, C.M., Heron, S.F., Hoey, A.S., Hoogenboom, M.O., Liu, G., 2018. Global warming transforms coral reef assemblages. Nature 556, 492.

Hughes, T.P., Kerry, J.T., Baird, A.H., Connolly, S.R., Chase, T.J., Dietzel, A., Hill, T., Hoey, A.S., Hoogenboom, M.O., Jacobson, M., 2019. Global warming impairs stock–recruitment dynamics of corals. Nature 568, 387–390.

Jackson, J., 1986. Modes of dispersal of clonal benthic invertebrates: consequences for species' distributions and genetic structure of local populations. Bull. Mar. Sci. 39, 588–606.

Jackson, J., Coates, A., 1986. Life cycles and evolution of clonal (modular) animals. Philos. Trans. R. Soc. Lond. B Biol. Sci. 313, 7–22.

Kayal, M., Lenihan, H.S., Brooks, A.J., Holbrook, S.J., Schmitt, R.J., Kendall, B.E., 2018. Predicting coral community recovery using multi-species population dynamics models. Ecol. Lett. 21, 1790–1799.

Kealoha, A.K., Shamberger, K.E., Reid, E.C., Davis, K.A., Lentz, S.J., Brainard, R.E., Oliver, T.A., Rappé, M.S., Roark, E.B., Rii, Y.M., 2019. Heterotrophy of oceanic particulate organic matter elevates net ecosystem calcification. Geophys. Res. Lett. 46, 9851–9860.

Knowlton, N., 2001. The future of coral reefs. Proc. Natl. Acad. Sci. U. S. A. 98, 5419–5425.

Kodera, S.M., Edwards, C.B., Petrovic, V., Pedersen, N.E., Eynaud, Y., Sandin, S.A., 2020. Quantifying life history demographics of the scleractinian coral genus Pocillopora at Palmyra Atoll. Coral Reefs 39, 1091–1105.

Leichter, J.J., Genovese, S.J., 2006. Intermittent upwelling and subsidized growth of the scleractinian coral Madracis mirabilis on the deep fore-reef slope of Discovery Bay, Jamaica. Mar. Ecol. Prog. Ser. 316, 95–103.

Lirman, D., Gracias, N.R., Gintert, B.E., Gleason, A.C.R., Reid, R.P., Negahdaripour, S., Kramer, P., 2007. Development and application of a video-mosaic survey technology to document the status of coral reef communities. Environ. Monit. Assess. 125, 59–73.

Lough, J., 2000. 1997–98: unprecedented thermal stress to coral reefs? Geophys. Res. Lett. 27, 3901–3904.

Loya, Y., Sakai, K., Yamazato, K., Nakano, Y., Sambali, H., van Woesik, R., 2001. Coral bleaching: the winners and the losers. Ecol. Lett. 4, 122–131.

Mackie, G., 1986. From aggregates to integrates: physiological aspects of modularity in colonial animals. Philos. Trans. R. Soc. Lond. B Biol. Sci. 313, 175–196.
Madin, J.S., Hoogenboom, M.O., Connolly, S.R., Darling, E.S., Falster, D.S., Huang, D., Keith, S.A., Mizerek, T., Pandolfi, J.M., Putnam, H.M., 2016. A trait-based approach to advance coral reef science. Trends Ecol. Evol. 31, 419–428.
Mangubhai, S., Nand, Y., Rotjan, R., 2019. Discovery of a recovering climax *Acropora* community in Kanton Lagoon in the remote Phoenix Islands Protected Area. Pac. Conserv. Biol. 25, 322–323.
McGill, B.J., Enquist, B.J., Weiher, E., Westoby, M., 2006. Rebuilding community ecology from functional traits. Trends Ecol. Evol. 21, 178–185.
McNamara, D.E., Cortale, N., Edwards, C., Eynaud, Y., Sandin, S.A., 2019. Insights into coral reef benthic dynamics from nonlinear spatial forecasting. J. R. Soc. Interface 16, 20190047.
Mollica, N.R., Cohen, A.L., Alpert, A.E., Barkley, H.C., Brainard, R.E., Carilli, J.E., DeCarlo, T.M., Drenkard, E.J., Lohmann, P., Mangubhai, S., Pietro, K.R., Rivera, H.E., Rotjan, R.D., Scott-Buechler, C., Solow, A.R., Young, C.W., 2019. Skeletal records of bleaching reveal different thermal thresholds of Pacific coral reef assemblages. Coral Reefs 38, 743–757.
Moritz, C., Vii, J., Lee Long, W., Tamelander, J., Thomassin, A., Planes, S. (Eds.), 2018. Status and Trends of Coral Reefs of the Pacific. Global Coral Reef Monitoring Network, IUCN.
Naughton, P., Edwards, C., Petrovic, V., Kastner, R., Kuester, F., Sandin, S., 2015. Scaling the annotation of subtidal marine habitats. In: WUWNet '15, October 22–24 2015, Washington DC, USA.
Nicosevici, T., Gracias, N., Negahdaripour, S., Garcia, R., 2009. Efficient three-dimensional scene modeling and mosaicing. J. Field Rob. 26, 759–788.
Pedersen, N.E., Edwards, C.B., Eynaud, Y., Gleason, A.C.R., Smith, J.E., Sandin, S.A., 2019. The influence of habitat and adults on the spatial distribution of juvenile corals. Ecography 42, 1–11.
Petrovic, V., Vanoni, D.J., Richter, A.M., Levy, T.E., Kuester, F., 2014. Visualizing high resolution three-dimensional and two-dimensional data of cultural heritage sites. Mediter. Archaeol. Archaeom. 20, 93–100.
Pianka, E.R., 1970. On r-and K-selection. Am. Nat. 104, 592–597.
Pinzón, J.H., LaJeunesse, T.C., 2011. Species delimitation of common reef corals in the genus *Pocillopora* using nucleotide sequence phylogenies, population genetics and symbiosis ecology. Mol. Ecol. 20, 311–325.
Pizarro, O., Eustice, R.M., Singh, H., 2009. Large area 3-D reconstructions from underwater optical surveys. IEEE J. Ocean. Eng. 34, 150–169.
Razak, T.B., Roff, G., Lough, J.M., Mumby, P.J., 2020. Growth responses of branching versus massive corals to ocean warming on the Great Barrier Reef, Australia. Sci. Total Environ. 705, 135908.
Riegl, B., Purkis, S., 2015. Coral population dynamics across consecutive mass mortality events. Glob. Chang. Biol. 21, 3995–4005.
Roff, G., Bejarano, S., Bozec, Y.-M., Nugues, M., Steneck, R.S., Mumby, P.J., 2014. *Porites* and the Phoenix effect: unprecedented recovery after a mass coral bleaching event at Rangiroa Atoll, French Polynesia. Mar. Biol. 161, 1385–1393.
Sebens, K.P., 1987. The ecology of indeterminate growth in animals. Annu. Rev. Ecol. Syst. 18, 371–407.
Smith, L.W., Wirshing, H., Baker, A., Birkeland, C., 2008. Environmental versus genetic influences on growth rates of the corals Pocillopora eydouxi and Porites lobata (Anthozoa: Scleractinia). Pac. Sci. 62, 57–69.
Szmant, A.M., 1986. Reproductive ecology of Caribbean reef corals. Coral Reefs 5, 43–53.

Tanner, J., Hughes, T., Connell, J., 2009. Community-level density dependence: an example from a shallow coral assemblage. Ecology 90, 506.
Tilman, D., 1994. Competition and biodiversity in spatially structured habitats. Ecology 75, 2–16.
Vardi, T., Williams, D.E., Sandin, S.A., 2012. Population dynamics of threatened elkhorn coral in the northern Florida Keys, USA. Endanger. Species Res. 19, 157–169.
Vermeij, M.J.A., Sandin, S.A., 2008. Density-dependent settlement and mortality structure the earliest life phases of a coral population. Ecology 89, 1994–2004.
Veron, J.E.N., 2000. Corals of the World. Australian Institute of Marine Science, Townsville, Australia.
Werner, P.A., Caswell, H., 1977. Population growth rates and age versus stage-distribution models for teasel (*Dipsacus sylvestris* Huds.). Ecology 58, 1103–1111.
Yeakel, K.L., Andersson, A.J., Bates, N.R., Noyes, T.J., Collins, A., Garley, R., 2015. Shifts in coral reef biogeochemistry and resulting acidification linked to offshore productivity. Proc. Natl. Acad. Sci. U. S. A. 112, 14512–14517.

CHAPTER EIGHT

Biophysical model of coral population connectivity in the Arabian/Persian Gulf

Geórgenes Cavalcante[a,b,*], Filipe Vieira[a], Jonas Mortensen[c], Radhouane Ben-Hamadou[d], Pedro Range[d], Elizabeth A. Goergen[d], Edmo Campos[a,f], and Bernhard M. Riegl[e]

[a]Department of Biology, Chemistry and Environmental Sciences, College of Arts and Sciences, American University of Sharjah, Sharjah, United Arab Emirates
[b]Instituto de Ciências Atmosféricas (ICAT), Universidade Federal de Alagoas, Maceió, Brazil
[c]DHI A/S. Agern Allé 5, Hørsholm, Denmark
[d]Department of Biological and Environmental Sciences, Qatar University, Doha, Qatar
[e]Department of Marine and Environmental Sciences, Halmos College of Arts and Sciences, Nova Southeastern University, Dania Beach, FL, United States
[f]Instituto Oceanografico, Universidade de São Paulo, São Paulo, Brazil
*Corresponding author: e-mail addresses: gcavalcante@aus.edu; georgenescavalcante@icat.ufal.br

Contents

Advances in Marine Biology, Volume 87
ISSN 0065-2881
https://doi.org/10.1016/bs.amb.2020.07.001

Abstract

The coral reef ecosystems of the Arabian/Persian Gulf (the Gulf) are facing profound pressure from climate change (extreme temperatures) and anthropogenic (land-use and population-related) stressors. Increasing degradation at local and regional scales has already resulted in widespread coral cover reduction. Connectivity, the transport and exchange of larvae among geographically separated populations, plays an essential role in recovery and maintenance of biodiversity and resilience of coral reef populations. Here, an oceanographic model in 3-D high-resolution was used to simulate particle dispersion of "virtual larvae." We investigated the potential physical connectivity of coral reefs among different regions in the Gulf. Simulations reveal that basin-scale circulation is responsible for broader spatial dispersion of the larvae in the central region of the Gulf, and tidally-driven currents characterized the more localized connectivity pattern in regions along the shores in the Gulf's southern part. Results suggest predominant self-recruitment of reefs with highest source and sink ratios along the Bahrain and western Qatar coasts, followed by the south eastern Qatar and continental Abu Dhabi coast. The central sector of the Gulf is suggested as recruitment source in a stepping-stone dynamics. Recruitment intensity declined moving away from the Straits of Hormuz. Connectivity varied in models assuming passive versus active mode of larvae movement. This suggests that larval behaviour needs to be taken into consideration when establishing dispersion models, and establishing conservation strategies for these vulnerable ecosystems.

1. Introduction

Significant fluctuations of population sizes are increasingly common in coral reef organisms. These can be outbreaks of predators, like Crown-of-Thorns Starfish (Shafir et al., 2008) or coral-eating snails (De'ath et al., 2012), or competitors such as brown algae (Mumby et al., 2007; Van den Hoek et al., 1995). Population collapses of fishes, corals or other organisms are increasing and frequently demonstrated (McClanahan et al., 2008; Robbins et al., 2006). An ever-growing human footprint leads to more resource-extraction by capture (fisheries, collection of ornamentals, etc.) or resource removal either by mortality events or by destruction of viable habitat (Cinner et al., 2018). Changes in populations of reproducing organisms have obvious repercussions on population dynamics and lead to feedbacks into upward- or downward cycles, or in the worst case, one-way trajectories of decline (Case, 2000; Caswell, 2001). Corals and fishes are organisms that show tendencies to get ever rarer on reefs as natural and man-made environmental impacts increase (Kayal et al., 2012).

If organismic populations on coral reefs are to survive in such difficult settings, first principles of population dynamics suggest that replenishment by natural recruitment will be key to survival (Case, 2000; Caswell, 2001; Hutchinson, 1978). Population fecundity, connectivity among populations, and levels of recruitment must, therefore, be known to understand and model the potential survival of coral reef organisms (Paris et al., 2007). This is no trivial task since recruits of virtually all organisms are small and difficult, if not impossible, to track unless sophisticated genetic tools are at hand (Baums et al., 2014; Berumen et al., 2019) to determine the provenance of propagules. While much progress has been made by using traditional field methods (settlement plates, larval capture, etc.; Bauman et al., 2012; Burt and Bauman, 2019), there remains a need to know more details and to better evaluate future scenarios. Also, some predictions regarding the likelihood of population connectivity based on demographic models (Riegl and Purkis, 2015; Riegl et al., 2017) can only be verified by sophisticated analyses of propagule flow among these populations.

Biophysical models of larval dispersion that integrate particle transport with physical oceanographic models have been used with much success to explain patterns in biodiversity (Cowen et al., 2000; Paris et al., 2007; Werner et al., 2007), invasion dynamics (Johnston and Akins, 2016; Johnston and Purkis, 2015), and potential pathways of larvae for population recovery (Cavalcante and Burt, 2016; Riegl et al., 2017, 2018). Such approaches bear much promise for understanding and predicting the future of marine resources in increasingly depleted coral reefs.

The Arabian/Persian Gulf (hereafter termed "Gulf") is a peripheral, epicontinental sea connected to the Indian Ocean and was, until relatively recently, home to large reef areas that have since been severely restricted (Riegl et al., 2018; Sale et al., 2011; Sheppard, 2016; Sheppard et al., 2010). A thriving economy and growing human population put massive pressure on the reefs of the region (Riegl and Glynn, 2020, in this volume), which is home to some of the world's largest coastal alteration (dredge-and-fill) and desalination projects (Sale et al., 2011; Sheppard et al., 2010; Van Lavieren et al., 2011). With habitat rapidly disappearing and the region being highly sensitive to climate extremes (Bauman et al., 2015; Burt et al., 2014; Riegl and Purkis, 2012), the future of most coral reef taxa and reef areas seems to depend on recruitment to replenish the frequent and severe losses (Pratchett et al., 2017; Riegl et al., 2017). The status, whether extant or soon extinct, of some ecosystem engineering species depends on the availability of upstream populations to seed and allow replenishment of depleted populations (Riegl et al., 2018).

This study uses a three-dimensional oceanographic model at 1 km resolution to model particle transport within the Gulf and to develop predictions of population connectivity. Previous biophysical models of settlement and population replenishment dynamics in this region were two-dimensional and considered only the ocean's surface layer with low resolution in the coastal areas (Cavalcante and Burt, 2016; Riegl et al., 2017, 2018). However, large areas of the Gulf harbour deeper (to >30 m) coral populations and the densest reef assemblages are close to the shoreline. This new model therefore represents an advancement in technical capability to develop much finer-grained scenario-models and predictions. In this paper we use a 3D high-resolution model that reproduces the circulation features of the Gulf, together with an agent-based model that reproduces the transport and biological processes of spawned eggs to first, (i) understand whether the larvae spawned at a certain area can travel to form coral colonies at other areas and, second (ii) to explore how distinct coral populations are connected within the Gulf, which is crucial to coral conservation efforts.

2. Material and methods

2.1 Study area

The Gulf is a shallow semi-enclosed marginal sea connected to the northern Indian Ocean through the narrow Straits of Hormuz (Chao and Al-Hajri, 1992; Sheppard et al., 1992; Yao and Johns, 2010). Although the Gulf is relatively wide (~810 km), its bathymetry is shallow and narrow, with an average depth (asymmetric along its axis) of 36 m, and an average width of 200 km (Emery, 1956; Kämpf and Sadrinasab, 2006). Due to its shallow nature and seafloor morphology, large-scale oceanographic circulation within the Gulf is driven by seasonally varying surface heat flux, freshwater flux and wind stress (Cavalcante and Burt, 2016; Kämpf and Sadrinasab, 2006; Reynolds, 1993; Swift and Bower, 2003).

2.2 Hydrodynamic modelling

A 3D flexible mesh model (DHI, 2013) was used to simulate hydrodynamic flow. This model was based on previous studies in the Gulf (Cavalcante and Burt, 2016; Cavalcante et al., 2012) and complex coastal regions (e.g. Jones et al., 2007; Lawrence et al., 2009; Sundelöf and Jonsson, 2012). A detailed description of the model can be found in Cavalcante and Burt (2016). Boundary conditions were generated by interpolating the predicted tidal constituents and u and v current speed, salinity and temperature from HYCOM (3-h intervals; https://www.hycom.org/) along the eastern limits of the model domain. Bathymetric data was derived from global bathymetry (SRTM30_PLUS; http://comlmaps.org/how-to/layers-and-resources/physical-environment/srtm30-plus-global-topography/) and produced a domain comprising of a horizontal grid at 1 km resolution, with a resulting mesh consisting of 18,990 nodes and 36,320 elements, and with 11 sigma layers in the vertical resolution. This domain extended northward from 23.7°N and westward from 58.0°E (Fig. 1A).

The model was driven by wind velocity, wind direction, and air pressure 2 m above sea level. For the relatively small, semi-enclosed basin of the Gulf, the higher spatial and temporal resolution allowed better representation of the local wind fields. Wind data from the Climate Forecast System Reanalysis (CFSR, Saha et al., 2010, 2014) were selected to force the MIKE model. These atmospheric variables were also applied within the model's calibration and validation period (April to May 2009). Wind and air pressure data had a resolution of $\sim 0.2^0$ ($\sim$20 km) and $\sim 0.5^0$ ($\sim$50 km), respectively. All atmospheric data were obtained at hourly intervals across the experimental period from the Research Data Archive held within the

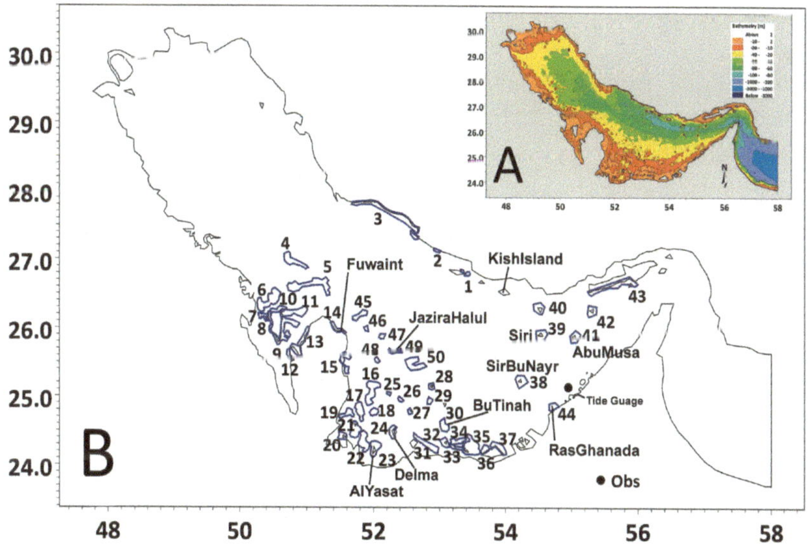

Fig. 1 Model domain and bathymetry (A), coral reef distribution within the model domain (B). Fifty reefs were selected covering areas depicted by the blue contours. The location of the moored station is indicated by a "black dot" ($\sim$17 m depth). Data from HYCOM model were extracted at the same location.

Computational and Information Systems Laboratory at the National Center for Atmospheric Research (Colorado; http://rda.ucar.edu).

2.3 Field data

Current speed and direction were collected during the period of 1 April to 31 May 2009, from a single 600 kHz Acoustic Doppler Current Profilers (ADCP). The ADCP was deployed at ~17 m depth within the offshore region of the Dubai coast on the southern Gulf area (25.179 N, 55.020E; Fig. 1B). The measured data were used to evaluate the performance of the 3D numerical model applied in this study. The comparison of observed and modelled results are presented in the model performance and validation section.

2.4 Model performance and validation

The sensitivity of the model to changes in bathymetry, wind stress and other physical forcings that may affect hydrodynamics was investigated in depth throughout 3D simulations. The model was calibrated and validated against in situ observations of (u) alongshore and (v) cross-shore currents measured at the moored station and tide level (Fig. 1B). Tide elevation measured at Jebel Ali Port (Fig. 1B) were used to further test and validate the model for the period April–May 2016. A demonstration of the model's performance against measured data at the offshore station (black circle; Fig. 1B), and tide elevation is provided in Fig. 2B.

The error metrics bias, root mean square error (RMSE), and correlation coefficient (r) were used to evaluate the (u) alongshore and (v) cross-shore currents, and tide level model results. The statistics are defined as:

$$Bias = \overline{y} - \overline{x} \tag{1}$$

$$RMSE = \sqrt{\frac{\sum_i (x_i - y_i)^2}{n}} \tag{2}$$

$$r = \frac{\sum_i ((x_i - \overline{x})(y_i - \overline{y}))}{\sqrt{\sum_i (x_i - \overline{x})^2 \sum_i (y_i - \overline{y})^2}} \tag{3}$$

where x_i and y_i represent the measured and modelled values, respectively, n the total number of observations, and $\overline{x}$, $\overline{y}$ are measured and modelled averages, respectively.

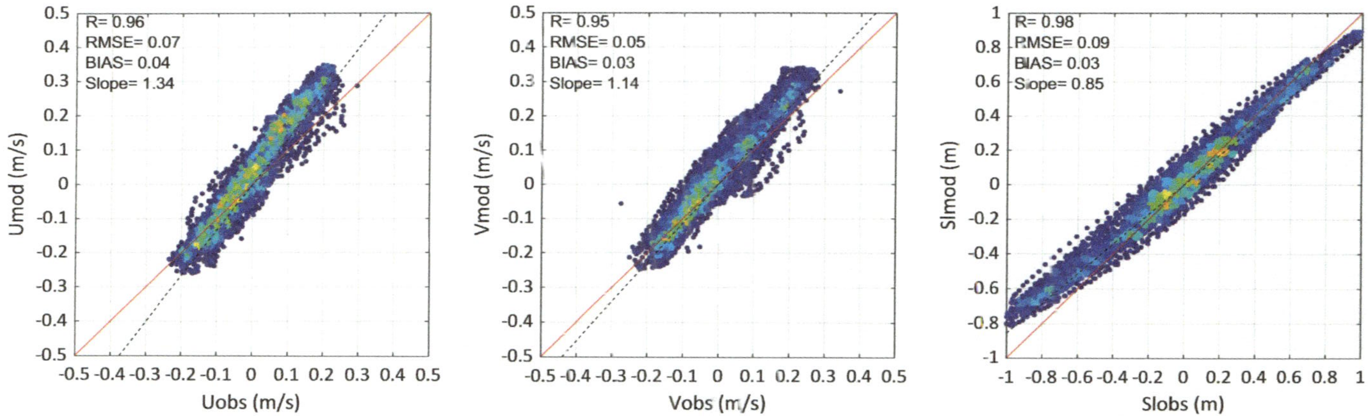

Fig. 2 Density scatter plots of (*u*) alongshore and (*v*) cross-shore currents for the offshore station, and tide level at Jebel Ali tide gauge against Mike 3D. ADCP data correspond to the period April–May 2009, and tide data is for April–May 2016.

2.5 Agent based model (ABM) description

In the model, transport by currents is calculated using a combination of advection and dispersion, the former determined by the current velocity predicted by the hydrodynamic model. The sub-scale mixing and dispersion of particles (larvae) is accounted for by including the MIKE built-in Scaled Eddy Viscosity formulation, with an applied scaling factor of 1 in the horizontal and vertical representing the movement of agents (larvae) in response to currents not resolved by the hydrodynamic model (MIKE 3 FM, 2013). The resulting transport of individual agents (larvae) is derived from a simple numerical solution to the Langevin equation which is typically applied for simulating advection-dispersion processes:

$$\Delta X = X_n - X_{n-1} = A(X_{n-1}, t_{n-1}) \cdot \Delta t + \sigma_L \cdot N(0, 1) \tag{4}$$

where n, $n-1$ (point in space), t is time, A is a drift term, $N(0, 1)$ is the standard normal distribution with a mean of 0 and a variance of 1, while σ_L is the standard deviation of the turbulent dispersion.

As the dispersal of broadcasting coral in its initial life stage is predominately determined by ocean currents and winds, the timing of the spawning event plays a key role in resulting distribution patterns. Regional differences in the timing of spawning exist, but in the Gulf it is well documented for *Acropora downingi*, *A.* cf. *clathrata*, *A.* cf. *valenciennesi*, *A. arabensis*, *Cyphastrea microphthalma*, *Platygyra daedalea*, *Dipsastraea pallida*, and *Porites harrisoni* (Al-Hashmi, 2012; Bauman et al., 2011; Fadlallah, 1996; Harrison, 1995; Howells et al., 2014, 2016), and usually occurs over a period of 1–3 days following a full moon in late April and May.

2.6 Model parameterization

All simulation runs were initiated with the same values of model parameters (Table 1). This was done to investigate the effect of the source location, substrate area and the effect of hydrodynamics between locations, to generate source and sink maps of distribution of competent larvae.

To settle successfully on suitable substrate in the model, each individual particle had to pass some conditional statements, such as:

1. Particle had to have gained competency (i.e. the capability to settle, defined in Table 1).
2. Particle x,y-coordinates in the given model time step had to coincide with a reef area.
3. Particle needed to be <0.5 m vertical distance above reef substrate.
4. Current velocity needed to be <0.5 m/s at the moment of settlement.

Table 1 List of the main user-definable constants in the model setup, and the respective value applied in the different model scenarios (Bode et al., 2019; Connolly and Baird, 2010; Ritson-Williams et al., 2009; Storlazzi et al., 2017).

Symbol	Description	Value	Unit
z_{min}	Minimum daily instantaneously mortality rate	0.01	day^{-1}
z_{max}	Maximum daily instantaneously mortality rat	0.3	day^{-1}
α_z	Shape parameter of the of the mortality curve	0.76	–
λ_z	Scale parameter of the mortality curve	0.1	–
v_h	Zygote positive buoyancy	0.0014	m/s
${}^{CG}max$	Maximum daily probability of competency gain	0.2	day^{-1}
CG_α	Shape parameter for the sigmoidal competency gain curve	2.5	–
CG_t	Time before a zygote experiences max. probability of competency gain.	5	days
${}^{I}min$	Minimum incubation time required before zygote can become a competent larvae	24	hours
${}^{CL}max$	Maximum daily probability of competency loss	0.3	day^{1}
CL_α	Shape parameter for the sigmoidal competency loss curve	0.2	–
CL_t	Time before a coral larvae experiences maximum probability of competency loss	45	days
v_s	Larvae settling speed	0.001	m/s
$dismax$	The maximum vertical distance from larvae to the seafloor (substrate) before settlement can occur	0.5	metre
${}^{v}max$	The maximum current velocity at which successful settlement can occur	0.5	m/s

2.7 Model setup

Based on available GIS dataset of reef extent in the Gulf, 50 reef groups were defined (Fig.1B). Each reef group acted as an individual larvae source and sink in each model scenario (50 individual simulations) to establish the local variation in dispersal and settlement patterns.

The area around Nayband Bay in Iran was chosen as the site of a first numeric larvae release experiment due to its northern-central position in the Gulf which made it ideal to study short- and long-distance travel (Fig. 1B—Reef 3). This area used to be rich in coral reefs but has lost virtually its entire coral cover (Fatemi and Shokri, 2001). Besides being instructive for understanding model behaviour, the model also illustrates loss of connectivity opportunity by elimination of this important central coral reef area.

Since mass spawning events in late April account for the majority of spawned larvae, this study concentrated on dispersal and connectivity patterns arising from the spawning event in April 2016 (22/04–24/04). Coincident with the modelled period, field investigations confirmed larvae settlement on artificial substrata (Burt and Bauman, 2019). The model assumed that most larvae will either settle or perish within the first ~45 days from spawning, despite a potential maximum lifespan of coral larvae >100 days (Graham et al., 2008). The chosen model period covered 17/04–10/06/2016 (45 days of larval competence period plus an initiation and termination buffer), which was sufficient to allow most larvae to settle on suitable substrata under suitable conditions. The drift profile of each particle was determined by data from the hydrodynamic model, with time steps of 300 s. A total of 10 million virtual particles were released per day at each source reef per simulation run, with a total of 30 million particles for the 3-day release period. While this may underestimate true larval productivity, it is a conservative estimate allowing to evaluate connectivity patterns with high confidence.

Larval exchange between reefs was logged to describe connectivity and to identify sources and sinks based on the basin-scale circulation from the Iranian coast towards Qatar and the southern Gulf (UAE). The model domain was restricted to the longitudes 50–56 E, the Gulf's eastern half. Hydrographic model and ABM simulation outputs were saved for every 60 min of model run.

3. Results

3.1 Hydrodynamic model validation

Comparison of model and validation data indicated good agreement for u and v parameters and tidal elevation. Statistical analysis demonstrated good performance of the model (Fig. 2). The correlation coefficients obtained were 0.96 (u), 0.95 (v), and 0.98 for tide elevation. RMSE for u, v, and tide elevation were 0.07 m/s, 0.05 m/s, and 0.09 m, bias were 0.04 m/s, 0.03 m/s and 0.03 m, respectively. The slopes were 1.34, 1.14, and 0.85, respectively.

3.2 Circulation patterns and sensitivity analysis (Passive versus active particles)

Coral larvae dispersed mainly along predominant flow. Based on the validated model, the typical circulation fields in the Gulf during the study period

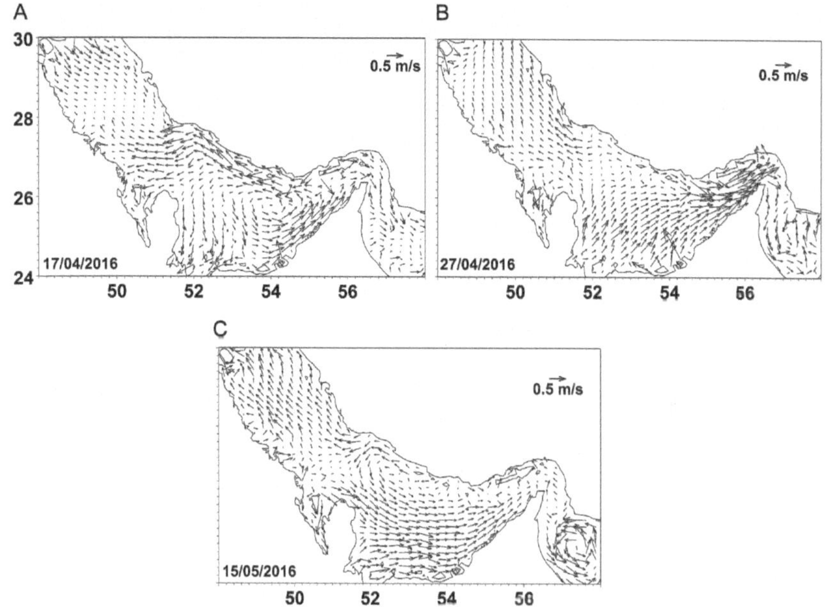

Fig. 3 Modelled current patterns for three distinct circulation features captured by the model during the coral larval simulation period that demonstrates the well represented eddy fields present in the Arabian Gulf during this time of the year, and the general basin-circulation. (A) Cyclonic eddy formation south of Kish Island on 14 April 2016 (core area at approximately 54°E and 26°N). (B) Small eddy field off Iranian coast on 27 April 2016 (core area at approximately 52.4°E and 27°N). (C) General circulation feature of the Gulf on 15 May 2016.

were identified, which allowed reliable estimates of larvae connectivity throughout the region. The model resolved the major circulation fields and the more variable eddy features for the period April–May–June (Fig. 3). Small and large variable eddy field patterns were observed throughout the simulation period. The first eddy had strong inflow (~0.5 m/s) along the Iranian coast and Kish Island (54.5°E, 26.3°N—Figs 1B and 3A), then flowing to the south towards SirBuNayr Island (54.2°E, 25.2°N—Figs 1B and 3A). A second eddy field with relatively weak currents (~0.15 m/s) was centred at 52.4°E and 27°N off the Iranian coast (Fig. 3B). The third pattern was characterized by the well-known general basin-circulation flowing from Iran to the south towards north-northeast Qatar and then easterly along Abu Dhabi (Fig. 3C).

A sensitivity analysis provided information on particle dispersion patterns in passive and active mode scenarios (Fig. 4A and B). In the passive mode (particles without biological attributes), the trajectories of particles spread

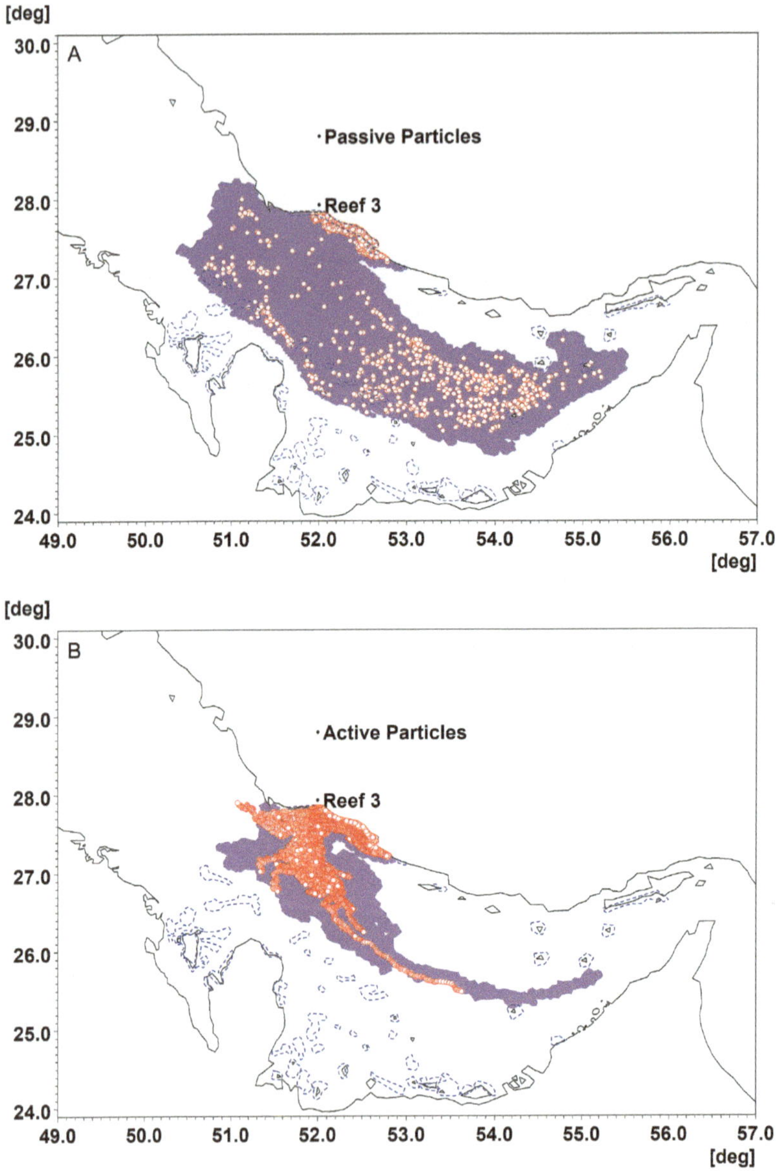

Fig. 4 Evolution of particle dispersion assuming no biological processes (passive particles) (A), and including biological processes (active particles—see Table 1 for details) (B) released from "Reef 3." The violet contour area represents the cumulative density of all virtual particles through time in the simulations demonstrating the larger spatial distribution in "larval location" under (A) only hydrodynamic forcings, than in (B) with life functions such as mortality, settlement and competency loss reducing larval density offshore the source area. The red/white dots represent locations of individual, randomly chosen particles.

throughout the Gulf initially towards Qatar, and then concentration of particles sparsened towards the southeast. The density pattern highlights the interactions between basin current patterns and the dynamics of larval trajectories, indicating the physical connectivity along different coral communities after 45 days of simulation (Fig. 4A).

In the active mode (AM; particles with biological attributes; Table 1), trajectories formed a denser plume with more localized distribution, particularly in the centre of the Gulf where larval movement decreased with increasing distance from the source area. Larvae then spread towards the southeast within a narrowing plume and particles exhibited less spread (Fig. 4B). Compared to the passive mode, trajectories affected by the biological parameters (AM) were substantially reduced. Despite the same particle release site at Reef 3, the mortality factor reduced the number of larvae in the water as well as their mobility, resulting in reduced transport distances (Fig. 4B). Hypothetical maximum potential dispersal range for the modelled spawning period is shown as a purple contour in Fig. 4. Allowing larvae to survive for longer periods may increase the number of days for hydrodynamic incorporation and lead to an increase in the average distance travelled by the particle even reaching regions in the southern Gulf.

3.3 Larvae dispersal patterns

Simulations showed that coral larvae could be transported distances up to 328 km in the central region of the Gulf (Fig. 5A—red line). The time and distance travelled depended on geographical location of the sources, hydrodynamic conditions, wind direction, and strength. For instance, particles released at Reef 3 (Fig. 5A) were generally pushed offshore by the semi-permanent anticlockwise basin-scale circulation that moves at an average speed of about 0.5 m/s, and then followed a generally narrow dispersion through the middle region of the Gulf (Fig. 5A). On the other hand, particles released along the east side of Qatar and near continental Abu Dhabi (Fig. 5A—Reef 23) tended to be affected by the weak residual tidal current with travel distances up to 52 km parallel to shore. The majority of the larvae were transported shorewards and towards reefs near the source areas, resulting in high self-seeding (recruitment on the natal reef). Similarly, simulations with source areas near Bahrain (reef 6—green in Fig. 5A) and western Qatar revealed that larvae did not travel far (~43 km). This was attributed to short drift time due to coastal geomorphology restricting the water circulation and dominant tidal currents from northwest causing the majority of larvae to be washed ashore.

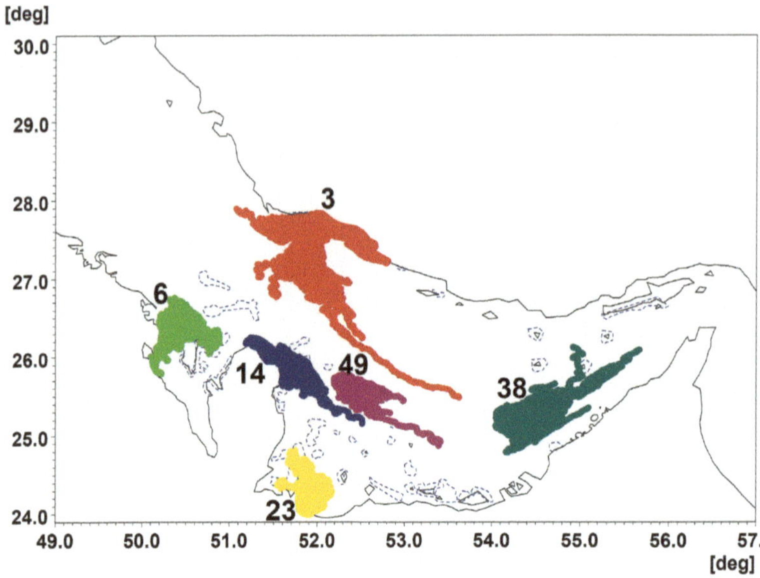

Fig. 5 Example modelled Lagrangian coral larval dispersal patterns from six reefs in the Persian/Arabian Gulf during the coral larval spawning event that demonstrates the effects of water circulation features and biological parameters shown in Figs 3 and 4B, respectively. Source reef numbers 14, 23, 38, and 49 represent Fuwaint, Al Yasat, SirbyNayr, and JaziraHalul reefs, respectively. The different colour lines show the modelled larval tracks of individuals that become competent, but did not necessarily recruit to a new reef by the end of the simulation. Note the range of dispersal patterns from a given reef under different oceanographic forcing, depending on their location and demonstrating the potential for a given reef to seed different regions and their respective reefs over relatively short timescales. Reef site 3 (Nayband Bay) now is largely dead—the serious implications of the loss of this area can be read from the large red area of larval coverage that is no longer realized in nature.

All released particles from the region of reefs 14 (Fuwaint—blue colour) and 49 (JaziraHalul—pink colour) were transported towards the southeast following the predominant water circulation (Fig. 5A). The larvae plume spread in the beginning only to restrict towards the end of the simulations. The average distance travelled by larvae from reefs 14, 49, and 38 in the cross-shore direction was shorter (~52, 55 and 76 km) than the alongshore distance (~133, 146, and 209 km). At reef 14 larvae were vulnerable to washing up on shore (Fig. 5A—light green colour) because coral patches were located nearshore on a wind-exposed coastline with low tidal current amplitude. Otherwise, larvae from reef 14 were affected by currents moving southward along the Qatar coast and the predominant eastward flow

(Fig. 5A—blue colour). At Reef 49, corals were situated on a wind-exposed shoreline with strong energy, coupled with generally moderate 0.4 m/s to high 0.8 m/s current speeds from the anticyclonic Gulf circulation (Fig. 5A—pink colour). Larvae from reef 38 dispersed towards the Straits of Hormuz corresponding with formation of small-scale eddies during May–June, which also made larvae vulnerable to washing up on shore (Fig. 5A—green colour). This area is also exposed to strong winds, the dominant outflow circulation from the Gulf towards the Indian Ocean, and the less intense tide-generated current velocity. Documented larval trajectories suggested that coral larvae could either (1) drift long distances and seed large areas, or (2) remain fairly close to the natural sources, with many washing up on shore. The dominant dispersal pattern was further investigated.

Although the model successfully simulated larval dispersal to several reefs (Fig. 6A), only few areas were indeed settled in the simulations (Fig. 6B). Larvae were predicted to gain competence on average 5 days after release and to perish after 45 days, thus allowing them to move long enough distances for patch connectivity. However, although the cumulative larvae density from reef source 14 indicated that 9 reefs could be reached excluding self-seeding (namely reefs 15, 16, 29, 30, 45, 46, 48, 49, 50—Fig. 6A), settlement occurred only on five reefs including self-seeding (reefs 14, 15, 16, 46 and 48—Fig. 6B). The longest distance travelled from reef 14 was ~260 km, potentially reaching as far as reef 30 (Bu Tinah—Fig. 6A), and the reef with highest settlement (reef 16—Fig. 6A) was located about 98 km from source reef 14. Such results are supported by the model's sensitivity analysis and reveal the substantial effect of choice of parameters on movement and the potential for settlement. The number of thresholds required for larvae to allow settlement once competent (whether within a model cell with substrate, maximum particle speed, minimum depth of substrate) and the daily mortality rate significantly influenced larvae loss and dispersal range in the model area.

3.4 Reef connectivity pathways

A connectivity matrix was developed to display the modelled larval connectivity among reefs (Fig. 7). A persistently high degree of self-seeding was observed (Fig. 7). This suggests that many reefs retain most larvae under the modelled conditions and coral populations are thus either sustained by their own larvae or those from nearby reefs. The potential for connectivity to distant reefs also existed (Figs 5 and 6A) but was the minority of cases.

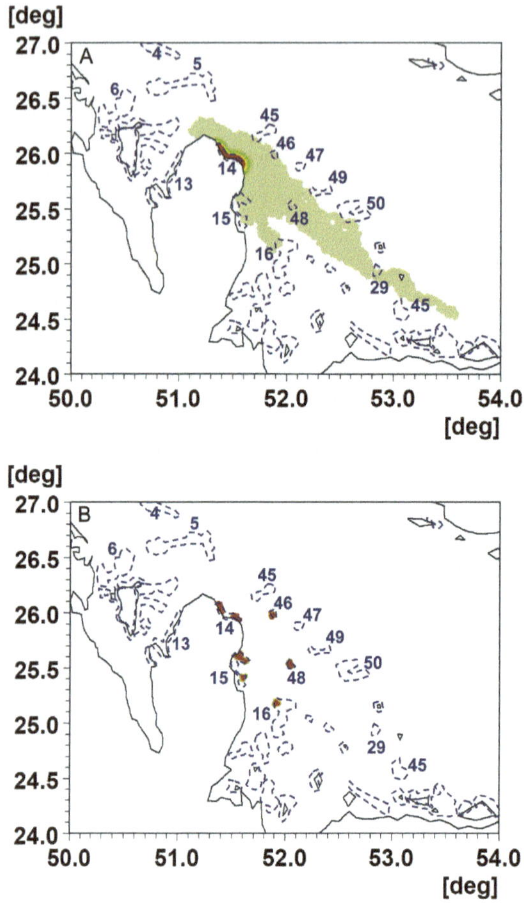

Fig. 6 Example modelled Lagrangian coral larval dispersal patterns from reef 14 (Fuwaint) on the northern Qatar coast (A) that demonstrate the potential range of dispersal and the settlement reefs (B). The green contour shows the cumulative area of all larval tracks (A), while the red contours denote successful larval recruitment to new reefs (B). Dispersal reached nine reefs not considering self-seeding (15, 16, 29, 30, 45, 46, 48, 49, 50), but coral larvae only successfully settled on five reefs, including self-seeding, which indicates the effects of biological parameters, especially the mortality rate, on success of seeding other reefs.

The downstream connectivity matrix showed 43 out of 50 source reefs retaining >50% of released larvae. Of these reefs, half retained up to 90% of their larvae (Fig. 7) and 6 retained 100% of their larvae (Iranian coast (3), north of Bahrain (4), AbuMusa (41), RasGhanada (44), northeast of Qatar (48) and central sector of the Gulf (50)—see Fig. 1B for reference). The main potential source areas for larvae travelling to other reefs were near the Bahrain coast and west of Qatar (highest source strength

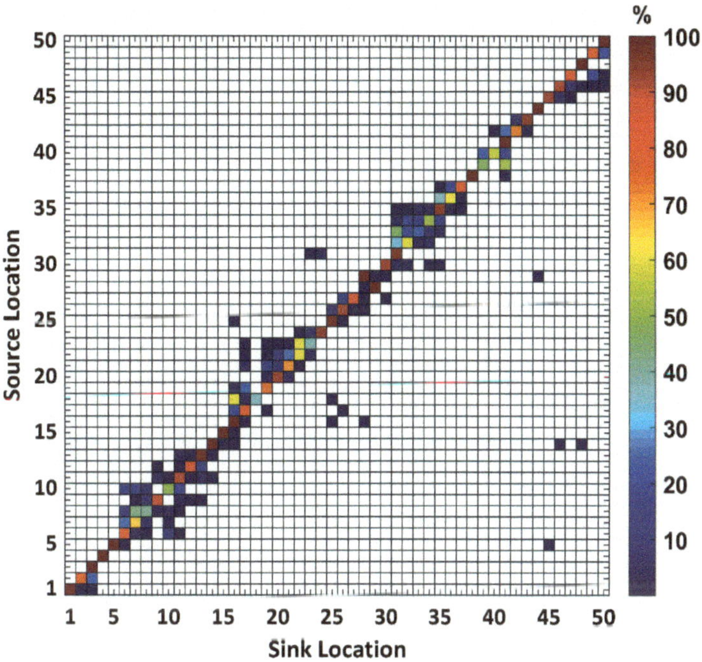

Fig. 7 Matrix representation of modelled coral larval connectivity between source reefs and reef sinks in the Arabian Gulf during the spawning event in April 2016. Colours represent the percentage of larvae spawned from a given source reef settling on another reef (note the non-linear scale); the diagonal dashed line from the bottom left to top right designates self-seeding reefs.

on reefs 6–13, including self-seeding), followed by the southeast Qatar coast and nearshore Abu Dhabi (reefs 19–23, including self-seeding), and the region south of BuTinah reef, where a total of 7 reefs (31–37), exhibited a marked source-sink dynamics with adjacent reefs (Fig. 7). The Iranian coast with reefs 1–3, 38 and 50, had the weakest source potential (Fig. 7). Similarly, the region north of Bahrain (reef 4–5) and near the Strait of Hormuz (reefs 40 and 42) provided few larvae to other reefs. Reef 29 in the central lower sector of the Gulf was the recruitment source to the most distant reef, its larvae travelled ~172 km to seed reef 44 (RasGhanada) on the coast of Abu Dhabi (Fig. 7). 52.9% of larvae originated at Siri reef (41) were transported to AbuMusa (41), while 47.1% remained in Siri, suggesting the influence of the anticlockwise circulation typically formed during this period of the year. Overall, reef 35 (seeded 7 reefs), and reefs 8, 9, 11, 22, 23 (AlYasat), with a total of 6 seeded reefs were the larval sources for the greatest numbers of reefs in the Gulf.

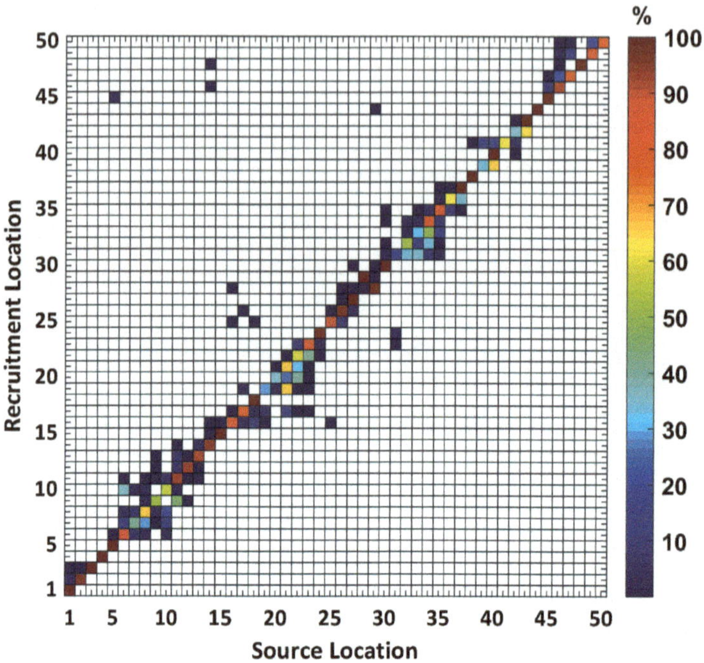

Fig. 8 Matrix representation of modelled coral larval connectivity between recruitment reefs and source reefs in the Gulf during the spawning event in April 2016. Colours represent the percentage of larvae recruited by a given reef from another reef (note the non-linear scale); the diagonal dashed line from the bottom left to top right designates self-recruiting reefs.

Variability of recruit origin at any given reef was evaluated by calculating percent contribution of each source to the total number of recruited larvae (upstream connectivity matrix, Fig. 8). Four reef clusters had high recruitment rates from neighbouring reefs within the same source areas. These same regions also had high values of self-seeding. The highest sink or recruitment concentration areas were along the Bahrain coast and west of Qatar (reefs 6–13, much self-seeding and recruiting to neighbouring reefs) (Fig. 8). The area of reefs 19–23 indicated the second densest recruitment concentration (southeast Qatar coast and nearshore Abu Dhabi). The continental Abu Dhabi coast (reefs 31–35) also received many larvae mostly from adjacent reefs. Reefs 49 and 50 can be considered a joint recruitment area (Fig. 8). Reefs located on the northeast coast of Bahrain (reef 11), and east of Qatar (reefs 16 and 17), were the larval sinks for the greatest numbers of reefs in the Gulf receiving larvae from seven different sources. Reef 19, along the southeast Qatar coast and reefs 31 and 35, on the Abu Dhabi coast received larvae from 6 different sources (Fig. 8). Some source reefs that did not seed other reefs were sink/recruitment areas receiving from other sites (Fig. 8). AbuMusa (41) seeded no other reefs but received larvae from five, including itself (38, 39, 40, 41 and 42). Similarly, reef 50 in the Gulf's central sector did not seed but received larvae from 4 reefs, including itself (46, 47, 49 and 50; Fig. 8; Table 2). Except for reef 41 (AbuMusa), recruitment intensity declined along reefs 38–43 in a direction moving away from the Straits of Hormuz (Fig. 8). The majority of the reefs were characterized by self-seeding, except AlYasat (23) which sent 60.06% of its larvae to settle on reef 22, and JaziraHalul (49) with sent 20.44% of its larvae to recruit on reef 50 (Table 2). Although self-seeding was high among sink reefs, AbuMusa also received recruits from reefs 39, 40 and 42 (Table 2).

Overall, settlement success was estimated by the percentage of larvae that settled in any given reef corresponding to the total amount released by each reef (Fig. 9). Due to the model's parameters, most larvae were lost during the simulation period and only reefs 6, 13 and 36 achieved a settlement success of approximately 11%, including self-seeding. The maximum percentage of settlement success dropped to 5.5% (reefs 8 and 10), 5.7% (reef 34), and 6.7% (reef 29) after removing the self-seeding contribution. The average success of delivery to reefs over these modelled competency periods and the effects of constants affecting movement and survival of larvae was 5.8% and 1.4% with and without self-seeding contribution, respectively. Altogether, this documents the prevalence of self-seeding for the maintenance of coral populations on Gulf reefs. Some level of mixing of populations was also achieved by successful recruitment to other, mostly nearby, reefs (Fig. 7).

4. Discussion

Coral connectivity in the Arabian/Persian Gulf was investigated using a coupled 3D Hydrodynamic and Agent Based Model to evaluate the degree of connectivity among reef areas. A predominant "self-seeding" dynamics with high rates of larval retention to natal reefs and some contribution from nearby reefs appear the primary mechanisms sustaining the Gulf's coral populations. But the study also found the potential for larvae to disperse over long distances, albeit rarely and dependent on specific sets of physical and survival conditions.

Coral species have variable strategies for reproduction and dispersal, however, most Gulf corals are broadcast spawners (Bauman et al., 2012).

Table 2 Connectivity matrix showing the percentage of source/recruitment dynamics from modelled coral larvae particles released from several known reefs with percentage of settled larvae at each reef group, and correspondent natal reef that were recruited by each known reef.

Reef	Reef Number (RN)	Source (sends larvae to) RN	Source (sends larvae to) %	Recruitment (receives larvae from) RN	Recruitment (receives larvae from) %
Fuwaint	14	14	99.32	11	1.21
		15	0.32	13	4.27
		16	0.05	14	94.52
		46	0.09	–	–
		48	0.23	–	–
AlYasat	23	17	0.47	22	10.87
		19	0.08	23	87.51
		20	0.04	24	1.59
		21	0.95	31	0.03
		22	60.05	–	–
		23	38.42	–	–
BuTinah	30	30	93.72	27	0.02
		31	3.34	29	0.06
		32	2.29	30	99.92
		34	0.46	–	–
		35	0.20	–	–
SirBuNayr	38	38	99.85	38	100
		41	0.15	–	–
AbuMusa	41	–	–	38	0.25
		–	–	39	14.41
		–	–	40	17.12
		–	–	41	62.21
		–	–	42	6.01
RasGhanada	44	44	100	29	0.02
		–	–	44	99.98
JaziraHalul	49	49	79.56	46	1.97
		50	20.44	47	11.75
		–	–	49	86.28

The settlement/recruitment percentage value is relative to the total number of coral larvae that settled at each reef and the number that were recruited at each reef group. RN=Reef Number.

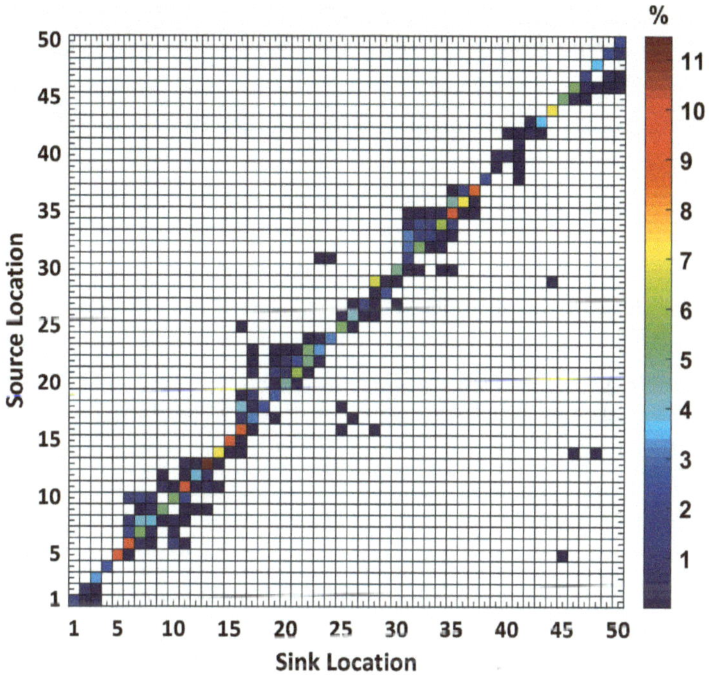

Fig. 9 Connectivity matrix showing the percentage of success modelled coral larvae particles released from 50 source reefs and correspondent sinking settling reefs. The settlement success value is relative to the total number of coral larvae particles released at each source reef and the number that settled at each reef group.

These corals release large quantities of gametes directly into the water column for fertilization (Lipcius et al., 2008). Bundles of positively buoyant gametes will then float to the surface where their dispersal, given their limited mobility, is almost entirely dependent on water currents (Black, 1993). Dynamics of the presented model compared well to known Gulf oceanography (Alosairi et al., 2011; Thoppil and Hogan, 2009, 2010a,b) and thus allowed a realistic focus on larval transport in realistic oceanographic model conditions. This model successfully expanded upon others previously used by the authors and significantly adds to the precision in the estimation of recruitment pathways in the Gulf (Cavalcante and Burt, 2016; Riegl et al., 2017, 2018).

The strong influence of hydrodynamics and atmospheric fields during the period of simulation (April–June 2016) was not unexpected and clearly suggests that strong inter-annual variability in recruitment dynamics must be expected. The shallow nature of the Gulf and the surrounding topography make its currents particularly sensitive to atmospheric forcings (Burt et al., 2019; Thoppil and Hogan, 2010a,b). Therefore, some variability in the observed reef connectivity patterns can be expected particularly in years with strong atmospheric anomalies. Winds in the Gulf are strongly seasonal and can increase dramatically during so-called Shamal events (north winds in Arabic) reaching 10–25 m/s, and lasting up to several days (El-Sabh and Murty, 1989; Riegl and Purkis, 2012). During April and July, the coastal flow along the Iranian coast increases and weakening of winds was regarded as one the factors for increasing inflow into the Gulf (Thoppil and Hogan, 2010a,b). Thus, variability of wind strength will impact the development of large/regional circulation features in the central Gulf and will play an important role on larval connectivity patterns. Further research on the role of wind-induced circulation on larvae transport and dispersion dynamics is warranted and underway.

Larvae dispersion plumes differed between passive and active mode scenarios. This indicated a potential for connectivity among distant area dependent on the combination of physical forcings (currents and winds) and larval biology. Differences between the passive and active dispersion models can be explained by the longer persistence of passive particles not subjected to the stringent biological criteria (daily mortality, settlement, etc., Table 1) in the active mode and an additional strong influence of daily tidal currents in passive mode. This caused the cumulative passive dispersion plume to occupy a larger area than predicted by just the prevailing Gulf circulation, forcing larvae towards the southeast region. In active mode, initial dispersion resembled that of passive mode, but as simulations progressed, the number of available particles in the dispersion plume were reduced. Larval density at the periphery of the plume was lower, and therefore the "daily mortality" parameter in active mode rapidly caused thinning in these peripheral areas and their eventual demise. The higher number of particles in the central plume allowed its persistence and transport by the prevailing Gulf currents for an extended period and longer distance before larvae settled or perished.

Not only oceanographic variability, but also choice of parameters in the ABM influenced model outcomes. Sensitivity analyses on the parameters "incubation period," "daily mortality rate" and "maximum competence period" indicated no major changes in connectivity patterns. A high degree of larvae retention at natal reefs was always observed which reaffirms the overriding role of physical forcing. Parameters characterizing larval life history were, however, fairly generic since based on relatively few studies (Bode et al., 2019; Connolly and Baird, 2010; Ritson-Williams et al.,

2009; Storlazzi et al., 2017) of which most were not specific to the Gulf. Since Gulf reefs harbour ~70 species of coral (Riegl and Purkis, 2012; Sheppard, 2016), differences in larval behaviour with resultant differences in dispersal and reef connectivity should be expected. Our results therefore represent fairly generic connectivity patterns based on a fairly generic coral larvae type. More specific results can be obtained once more species-specific data about larval survival rates, competency, and settlement dynamics become available.

Despite all caveats and necessary shortcomings, the model results resonate well with predictions and hypotheses from other studies that could so far not be verified. Riegl and Purkis (2015) observed that populations in five dominant Gulf coral species at Ras Ghanada (UAE) appeared to have reached a stable size distribution defined by a matrix population model after ~10 disturbance-free years. Since stable state can only be reached in a closed population, this predicted the Ras Ghanada population to be largely self-seeding which was now verified by the present model. Ras Ghanada did not receive propagules from any other reef in the study period, and only provided larvae to the reefs immediately nearby (Table 2). Such new knowledge allows to draw important conclusions about patterns of coral population dynamics and degradation in the Gulf. First, the high levels of self-seeding suggest that regeneration after mass mortality events must be primarily from survivors. This helps explain how demonstrated changes in community structure, driven by a population spin-down of formerly ecosystem engineering species (Riegl et al., 2018), could happen in the space of only two decades. Strongly localized larval dynamics would favour recruitment by species surviving disturbances with a remaining large population. Regular and intense recruitment by these species into their natal reef would allow them to reach stable state population structure relatively quickly. The previously dominant species which suffered far greater mortality and thus greater (or even complete) depression in local larvae availability, *Acropora downingi*, would have to await one of the rarer long-range recruitment events. That species would not only recruit later and in lower density, due to the long travel distance of its larvae, but also less recruitment space would be available due to increased space competition by the survivors. Such dynamics may explain an observed lack of regeneration in tabular *Acropora* in the Gulf, which contrasts to other regions with different recruitment dynamics (Pratchett et al., 2020).

Physical connectivity between more distant regions depended on larval mortality rate in the model. Broadcast spawners with long-lived larvae, like most Gulf corals (Bauman et al., 2011; Howells et al., 2014), should have the potential of long-range dispersal. The relatively low expected rates of successful long-distance dispersal would increase if longer larval survival than assumed in the model could be demonstrated. At present, arguments for long-range connectivity need to invoke "stepping-stone dynamics." Corals may settle on an easily reachable reef, and in later generations, their offspring may move to reefs further away (e.g. in Fig. 5, corals could step from region 14 to 39 and from there to 48). Via several recruitment events across several stepping stones, the present study suggests that reefs in Iran may connect to Qatar, and those via the Abu Dhabi offshore islands and banks to the UAE coast (Fig. 4–6). This may be critical to maintaining populations that suffered heavy mortality in the southern Gulf, such as in *Acropora* species. Such dynamics may have assisted *Acropora* to regenerate rapidly after widespread mass mortality in 1964 (Shinn, 1976), 1981/2 and 1996/8 (Riegl, 1999).

Global change may interfere with long-range dispersal. Larval pelagic phase shortens and survival declines with rising ocean temperature (Figueiredo et al., 2014) suggesting that assumptions of mostly local recruitment are likely to be ever more important. Long-range dispersal in the Gulf may lead to connectivity across temperature gradients which generates its own challenges (Kleypas et al., 2016). Particularly if Iranian corals recruit to the southern Gulf, the parent colonies may have been adapted to generally more benign environmental conditions and larvae may not carry suitable genetic adaptation to survive at their place of settlement. The Gulf therefore is an open-air laboratory to study the positive, neutral, or negative effects of long-range versus local connectivity. Models like ours quantify the pathways, giving ecologists a better framework for explaining observations and designing experiments.

5. Conclusions

In this study, the developed 3D hydrodynamic model performed as expected for the simulated period. The model was able to simulate the basin-scale circulation and reproduce the observed characteristics of the eddies typical for the period of study, and agreed with in situ observations, the HYCOM database, and other studies on the circulation-driven mechanisms in the Gulf. The 3D model highlighted the importance of atmospheric fields on circulation patterns and the understanding of local circulation characteristics. The hydrodynamic simulations demonstrated the potential for interpreting the key identified physical connectivity pathways of the distinct regions with regards to coral reefs in the Gulf. Coral larvae have the potential to be dispersed long distances under basin-scale currents and eddies, but tend to shorten the travel distance under localized tidal-driven currents.

Acknowledgements

This work was made possible by NPRP grant # [10-0205-170342] from the Qatar National Research Fund (a member of Qatar Foundation). The findings achieved herein are solely the responsibility of the authors. This work was partially supported by the American University of Sharjah [grant number FRG19-M-G74]. The authors would also like to thank the anonymous reviewers for their valuable comments that contributed to the final article. Wind and atmospheric pressure series data were sourced from the Research Data Archive (RDA) which is maintained by the Computational and Information Systems Laboratory (CISL) at the National Center for Atmospheric Research (NCAR). NCAR is sponsored by the National Science Foundation (NSF). We are grateful to DHI Water & Environment, Brazil, for allowing us to use their software MIKE 3.

References

Al-Hashmi, A., 2012. Coral reef reproduction and its features in the Arabian gulf (Jebel Ali, UAE). Int. J. Environ. Sustain. 1 (3), 12–21.

Alosairi, Y., Imberger, J., Falconer, R., 2011. Mixing and flushing in the Persian Gulf (Arabian Gulf). J. Geophys. Res. 116, C03029.

Bauman, A.G., Baird, A.H., Cavalcante, G.H., 2011. Coral reproduction in the world's warmest reefs: southern Persian Gulf (Dubai, United Arab Emirates). Coral Reefs 30, 405–413.

Bauman, A.G., Feary, D.A., Heron, S.F., Pratchett, M.S., Burt, J.A., 2012. Multiple environmental factors influence the spatial distribution and structure of reef communities in the northeastern Arabian Gulf. Mar. Pollut. Bull. 72, 302–312.

Bauman, A.G., Guest, J.R., Dunshea, G., Low, J., Todd, P.A., 2015. Coral settlement on a highly disturbed equatorial reef system. PLoS One 10 (5), e0127874. https://doi.org/10.1371/journal.pone.0127874.

Baums, I.B., Devlin-Durante, M., Laing, B.A.A., Feingold, J.S., Smith, T.B., Bruckner, A., Monteiro, J., 2014. Marginal coral populations: the densest known aggregation of *Pocillopora* in the Galapagos Archipelago is of asexual origin. Front. Mar. Sci. 1, 1–11.

Berumen, M.L., Roberts, M., Sinclair-Taylor, T., Dibattista, J., Saenz-Agudelo, P., Isari, S., He, S., Khalil, M., Hardenstine, R., Tietbohl, M., Priest, M., Kattan, A., Coker, D., 2019. Fishes and connectivity of Red Sea coral reefs. In: Voolstra, C.R., Berumen, M.L. (Eds.), Coral Reefs of the Red Sea. Springer, pp. 157–179. https://doi.org/10.1007/978-3-030-05802-9_8.

Black, K.P., 1993. The relative importance of local retention and inter-reef dispersal of 553 neutrally buoyant material on coral reefs. Coral Reefs 12, 43–53.

Bode, M., Leis, J.M., Mason, L.B., Williamson, D.H., Harrison, H.B., 2019. Successful validation of a larval dispersal model using genetic parentage data. PLoS Biol. 17 (7), e3000380.

Burt, J.A., Bauman, A.G., 2019. Suppressed coral settlement following mass bleaching in the southern Persian Gulf. Aquat. Ecosyst. Health Manage. 1–9.

Burt, J.A., Van Lavieren, H., Feary, D.A., 2014. Persian Gulf reefs: an important asset for climate science in urgent need of protection. Ocean Challenge 20, 49–56.
Burt, J.A, Paparella, F., Al-Mansoori, N., Al-Mansoori, A., Al-Jailani, H. 2019. Causes and consequences of the 2017 coral bleaching event in the southern Persian/Arabian Gulf. Coral Reefs. 38(4), pages567–589. https://doi.org/10.1007/s00338-00019-01767-y.
Case, T., 2000. An Illustrated Guide to Theoretical Ecology. Oxford University Press, 449 pp.
Caswell, H., 2001. Matrix Population Models. Sinauer, Sunderland.
Cavalcante, G.H., Feary, D.A., Burt, J.A., 2016. The influence of extreme winds on coastal oceanography and its implications for coral population connectivity in the southern Arabian Gulf. Mar. Pollut. Bull. 105, 489–497.
Cavalcante, G.H., Kjerfve, B., Feary, D.A., 2012. Examination of residence time and its relevance to water quality within a coastal mega-structure: the Palm Jumeirah Lagoon. J. Hydrol. 468–469, 111–119.
Chao, S.Y., Kao, T.W., Al-Hajri, K.R., 1992. A numerical investigation of circulation in the Arabian Gulf. J. Geophys. Res. 97, 219–236.
Cinner, J.E., Maire, E., Hutchery, C., MacNeil, A., Graham, N.A.J., 2018. Gravity of human impacts mediates coral reef conservation gains. Proc. Natl. Acad. Sci. U. S. A. 115 (27), E6116–E6125.
Connolly, S.R., Baird, A.H., 2010. Estimating dispersal potential for marine larvae: dynamic models applied to scleractinian corals. Ecology 91 (12), 3572–3583.
Cowen, R.K., Paris, C.B., Srinivasan, A., 2000. Scaling of connectivity in marine populations. Science 311 (5760), 522–527.
De'ath, G., Fabricius, K.E., Sweatman, H., Puotinen, M., 2012. The 27-year decline of coral cover on the great barrier reef and its causes. Proc. Natl. Acad. Sci. U. S. A. 109, 17995–17999.
DHI, 2013. Water & Environment, MIKE 21 and MIKE 3 Flow Model FM Hydrodynamic Module. DHI Group.
El-Sabh, M.I., Murty, T.S., 1989. Storm surges in the Arabian Gulf. Nat. Hazards 1 (4), 371–385.
Emery, K.O., 1956. Sediments and water of the Persian Gulf. AAPG Bull. 40, 2354–2383.
Fadlallah, Y.H., 1996. Synchronous spawning of *Acropora clathrata* coral colonies from the western Arabian Gulf (Saudi Arabia). Bull. Mar. Sci. 59, 209–216.
Fatemi, S.M.R., Shokri, M.R., 2001. Iranian coral reefs status with particular reference to Kish Island, Persian Gulf. In: Proceedings of the International Coral Reef Initiative Indian Ocean Regional Workshop, November 26–28, Mozambique, pp. 1–13.
Figueiredo, J., Baird, A.H., Harii, S., Conolly, S.R., 2014. Increased local retention of coral reef larvae as a result of ocean warming. Nat. Clim. Chang. 4, 498–502.
Graham, E.M., Baird, A.H., Connolly, S.R., 2008. Survival dynamics of scleractinian coral larvae and implications for dispersal. Coral Reefs. 27, 529–539.
Harrison, P.L., 1995. Status of the Coral Reefs of Kuwait. Final Rep. U. N. Ind. Dev. Organ. U. N. Dev. Programme UNIDO UNDP Vienna.
Howells, E.J., Abrego, D., Vaughan, G.O., Burt, J.A., 2014. Coral spawning in the Gulf of Oman and relationship to latitudinal variation in spawning season in the Northwest Indian Ocean. Sci. Rep. 4, 7484.
Howells, E.J., Ketchum, R.N., Bauman, A.G., Mustafa, Y., Watkins, K.D., Burt, J.A., 2016. Species-specific trends in the reproductive output of corals across environmental gradients and bleaching histories. Coral Reefs Arab. 105, 532–539.
Hutchinson, G.E., 1978. An Introduction to Population Ecology. Yale University Press, 260 pp.
Johnston, M.W., Akins, J.L., 2016. The non-native royal damsel (*Neopomacentrus cyanomos*) in the southern Gulf of Mexico; an invasion risk? Mar. Biol. 163 (1), 1–14.

Johnston, M.W., Purkis, S.J., 2015. A coordinated and sustained international strategy is required to turn the tide on the Atlantic lionfish invasion. Mar. Ecol. Prog. Ser. 533, 219–235. https://doi.org/10.3354/meps11399.

Jones, O.P., Petersen, O.S., Kofoed-Hansen, H., 2007. Modelling of complex coastal environments: some considerations for best practice. Coast Eng. 54, 717–733.

Kämpf, J., Sadrinasab, M., 2006. The circulation of the Persian Gulf: a numerical study. Ocean science, 2, 1-15.Lobel, P.S. 2011. Transport of reef larvae by an ocean eddy in Hawaiian waters. Dyn. Atmos. Oceans 52, 119–130.

Kayal, M., Vercelloni, J., Lison de Loma, T., 2012. Predator crown-of-thorns starfish (*Acanthaster planci*) outbreak, mass mortality of corals, and cascading effects on reef fish and benthic communities. PLoS One 7 (10), e47363.

Kleypas, J.A., Thompson, D.M., Castruccio, F.S., Curchitser, E.N., Pinsky, M., Watson, J/R., 2016. Larval connectivity across temperature gradients and its potential effect on heat tolerance in coral populations. Glob. Chang. Biol. 22, 3539–3549.

Lawrence, J., Kofoed-Hansen, H., Chevalier, C., 2009. High resolution metocean modelling at EMEC's (UK) marine energy test sites. In: Proc 8th Eur Wave and Tidal Energy Conf, Uppsala. www.emec.org.uk/pdf.

Lipcius, R.N., Eggleston, D.B., Schreiber, S.J., Seitz, R.D., Shen, J., Sisson, M., Stockhausen, W.T., 2008. Importance of metapopulation connectivity to restocking and restoration of marine species. Rev. Fish. Sci. 16, 101–110.

McClanahan, T.R., Hicks, C.C., Darking, E.S., 2008. Malthusian overfishing and efforts to overcome it on Kenyan coral reefs. Ecol. Appl. 18 (6), 1516–1529.

Mumby, P., Hastings, A., Edwards, H.J., 2007. Thresholds and the resilience of Caribbean coral reefs. Nature 450, 98–101.

Paris, C.B., Cherubin, L.M., Cowen, R.K., 2007. Surfing, spinning, or diving from reef to reef: effects of population connectivity. Mar. Ecol. Prog. Ser. 347, 285–300.

Pratchett, M.S., Baird, A.H., Bauman, A.G., Burt, J.A., 2017. Abundance and composition of juvenile corals reveals divergent trajectories for coral assemblages across the United Arab Emirates. Mar. Pollut. Bull. 114, 1031–1035.

Pratchett, M., McWilliam, M.J., Riegl, B., 2020. Contrasting shifts in coral assemblages with increasing disturbances. Coral Reefs 39, 783–793. https://doi.org/10.1007/s00338-020-01936-4.

Reynolds, R.M., 1993. Physical oceanography of the Gulf, strait of Hormuz and the Gulf of Oman-results from the Mt Mitchell expedition. Mar. Pollut. Bull. 27, 35–59.

Riegl, B., 1999. Coral communities in a non-reef setting in the southern Arabian Gulf (Dubai, UAE): fauna and community structure in response to recurrent mass mortality. Coral Reefs 18 (1), 63–73.

Riegl, B., Glynn, P.W., 2020. Population dynamics of the reef crisis: consequences of the growing human population. In: Riegl, B. (Ed.), Population Dynamics of the Reef Crisis, vol. 87. Advances in Marine Biology, Elsevier, pp. 1–30 (in this volume). https://doi.org/10.1016/bs.amb.2020.07.004.

Riegl, B., Purkis, S.J., 2012. Coral Reefs of the Gulf: Adaptation to Climatic Extremes. Springer, 379 pp.

Riegl, B., Purkis, S.J., 2015. Coral population dynamics across consecutive mass mortality events. Glob. Chang. Biol. 21 (9), 3995–4005.

Riegl, B., Cavalcante, G., Bauman, A., Feary, D., Steiner, S., Purkis, S., 2017. Demographic mechanisms of reef coral species winnowing from communities under increased environmental stress. Front. Mar. Sci. 4, 344. https://doi.org/10.3389/fmars.2017.00344.

Riegl, B., Johnston, M., Purkis, S., Howells, E., Burt, J., Steiner, S.C.C., Sheppard, C.R.C., Bauman, A., 2018. Population collapse dynamics in *Acropora downingi*, an Arabian/Persian gulf ecosystem-engineering coral, linked to rising temperature. Glob. Chang. Biol. 24 (6), 2447–2462. https://doi.org/10.1111/gcb.14114.

Ritson-Williams, R., Arnold, S., Fogarty, N., Steneck, R.S., Vermeij, M., Paul, V.J., 2009. New perspectives on ecological mechanisms affecting coral recruitment on reefs. Smithson. Contrib. Mar. Sci. 38, 437–457.

Robbins, W.D., Hisano, M., Connolly, S., Choat, H., 2006. Ongoing collapse of coral-reef shark populations. Curr. Biol. 16 (23), 2314–2319.

Saha, S., Moorthi, S., Pan, H.-L., Wu, X., Wang, J., Nadiga, S., Tripp, P., Kistler, R., Woollen, J., Behringer, D., Liu, H., Stokes, D., Grumbine, R., Gayno, G., Wang, J., Hou, Y.-T., Chuang, H., Juang, H.-M.H., Sela, J., Iredell, M., Treadon, R., Kleist, D., Delst, P.V., Keyser, D., Derber, J., Ek, M., Meng, J., Wei, H., Yang, R., Lord, S., van den Dool, H., Kumar, A., Wang, W., Long, C., Chelliah, M., Xue, Y., Huang, B., Schemm, J.-K., Ebisuzaki, W., Lin, R., Xie, P., Chen, M., Zhou, S., Higgins, W., Zou, C.-Z., Liu, Q., Chen, Y., Han, Y., Cucurull, L., Reynolds, R.W., Rutledge, G., Goldberg, M., 2010. The NCEP climate forecast system reanalysis. Bull. Am. Meteorol. Soc. 91 (8), 1015–1058.

Saha, S., Moorthi, S., Wu, X., Wang, J., Nadiga, S., Tripp, P., Behringer, D., Hou, Y.-T., Chuang, H., Iredell, M., Ek, M., Meng, J., Yang, R., Mendez, M.P., van den Dool, H., Zhang, Q., Wang, W., Chen, M., Becker, E., 2014. The NCEP climate forecast system version 2. J. Climate 27 (6), 2185–2208. https://doi.org/10.1175/jcli-d-12-00823.1.

Sale, P.F., Feary, D.A., Burt, J.A., Bauman, A.G., Cavalcante, G.H., Drouillard, K.G., Kjerfve, B., Marquis, E., Trick, C.G., Usseglio, P., van Lavieren, H., 2011. The growing need for sustainable ecological management of marine communities of the Persian Gulf. Ambio 40, 4–17.

Shafir, S., Gur, O., Rinkevich, B., 2008. A *Drupella cornus* outbreak in the northern Gulf of Eilat and changes in coral prey. Coral Reefs 27, 379.

Sheppard, C.R.C., 2016. Coral reefs in the Gulf are mostly dead now, but can we do anything about it? Mar. Pollut. Bull. 105 (2), 593–598.

Sheppard, C., Price, A., Roberts, C., 1992. Marine Ecology of the Arabian Region: Patterns and Processes in Extreme Tropical Environments. Academic Press, Toronto.

Sheppard, C.R.C., Al-Hoseine, M., Al-Jamali, F., Al-Yamani, F., Baldwin, R., Bishop, J., Benzoni, F., Dutrieux, E., Dulvy, N.K., Durvasula, S.R.V., Jones, D.A., Loughland, R., Medio, D., Nithyanandan, M., Pilling, G.M., Polikarpov, I., Price, A.R.G., Purkis, S., Riegl, B., Saburova, M., Samimi Namin, K., Taylor, O., Wilson, S., Zainal, K., 2010. The Persian/Arabian Gulf: a young sea in decline. Mar. Pollut. Bull. 60, 13–38.

Shinn, E.A., 1976. Coral recovery in Florida and the Perisan Gulf. Environ. Geol. 1, 241–254.

Storlazzi, C.D., van Ormondt, M., Chen, Y.-L., Elias, E.P.L., 2017. Modeling fine-scale coral larval dispersal and interisland connectivity to help designate mutually-supporting coral reef marine protected areas: insights from Maui Nui, Hawaii. Front. Mar. Sci. 4 (381), 1–14.

Sundelöf, A., Jonsson, P.R., 2012. Larval dispersal and vertical migration behaviour—a simulation study for short dispersal times. PSZN I: Mar. Ecol. 33, 183–193.

Swift, S.A., Bower, A.S., 2003. Formation and circulation of dense water in the Persian/Arabian Gulf. J. Geophys. Res. 108, 3004.

Thoppil, P.G., Hogan, P.J., 2009. On the mechanisms of episodic salinity outflow events in the Strait of Hormuz. J. Phys. Oceanogr. 39 (6), 1340–1360.

Thoppil, P.G., Hogan, P.J., 2010a. A modeling study of circulation and eddies in the Persian Gulf. J. Phys. Oceanogr. 40 (9), 2122–2134.

Thoppil, P.G., Hogan, P.J., 2010b. Persian gulf response to a wintertime shamal wind event. Deep-Sea Res. I 57, 946–955.

Van den Hoek, C., Mann, D.G., Jahns, H.M., 1995. Algae: An Introduction to Phycology. Cambridge University Press, Cambridge.

Werner, F.E., Cowen, R.K., Paris, C.B., 2007. Coupled biological and physical models: present capabilities and necessary developments for future studies of population connectivity. Oceanography 20 (3), 54–69.

Van Lavieren, H., Burt, J., Feary, D.A., Cavalcante, G., Marquis, E., Benedetti, L., Trick, C., Kjerfve, B., Sale, P.F., 2011. Managing the Growing Impacts of Development on Fragile Coastal and Marine Ecosystems: Lessons From the Gulf, A policy report. UNU-INWEH, Hamilton, ON, Canada.

Yao, F., Johns, W.E., 2010. A HYCOM modeling study of the Persian Gulf: 1. Model configurations and surface circulation. J. Geophys. Res. 115, C11017. https://doi.org/10.1029/2009JC005781.

CHAPTER NINE

Larval connectivity and water quality explain spatial distribution of crown-of-thorns starfish outbreaks across the Great Barrier Reef

S.A. Matthews[a,b,*], C. Mellin[c], and Morgan S. Pratchett[a]
[a]Australian Research Council Centre of Excellence for Coral Reef Studies, James Cook University, Townsville, QLD, Australia
[b]Australian Institute of Marine Science, Townsville, QLD, Australia
[c]The Environment Institute and School of Biological Sciences, University of Adelaide, Adelaide, SA, Australia
*Corresponding author: e-mail address: sammatthews990@gmail.com

Contents

Abstract

Outbreaks of the coral eating crown-of-thorns starfish (COTS; *Acanthasts* cf. *solaris*) occur in cyclical waves along the Great Barrier Reef (GBR), contributing significantly to the decline in hard coral cover over the past 30 years. One main difficulty faced by scientists and managers alike, is understanding the relative importance of contributing factors to COTS outbreaks such as increased nutrients and water quality, larval connectivity, fishing

Advances in Marine Biology, Volume 87
ISSN 0065-2881
https://doi.org/10.1016/bs.amb.2020.08.007

pressure, and abiotic conditions. We analysed COTS abundances from the most recent outbreak (2010–2018) using both boosted regression trees and generalised additive models to identify key predictors of COTS outbreaks. We used this approach to predict the suitability of each reef on the GBR for COTS outbreaks at three different levels: (1) reefs with COTS present intermittently (Presence); (2) reefs with COTS widespread and present in most samples and (Prevalence) (3) reefs experiencing outbreak levels of COTS (Outbreak). We also compared the utility of two auto-covariates accounting for spatial autocorrelation among observations, built using weighted inverse distance and weighted larval connectivity to reefs supporting COTS populations, respectively. Boosted regression trees (BRT) and generalised additive mixed models (GAMM) were combined in an ensemble model to reduce the effect of model uncertainty on predictions of COTS presence, prevalence and outbreaks. Our results from best performing models indicate that temperature (Degree Heating Week exposure: relative importance = 13.1%) and flood plume exposure (13.0%) are the best predictors of COTS presence, variability in chlorophyll concentration (12.6%) and flood plume exposure (8.2%) best predicted COTS prevalence and larval connectivity potential (22.7%) and minimum sea surface temperature (8.0%) are the best predictors of COTS outbreaks. Whether the reef was open or closed to fishing, however, had no significant effect on either COTS presence, prevalence or outbreaks in BRT results (<0.5%). We identified major hotspots of COTS activity primarily on the mid shelf central GBR and on the southern Swains reefs. This study provides the first empirical comparison of the major hypotheses of COTS outbreaks and the first validated predictions of COTS outbreak potential at the GBR scale incorporating connectivity, nutrients, biophysical and spatial variables, providing a useful aid to management of this pest species on the GBR.

1. Introduction

Outbreaks of the corallivorous crown-of-thorns starfish (*Acanthaster* cf. *solaris*) occur on the Great Barrier Reef (GBR) approximately every 15 years (James and Scandol, 1992; Kenchington, 1977; Lane, 1996; Reichelt et al., 1990; Seymour and Bradbury, 1999) and have been responsible for up to 40% of the loss of hard coral cover on the GBR over the last 30 years (De'ath et al., 2012). Long term data suggest that outbreaks originate in the northern GBR within Cairns-Lizard Island region, known as the "initiation zone" (Vanhatalo et al., 2017; Wooldridge and Brodie, 2015). These are known as primary outbreaks, and are characterised by the build-up of COTS densities through local recruitment during repeated successful spawning seasons (Endean, 1974; Johnson et al., 1992; Pratchett et al., 2014; Stump, 1996). Inevitably, larger population sizes of adult COTS in the "initiation zone" overcome Allee thresholds (Rogers et al., 2017), enhancing fertilisation and increasing propagule pressure, seeding downstream secondary outbreaks (Endean, 1974; Pratchett et al., 2014). As the outbreak progresses, aided by the prevailing south-easterly flow of the East Australian Current, these secondary outbreaks spread southward down the GBR in "waves" (Kenchington, 1977; Pratchett et al., 2014). Although the spatial patterns in the progression of an outbreak are somewhat consistent, predicting which reefs will be affected by an outbreak remains elusive.

While oceanographic patterns of larval dispersal (Hock et al., 2014; Scandol and James, 1992; Uthicke et al., 2015a), the anthropogenic erosion of natural population regulation ('terrestrial run-off hypothesis' (Birkeland, 1982) and 'predator removal hypothesis' (Endean, 1969)) are commonly used to explain the initiation of primary COTS outbreaks, the majority of reefs affected by COTS on the GBR are subject to the waves of secondary outbreaks. Understanding the spatial patterns of these secondary outbreaks and the role of their environmental drivers remains a key research and management knowledge gap (Pratchett et al., 2017a). While recent research has used estimated larval connectivity networks to explain which reefs might disproportionately be effective sources or sinks of COTS larvae across the GBR (Hock et al., 2014), it did not account for other important environmental drivers of COTS outbreaks (e.g. chlorophyll *a* (Fabricius et al., 2010)), nor did it allow for the distinction between primary and secondary outbreaks.

The density of COTS can vary greatly both within and among reefs over short time scales (Birkeland and Lucas, 1990; Pratchett et al., 2014). This spatial and temporal variability of COTS populations indicates that there may be important differences between the drivers of COTS presence on a reef, and the drivers of COTS outbreaks (>0.22 individuals per manta tow (De'ath, 2003; Sweatman et al., 2008)). Additionally, at a reef level, COTS are often found in patchy aggregations presumably linked to coral prey availability (Chesher, 1969; Pratchett et al., 2017a) and to enhance fertilisation potential in spawning aggregations (Babcock et al., 1994; Rogers et al., 2017). Such patchiness suggests that the prevalence (herein referred to as the proportion of tows in which COTS are observed) of a COTS outbreak on a reef may again be linked to subtly different environmental and ecological drivers than COTS presence or outbreaks.

Although several major hypotheses have been put forward to explain COTS population fluctuations, they generally fall into three major categories (1) larval-nutrient supply hypotheses, (2) predatory release hypotheses and (3) larval connectivity hypotheses. All of these hypotheses are underpinned by the immense fecundity and fast maturation of COTS individuals, making them naturally predisposed to large population fluctuations (Antonelli and Kazarinoff, 1984; Babcock et al., 2016b; Mellin et al., 2016; Uthicke et al., 2009; Vine, 1973). Enhanced nutrient levels since European settlement have been shown across multiple laboratory studies to have positive effects on COTS larval survival (Fabricius et al., 2010; Pratchett et al., 2017b; Uthicke et al., 2015b; Wolfe et al., 2015, 2017), while others have highlighted the confluence between elevated nutrient supply major and COTS outbreaks (Brodie et al., 2017; Fabricius et al., 2010; Wooldridge and Brodie, 2015). Some support for predatory release hypotheses has been provided by the reduced likelihood and intensity of COTS outbreaks on reefs with "no-take" zones closed to fishing (Sweatman, 2008; Vanhatalo et al., 2017). These studies indicate that increased fishing pressure on large predatory fish, diminishes their ability to regulate COTS populations, thereby increasing of the risk of outbreaks, but these studies did not account for any other influential covariates other than latitudinal and cross-shelf location. While some progress has been made to identify potential predators of adult and juvenile COTS (Cowan et al., 2017), it is still unclear how much changes to fish populations may affect the initiation and spread of COTS outbreaks. Finally, the recently developed hydrodynamic models for the GBR have been used to identify important source and sink reefs across the GBR network (Hock et al., 2014) and to help design decision frameworks to increase the effectiveness of COTS control on the GBR (Hock et al., 2016). These predictions are based entirely upon the potential larval connectivity estimates and assume all reefs have the same suitability for populations of COTS. An important extension of these models is to explicitly incorporate known (and potential) drivers of COTS outbreaks and spatial patterns. Most importantly these predictions should be validated against the available observation data to understand the associated uncertainty.

Although COTS dynamics on the GBR are clearly influenced by a combination of factors, there has yet to be an attempt to empirically assess the importance of each of these competing hypotheses in a single study. Babcock et al. (2016a) provided the first qualitative assessment of competing hypotheses, focusing on the role of nutrients and predation in mediating COTS population dynamics. While being a useful theoretical framework to understand COTS population dynamics, their model was not validated against empirical data and did not explicitly include larval connectivity. During the current outbreak cycle (2010–2019) there has been a vast expansion in the COTS control program, providing unprecedented monitoring data on COTS populations across a large extent of the GBR (Fig. 1).

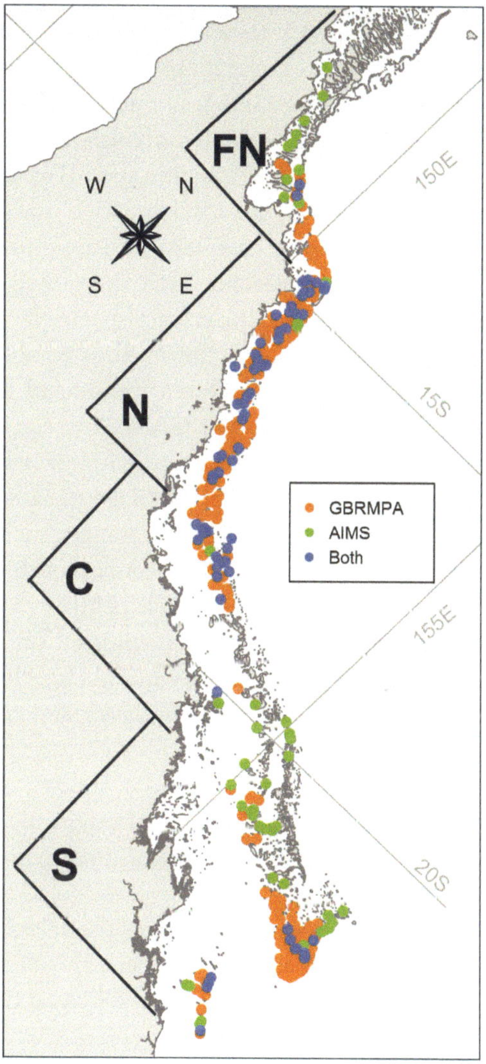

Fig. 1 Reefs with available COTS observation data collected between 2012 and 2018, surveyed by either GBRMPA (Orange: COTS Control Program and the joint Field Management/Queensland Parks and Wildlife Services COTS Response Program) or Australian Institute of Marine Science Long Term Monitoring Program (Green) across the management areas of the GBRMP (FN, Far Northern; N, Northern; C, Central; S, Southern). Reefs surveyed by both institutions are shown in blue.

This increased spatial and temporal resolution of observation data and the availability of GBR-wide estimates of environmental variables and connectivity estimates provides a unique opportunity to compare competing hypotheses in a spatially explicit manner. Importantly, they also provide an opportunity to investigate how these variables drive the presence, outbreaks and prevalence of COTS populations.

Species distribution modelling (SDM) is a powerful tool for predicting the spatial distribution of a species and identifying the relative importance of their environmental drivers across landscapes (Franklin and Miller, 2010; Guisan and Thuiller, 2005; Leathwick and Elith, 2009; Robinson et al., 2011). The nature of many ecological relationships are often non-linear and machine learning approaches such as boosted regression trees (BRT) have been increasingly used to predict species ranges and identify important predictor variables (De'ath, 2007; Elith et al., 2008; Sutcliffe et al., 2014). These approaches are often conducted in an ensemble model with generalised additive models (GAM) that can also account for non-linear trends by applying a smoother, providing useful flexibility for fitting ecologically realistic relationships in SDM (Leathwick and Elith, 2009). Importantly, for species with dynamic ranges such as COTS it is important to include covariates pertaining to important biotic interactions and dispersal pathways in an SDM approach (Elith et al., 2010; Mellin et al., 2016). SDM provides empirical and validated predictions of reefs most likely to experience COTS outbreaks while also modelling the relative importance of each of the competing hypotheses and their relationship to COTS.

In this study, we:

(1) Identify key environmental, spatial and hydrodynamic drivers that may explain spatial patterns of COTS presence, prevalence and outbreaks across the extent of the GBR;

(2) Construct an ensemble model to compare the relative influence and relationships between candidate predictors and COTS populations;

(3) Compare the utility of contemporary COTS larval connectivity estimates over simple distance estimates of connectivity for accounting for spatial autocorrelation and

(4) Derive predictive maps of COTS presence, outbreaks and prevalence to help identify potentially important reefs for COTS control operations.

2. Methods

2.1 Crown-of-thorns starfish observation data

Two independent sets of COTS observation data were gathered: (1) manta tow data from the Australian Institute of Marine Science Long Term Monitoring Program (AIMS LTMP) and (2) combined manta tow observations collected by the Great Barrier Reef Marine Park Authority's (GBRMPA) COTS control program and the Joint Field Management Program's COTS Response Program (collected by the Queensland Parks and Wildlife Services) (Fig. 1). The data used here represent the most recent, spatially and temporally extensive observational data for COTS. The AIMS dataset spans from 1983 to 2019, whereas the other datasets were collected between 2012 and 2019 and thus the temporal extent of this study has been limited to the most recent outbreaks of COTS (2012–2019). Due to the culling activity and tourism in the Northern management area, sampling intensity is greatest in these regions, however substantial data has been collected in the Central and Southern GBR (Fig. 1). Manta tow data from the FMP and the Control program was collated to calibrate the models, and the AIMS LTMP data was used to independently validate model predictions. Using the most recent outbreak however may reduce the model's performance in predicting the historical spatial distribution as this can vary between outbreak cycles (Fig. 2). This is addressed partially by validating model predictions against the 30 year AIMS LTMP dataset to test performance against historical trends and also by resampling observations to generate a distribution of effect sizes for best fit models (Fig. S5 in Supplementary material in the online version at https://doi.org/10.1016/bs.amb.2020.08.007). At least two complete manta-tows during the 2012–2019 period were required for a reef to be included in the analysis. In total 335 reefs were included for model fitting with observations for 113 reefs used for model validation.

2.2 Response variables

Survey data were collated at the reef level and presence of COTS was defined as any observation of COTS during manta tow observations of that reef (0–1; Present/Absent). Outbreaks at a reef were defined as per AIMS guidelines (De'ath, 2003; Sweatman et al., 2008) for manta tow data (>0.22 COTS/manta tow) (0–1; whether or not COTS had reached outbreak levels) and fitted using a binomial distribution. Prevalence of COTS at a reef was defined in this study as the proportion of tows where COTS were observed during a set of manta tow observations (a survey: observer towed around the perimeter of a reef (Miller et al., 2009)). As reefs are towed multiple times throughout the study period, the maximum recorded prevalence was used as the response variable to reflect the upper limits of COTS prevalence per reef (Fig. S3 in Supplementary material in the online version at https://doi.org/10.1016/bs.amb.2020.08.007). Using the maximum prevalence is biased towards high density observations of COTS. This is by design as the aim is to identify predictors of reefs that are susceptible to populations of COTS that are distributed widely across the focal reef. This approach does make the model susceptible to sampling intensity which was partially accounted for by using only reefs that had been surveyed at least twice over the study period. The prevalence proportion was logit transformed prior to model fitting using a Gaussian distribution (Warton and Hui, 2011). For COTS prevalence a two-step modelling approach was used to accommodate for the zero-inflated nature of the proportion data (Cragg, 2006; Potts and Elith, 2006). Also called the hurdle model, this approach uses the results from the COTS presence model to define suitable reefs and then models the prevalence on suitable reefs only (e.g. Mellin et al., 2012). The dataset was thus restricted to contain only reefs that have been observed to host COTS at least once before modelling the major predictors of COTS prevalence.

2.3 Candidate predictor variables

All predictor variables used within this analysis were originally compiled on a 0.01° (~1 km) grid across the GBR (Matthews et al., 2019) over various time scales (Table 1). These data were then averaged at the reef level to match the spatial scale of the observation data.

2.3.1 *Water quality variables*

Estimates of water quality were included as potential predictors for COTS distribution patterns due to the suggested effect of water quality on COTS larval development, survival and outbreaks. Water quality was considered in the study using satellite derived estimates of flood plume exposure (Devlin et al., 2012; Matthews et al., 2019) and annual model estimates of mean and standard deviation of chlorophyll *a* concentration (CSIRO, 2019). The average frequency of exposure to primary (turbid, sediment dominated plumes) and secondary (chlorophyll dominated plumes) flood plumes was calculated across the GBR as the proportion of weeks within the Australian wet season where exposure to the plume type occurred between 2000 and 2014 (Devlin et al., 2012). Daily estimates of chlorophyll concentrations were obtained from the eReefs 4 km biogeochemical model (CSIRO, 2019) 2013–2018 (all available data), and then annual mean and standard deviation were computed for each reef.

Table 1 Predictor variable definitions, units, code used in SDM analyses: Env, environmental; WQ, water quality; Spat, Spatial; Cor, coral; Dist, disturbance; Conn, connectivity; = AutoCov, =auto-covariate.

Type	Code	Description	Units	Years	Source	COTS Reference[a]
Env	O2_AV	Average oxygen	$mL\,L^{-1}$	1960–2006	Huang et al. (2010), Matthews et al. (2019)	Lamare et al. (2014), Hardy et al. (2014)
Env	NO3_AV	Average nitrate	μM	1960–2006		Birkeland (1982)
Env	S_AV	Average salinity	PSU	1960–2006		Lucas (1973), Allen et al. (2017), Caballes et al. (2017b)
Env	S_SR	Seasonal range salinity	PSU	1960–2006		
Env	BATHY	Seafloor depth	m	1963–2009		Johnson (1992)
Env	STRESS	Percentage of time for which the bed shear stress was >0.4 Pa (Wave exposure proxy)	%	2010		Moran (1986)
Env	SST_MIN	Minimum sea surface temperature	°C	1999–2008		Kamya et al. (2014), Lamare et al. (2014)
Env	MUD	percentage of a seabed sediment sample that is <63 μm in diameter	%	1960–2009		Wolanski et al. (2003)[a]
WQ	WQ_1	Primary (representing turbid, sediment dominated plume) flood plume frequency (weeks occurred/total weeks) during wet season (max = 26).—relative frequency	0–1	2000–2014	Devlin et al. (2012), Matthews et al. (2019)	Fabricius et al. (2010), Wolfe et al. (2015), Wooldridge and Brodie (2015), Pratchett et al. (2017b), Brodie et al. (2017)
WQ	WQ_2	Secondary chlorophyll dominated plume—relative frequency	0–1	2000–2014		

Continued

Table 1 Predictor variable definitions, units, code used in SDM analyses: Env, environmental; WQ, water quality; Spat, Spatial; Cor, coral; Dist, disturbance; Conn, connectivity; = AutoCov, =auto-covariate.—cont'd

Type	Code	Description	Units	Years	Source	COTS Reference[a]
WQ	WQ_3	Further extent of plume, as delineated by salinity <34 ppt—relative frequency	0–1	2000–2014		
WQ	CHL_AV	Wet season mean chlorophyll	$mg\,m^{-3}$	2012–2018	CSIRO (2019)	
WQ	CHL_SD	Wet season standard deviation chlorophyll	$mg\,m^{-3}$	2012–2018	CSIRO (2019)	
Spat	SHELF	Cross shelf location (Inner, Middle, Outer)	Factor (3 Levels)	–	AIMS (2019)	Moran (1986)
Spat	SECTOR	Marine Park Latitudinal Sector (Southern, Central, Northern, Far Northern)	Factor (4 Levels)	–	GBRMPA (2014a)	Moran (1986), Fabricius et al. (2010)
Spat	ZONE	Open or closed to fishing	Factor (2 Levels)	–	GBRMPA (2014b)	Sweatman et al. (2008), Vanhatalo et al. (2017)
Cor	BENT_CL	Benthic coral community type	Factor (6 Levels)	Estimated 2018	Mellin et al. (2019a)	Lucas (1984), Pratchett (2007), Caballes et al. (2017a)
Cor	HC_MAX	Predicted maximum hard coral cover	0–100%	Estimated 2018		Lucas (1984), Caballes et al. (2017a)
Dist	DHW_EXP	Mean exposure to degree heating weeks 1985–2012	Num	1985–2012	Heron et al. (2016), Matthews et al. (2019)	Kamya et al. (2014), Lamare et al. (2014)

Dist	CYC_EXP	Mean exposure to cyclonic waves (>4 m) 1985–2012	Num	1985–2012	Puotinen et al. (2016), Matthews et al. (2019)	Osborne et al. (2011), Sato et al. (2018)[a]
Dist	BL_EXP	Mean exposure to 1998 and 2002 bleaching events	1–5	1998, 2002	Berkelmans et al. (2004), Matthews et al. (2019)	Hughes et al. (2018)[a]
Conn	IN_STR	Weighted sum of strength of incoming connections	Num	2009–2013	Hock et al. (2017)	Hock et al. (2014, 2017), Wooldridge and Brodie (2015)
Conn	LAR_RET	Estimated proportion of larval retention	%	2009–2013		
Auto Cov	AC_DIST	Auto-covariate weighted by distance to reefs with COTS present	Num	–	Dormann et al. (2007)	
Auto Cov	AC_CONN	Auto-covariate weighted by estimated larval connectivity to reefs with COTS present	Num	–	Dormann et al. (2007), Hock et al. (2017)	

[a]Indicates references that relate to the effect of the variable on COTS' coral prey and not COTS directly.
References are provided for the source of the data and how these variables relate to COTS populations.

2.3.2 Disturbance variables

GBR-wide estimates of mean annual exposure to potentially damaging waves generated by tropical cyclones and Degree Heating weeks (as an index of bleaching risk) (1986–2012) alongside exposure to the 1998 and 2002 bleaching events (Matthews et al., 2019). COTS typically prefer fast growing *Acropora* spp. corals (Pratchett et al., 2014), which despite being particularly susceptible to acute disturbances (e.g. Hughes et al., 2018), recruit abundantly and often recover rapidly following disturbances (Linares et al., 2011; Osborne et al., 2011; Sato et al., 2018). COTS populations are thus presumably mediated by the disturbance history of a reef, which are broadly captured by our indices of historical exposure to cyclonic waves, thermal stress and bleaching.

2.3.3 Environmental variables

Static estimates of environmental variables, originally collated from the Commonwealth of Australia's Environment Research Facility (CERF) Marine Biodiversity Hub (Huang et al., 2013), were obtained from Matthews et al. (2019) over varying time scales (Table 1). These environmental variables include: annual mean estimates of nitrate, oxygen, phosphate, silicate, temperature and salinity, bathymetry, percentage cover of sediment components, multiple indices of the strength and frequency of the combined wave–current bed shear stress and spatial variables of minimum distances to the coast and the ocean at the edge of the GBR lagoon (Table 1). These variables have been successfully used to predict benthic community composition (Mellin et al., 2019a), fish assemblages (Mellin et al., 2010) and other inter-reef species richness and abundance on the GBR (Sutcliffe et al., 2014). Moreover the dataset includes estimates of a number of variables shown to be important in affecting different life history stages of COTS such as salinity, oxygen and water temperature (Table 1).

2.3.4 Coral variables

Estimated maximum coral cover and benthic community were obtained from Mellin et al. (2019a). Benthic communities were identified using multivariate regression trees (De'ath, 2002), modelling the relationship between spatial and environmental covariates (Matthews et al., 2019) with the relative cover of benthic groups and coral taxa. Clusters of reefs were then defined by the multivariate regression tree corresponding to distinct taxonomic assemblages, characterised by its indicator taxa. The six benthic communities identified were: (i) outer reef—soft coral dominated; (ii) outer reef—digitate coral dominated; (iii) outer reef—tabulate coral dominated; (iv) mid shelf reef—mixed assemblage; (v) inner shelf reef—*Porites* dominated; (vi) inner shelf reef—macroalgae dominated (Fig. S1 in Supplementary material in the online version at https://doi.org/10.1016/bs.amb.2020.08.007).

Maximum coral cover, an index of the amount of substrate available or reef carrying capacity, was estimated from a 22-year time series at the 46 long term monitoring sites on the GBR based on a Gompertz model of coral growth (MacNeil et al., 2019; Mellin et al., 2019a). This model assumes that in the absence of disturbance, coral cover increases from its initial value (HC_{ini}, in 1996 in this case) to its asymptote (HC_{max}, determined by the reef carrying capacity or amount of available substrate).

2.3.5 Spatial variables

Outbreaks of COTS are spatially patchy yet are more common within the mid-shelf reefs on the GBR and with some evidence suggesting higher outbreak probabilities at reefs open to fishing (Sweatman, 2008; Vanhatalo et al., 2017). To account for latitudinal and longitudinal patterns we included the cross-shelf location (inner, middle, outer) and four latitudinal sectors (Southern, Central, Central-Northern, Far Northern). Additionally, the effect of zoning on COTS populations was included as a categorical variable indicating whether a reef was open or closed for fishing (i.e. no-take area).

2.3.6 Larval connectivity variables

On the GBR the development of advanced hydrodynamic and larval connectivity models (Condie et al., 2012; CSIRO, 2019) has facilitated the development of larval connectivity networks for COTS (Hock et al., 2014). These networks are constructed from 4 years of hydrodynamic modelled data and estimate the potential connectivity between all reefs on the GBR (2009–2013). Potential connectivity is a simplification of real-world processes which essentially represents maximal physically possible advective transport among reefs as implied by the model and expressed as a proportion of simulated larvae reaching destination reefs (Hock et al., 2014). As our COTS observational dataset spans 9 years, and our goal is to highlight the most important factors promoting COTS outbreaks, we used an averaged connectivity network that identifies the most consistent between-reef larval connections across years. From this network we computed the weighted in-strength for every reef, defined at the reef level as the sum of the potential downstream connectivity from every other source reef on the GBR.

A reef with high in-strength will be characterised by high potential connectivity from many source reefs, thus potentially representing an important sink for COTS larvae. The in-strength s of a reef i is defined as

$$s_i = \sum_j w_{ij}$$

where w_{ij} is the potential connectivity from reef i to reef j (Barrat et al., 2004).

2.4 Spatial autocorrelation

Outbreaks of COTS on individual reefs are not independent of one another, as an outbreak at one reef will likely seed an outbreak at a neighbouring reef if there is sufficient larval supply via ocean currents (hereafter connectivity). Connectivity between reefs and spatial correlation of environmental drivers means there will be spatial autocorrelation within our response variables. To account for this autocorrelation we adopted the autologistic model (Dormann et al., 2007) where a distance-weighted term (the auto-covariate) is added to the predictor set, to represent the influence of neighbouring observations.

The auto-covariate A at any site i is calculated as:

$$A_i = \sum_{j \in k_i} v_{ij} y_j$$

where y_j is the response value of $\boldsymbol{y}$ at site j among site i's set of k_i neighbours; and v_{ij} is the weight given to site j's influence over site i (Augustin et al., 1996; Gumpertz et al., 1997). We calculated two competing auto-covariates: (i) a distance auto-covariate, whereby observations at site j were weighted by their inverse geographical distance to site i, and (ii) a connectivity auto-covariate, whereby observations were weighted by the estimated larval connectivity from site j to site i. Including the spatial auto-covariates allows the model to account for autocorrelation, but also allows the comparison of estimates of larval connectivity for COTS (Hock et al., 2014) to estimates based solely on distance. Additionally the connectivity auto-covariate models the larval connectivity to reefs that have experienced outbreaks, compared to the in-strength variable which measures how well connected the focal reefs is to all others within the network, regardless of the outbreak status of those reef.

2.5 Model fitting

Boosted regression trees (BRT) were used to fit our candidate predictors set to COTS observation data. BRT is a machine learning algorithm that uses many simple decision trees to iteratively boost the predictive performance of the final models (Elith et al., 2008). Model settings include the learning rate (lr) that controls the contribution of each tree to the final model and tree complexity (tc) that determines the extent to which interactions were fitted. The number of trees that achieved minimal predictive deviance (i.e. the loss in predictive performance due to a suboptimal model) was determined using cross-validation (function gbm.step with tc = 3, lr = 0.001, bag fraction = 0.5) (Elith et al., 2008; R Development Core Team, 2017). Multicollinearity was assessed between candidate variables, however none were omitted with Pearson's correlation values <0.7 (Fig. S2 in Supplementary material in the online version at https://doi.org/10.1016/bs.amb.2020.08.007). To investigate the effect of spatial autocorrelation and the utility of connectivity- and distance-based auto-covariates, each BRT was fitted to our three response variables (Presence, Prevalence and Outbreak) three times: (1) with no auto-covariate; (2) with the spatial auto-covariate; (3) with the connectivity auto-covariate, to give a total of 9 BRT models. One of the useful outputs of BRT is the relative influence (variable importance, in %) each predictor had in the classification process, allowing each predictor to be ranked in terms of their utility in predicting presence, prevalence or outbreaks of COTS.

Generalised additive models (GAM) were also fit using the same candidate predictor set to potentially identify more parsimonious models. GAM provide a flexible middle ground between traditional generalised linear model models and machine learning approaches. All potential combinations of variables were fitted in a full subsets theoretic approach (Johan, 2015) restricting the GAM to include a maximum of six variables and maximum five knots per variable (*n* models = 40,815). Full subset models were fitted to each of our three responses in the same categories as the BRT: (1) with no auto-covariate; (2) with the spatial auto-covariate; (3) with the connectivity auto-covariate, for a total of 9 GAM model categories. Each of the variables were then assigned a relative importance indicating the frequency of their inclusion in best-fit models (Fisher et al., 2018). From the full subset of models for each of the nine model categories the single best performing model was chosen via Bayesian Information Criterion (BIC). Partial dependency plots (which reveal the effect of a variable on the response after accounting for the average effects of all other variables (Elith et al., 2008)) were produced for the top four most influential predictors for both BRT and GAM models, to investigate the effects of predictors independently. BRT and GAM models were fitted and validated in R 3.5.3 (R Development Core Team, 2017) using the "gbm" and "FSSgam" packages respectively.

2.6 Model validation

Each of the 9 BRT and 9 GAM models were subsequently validated against the independent test data and using 10-fold cross validation to estimate prediction accuracy. Validation statistics were combined to produce a single validation metric V_m, calculated for the mth model as:

$$V_m = \frac{DE_m + (CV.ACC_m) + (IND.ACC_m)}{3} \quad (1)$$

where DE is the proportion of deviance explained; CV.ACC is the 10-fold cross-validation accuracy and IND.ACC is accuracy when validated against the AIMS Long Term Monitoring Program dataset. For binary response variables (presence, outbreak), accuracy refers to the proportion of validation observations that were correctly classified, assuming a 0.5 threshold, for Gaussian responses (prevalence), accuracy refers to the mean prediction accuracy (1—mean prediction error). Additionally for the best performing model, empirical data was resampled 100 times over a 0.2° (~20 km) grid selecting 1 representative reef per grid cell. Models were then refit for these resampled datasets, to create a distribution of effect sizes (variable importance values) for each of the candidate predictor variables to assess the influence of sampling bias (Fig. S5 in Supplementary material in the online version at https://doi.org/10.1016/bs.amb.2020.08.007).

2.7 Ensemble prediction

The boosted regression tree and generalised additive models for each response were then combined to form weighted ensemble predictions based on the method below (Marmion et al., 2009):

$$WA_i = \frac{\sum_m (V_m \times PRED_{m_i})}{\sum_m V_m} \quad (2)$$

where WA_i is the weighted ensemble prediction for the ith location, $PRED_{m_i}$ is the ith prediction for model m, and V_m is the validation statistic for model m (Eq. 2).

A second ensemble approach was also made by "stacking" the predictions from all boosted regression tree and GAMs into a final boosted regression tree model (Wolpert, 1992; Ting and Witten, 1999). This is a machine learning approach to ensembles, allowing the boosted regression tree algorithm to choose which model predictions are the most useful in explaining the response variable. The weighted and "stacked" ensemble were then validated against an independent dataset (AIMS Long Term Monitoring Program) to test the generalisability of the two approaches. Final predictions were then made across all reef locations on the GBR using the best performing ensemble model to indicate the reefs most suitable to COTS presence, prevalence and outbreaks.

3. Results

3.1 Presence model

Presence of COTS individuals on a reef was best explained by the model without auto-covariate structure for both BRT and GAM approaches according to both model fit (68.5% and 61.8% deviance explained respectively) and validation against independent data (78.8% and 71.7% accuracy respectively) (Table 2). For the best fitting BRT, mean annual exposure to degree heating weeks (DHW_EXP = 13.1% relative influence); exposure to chlorophyll-rich secondary plumes (Secondary = 13.0%) and minimum sea surface temperature (SST_MIN = 10.8%) were the most influential predictors (Fig. 2A). Partial dependency plots indicated a strong negative threshold response at ~1.1 DHW/year and minimum sea surface temperature of ~24 °C. Conversely, increasing variability in chlorophyll concentration (CHL_SD) had a relatively linear positive effect of COTS presence, while the frequency of exposure to secondary plumes variables and COTS presence indicated a positive threshold type response at low levels, highlighting the importance of availability of larval nutrition for the establishment of COTS populations.

For GAM models, exposure to cyclonic waves (included in 96.0% of top performing models), primary flood plume exposure (WQ_1 = 92.8%), variation in chlorophyll concentration (CHL_SD = 87.9%) were the most important predictors, followed by average nitrate concentration (NO3_AV = 20.4%) and minimum temperature (SST_MIN = 11.4%).

Table 2 Validation metrics for each model and ensemble for each of the three response variables: presence, outbreaks and prevalence.

Response	Framework	Model	DE	CV.ACC[a]	IND.ACC[a]	Vm
Presence	BRT	PA	0.68	0.87	0.79	0.78
		$PA_{AC.Dist}$	0.68	0.87	0.75	0.77
		$PA_{AC.Conn}$	0.66	0.85	0.74	0.75
	GAM	PA	0.62	0.84	0.72	0.73
		$PA_{AC.Dist}$	0.58	0.83	0.73	0.71
		$PA_{AC.Conn}$	0.62	0.84	0.72	0.73
	ENSEMBLE	Weighted	NA	0.93	0.72	–
		Stacked	0.86	0.96	0.73	–
Prevalence	BRT	PREV	0.85	0.85	0.87	0.86
		$PREV_{AC.Dist}$	0.80	0.85	0.89	0.84
		$PREV_{AC.Conn}$	0.88	0.85	0.88	0.87
	GAM	PREV	0.59	0.84	0.85	0.75
		$PREV_{AC.Dist}$	0.67	0.85	0.87	0.79
		$PREV_{AC.Conn}$	0.57	0.83	0.85	0.75
	ENSEMBLE	Weighted	NA	0.89	0.87	–
		Stacked	0.9192	0.94	0.86	–
Outbreak	BRT	OUT	0.75	0.85	0.77	0.79
		$OUT_{AC.Dist}$	0.74	0.87	0.77	0.79
		$OUT_{AC.Conn}$	0.77	0.86	0.78	0.80
	GAM	OUT	0.61	0.81	0.71	0.70
		$OUT_{AC.Dist}$	0.65	0.82	0.75	0.76
		$OUT_{AC.Conn}$	0.61	0.82	0.71	0.71
	ENSEMBLE	Weighted	NA	0.94	0.77	–
		Stacked	0.87	0.97	0.80	–

[a]Accuracy for prevalence is calculated as 1-mean prediction error.
Models presented in bold represent the best performing model for each response and model framework.

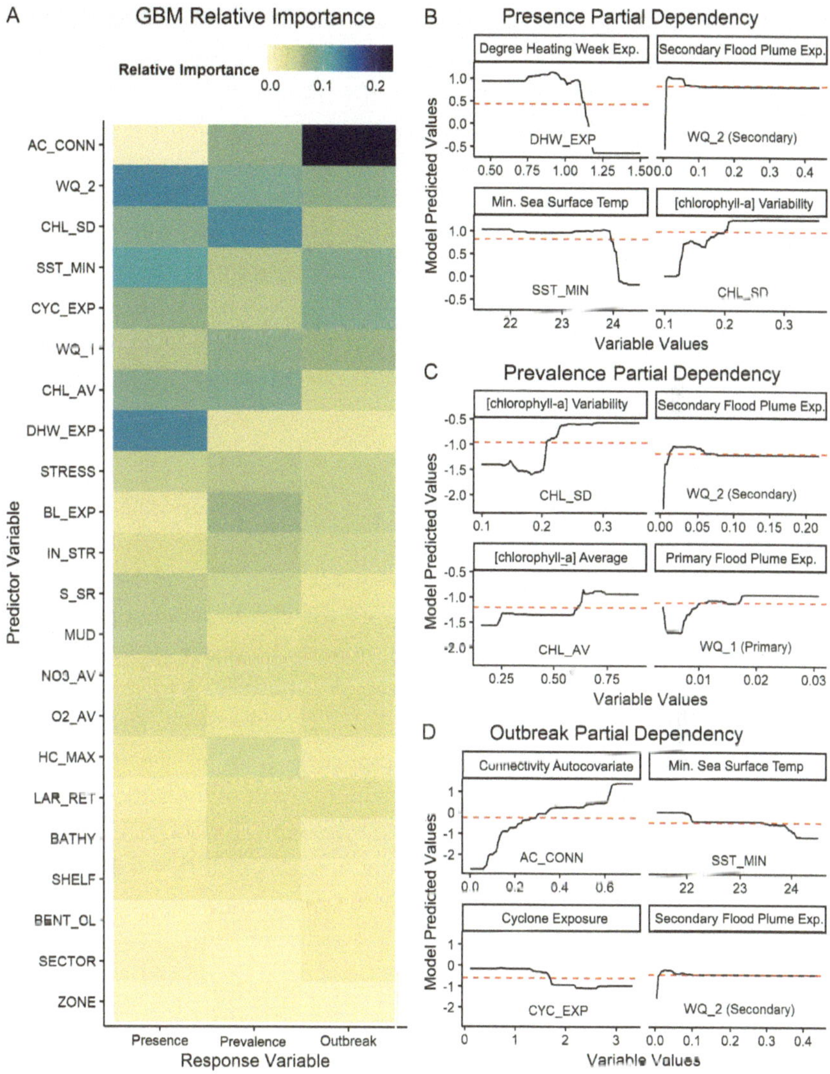

Fig. 2 Proportion relative importance (Left hand side) (A) and partial dependency plots of the four most influential variables (Right hand side) for best fit boosted regression tree models of COTS Presence (B), Prevalence (C) and Outbreaks (D). Best fit models (Table 2) are presented for each COTS response variable.

Marginal effects plots of the top performing model reveal a strong positive relationship of both chlorophyll variability (CHL_SD) and in-strength (IN_STR) on the probability of COTS presence (Fig. 3B). Exposure to primary flood plumes (WQ_1) indicated high probabilities of COTS presence in the absence of primary plumes and a severe dip at low levels before dramatically dropping off again at high exposure to these turbid flood plumes. This finding was relatively consistent for reefs either open or closed to fishing, although sample size was low at the upper extremes of primary flood plumes. Cyclonic wave exposure (CYC_EXP) had a generally positive relationship with COTS presence, with a major dip apparent in closed fishing sites, however the confidence limits overlap substantially.

3.2 Prevalence model

In contrast to COTS presence, the prevalence of COTS populations on a reef was best explained by the connectivity-based auto-covariate model according to both model fit (89.8% deviance explained) and 10 fold cross validation (14.9% mean prediction error) for BRT models (Table 2). Predictions for COTS prevalence were dominated by water quality variables (CHL_SD = 12.6%; WQ_2 = 8.2%; CHL_AV = 8.0%, WQ_1 = 7.4%) as well as the connectivity auto-covariate (AC_CONN = 7.4%) and exposure to bleaching events (BL_EXP = 6.5%). Inspection of the partial dependency plots indicates generally positive relationships for each of the nutrient related variables (Fig. 2D). Again, benthic community type (BENT_CL = 0.55%), latitudinal sector (SECTOR = 0.21%) and zoning (ZONE = 0.42%) were poor predictors of COTS prevalence across a reef.

For GAM models, using the distance based auto-covariate yielded the best results for model fit (66.7% deviance explained) and validation against independent data (86.6% accuracy). The distance based auto-covariate (AC_DIST included in 98.4% of top performing models), bed sheer stress (wave exposure proxy) (STRESS = 98.0%), and exposure to cyclonic waves (CYC_EXP = 49.6%) were the most reliable variables, followed by zoning (ZONE = 26.5%), exposure to secondary (WQ_2 = 25.9%) and primary (WQ_2 = 11.8%) flood plumes. Although models containing the distance based auto covariate outperformed the larval connectivity auto-covariate, the in-strength of a reef was included in a large portion of top performing models (IN_STR = 88.1%). Marginal effects plots of the top performing GAM model reveal strong positive effects of both the distance-based auto-covariate and the in-strength of a reef with a negative effect of bed sheer stress and a parabolic relationship with cyclone exposure, with prevalence increasing at higher levels of exposure (Fig. 3D).

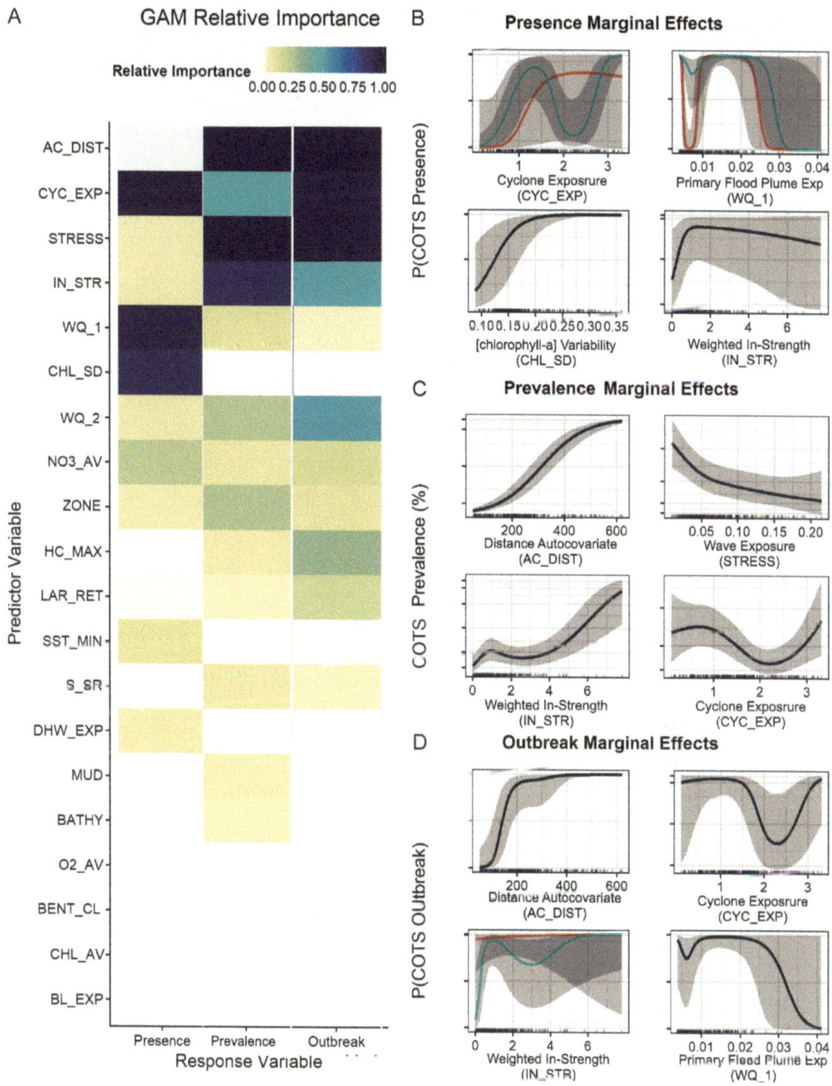

Fig. 3 Proportion relative importance (A) and marginal effects plots of the four most influential (Right hand side) for best fit Generalised Additive models of COTS Presence (B), Prevalence (C), and Outbreaks (D). Variables with no coloration were not used in any of the best fit models as per the FSSgam procedure. For models including an interaction with zoning, red lines represent reefs closed to fishing and green lines are those that are open for fishing. NB The partial dependencies shows the single best fit model. The single best fit model does not necessarily include the all the top predictors.

3.3 Outbreak model

The outbreak potential of a reef was best explained by the connectivity based auto-covariate model according to both model fit (76.6% deviance explained) and validation with independent data (78.0% accuracy) for BRT. Importantly, the connectivity derived auto-covariate was by far the most influential predictor (AC_CONN=22.7% variable importance), exhibiting a strong positive effect outbreak potential (Fig. 2C). Cyclonic wave exposure (8.0%—negative relationship), minimum sea surface temperature (8.0%—negative relationship), secondary (7.4%—threshold positive relationship) and primary flood plume exposure (6.8%—positive relationship) were also relatively useful predictors of COTS outbreak potential.

GAM models performed best when including the distance based auto-covariate (64.9% deviance explained; 75.6% independent validation accuracy). The distance based auto-covariate (AC_DIST=included in 98.7% of top performing models), bed-sheer stress (wave exposure proxy) (STRESS=95.9%), exposure to cyclonic waves (CYC_EXP=94.7%) were the most reliable variables, with exposure to secondary flood plumes (WQ_2=53.5%) and estimated maximum coral cover (HC_MAX=30.9%) also included in a large portion of top performing models (Fig. 3C). As for outbreaks, although models containing the distance based auto covariate outperformed the larval connectivity auto-covariate, the in-strength of a reef was included in a large portion of top performing models (IN_STR=51.3%). Similarly, marginal effects plots for the top performing model reveal a strong positive relationship for the distance auto-covariate, however the sharper increase suggests that reaching outbreak densities is easier than establishing a truly widespread and prevalent population. Outbreaks appear to occur most frequently at reefs that have experienced lower levels of cyclone exposure, however there is another increase in outbreak probability at the maximum levels of exposure. COTS outbreaks also occur at low to moderate exposure to primary flood plumes. Interestingly, there is a sharp increase in the probability of outbreak at relatively low levels of in-strength (i.e. potential larval supply) for reefs closed to fishing, whereas reefs open to fishing have high chances of having COTS outbreaks regardless of their larval supply.

3.4 Ensemble predictions

Ensemble predictions were made for each of the three response variables by combining predictions from all six models (three GAM, three BRT) into both a validation-weighted and a stacked ensemble (Fig. 4).

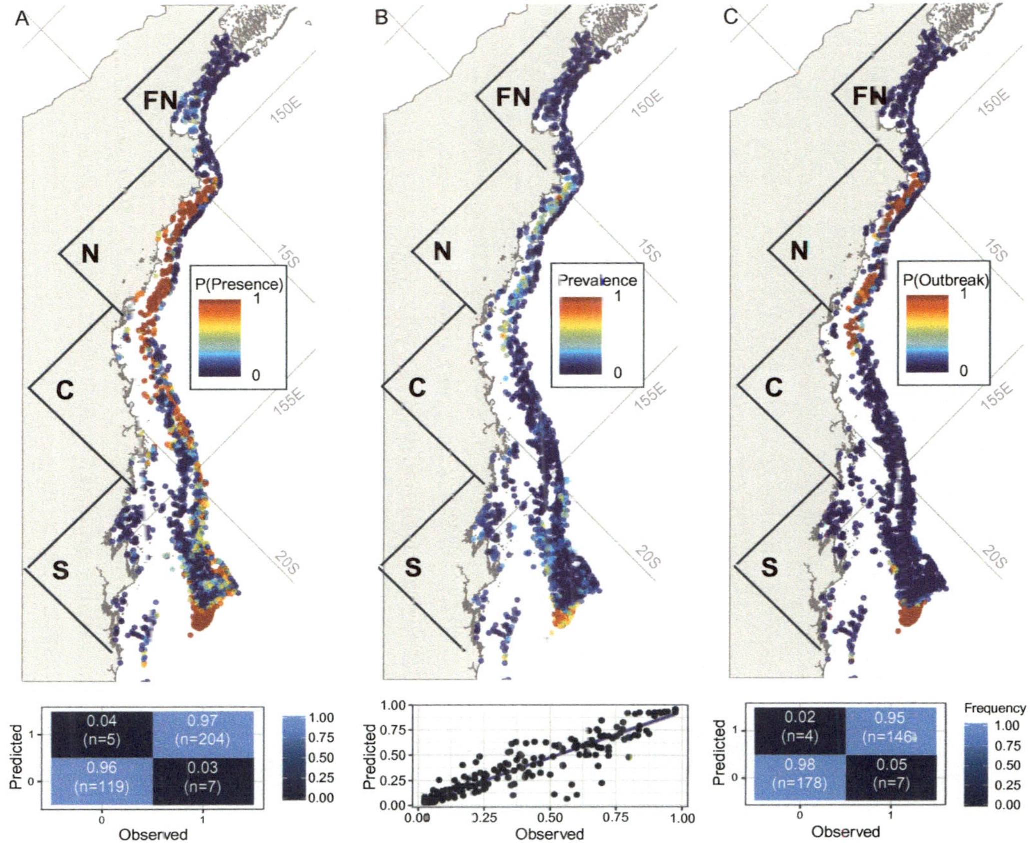

Fig. 4 Stacked ensemble predictions and cross-validation for COTS presence (A), prevalence (B) and outbreaks (C) across the management areas of the GBRMP (FN, Far Northern; N, Northern; C, Central; S, Southern). Validation statistics are presented as a confusion matrix for presence and outbreaks (0.5 classification threshold) and as linear regression for prevalence.

While dramatically improving model fit and accuracy with the training data, the stacked ensemble method did not improve the accuracy of predictions made by the either best fit model alone or the weighted ensemble approach for the validation dataset (Table 2). For binary outcomes validated classification was more prone to Type II errors (Presence: false negative rate = 0.31; Outbreak: false negative rate = 0.30) than Type I errors (Presence: false positive rate = 0.21; Outbreak: false positive rate = 0.15).

While the upper limits of each of the response variables were generally located in the Swains reefs of the Southern GBR and mid-shelf reefs of the Northern GBR, distinct spatial patterns were observable for each response. Relatively high probabilities (>0.5) of COTS presence were predicted offshore of the Whitsundays and Mackay Regions, where little empirical data exists (Fig. 4A). While COTS may be present along inshore reefs (Fig. 4), our results highlight that these regions are not suitable for COTS outbreaks or prevalent (i.e. widespread) populations. Although a number of reefs in the central northern region of the GBR were predicted to be suitable for COTS outbreaks, few in this region were predicted to have high levels of prevalence.

Ensemble predictions estimated that at the GBR scale, 32% of reefs are suitable for COTS (>50% probability), while 12% were susceptible to outbreaks of COTS (>0.22 COTS/Manta Tow) and only 7% were likely to exhibit widespread prevalent population (>50% tows observing a COTS), however as this dataset has limited data in the central and southern regions, this is likely an underestimate of reefs susceptible to COTS. There is significant variation among cross-shelf locations and latitudinal sectors (Fig. 4), with COTS outbreaks and prevalent populations primarily on mid-shelf reefs in the Southern and Central Northern regions of the GBR (Fig. S4 in Supplementary material in the online version at https://doi.org/10.1016/bs.amb.2020.08.007). Additionally, while sampling was biased towards the Northern region, results from resampling indicate that the general patterns of variable importance are robust. For prevalence and outbreak responses however, the relative importance of temperature variables (SST_MIN) and storm exposure (CYC_EXP, STRESS) are generally higher than when considering the full dataset (Fig. S5 in Supplementary material in the online version at https://doi.org/10.1016/bs.amb.2020.08.007), similar to the results from the GAM's (Fig. 3).

4. Discussion

COTS presence, outbreaks and prevalence were explained, to varying degrees, by larval connectivity, water quality and wave exposure (cyclonic and bed shear stress). For both modelling frameworks, predicting COTS presence on a reef was best achieved by models that did not include the spatial or connectivity based auto-covariate. Importantly however, for COTS to establish secondary outbreaks and/or prevalent populations, our results indicate that there needs to be a substantial larval input from connected reefs, as exemplified by the importance of the connectivity based auto-covariate variables for these responses. Respectively, the connectivity and distance-based auto-covariates act as modelled estimates and proxy indicators of COTS larval supply to a reef, and indeed both have been shown to be equally useful predictors of outbreak probability at a reef (Hock et al., 2014). Together these results indicate that reliable connectivity pathways with COTS source reefs may not necessarily be a prerequisite for COTS presence, as intermittent larval supply may be sufficient to establish low-density COTS populations. However, the establishment of widespread prevalent and/or outbreaking COTS populations are seemingly limited by larval supply from COTS source reefs, with continued propagule pressure promoting COTS outbreak populations above ecologically sustainable levels (Black and Moran, 1991; Hock et al., 2014; Pratchett et al., 2017a). These findings add empirical support for larval supply hypotheses and use of connectivity models to predict patterns of COTS secondary outbreaks.

Proxy indicators of water quality (relative flood plume exposure frequency) and larval nutrient supply (annual maximum chlorophyll concentration) had strong positive effects on COTS presence, prevalence and COTS outbreaks. These findings are consistent with larval nutrient supply hypotheses and numerous recent studies highlighting increased larval survival at elevated food concentrations (Pratchett et al., 2017b; Uthicke et al., 2015b; Wolfe et al., 2015, 2017) and correlations between outbreak cycles and chlorophyll *a* peaks (Brodie et al., 2017; Wooldridge and Brodie, 2015). Particularly important were the results from the BRT models that indicate sharp threshold-type responses at low levels of chlorophyll dominated secondary plume exposure for all three COTS response variables and a spike in COTS prevalence at ~0.25 and 0.6 μg chl *a* L^{-1} (Fig. 2C). These results provide some support to the contentious threshold responses to chlorophyll concentrations identified by Fabricius et al. (2010), which occur between 0.25 and 1 μg chla *a* L^{-1}. Influential positive relationships were also observed for mean and standard deviation of chlorophyll concentration, indicating that both baseline nutrient concentrations (Fabricius et al., 2010; Pratchett et al., 2017b; Wolfe et al., 2015) and exposure to nutrient pulses are indeed important predictors of COTS outbreak dynamics (Brodie et al., 2017; Wooldridge and Brodie, 2015). Low levels of exposure to the more turbid primary plumes was an important predictor, particularly for COTS outbreaks and prevalence, yet high levels of exposure appear to be deleterious to COTS populations, presumably due to decreased salinity. Reduced salinity (30% of ambient conditions) has been shown to increase larval survival (Lucas, 1973), yet Caballes et al. (2017b) observed a sharp decline in fertilisation, gastrulation and cleavage rates below 30%, indicating the importance of moderate exposure to plume conditions. These relationships are in accordance with Pratchett et al.'s (2014) assertion that the spread of outbreaks is conditional on years of high larval survivorship, which is facilitated by major flood events (i.e. infrequent exposure to primary and secondary plumes) that enhance food availability, and promote larval survival. Indeed, our results suggest that regulating water quality on the GBR remains one of the most important management actions for early intervention to reduce or mitigate damage caused by COTS outbreaks.

Most importantly, our results highlight the importance of moving beyond simple (single or dual variable) approaches to modelling COTS distribution patterns. While there is merit in creating easy-to-interpret and parsimonious models, there is an inherent trade-off with model accuracy. For COTS, prediction accuracy is key to determining the allocation of limited resources for COTS control. For example our full model predictions (~95% Accuracy) were far more effective than using solely the potential supply of COTS larvae to a reef (in-strength) (61.1%), exposure to flood plumes (68.1%) or combining both (68.1%). Previous research has used solely simulated connectivity networks to estimate the most important reefs for spreading waves of COTS outbreaks (Hock et al., 2014), planning optimised control procedures (Hock et al., 2016) and identifying important coral source reefs which could help replenish depleted broodstock (Hock et al., 2017). These estimates of COTS and coral larval dispersal are also used by the GBRMPA to prioritise reefs for COTS control. Our findings do not undermine the utility of these connectivity estimates, but combining them with ecological and environmental parameters can improve predictions of

COTS distribution patterns. Moreover, improvements to the underlying assumptions of the connectivity models, particularly the detail of maps (Roelfsema et al., 2018) used to determine reef habitats, and calibration with finer scale hydrodynamic models (Bode et al., 2018; Mumby et al., 2018; Thomas et al., 2014) will further improve modelling estimates and decision-making regarding the allocation of COTS control resources.

One of the most persistent hypotheses to explain the differential occurrence of COTS outbreaks at individual reefs is the "predator removal hypothesis". Most studies, however, do not focus on specifically on the abundance of the COTS high profile natural predator (the giant triton, *Charonia tritonis*) due to its scarcity on the GBR, and instead use fishing pressure (i.e. zoning regulation) as a proxy indicator for predator abundance (Sweatman, 2008; Vanhatalo et al., 2017). Our study found that zoning regulation was the lowest ranked predictor within the BRT framework, however a few interesting patterns were identified in the GAM models (Fig. 3B and C). Most significant was the apparent interaction between potential larval supply (IN_STR) and zoning. The best-fit GAM model indicated low outbreak probability at low levels of larval supply for reefs closed to fishing, but outbreak probabilities remained high at all levels of potential larval supply for reefs open to fishing (Fig. 3C). This suggests that reefs with lower larval supply may still be able to reach outbreak densities if predation pressure has been reduced. Therefore, our results provide some tentative support for predator removal hypotheses, however it must be noted that when compared to the importance of other predictors (namely connectivity and water quality) the effect of zoning was small. Furthermore, it must be kept in mind that our analysis, like most other studies, does not account for the effect of the giant triton, as even at pre-harvest densities it is uncertain that they would have been effective at controlling COTS outbreaks (Pearson and Endean, 1969; Pratchett et al., 2014). Contemporary research is aimed at rearing giant tritons and synthesising their predator cues (Bose et al., 2017; Hall et al., 2017a,b) in an attempt to control COTS populations and thus future scenario modelling should aim to identify the potential utility of such approaches.

Understanding and managing for the cumulative impact of disturbance on the GBR, and coral reefs globally, has been a key research objective in recent years (Burke et al., 2011; Hughes et al., 2017a; Mellin et al., 2019a) and our study highlights complex links between disturbance regimes (namely thermal stress and tropical cyclones) and COTS outbreaks. Cyclone exposure was also an influential predictor for all three responses across both modelling frameworks but was more influential in GAM models (Figs 2 and 3). GAM models revealed interesting marginal effects with a reduction of outbreak, prevalence and presence potential as exposure increased, before an increase at the highest levels of exposure. This can be partially explained by COTS feeding preference for fast growing corals (e.g. *Acroporids, Pocilloporids*) (Pratchett, 2007) which are both the most acutely affected by and first to recolonise from severe disturbance events (Mellin et al., 2019b). In this way, recent exposure to damaging cyclonic waves may eliminate COTS preferred food source, yet after some years of recovery, these reefs may become ideal locations for COTS outbreaks. There was a clearer negative trend for wave exposure (i.e. non-cyclonic), indicating COTS preference for more protected mid-shelf reefs, where they are less likely to be disrupted from feeding via oceanic wave action.

In our study, the impact of bleaching and thermal stress was less clear cut, likely because enhanced COTS larval survival in higher temperatures (Uthicke et al., 2015b) competes with the reduction of coral food supply for adults following severe thermal stress events. This means that temperature and thermal stress, while being important to COTS life history and prey availability, become difficult to interpret with regards to their effect upon COTS spatial distribution. Exemplifying this, although we found DHW exposure and minimum sea surface temperature to be useful predictors of COTS presence (Figs 2 and 3), these more likely represent the latitudinal gradient that exists due to numerous severe outbreaks observed in the southern tip of the GBR and none observed in the warmer, most northern sections. Additionally, there is limited data in the northern region where the greatest impact of the 2016–2017 mass bleaching events was recorded, making it difficult to draw solid conclusions about the effect of thermal stress and bleaching on COTS. It can be postulated however, that the impact of severe bleaching episodes may be similar to that of tropical cyclones, during which COTS preferred food sources (*Acropora sp.*) experience the most severe mortality (Hughes et al., 2018), but are also the first to recolonise (Mellin et al., 2019b) given sufficient broodstock remains. Although reefs severely affected by bleaching may not be prone to COTS outbreaks in the short term, it is assumed that if they recover, they will once again become vulnerable to COTS outbreaks. It is important to note that as the incidence of severe bleaching events increases (Hughes et al., 2017b; King et al., 2017), coral cover may decline to a point where no self-sustaining COTS populations can be supported, altering the system entirely.

A few important caveats need to be considered when interpreting our results. Primarily, it is important to note that sampling density plays an influential role in the predictability of COTS distribution patterns. Given that the major predictors of COTS prevalence and outbreaks are auto-covariates which relate either the distance or connectivity to reefs with outbreaks, scarce empirical data result in uncertain predictions. For example, high probabilities (>0.5) of COTS presence (Fig. 3A) were predicted offshore of the Whitsundays and Mackay (Central-Southern) yet outbreaks or prevalent populations were not predicted in this region. Due to the lack of empirical data in this region, it is unclear if these predictions indicate connectivity between high density outbreak reefs in the Central GBR to the Southern region. It is more likely that our model underestimates the probability of outbreaks and prevalent populations in this region due to the high relative influence of the auto-covariates for these responses. However, as the COTS control program has expanded, there will be unprecedented levels of observational data for the GBR, particularly in the traditionally data poor regions of the far North and Central regions (Fig. 1). We suggest that as the data from the expanded COTS control program continues to be collected, filling in spatial and temporal gaps, such modelling exercises should be updated to include all available data, or potentially automated as part of the GBRMPA's planned Reef Integrated Monitoring and Reporting program to keep predictions up to date and continually improving. Additionally, while the relative importance between modelling frameworks were generally similar, the relative importance of wave and cyclone exposure (STRESS and CYC_EXP) was markedly higher in the GAM framework. This is most likely due to the restriction placed (by design) on the GAM approach, limiting the number of variables (6) and degree of non-linearity allowed (5 knots). This approach favours models with less complex non-linear trends such as wave and cyclone exposure (Fig. 3B–D) compared to the water quality and chlorophyll variables favoured by the BRT approach (Fig. 2B–D).

Our study is the first to compare the relative influence of these variables alongside a host of other potentially influential factors (i.e. salinity, temperature, wave exposure) on the presence of COTS individuals, the probability of COTS populations reaching outbreak densities and the prevalence of these populations. This highlights the importance of incorporating all available data (namely connectivity and environmental variables) for both the assessment of contemporary theory and also for embedding data into the decision making process. In a time where extensive observational and modelled environmental estimates exist for the entire extent of the GBR and reefs worldwide, it is imperative to build our predictions from all available data and using modelling frameworks that can accommodate them. Given that 30 years of research have not provided a satisfactory simple model to explain spatial patterns of COTS outbreaks (Babcock et al., 2016a; Pratchett et al., 2017a), it is vital to include as much relevant information as possible in our estimates. This study successfully integrates data from multiple disparate sources into a cohesive framework, comparing the relative influence of competing hypotheses for COTS population dynamics while also allowing for the nature of individual relationships to be interrogated. Continuing to build data-driven platforms for decision making and embracing emerging technologies for detection will prove crucial to the effective control of COTS on the GBR and globally.

References

AIMS, 2019. Reef Monitoring. https://www.aims.gov.au/docs/research/monitoring/reef/reef-monitoring.html. Accessed on 2019-03-02.

Allen, J.D., Schrage, K.R., Foo, S.A., Watson, S.A., Byrne, M., 2017. The effects of salinity and pH on fertilization, early development, and hatching in the crown-of-thorns seastar. Diversity 9, 13.

Antonelli, P.L., Kazarinoff, N.D., 1984. Starfish predation of a growing coral reef community. J. Theor. Biol. 107, 667–684.

Augustin, N.H., Mugglestone, M.A., Buckland, S.T., 1996. An autologistic model for the spatial distribution of wildlife. J. Appl. Ecol. 33, 339.

Babcock, R.C., Mundy, C.N., Whitehead, D., 1994. Sperm diffusion models and in situ confirmation of long-distance fertilization in the free-spawning asteroid *Acanthaster planci*. Biol. Bull. 186, 17–28.

Babcock, R.C., Dambacher, J.M., Morello, E.B., Plagányi, É.E., Hayes, K.R., Sweatman, H.P.A., Pratchett, M.S., 2016a. Assessing different causes of crown-of-thorns starfish outbreaks and appropriate responses for management on the great Barrier Reef. PLoS One 11, e0169048.

Babcock, R.C., Milton, D.A., Pratchett, M.S., 2016b. Relationships between size and reproductive output in the crown-of-thorns starfish. Mar. Biol. 163, 234.

Barrat, A., Barthélemy, M., Pastor-Satorras, R., Vespignani, A., 2004. The architecture of complex weighted networks. Proc. Natl. Acad. Sci. U. S. A. 101, 3747–3752.

Berkelmans, R., De'ath, G., Kininmonth, S., Skirving, W.J., 2004. A comparison of the 1998 and 2002 coral bleaching events on the Great Barrier Reef: spatial correlation, patterns, and predictions. Coral Reefs 23, 74–83.

Birkeland, C., 1982. Terrestrial runoff as a cause of outbreaks of *Acanthaster planci* (Echinodermata: Asteroidea). Mar. Biol. 69, 175–185.

Birkeland, C., Lucas, J.S., 1990. *Acanthaster planci*: Major Management Problem of Coral Reefs. CRC Press, Boca Raton, FL.

Black, K.P., Moran, P.J., 1991. Influence of hydrodynamics on the passive dispersal and initial recruitment of larvae of *Acanthaster planci* (Echinodermata: Asteroidea) on the Great Barrier Reef. Mar. Ecol. Prog. Ser. 69, 55–65.

Bode, M., Bode, L., Choukroun, S., James, M.K., Mason, L.B., 2018. Resilient reefs may exist, but can larval dispersal models find them? PLoS Biol. 16, e2005964.

Bose, U., Suwansa-ard, S., Maikaeo, L., Motti, C.A., Hall, M.R., Cummins, S.F., 2017. Neuropeptides encoded within a neural transcriptome of the giant triton snail Charonia tritonis, a Crown-of-Thorns Starfish predator. Peptides 98, 3–14.

Brodie, J., Devlin, M., Lewis, S., 2017. Potential enhanced survivorship of crown of thorns starfish larvae due to near-annual nutrient enrichment during secondary outbreaks on the central mid-shelf of the Great Barrier Reef, Australia. Diversity 9, 17.

Burke, L., Reytar, K., Spalding, M., Perry, A., 2011. Reefs at Risk Revisited: Technical Notes on Modeling Threats to the World's Coral Ree. pp. 1–19.

Caballes, C.F., Pratchett, M.S., Buck, A.C.E., 2017a. Interactive effects of endogenous and exogenous nutrition on larval development for crown-of-thorns starfish. Diversity 9, 15.

Caballes, C.F., Pratchett, M.S., Raymundo, M.L., Rivera-Posada, J.A., 2017b. Environmental tipping points for sperm motility, fertilization, and embryonic development in the crown of thorns starfish. Diversity 9, 10.

Chesher, R.H., 1969. Destruction of the Pacific corals by the sea star *Acanthaster planci*. Science 165, 280–283.

Condie, S., Hepburn, M., Mansbridge, J., 2012. In: Modelling and Visualisation of Connectivity of the Great Barrier Reef. Proceedings of the 12th International Coral Reef Symposium, Cairns, Australia 9–13 July.

Cowan, Z.-L., Pratchett, M., Messmer, V., Ling, S., Cowan, Z.-L., Pratchett, M., Messmer, V., Ling, S., 2017. Known predators of crown-of-thorns starfish (*Acanthaster* spp.) and their role in mitigating, if not preventing, population outbreaks. Diversity 9, 7.

Cragg, J.G., 2006. Some statistical models for limited dependent variables with application to the demand for durable goods. Econometrica 39, 829.

CSIRO. 2019. eReefs Research. https://research.csiro.au/ereefs/, Accessed on 20/4/2019.

De'ath, G., 2002. Multivariate regression trees: a new technique for modeling species-environment relationships. Ecology 83, 1105–1117.

De'ath, G., 2003. Analyses of Crown-of-Thorns Starfish Data From the Fine-Scale Surveys and the Long-Term Monitoring Program Manta Tow Surveys. CRC Reef Research Centre Technical Report. CRC Reef Research Centre, Townsville.

De'ath, G., 2007. Boosted trees for ecological modeling and prediction. Ecology 88, 243–251.

De'ath, G., Fabricius, K.E., Sweatman, H., Puotinen, M., 2012. The 27-year decline of coral cover on the Great Barrier Reef and its causes. Proc. Natl. Acad. Sci. U. S. A. 109, 17995–17999.

Devlin, M.J., McKinna, L.W., Álvarez-Romero, J.G., Petus, C., Abott, B., Harkness, P., Brodie, J., 2012. Mapping the pollutants in surface riverine flood plume waters in the Great Barrier Reef, Australia. Mar. Pollut. Bull. 65, 224–235.

Dormann, C.F., Mcpherson, J.M., Araújo, M.B., Bivand, R., Bolliger, J., Carl, G., Davis, R., Hirzel, A., Jetz, W., Kissling, W.D., Kühn, I., Ohlemüller, R., Peres-Neto, P.R., Reineking, B., Oder, S., Schurr, F.M., Wilson, R., 2007. Methods to account for spatial autocorrelation in the analysis of species distributional data: a review. Ecography 30, 609–628.

Elith, J., Leathwick, J.R., Hastie, T., 2008. A working guide to boosted regression trees. J. Anim. Ecol. 77, 802–813.

Elith, J., Kearney, M., Phillips, S., 2010. The art of modelling range-shifting species. Methods Ecol. Evol. 1, 330–342.

Endean, R., 1969. Report on Investigations Made Into Aspects of the Current *Acanthaster planci* (Crown-of-Thorns) Infestations of Certain Reefs of the Great Barrier Reef. Queensland Department of Primary Industries (Fisheries branch), Brisbane, Australia.

Endean, R., 1974. In: *Acanthaster planci* on the Great Barrier Reef. Proceedings of the Second International Symposium on Coral Reefs. vol. 1, pp. 563–576.

Fabricius, K.E., Okaji, K., De'ath, G., 2010. Three lines of evidence to link outbreaks of the crown-of-thorns seastar *Acanthaster planci* to the release of larval food limitation. Coral Reefs 29, 593–605.

Fisher, R., Wilson, S.K., Sin, T.M., Lee, A.C., Langlois, T.J., 2018. A simple function for full-subsets multiple regression in ecology with R. Ecol. Evol. 8, 6104–6113.

Franklin, J., Miller, J.A., 2010. Mapping Species Distributions: Spatial Inference and Prediction. Cambridge University Press.

GBRMPA. 2014a. Management Areas of the GBRMP. http://www.gbrmpa.gov.au/geoportal/, Accessed on 23/1/2019.

GBRMPA, 2014b. Great Barrier Reef Marine Park Zoning. http://www.gbrmpa.gov.au/geoportal/. Accessed on 2019-02-03.

Guisan, A., Thuiller, W., 2005. Predicting species distribution: Offering more than simple habitat models. Ecol. Lett. 8, 993–1009.

Gumpertz, M.L., Graham, J.M., Ristaino, J.B., 1997. Autologistic model of spatial pattern of phytophthora epidemic in bell pepper: effects of soil variables on disease presence. J. Agric. Biol. Environ. Stat. 2, 131.

Hall, M.R., Kocot, K.M., Baughman, K.W., Fernandez-Valverde, S.L., Gauthier, M.E.A., Hatleberg, W.L., Krishnan, A., McDougall, C., Motti, C.A., Shoguchi, E., Wang, T., Xiang, X., Zhao, M., Bose, U., Shinzato, C., Hisata, K., Fujie, M., Kanda, M., Cummins, S.F., Satoh, N., Degnan, S.M., Degnan, B.M., 2017a. The crown-of-thorns starfish genome as a guide for biocontrol of this coral reef pest. Nature 544, 231–234.

Hall, M.R., Motti, C.A., Kroon, F., 2017b. The potential role of the giant triton snail, Charonia tritonis (Gastropoda: Ranellidae) in mitigating populations of the crown-of-thorns starfish. Report to the National Environmental Science Programme. Cairns.

Hardy, N.A., Lamare, M., Uthicke, S., Wolfe, K., Doo, S., Dworjanyn, S., Byrne, M., 2014. Thermal tolerance of early development in tropical and temperate sea urchins: inferences for the tropicalization of eastern Australia. Mar. Biol. 161, 395–409.

Heron, S.F., Johnston, L., Liu, G., Geiger, E.F., Maynard, J.A., De La Cour, J.L., Johnson, S., Okano, R., Benavente, D., Burgess, T.F.R., Iguel, J., Perez, D.I., Skirving, W.J., Strong, A.E., Tirak, K., Eakin, C.M., 2016. Validation of reef-scale thermal stress satellite products for coral bleaching monitoring. Remote Sens. (Basel) 8, 1–16.

Hock, K., Wolff, N.H., Condie, S.A., Anthony, K.R.N., Mumby, P.J., 2014. Connectivity networks reveal the risks of crown-of-thorns starfish outbreaks on the Great Barrier Reef. J. Appl. Ecol. 51, 1188–1196.

Hock, K., Wolff, N.H., Beeden, R., Hoey, J., Condie, S.A., Anthony, K.R.N., Possingham, H.P., Mumby, P.J., 2016. Controlling range expansion in habitat networks by adaptively targeting source populations. Conserv. Biol. 30, 856–866.

Hock, K., Wolff, N.H., Ortiz, J.C., Condie, S.A., Anthony, K.R.N., Blackwell, P.G., Mumby, P.J., 2017. Connectivity and systemic resilience of the Great Barrier Reef. PLoS Biol. 15, e2003355.

Huang, Z., Brooke, B., Whitta, N., Potter, A., Fuller, M., Dunn, J., 2010. Australian marine physical environmental data-description and metadata. Geosci. Australia Record 2010 (32), 141.

Huang, Z., Brooke, B., Whitta, N., Potter, A., 2013. Australian marine physical environmental data—descriptions and metadata. Geosci. Australia Record 2010, 32.

Hughes, T.P., Barnes, M.L., Bellwood, D.R., Cinner, J.E., Cumming, G.S., Jackson, J.B.C., Kleypas, J., Van De Leemput, I.A., Lough, J.M., Morrison, T.H., Palumbi, S.R., Van Nes, E.H., Scheffer, M., 2017a. Coral reefs in the Anthropocene. Nature 546, 82–90.

Hughes, T.P., Kerry, J.T., Álvarez-Noriega, M., Álvarez-Romero, J.G., Anderson, K.D., Baird, A.H., Babcock, R.C., Beger, M., Bellwood, D.R., Berkelmans, R., Bridge, T.C., Butler, I.R., Byrne, M., Cantin, N.E., Comeau, S., Connolly, S.R., Cumming, G.S., Dalton, S.J., Diaz-Pulido, G., Eakin, C.M., Figueira, W.F.,

Gilmour, J.P., Harrison, H.B., Heron, S.F., Hoey, A.S., Hobbs, J.P.A., Hoogenboom, M.O., Kennedy, E.V., Kuo, C.Y., Lough, J.M., Lowe, R.J., Liu, G., McCulloch, M.T., Malcolm, H.A., McWilliam, M.J., Pandolfi, J.M., Pears, R.J., Pratchett, M.S., Schoepf, V., Simpson, T., Skirving, W.J., Sommer, B., Torda, G., Wachenfeld, D.R., Willis, B.L., Wilson, S.K., 2017b. Global warming and recurrent mass bleaching of corals. Nature 543, 373–377.

Hughes, T.P., Kerry, J.T., Baird, A.H., Connolly, S.R., Dietzel, A., Eakin, C.M., Heron, S.F., Hoey, A.S., Hoogenboom, M.O., Liu, G., McWilliam, M.J., Pears, R.J., Pratchett, M.S., Skirving, W.J., Stella, J.S., Torda, G., 2018. Global warming transforms coral reef assemblages. Nature 556, 492–496.

James, M.K., Scandol, J.P., 1992. Larval dispersal simulations: correlation with the crown-of-thorns starfish outbreaks database. Aust. J. Mar. Freshwat. Res., 569–582.

Johan, R., 2015. Penggunaan Kortikosteroid Topikal yang Tepat. Contin. Prof. Dev. 42, 308–312.

Johnson, C., 1992. Settlement and recruitment of *Acanthaster planci* on the Great Barrier Reef: questions of process and scale. Aust. J. Mar. Freshwat. Res. 43, 611–627.

Johnson, D., Moran, P., Baker, V., Christie, C., 1992. An attempt to locate high density populations of juvenile crown-of-thorns starfish (*Acanthaster planci*) on the central Great Barrier Reef. Coral Reefs 11, 122.

Kamya, P.Z., Dworjanyn, S.A., Hardy, N., Mos, B., Uthicke, S., Byrne, M., 2014. Larvae of the coral eating crown-of-thorns starfish, *Acanthaster planci* in a warmer-high CO2 ocean. Glob. Chang. Biol. 20, 3365–3376.

Kenchington, R.A., 1977. Growth and recruitment of *Acanthaster planci* (L.) on the great barrier reef. Biol. Conserv. 11, 103–118.

King, A.D., Karoly, D.J., Henley, B.J., 2017. Australian climate extremes at 1.5 °C and 2 °C of global warming. Nat. Clim. Chang. 7, 412–416.

Lamare, M., Pecorino, D., Hardy, N., Liddy, M., Byrne, M., Uthicke, S., 2014. The thermal tolerance of crown-of-thorns (*Acanthaster planci*) embryos and bipinnaria larvae: implications for spatial and temporal variation in adult populations. Coral Reefs 33, 207–219.

Lane, D.J.W., 1996. A crown-of-thorns outbreak in the eastern Indonesian Archipelago, February 1996. Coral Reefs 15, 209–210.

Leathwick, J.R., Elith, J., 2009. Species distribution models: ecological explanation and prediction across space and time. Annu. Rev. Ecol. Evol. Syst. 40, 677–697.

Linares, C., Pratchett, M.S., Coker, D.J., 2011. Recolonisation of Acropora hyacinthus following climate-induced coral bleaching on the Great Barrier Reef. Mar. Ecol. Prog. Ser. 438, 97–104.

Lucas, J.S., 1973. Reproductive and larval biology of *Acanthaster planci* (L.) in Great Barrier Reef waters. Micronesica 9, 197–203.

Lucas, J.S., 1984. Growth, maturation and effects of diet in *Acanthaster planci* (L.) (Asteroidea) and hybrids reared in the laboratory. J. Exp. Mar. Biol. Ecol. 79, 129–147.

MacNeil, M.A., Mellin, C., Matthews, S., Wolff, N.H., McClanahan, T.R., Devlin, M., Drovandi, C., Mengersen, K., Graham, N.A.J., 2019. Water quality mediates resilience on the Great Barrier Reef. Nat. Ecol. Evol. 3, 620–627.

Marmion, M., Parviainen, M., Luoto, M., Heikkinen, R.K., Thuiller, W., 2009. Evaluation of consensus methods in predictive species distribution modelling. Divers. Distrib. 15, 59–69.

Matthews, S.A., Mellin, C., MacNeil, A., Heron, S.F., Skirving, W., Puotinen, M., Devlin, M.J., Pratchett, M., 2019. High-resolution characterization of the abiotic environment and disturbance regimes on the Great Barrier Reef, 1985–2017. Ecology 100, e02574.

Mellin, C., Bradshaw, C.J.A., Meekan, M.G., Caley, M.J., 2010. Environmental and spatial predictors of species richness and abundance in coral reef fishes. Glob. Ecol. Biogeogr. 19, 212–222.

Mellin, C., Russell, B.D., Connell, S.D., Brook, B.W., Fordham, D.A., 2012. Geographic range determinants of two commercially important marine molluscs. Divers. Distrib. 18, 133–146.
Mellin, C., Lurgi, M., Matthews, S., MacNeil, M.A., Caley, M.J., Bax, N., Przeslawski, R., Fordham, D.A., 2016. Forecasting marine invasions under climate change: Biotic interactions and demographic processes matter. Biol. Conserv. 204, 459–467.
Mellin, C., Matthews, S., Anthony, K.R.N., Brown, S.C., Caley, M.J., Johns, K.A., Osborne, K., Puotinen, M., Thompson, A., Wolff, N.H., Fordham, D.A., MacNeil, M.A., 2019a. Spatial resilience of the Great Barrier Reef under cumulative disturbance impacts. Glob. Chang. Biol. 25, 2431–2445.
Mellin, C., Thompson, A., Jonker, M.J., Emslie, M.J., Mellin, C., Thompson, A., Jonker, M.J., Emslie, M.J., 2019b. Cross-shelf variation in coral community response to disturbance on the Great Barrier Reef. Diversity 11, 38.
Miller, I.R., Jonker, M., Coleman, G., 2009. Crown-of-thorns starfish and coral surveys using the manta tow and SCUBA search techniques. Long-term Monitoring of the Great Barrier Reef Standard Operation Procedure Number 9 Edition 3. Page Standard Operation Procedure, AIMS.
Moran, P.J., 1986. The acanthaster phenomenon. Oceanogr. Mar. Biol. Annu. Rev. 24, 379–480.
Mumby, P.J., Hock, K., Condie, S.A., Ortiz, J.C., Wolff, N.H., Anthony, K.R.N., Blackwell, P.G., 2018. Response to Bode and colleagues: 'Resilient reefs may exist, but can larval dispersal models find them?'. PLoS Biol. 16, e2007047.
Osborne, K., Dolman, A.M., Burgess, S.C., Johns, K.A., 2011. Disturbance and the dynamics of coral cover on the Great Barrier Reef (1995-2009). PLoS One 6, e17516.
Pearson, R.G., Endean, R.G., 1969. A preliminary study of the coral predator *Acanthaster planci* (L.) (Asteroidea) on the Great Barrier Reef. Fish. Notes Queensland Dep. Harb. Mar. 3, 27–55.
Potts, J.M., Elith, J., 2006. Comparing species abundance models. Ecol. Model. 199, 153–163.
Pratchett, M.S., 2007. Feeding preferences of *Acanthaster planci* (Echinodermata: Asteroidea) under controlled conditions of food availability. Pac. Sci. 61, 113–120.
Pratchett, M.S., Caballes, C.F., Rivera-Posada, J.A., Sweatman, H.P.A., 2014. Limits to understanding and managing outbreaks of crown of thorns starfish (Ancathanster spp). Oceanogr. Mar. Biol. Annu. Rev. 52, 133–200.
Pratchett, M., Caballes, C., Wilmes, J., Matthews, S., Mellin, C., Sweatman, H., Nadler, L., Brodie, J., Thompson, C., Hoey, J., Bos, A., Byrne, M., Messmer, V., Fortunato, S., Chen, C., Buck, A., Babcock, R., Uthicke, S., 2017a. Thirty years of research on crown-of-thorns starfish (1986–2016): scientific advances and emerging opportunities. Diversity 9, 41.
Pratchett, M., Dworjanyn, S., Mos, B., Caballes, C., Thompson, C., Blowes, S., 2017b. Larval survivorship and settlement of crown-of-thorns starfish (*Acanthaster* cf. *solaris*) at varying algal cell densities. Diversity 9, 2.
Puotinen, M., Maynard, J.A., Beeden, R., Radford, B., Williams, G.J., 2016. A robust operational model for predicting where tropical cyclone waves damage coral reefs. Sci. Rep. 6, 26009.
R Development Core Team, 2017. R: A language and environment for statistical computing. Vienna, Austria.
Reichelt, R.E., Bradbury, R.H., Moran, P.J., 1990. Distribution of *Acanthaster planci* outbreaks on the Great Barrier Reef between 1966 and 1989. Coral Reefs 9, 97–103.
Robinson, L.M., Elith, J., Hobday, a.J., Pearson, R.G., Kendall, B.E., Possingham, H.P., Richardson, a.J., 2011. Pushing the limits in marine species distribution modelling: Lessons from the land present challenges and opportunities. Glob. Ecol. Biogeogr. 20, 789–802.

Roelfsema, C., Kovacs, E., Ortiz, J.C., Wolff, N.H., Callaghan, D., Wettle, M., Ronan, M., Hamylton, S.M., Mumby, P.J., Phinn, S., 2018. Coral reef habitat mapping: A combination of object-based image analysis and ecological modelling. Remote Sens. Environ. 208, 27–41.

Rogers, J.G.D., Plágányi, É.E., Babcock, R.C., 2017. Aggregation, Allee effects and critical thresholds for the management of the crown-of-thorns starfish *Acanthaster planci*. Mar. Ecol. Prog. Ser. 578, 99–114.

Sato, Y., Bell, S.C., Nichols, C., Fry, K., Menéndez, P., Bourne, D.G., 2018. Early-phase dynamics in coral recovery following cyclone disturbance on the inshore Great Barrier Reef, Australia. Coral Reefs 37, 431–443.

Scandol, J.P., James, M.K., 1992. Hydrodynamics and larval dispersal: a population model of *Acanthaster planci* on the Great Barrier Reef. Aust. J. Mar. Freshwat. Res. 43, 583–596.

Seymour, R.M., Bradbury, R.H., 1999. Lengthening reef recovery times from crown-of-thorns outbreaks signal systemic degradation of the Great Barrier Reef. Mar. Ecol. Prog. Ser. 176, 1–10.

Stump, R., 1996. An investigation to describe the population dynamics of *Acanthaster planci* (L.) around Lizard Island, Cairns section, Great Barrier Reef Marine Park. CRC Reef Research Centre Technical Report No. 10. Townsville, Australia.

Sutcliffe, P.R., Mellin, C., Pitcher, C.R., Possingham, H.P., Caley, M.J., 2014. Regional-scale patterns and predictors of species richness and abundance across twelve major tropical inter-reef taxa. Ecography 37, 162–171.

Sweatman, H., 2008. No-take reserves protect coral reefs from predatory starfish. Curr. Biol. 18, 598–599.

Sweatman, H.P.A., Cheal, A., Coleman, G., Emslie, M.J., Johns, K.A., Jonker, M., Miller, I., Osborne, K., 2008. Long-term monitoring of the Great Barrier Reef, status report number 8. AIMS, Townsville, Australia.

Thomas, C.J., Lambrechts, J., Wolanski, E., Traag, V.A., Blondel, V.D., Deleersnijder, E., Hanert, E., 2014. Numerical modelling and graph theory tools to study ecological connectivity in the Great Barrier Reef. Ecol. Model. 272, 160–174.

Ting, K.M., Witten, I.H., 1999. Issues in stacked generalization. J. Artif. Intell. Res. 10, 271–289.

Uthicke, S., Schaffelke, B., Byrne, M., 2009. A boom–bust phylum? Ecological and evolutionary consequences of density variations in echinoderms. Ecol. Monogr. 79, 3–24.

Uthicke, S., Doyle, J., Duggan, S., Yasuda, N., McKinnon, A.D., 2015a. Outbreak of coral-eating crown-of-thorns creates continuous cloud of larvae over 320 km of the Great Barrier Reef. Sci. Rep. 5, 16885.

Uthicke, S., Logan, M., Liddy, M., Francis, D., Hardy, N., Lamare, M., 2015b. Climate change as an unexpected co-factor promoting coral eating seastar (*Acanthaster planci*) outbreaks. Sci. Rep. 5, 8402.

Vanhatalo, J., Hosack, G.R., Sweatman, H., 2017. Spatiotemporal modelling of crown-of-thorns starfish outbreaks on the Great Barrier Reef to inform control strategies. J. Appl. Ecol. 54, 188–197.

Vine, P.J., 1973. Crown of thorns (*Acanthaster planci*) plagues: the natural causes theory. Atoll Res. Bull. 166, 1–10.

Warton, D.I., Hui, F.K.C., 2011. The arcsine is asinine: the analysis of proportions in ecology. Ecology 92, 3–10.

Wolanski, E., Richmond, R., Mccook, L., Sweatman, H., 2003. Mud, marine snow and coral reefs the survival of coral reefs requires integrated watershed-based management activities and marine conservation. Am. Sci. 91, 44–51.

Wolfe, K., Graba-Landry, A., Dworjanyn, S.A., Byrne, M., 2015. Larval starvation to satiation: Influence of nutrient regime on the success of *Acanthaster planci*. PLoS One 10, 1–17.

Wolfe, K., Graba-Landry, A., Dworjanyn, S.A., Byrne, M., 2017. Superstars: assessing nutrient thresholds for enhanced larval success of *Acanthaster planci*, a review of the evidence. Mar. Pollut. Bull. 116, 307–314.

Wolpert, D.H., 1992. Stacked generalization. Neural Netw. 5, 241–259.

Wooldridge, S.A., Brodie, J.E., 2015. Environmental triggers for primary outbreaks of crown-of-thorns starfish on the Great Barrier Reef, Australia. Mar. Pollut. Bull. 101, 805–815.

CHAPTER TEN

COTSMod: A spatially explicit metacommunity model of outbreaks of crown-of-thorns starfish and coral recovery

S.A. Matthews[a,b,*], K. Shoemaker[c], Morgan S. Pratchett[a], and C. Mellin[d]
[a]Australian Research Council Centre of Excellence for Coral Reef Studies, James Cook University, Townsville, QLD, Australia
[b]Australian Institute of Marine Science, Townsville, QLD, Australia
[c]Department of Natural Resources and Environmental Science, University of Nevada Reno, Reno, NV, United States
[d]The Environment Institute and School of Biological Sciences, University of Adelaide, Adelaide, SA, Australia
[*]Corresponding author: e-mail address: sammatthews990@gmail.com

Contents

Abstract

Outbreaks of the Pacific crown-of-thorns starfish (COTS; *Acanthaster* cf. *solaris*) have been responsible for 40% of the decline in coral cover on the GBR over the last 35 years. With the intensity and frequency of bleaching and cyclonic disturbances increasing, effectively managing these outbreaks may allow reefs an opportunity to recover from these cumulative impacts. Significant research effort has been directed toward developing regional scale models for COTS outbreaks, but these have yet to be fit explicitly to long term time series at the scale of the entire GBR, nor do previous research efforts incorporate explicit estimates of cumulative disturbance history. We developed a

Advances in Marine Biology, Volume 87
ISSN 0065-2881
https://doi.org/10.1016/bs.amb.2020.09.001

stage-based metapopulation model for COTS at a 1 × 1 km resolution using long-term time series and modelled estimates of COTS larval connectivity, nutrient concentrations and important vital rates estimated from the literature. We coupled this metapopulation model to an existing spatially explicit model of coral cover growth, disturbance and recovery across the GBR from 1996 to 2017 to create a metacommunity model. Our results were validated against a spatially and temporally extensive dataset of COTS and coral cover across the GBR, predicting an average coral decline of 1.3% p.a. across the GBR, and accurately recreating coral cover trajectories (mean prediction error = 7.1%) and COTS outbreak classification (accuracy = 80%). Sensitivity analyses revealed that overall model accuracy was most sensitive to larval predation (boosted regression tree; relative importance = 46.7%) and two parameters defining juvenile density dependent mortality (21.5% and 17.5%). The COTS model underestimated peak COTS densities particularly in the Swains and Townsville sectors of the reef, while overestimating COTS density during non-outbreak years. A better understanding of inter-annual variability in larval connectivity, and regionally variable density dependence for adult COTS life stages may improve model fit during these extreme outbreak events. Our model provides a platform to develop upon, and with improvements to estimates of larval connectivity and larval predation could be used to simulate the effects of implementing varying combinations of COTS interventions. This research highlights the importance of the early life history stages of COTS as drivers of outbreak dynamics, emphasizing the need for further empirical research to estimate these parameters.

1. Introduction

Crown of thorns starfish (COTS; *Acanthaster* spp.) are a prominent coral predator on coral reefs in the Indo-Pacific. COTS are characterized by an immense fecundity (Babcock et al., 2016b; Kettle and Lucas, 1987), combined with synchronous spawning and consequently high fertilization rates (Babcock et al., 1994; Babcock and Mundy, 1992), making them predisposed to rapid increases in density, or outbreaks (reaching up to >1000 individuals Ha^{-1}) (Chesher, 1969; De'ath, 2003; Kayal et al., 2012). These factors combined with increased nutrient availability for larval COTS (Brodie et al., 2005), reduced predation on both juvenile and adult life stages (Endean, 1969; Sweatman, 2008), and favourable hydrodynamic conditions (Hock et al., 2014; Wooldridge and Brodie, 2015) are hypothesized to be responsible for the initiation and spread of COTS outbreaks. On the Great Barrier Reef, outbreaks of COTS are a major cause of coral loss, responsible for ~40% of the coral declines over the last 30 years (De'ath et al., 2012; Mellin et al., 2019). Moreover, among the major causes of coral loss (e.g., cyclones and bleaching), COTS outbreaks are the only disturbance that may be readily amenable to direct interventions (De'ath et al., 2012). Although outbreaks of COTS has been a key research and management focus (GBRMPA, 2017; Pratchett et al., 2014, 2017a; Westcott et al., 2016), there remains considerable uncertainty regarding the exact mechanisms by which COTS outbreaks initiate, spread and halt (Pratchett et al., 2017a) as well as the efficacy of COTS management more broadly (Westcott et al., 2020). Creating predictive modelling frameworks that account for such uncertainties are fundamental in improving management of COTS and the desired outcome of protecting coral.

Primary outbreaks of COTS build up over a number of years within the Northern management region (Fig. 1: Cairns and Cooktown/Lizard Island sectors) (Endean, 1974; Johnson, 1992; Pratchett, 2005; Stump, 1996; Vanhatalo et al., 2017). These outbreak densities of COTS (>0.22 COTS per 2 min manta tow) are potentially initiated by elevated nutrient conditions resulting from riverine input during the wet season, which is hypothesized to dramatically increase the survival of COTS larvae (Brodie et al., 2005; Fabricius et al., 2010; Wolfe et al., 2017). However, elevated nutrient conditions occur quite frequently (3 5 years) in this region and do not always result in primary outbreaks (Wooldridge and Brodie, 2015). Wooldridge and Brodie (2015) highlighted how the initiation of primary outbreak occurred when elevated nutrients coincided with increased larval connectivity between reefs during neutral El Niño Southern Oscillation (ENSO) phases. Due to the preference of COTS for fast growing corals (Pratchett et al., 2014), and the relationship between maternal nutrition and fertilization success (Caballes et al., 2016), the time lag between outbreak cycles is also likely linked to the recovery period required for fast growing corals following a COTS outbreak.

Once initiated, secondary outbreaks are generally observed on the mid shelf reefs south of the "initiation box" and are exemplified by large larval influx and more dramatic increases in COTS densities, aided by near-annual nutrient enrichment (Brodie et al., 2017). Primary outbreaks that occur in the southern Swain sector of the GBR are suggested to be somewhat independent from and concurrent with the northern outbreaks (Brodie et al., 2017) (Fig. 1). Following an outbreak, COTS populations decline dramatically, and adult starfish are often not detected in surveys the following year (Pratchett, 2005; Sweatman et al., 2008; Vanhatalo et al., 2017). These declines are presumably due to density dependent mortality processes such as starvation or disease (Pratchett, 1999), although this process is poorly

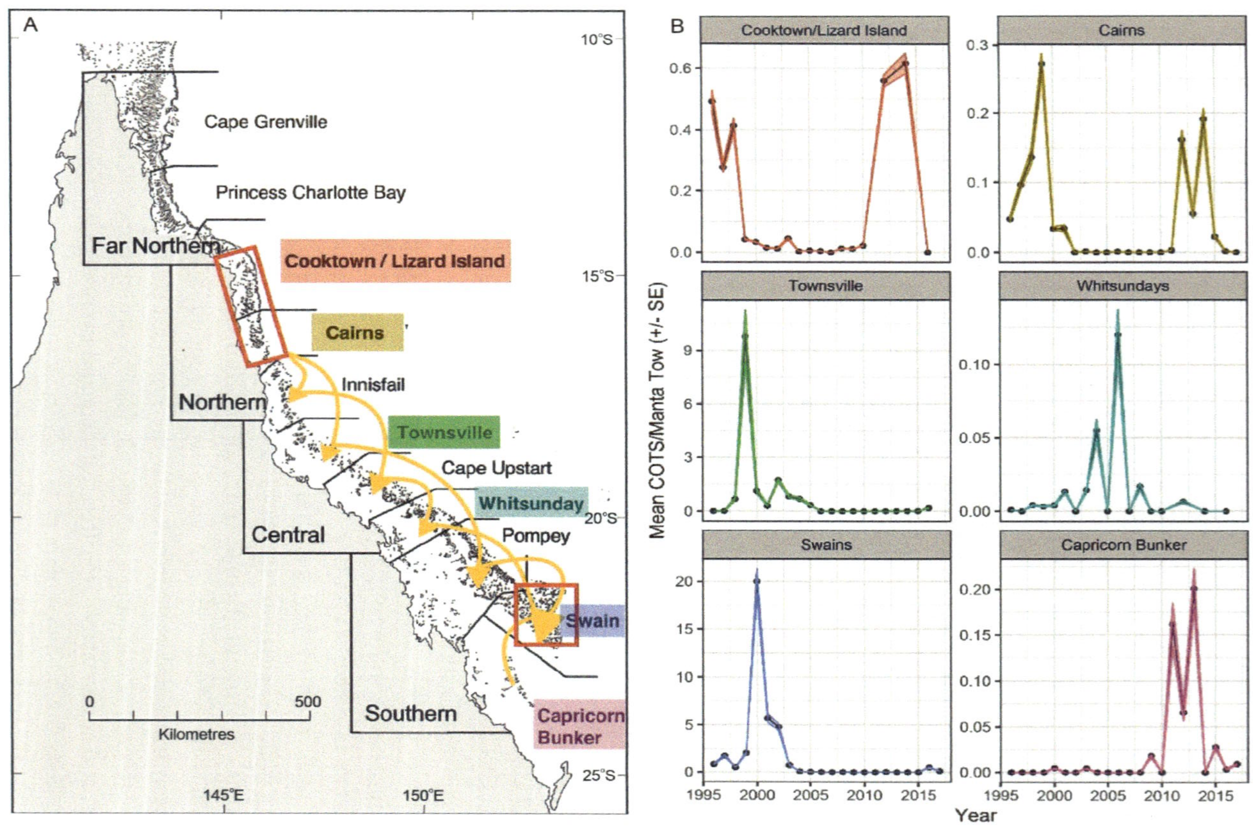

Fig. 1 (A) Representation of initiation zones for primary outbreaks (red boxes) and the subsequent propagation of secondary outbreaks at downstream reefs. (B) Empirical estimates of COTS densities aggregated for six major sectors of the GBR from the AIMS Long Term Monitoring Program (AIMS LTMP). (A) Also presents the four management areas of the GBR.

resolved and has not been well documented in the field or recreated in laboratory conditions (Mills, 2012; Pratchett et al., 2017a). After populations at a reef collapse, COTS are usually not observed again in significant densities for around 10 years, often allowing coral to recover (pending no other disturbance) before the next outbreak cycle begins (Fig. 1).

COTS rapid expansion in range and boom-bust dynamics during outbreak cycles (Uthicke et al., 2009) mean significant care must be taken when modelling their population dynamics (Mellin et al., 2016). Density dependent or resource dependent vital rates (e.g., mortality, fecundity) are key parameters for simulating boom and bust dynamics in a demographic population model (Caswell, 2006). For COTS, predator prey dynamics (Lotka, 1910; Volterra, 1928) have traditionally been used to model outbreak cycles (McCallum, 1993; Mellin et al., 2016; Morello et al., 2014), focusing primarily on coral prey availability as well as trophic interactions with COTS predators to determine COTS mortality rates. However, for COTS, it is not apparent that all "bust" phases of the outbreaks are necessarily linked to comprehensive depletion of coral resources (Pratchett, 2010), and density-dependent pathogenesis is also hypothesized to play an important role in COTS population collapse (Birkeland and Lucas, 1990; Pratchett, 1999; Zann et al., 1987). Ratio-dependent processes, where vital rates are linked to the ratio between available prey and predator abundance (Abrams and Ginzburg, 2000; Arditi and Ginzburg, 1989; Haque, 2009), are an alternative approach to traditional predator-prey type dynamics, aiming to represent both density and resource dependent processes. Incorporating both predator and prey dependent processes may offer a solution to simulating boom-bust dynamics for COTS, particularly to capture a "bust" mechanism that does not require total resource depletion.

Recent advances in hydrodynamic, COTS population and coral growth modelling were important precursors to the development of a modelling framework for a COTS-Coral metacommunity model at the GBR scale. MacNeil et al. (2019) developed a Gompertz-based coral growth model for reefs monitored by the Australian Institute of Marine Sciences Long Term Monitoring Program (AIMS LTMP). This model was extended by Mellin et al. (2019), who recreated the trajectories of coral cover across the entire GBR at a 1×1 km resolution using a dataset defining the disturbance history and environmental characteristics of reef locations (Matthews et al., 2019). Furthermore, significant advances in larval connectivity modelling (Condie et al., 2012; Hock et al., 2014) have allowed connectivity networks for COTS to be derived over four spawning seasons (2012–2016)

(Hock et al., 2017), while the development of the eReefs biogeochemical model allows estimates of important environmental conditions such as chlorophyll-a concentration to be estimated across the GBR for recent years (Chen et al., 2011; CSIRO, 2019; Johnson, 1992). Additionally, recent studies have developed population models for COTS at small scales, highlighting the importance of trophic interactions in COTS population dynamics (Mellin et al., 2016; Morello et al., 2014) and identifying important ecologically relevant thresholds for COTS control above which coral cover is reduced (Babcock et al., 2014) and fertilization rates (Rogers et al., 2017). Combined with the extensive literature surrounding COTS biology outbreaks (reviewed by Caballes and Pratchett, 2014; Pratchett et al., 2014, 2017a,b) these advances provide the platform necessary to build a COTS-Coral metacommunity model that will allow researchers and managers to investigate the likely effect of management actions.

This study builds upon an existing model of coral growth and disturbance across the GBR (MacNeil et al., 2019; Matthews et al., 2019; Mellin et al., 2019) by incorporating a COTS population model at each reef, linked by larval connectivity estimates (Hock et al., 2014, 2017) to create a COTS-Coral metacommunity model for the GBR at a 1 × 1 km resolution. This study aims to: (1) use the best available data to recreate and validate the trajectories of COTS and coral populations across the entire GBR; (2) investigate which parameters are most important for recreating COTS outbreak patterns and (3) provide a platform for future development to compare COTS management strategies.

2. Methods

A COTS-Coral metacommunity model (https://github.com/sammatthews990/COTS_Model) was constructed at a 1 × 1 km resolution to recreate the trajectories of coral and COTS across the GBR between 1996 and 2017. The metacommunity model framework can be summarized according to four subcomponents: (i) the coral dynamics model governing coral growth and impacts from other disturbances (cyclones, bleaching, disease); (ii) COTS population model recreating the life history and density dependent processes for each 1 × 1 km grid cell; (iii) a larval survival model which controls the proportion of larvae surviving from each spawning event based upon the nutrient conditions, and (iv) the larval connectivity model which disperses COTS larvae among reefs, linking individual reef population into a metapopulation model. COTS populations are initiated in the model using GBR-wide estimates of COTS density in 1996, derived from the AIMS LTMP (Matthews et al., 2019). Model simulations are then run to recreate trajectories between 1996 and 2017. The model is calibrated to manta tow observations for coral cover and COTS for 46 reefs that were surveyed at least 10 times providing reef-wide estimates of coral cover and COTS densities. Validation statistics were calculated on a further 91 reefs that were surveyed 5–10 times between 1996 and 2017.

2.1 Coral dynamics model

Coral growth was parameterized across a standardized 1 × 1 km grid for reef locations, as defined by a Gompertz-based growth curve (Table 1). Previous research defined the growth curves and disturbance coefficients for 47 AIMS LTMP reefs (MacNeil et al., 2019). These growth parameters were then modelled for all reef locations across the GBR based on the predicted coral community composition (Mellin et al., 2019). At each yearly time step, coral cover was affected by disturbance (cyclones, bleaching, disease) and then recovered as per the spatially explicit estimates for coral growth. Annual disturbance exposure for the model was provided from a dataset compiling the disturbance history for the GBR (1986–2017) on the standardized 1 × 1 km grid (Matthews et al., 2019). Although this coral model was originally calibrated to include disturbance from COTS outbreaks as predicted from the AIMS LTMP, in our framework this portion of the coral dynamics model was replaced in the present study by a stage-based COTS population model.

2.2 COTS population model

To represent COTS demography, we developed a stage-based metapopulation model with a larval, two juvenile and an adult stage (Fig. 2). Basic life history parameters include rates of density dependent mortality across age cohorts, density-dependent fertilization and feeding rates on corals. Stage-specific mortality rates and fertilization rates were modelled as a function of conspecific densities and resource availability within each 1 × 1 km grid cell. Starting values for these parameters were based on estimates from the literature where possible, and otherwise from expert opinion (Pratchett, unpublished data) (Table 1). Stage-specific COTS consumption rates, also

Table 1 Equations and sources for the population dynamics of COTS and Coral populations and vital rates for the demographic components of the COTS model where COTS abundances and coral cover are defined for the *i*th grid cell at the *r*th reef in year *y*.

Name	Equation	Source
Stage transition rates		
COTSL	$N_y^{L_i} = N_{y,Ai} \times Fec \times FvD \times (1 - Pred_L) \times Surv_y^{L_i}$	NA
COTSJ1	$N_{y+1}^{J1_i} = \left(\sum_i^r N_y^{L_i} \times CONN_{ri,y} \right) \times (1 - Mort_{J1})$	
COTSJ2	$N_{y+1}^{J2_i} = N_{y+1}^{J1_i} \times (1 - Mort_{J2})$	
COTSA	$N_{y+1}^{A_i} = N_{y+1}^{J2_i} \times (1 - Mort_A)$	
Coral growth		
Coral	$log(HC_{y+1,i}) = r_{s,i} + (1 - \alpha_i)\ log(HC_{yi}) + \sum_j \beta_j\ Disturb_{j,y,i} + \sum_j \beta_j\ Disturb_{j,y,i} \times PF_i$	MacNeil et al. (2019) and Mellin et al. (2019)
Vital rates		
FvD	$L_\infty (1 - e^{-K(N_{y,A})})$; $L_\infty = 0.8,\ K = 0.0007$	Babcock et al. (1994) and Rogers et al. (2017)
$Pred_L$	$N(\mu_P, \sigma_P)$; $N(0.98, 0.01)$	Scandol (1999)
$Surv_L$	$logit(Surv_y^{L_i}) = \beta_0 + \beta_1 log_2([Chl_y^i])$; $\beta_0 = [-0.4, 10]$, $\beta_1 = 2.02$	Fabricius et al. (2010) and Wolfe et al. (2017)
$Mort_{J1}$	$\frac{L}{1 + e^{-k\left(N_y^{J1_i} - x_0\right)}}; L = 1; k \sim N(-2e^7,\ 0.5e^7); x_0 \sim N(1e^{-7}, 0.2e^{-7})$	NA
$Mort_{J2}$	$\frac{L}{1 + e^{-k\left(N_y^{J2_i} - x_0\right)}}; L = 1; k \sim N(-7e^4,\ 2e^4); x_0 \sim N(2.5e^{-5}, 0.5e^{-5})$	NA

$Mort_A$	$f\left(\frac{HC}{COTS_y^{A_i}}\right) = \begin{cases} Mort_{A,\max}, & for \frac{HC}{COTS_y^{A_i}} < CCRatio_2 \\ b_{mort} \times \frac{HC}{COTS_y^{A_i}} + 1, & for\ CCRatio_2 < \frac{HC}{COTS_y^{A_i}} < CCRatio_1 \\ Mort_{A,\min}, & for \frac{HC}{COTS_y^{A_i}} \geq CCRatio_1 \end{cases}$	Scandol (1999) and Sweatman et al. (2008)
b_{mort}	$b_{mort} = {}^{Mort_{A,max} - 1}/_{CCRatio_1 - CCRatio_2}$	Slope of linear function, see Fig. 3C
Fec	$f\left(\frac{HC_y^i}{COTS_y^{A_i}}\right) = \begin{cases} Fec_{max}, for \frac{HC_y^i}{COTS_y^{A_i}} \geq CCRatio_1 \\ \left(b_y^{fec_i} \times \frac{HC_y^i}{COTS_y^{A}} + 0.1\right) Fec_{\max}, for \frac{HC_y^i}{COTS_y^{A_i}} < CCRatio_1 \end{cases}$	Babcock et al. (2016a) and Caballes et al. (2016)
b_{fec}	$b_{fec} = {}^{1 - Fec_{min}}/_{CCRatio_1}$	Slope of linear function, see Fig. 3D
Fec_{max}	$N(\mu_F, \sigma_F)$; $N(2e^7, 1e^7)$	Babcock et al. (2016a)
Constants		
Fec_{min}	Proportion of Fec_{max} at lowest levels of Coral-COTS ratio (see Fig. 3D) Starting estimate = 0.2	Scandol (1999)
$Mort_{A,max}$	Maximum mortality rate of $COTS_A$ at lowest levels of Coral-COTS ratio (see Fig. 3C) Starting estimate = 0.99	NA
$Mort_{A,min}$	Background mortality rate of $COTS_A$ at higher levels of Coral-COTS ratio (see Fig. 3C) Starting Estimate = 0.65	Scandol (1999) and Zann et al. (1990)

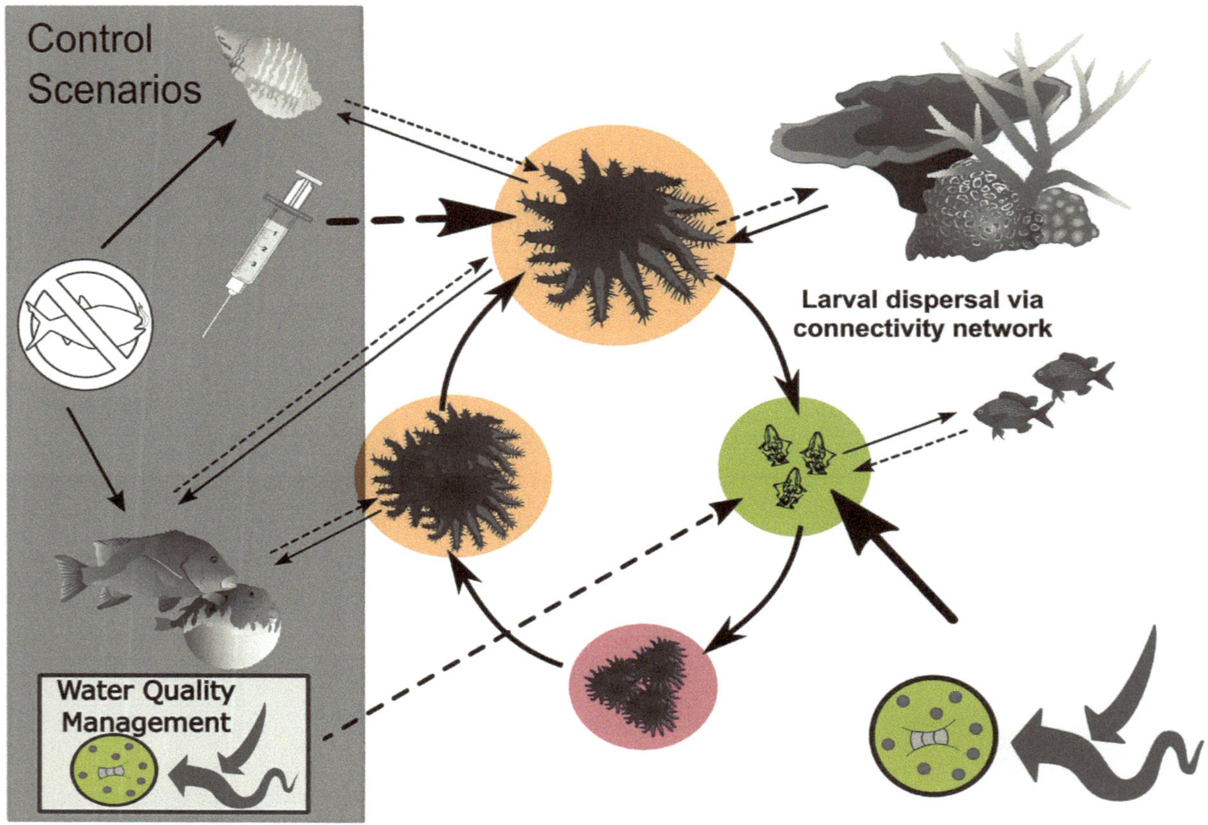

Fig. 2 Conceptual diagram of the main factors involved in COTS outbreaks on the GBR and their relationship to various stages of the COTS life cycle. The white area represents the interactions currently included in the COTS-Coral metacommunity model, the shaded portion of the diagram represents potential management scenarios that could be incorporated into the modelling framework for projections. Coloured circles represent COTS food source for that life stage (green = phytoplankton/chlorophyll; pink = crustose coralline algae; coral = scleractinian corals). Solid lines represent a positive effect and dashed lines indicate a negative effect. Symbols are courtesy of the Integration and Application Network, University of Maryland Centre for Environmental Science (ian.umces.edu/symbols/).

drawn from the literature, enable the COTS population model to link with the coral growth model as an additional disturbance factor. Importantly all the starting parameter estimates discussed here were simply "best guess estimates" and were allowed to vary within the model framework (via Latin hypercube sampling) to find the best model fit and test the sensitivity of model predictions to changes in estimates used. Hereafter, "starting values" are used to describe point estimates which were drawn from the literature and subsequently used to define an uncertainty interval extending the range of plausible values by 20% in both directions.

Adult mortality and fecundity rates are assumed to be linearly dependent on the ratio of available coral prey for each adult COTS (Fig. 3C and D). Coral cover-COTS ratio (*CCRatio*) was defined as *%Hard Coral Cover/ COTS per manta tow*. Two thresholds were defined with respect to *CCRatio*, $CCRatio_1$ aims to capture the ratio at which coral cover is expected to decline as a result of consumption of COTS (20–40% Hard Coral Cover/COTS per manta tow, depending on coral cover levels (Babcock et al., 2014)). At this threshold it is assumed that coral prey resources become scarce—increasing mortality, decreasing maternal nutrition and therefore reducing fecundity (Caballes et al., 2016). In the model, fecundity (mortality) is highest (lowest) at large values of *CCRatio* and decreases (increases) linearly towards its minimum (maximum) (Fig. 3). The maximum and minimum fecundity also use starting estimates from the literature (Babcock et al., 2016b). A second threshold $CCRatio_2$ was defined to capture the recurring dramatic collapse of COTS populations observed in the AIMS LTMP data (Fig. 1B), below which the mortality rate was set to its maximum ($Mort_{A,\ max}$; Table 1). This starting threshold was determined using empirical observation data from the AIMS LTMP (Fig. 3A and B). We used two approaches to identify starting values $CCRatio_1$ and $CCRatio_2$. Firstly, we fit a generalized additive mixed model (GAMM) to the proportional change in COTS populations (−1 to 1) in the year following a recorded COTS outbreak (>0.22 individuals Manta Tow^{-1}) as a function of *CCRatio*, using cross shelf location as random factor (Fig. 3A). While this model did not explain a significant amount of variation (GAMM: $R^2=0.03$, $P=0.11$), it suggests a threshold for *CCRatio* between 5 and 12 where COTS populations begin to decline drastically. The second approach was to model change in COTS population size (in the following year) for populations exceeding severe outbreak threshold (>1 individuals Manta Tow^{-1}) as a binary outcome in a generalized linear mixed model framework (increase in COTS density = 1; >50% decrease in COTS

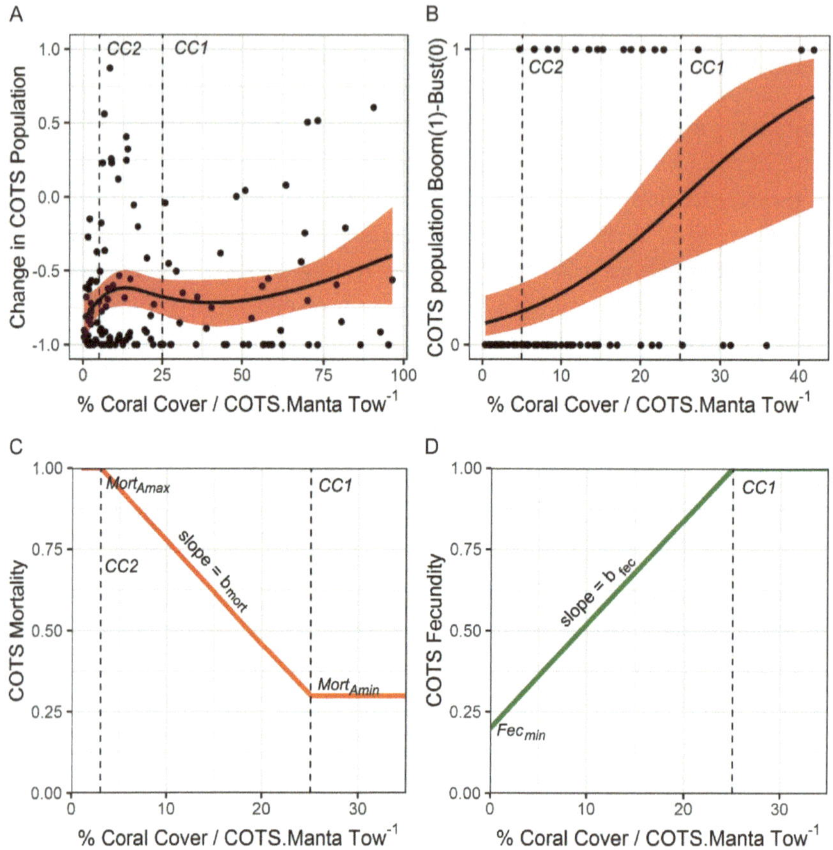

Fig. 3 (A) GAMM model for proportional COTS population declines following extreme COTS outbreak densities, dotted lines represent *CCRatio*$_1$ *and CCRatio*$_2$ which are used to define the onset of ratio-dependent processes (see Table 1); (B) binomial Relationship between Coral Cover-COTS ratio and the change to COTS population in the following survey year from AIMS LTMP data to provide initial estimates for ratio dependent thresholds used in the metacommunity model. Events were scored as 1 = COTS populations increased; 0 COTS populations declined sharply (>50% decline); (C) schematic of potential initial values for ratio-dependent thresholds for COTS mortality (MortA, Table 1), were 1 represents 100% COTS mortality, and (D) fecundity (Fec, Table 1) in relation to coral cover-COTS ratios, where 1 equals Fec_{max} (Table 1). The slope of these linear relationships are defined in Table 1.

density = 0) to identify the *CCRatio* at which increased mortality is triggered (Fig. 3). Despite a high level of dispersion in the data, results suggest that a declining *CCRatio* is significantly related to COTS population collapse (GLMM; $R^2 = 0.11$, $P < 0.01$). Importantly, there is no recorded case of

COTS population increase below a Coral-COTS ratio, *CCRatio* of 4.6 (Fig. 3B). From these results and the work of Babcock et al. (2014) we chose starting values of 5 for $CCRatio_2$ (threshold below which COTS populations collapse) and 25 for $CCRatio_1$ (threshold below which COTS mortality and fecundity are negatively affected). These starting values however were allowed to vary during model tuning to find the most optimum thresholds.

We modelled juvenile and sub-adult COTS mortality as a function of COTS density (we chose not to use a ratio-dependent model for these life stages as juveniles feed on crustose coralline algae, and there is no data on sub-adult densities to estimate ratio-dependent thresholds) within each 1 km grid cell according to a standard logistic growth model. We initially parameterized these relationships on the basis of estimates of COTS densities at critical outbreak thresholds (De'ath, 2003) (Table 2). Using these thresholds we estimated the number of juveniles and subadults present via an approximate estimate of COTS stable stage distribution generated using the R package "popbio" (Stubben and Milligan, 2007) and vital rates from Table 1.

Table 2 Estimated densities of different life stages of COTS at varying levels of Outbreaks to provide starting values for density dependent mortality curves.

COTS/Manta Tow	Outbreak status	Approx. adult density (adult ind km^{-2})	Subadult density (J2 ind km^{-2})[a]	Juvenile density (J1 ind km^{-2})[a]
0.01	No Outbreak	1000	6900	3,90,000
0.1	Potential	3500	23,600	13,54,000
0.22	Established	5100	33,500	19,23,000
1	Severe	11,100	71,900	41,22,000

[a]While earlier life stages are estimated from an approximate stable stage distribution.
There are four **COTS life stages** in the model (A = Adults, J2 = Subadult; J1 = Juveniles, L = Larval), which are defined by: $Pred_L$ = larval predation rate; $Surv_L$ = larval survival as predicted by the logit chlorophyll model derived from Fabricius et al (2012); *CONN* = average connectivity potential between reefs on the GBR; $Mort_{J1}$ = logistic density dependent mortality juvenile COTS; $Mort_{J2}$ = logistic density dependent mortality sub-adult (J2) COTS. $Mort_A$ = ratio dependent mortality for adult COTS (A); *FvD* = Von Bertalanffy growth curve for fertilization by density; *Fec* = ratio-dependent per capita fecundity for female COTS; Fec_{max} = per capita fecundity without resource limitation; $CCRatio_1$ = ratio dependent thresholds onset of ratio-dependent fecundity and adult mortality; $CCRatio_2$ = threshold below which COTS populations collapse; **Hard Coral Cover:** HC = hard coral cover; r_s = intrinsic coral growth rate; α = strength of density dependence for sessile species; β_j $Disturb_{j,\,y}$ = effect size and Estimate for *j*th disturbance (bleaching, cylones, disease, other); *PF* = combined exposure to flood plumes.
Adult densities are estimated using the calibration of De'ath (2003).

2.3 Larval survival model

To incorporate the well-established positive effects of prey availability on COTS larval survival (Fabricius et al., 2010; Pratchett et al., 2017b; Uthicke et al., 2015; Wolfe et al., 2015), a logistic model of larval survival in relation to chlorophyll concentrations ([chl-a]) was developed based on the results from Fabricius et al. (2010). Other research has shown COTS ability to survive in oligotrophic conditions to be much higher than those estimated by this model (Pratchett et al., 2017b, Wolfe et al., 2015) and thus the intercept and slope term of this relationship were included as a model parameter for tuning via Latin hypercube sampling (see Section 2.5). This provides the opportunity to test the sensitivity of model predictions to this crucial relationship, within empirically supported bounds.

In order to model larval survival, [chl-a] was estimated for the period 1996–2017 using the 4×4 km eReefs biogeochemical model for the GBR (CSIRO, 2019). This model has produced [chl-a] estimates between 2010 and 2017 that we used to determine the reef-level relationship between [chl-a] and ENSO cycles. On the GBR, ENSO is commonly linked to rainfall, with strong La Niña events increasing rainfall and thus nutrient enrichment (Devlin et al., 2001; Lough, 2001) and larval survival. Additionally, La Niña phases of the ENSO cycle are broadly linked to upwelling and phytoplankton blooms (Steinberg, 2011) and thus using ENSO for hindcasting offers an opportunity to incorporate variation in [chl-a] from both rainfall and upwelling events. For each reef, a GAM was fit to determine the relationship between ENSO (using the September-November Nino 3.4 index (Rayner et al., 2003)) and [chl-a], and thus hindcast to 1996 using historical Niño 3.4 data (Tables S1 and S2 in Supplementary materials in the online version at https://doi.org/10.1016/bs.amb.2020.09.001) (Fig. 4A and B). For each year of available eReefs data a matern variogram model (Matérn, 1960; Rossi et al., 1992) was fit to determine the spatial autocorrelation in the model residuals, and a mean psill (partial sill = sill − nugget) and range were calculated to determine the variogram model for La Niña, El Niño and neutral years. A total of 100 spatially correlated random replicate sets of chlorophyll predictions were then generated for each year based on the ENSO estimate to provide sub-reef (1×1 km grid cell) level variability and incorporate uncertainty in model estimates.

2.4 Larval connectivity model

Estimates for larval connectivity between reefs on the GBR were obtained from Hock et al. (2017). These connectivity networks are constructed from 4 years of hydrodynamic modelled data (2009–2013) and estimate the mean

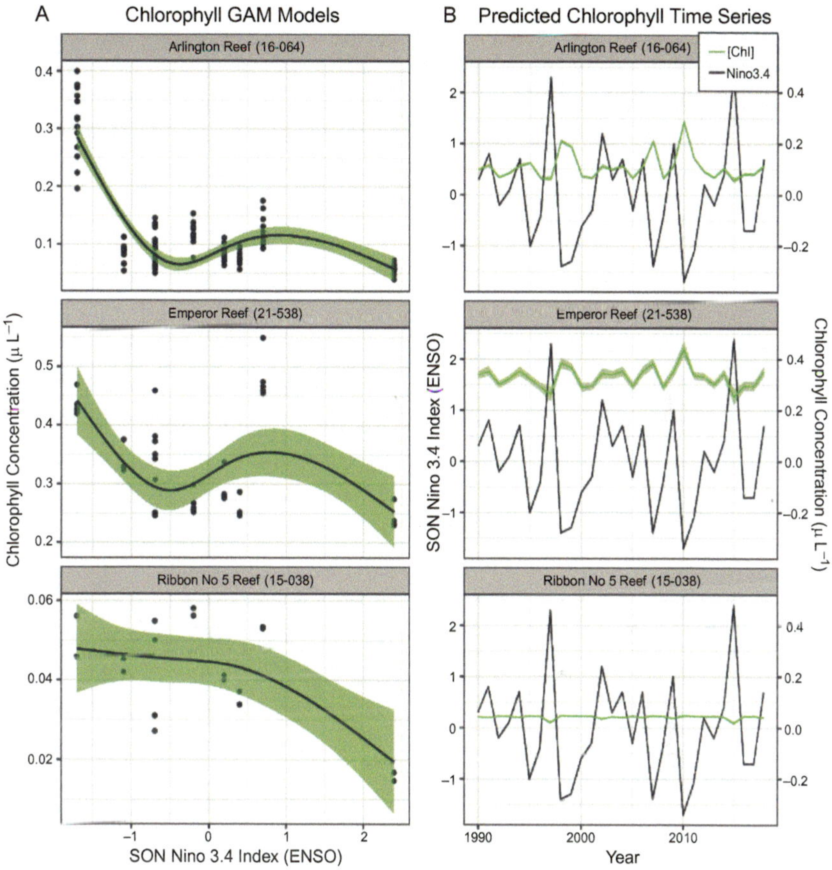

Fig. 4 (A) GAM models fitting ENSO Nino 3.4 index to eReefs predicted chlorophyll concentrations for 3 reefs with variable background concentrations and (B) the predicted chlorophyll concentrations for the period 1990–2018 (Green) overlaying the Nino 3.4 index (black line).

potential connectivity between all reefs on the GBR. Potential connectivity (expressed as a proportion) is a simplification of real-world processes which essentially represents maximal physically possible advective transport among reefs as implied by the model (Hock et al., 2014). During the dispersal phase of our model, surviving larvae from each source reef were pooled at the reef level and then distributed among connected reefs using the strength of each connection averaged from the 4 years of modelled estimates. Larvae were then evenly distributed among the grid cells at the target reef, and thus did not take into account the potential for larvae to be attracted to either prey or conspecific aggregations.

2.5 Model validation and sensitivity analysis

Model predictions were calibrated using AIMS LTMP Manta Tow to generate a base case model that provided a good fit to the data. Reefs chosen for calibration were surveyed at least 10 times providing reef wide estimates of coral cover and COTS densities ($N=46$). Predictions were then validated against reefs within the AIMS LTMP dataset that were not included in the 46 reefs used for calibration, yet included over 5 years of data ($N=91$). Using the starting parameter variables (Table 1), we used Latin hypercube sampling (Carnell, 2009; McKay et al., 1978) to sample evenly across the parameter space, where starting values were varied by ±20%. For each parameter combination 100 simulated model runs were conducted to provide an estimate of the uncertainty of the stochastic (disturbance, chlorophyll concentrations and larval connectivity estimates), density and ratio-dependent elements of the model. Mean prediction error (MPE) for both COTS and Coral were estimated at each reef. Additionally for COTS, model accuracy (ACC) and kappa (KAP) statistics (Cohen, 1960; Hossin and Sulaiman, 2015) for predicting presence or absence of COTS outbreaks (OUT) (>0.22 COTS/Manta Tow) were calculated. ACC and KAP were also used to classify COTS across the four outbreak categories (Table 2).The parameter set which reduced prediction error and maximized prediction ACC and KAP (after ACC and KAP standardization to a 0–1 range) across all reefs was thus identified as the base-case model (Eq. 1).

$$V_m = \frac{(1 - MPE.HC_m) + (OUT.ACC_m) + (OUT.KAP_m) + (CL.ACC_m) + (CL.KAP_m)}{5} \tag{1}$$

where V_m is the validation metric for each model run m, $MPE.\ HC_m$ is the mean prediction error for hard coral estimates; $OUT.\ ACC_m$ and $OUT.\ KAP_m$ are the model accuracy and kappa statistic for the binomial classification of COTS outbreaks and $CL.\ ACC_m$ and $CL.\ KAP_m$ are the model accuracy and kappa statistic for multilevel classification of the four outbreak categories (Table 2). Additionally, mean outbreak density was calculated for each latitudinal sector (Fig. 1) as the mean COTS density (COTS/Manta Tow) for model estimates that exceeded the COTS outbreak threshold. This was compared to the values calculated for our validation reefs from the AIMS LTMP to investigate spatial variability in model performance in predicting COTS outbreak densities. It is important to note that the above methods defined how the model was tuned and calibrated to find sensible parameters settings, however due to the complexity of the model and times series to which it was fit, there was no formal process to obtain a globally optimized parameter set.

To identify the sensitivity of predictions to parameter estimates, 200 Latin hypercube samples were drawn with each parameter varying within ±15% of the bounds identified by the base case model. These parameter values were then scaled and used as explanatory variables for our validation metric V_m (Eq. 1) using boosted regression trees (BRT) (Elith et al., 2008) to determine the most influential parameters and interactions between them for determining model accuracy (Norton, 2015). Additionally, these scaled variables were also used to predict annual coral loss and COTS accuracy in multilevel classification (No COTS, No Outbreak = <0.11, Potential Outbreak = <0.22, Established Outbreak = <1, Severe Outbreak >1 COTS/Manta Tow) to investigate the variables most influential for driving COTS densities and coral loss in our model. All models were constructed using R v3.4.1 (R Development Core Team, 2017), with tools for sensitivity analyses provided by the "lhs" (Carnell, 2009) and "dismo" (Hijmans et al., 2017) packages.

3. Results

The base-case model calibrated against the long-term training dataset (1995–2017) had a mean prediction error (MPE) of 7.19% for coral cover across the entire time series, and a MPE of 0.01 COTS/Manta Tow and mean prediction accuracy of 80.6% for COTS Outbreaks (V_m = 0.83). There was substantial spatial variation to the accuracy of model predictions (Table 3) most notably for COTS outbreaks. COTS densities were overestimated in both the Capricorn Bunker (CB) and Whitsunday (WH) sectors, with peak densities, particularly in the Townsville (TO) and Swains (SW) sectors, being underestimated.

3.1 Spatial patterns in COTS outbreaks

Outbreaks were initiated in 1996 in the "initiation box", peaking between 1997 and 1998. Outbreaks in northern regions of the GBR peaked at slightly lower densities than those predicted for secondary outbreaks in the Innisfail, Townsville and Swain regions, however these peak densities were much lower than observed in the monitoring data (Table 3). Patterns for individual calibration reefs generally followed peaks and troughs in COTS densities, although the model tended to overestimate COTS abundance during the ~10–15 year break between outbreak cycles (Fig. 5). Importantly, the second outbreak cycle beginning around 2010 was predicted by the model in

Table 3 Validation statistics for best-fit model parameters across latitudinal sectors of the GBR (Fig. 1A PC = Prince Charlotte Bay; CL = Cooktown/Lizard Island; CA = Cairns; IN = Innisfail; TO = Townsville; CU = Cape Upstart; WH = Whitsundays; PO = Pompeys; SW = Swains; CB = Capricorn Bunker) where MPE = mean prediction error, ACC = model accuracy, KAP = kappa and outbreak density = model predicted mean COTS/Manta Tow when density exceeds 0.22 COTS/Manta Tow and Δ Outbreak Density is the % difference between model predicted outbreak density and AIMS LTMP Manta Tow observation.

Sector	MPE Hard coral (%)	MPE COTS (COTS/MT)	Acc COTS presence	Acc COTS outbreak	Kap COTS outbreak	Outbreak density	Δ Outbreak density
PC	2.81	0	0.36	0.82	0	NA	−100%
CL	7.60	0.066	0.69	0.91	0.51	0.50	2%
CA	2.52	0.010	0.80	0.98	0	0.46	35%
IN	4.29	0.057	0.29	0.85	0.17	0.53	−80%
TO	3.94	0.19	0.61	0.58	−0.074	0.65	−80%
CU	4.79	0.061	0.43	0.78	−0.12	0.67	46%
WH	6.94	0.049	0.78	0.92	−0.04	0.81	113%
PO	7.64	0.068	0.32	0.82	−0.10	0.67	−78%
SW	4.81	0.16	0.55	0.55	0.21	0.70	−92%
CB	7.29	0.15	0.82	0.76	0.18	0.64	94%
Overall	7.20	0.01	0.627	0.806	0.219	0.63	−81%

the northern reefs (see Fig. 5B, Martin Reef and Linnet Reef) but was smaller in magnitude and began a year earlier than observed. In southern regions however, the model produced a multiple boom-and-bust dynamic, where COTS densities collapsed and increased again multiple times throughout the period, which was not supported by the empirical observations.

3.2 Spatial patterns in coral cover

As coral cover was predominantly predicted from a previously calibrated model, the trajectories more closely match the empirical data (Fig. 6B). Overall reefs in the Southern-Central (Pompeys and Whitsunday sectors) region of the GBR had the highest rates of coral decline, with the Northern region (Cairns and Cooktown/Lizard Island sectors) also experiencing annual declines around 1% per year. However, some pockets of the GBR were predicted to experience annual increases in coral cover,

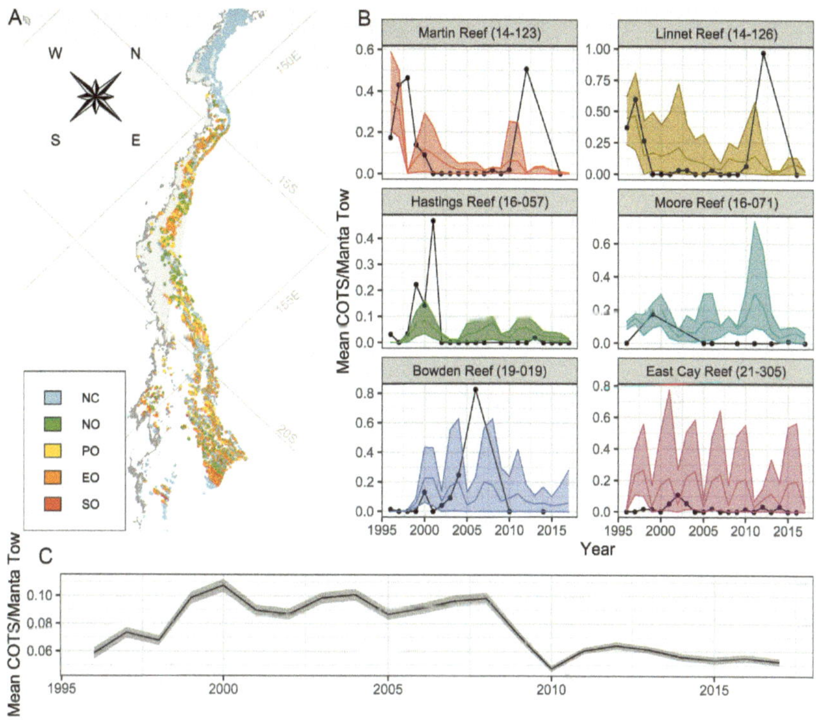

Fig. 5 (A) Predicted mean COTS abundance for each GBR reef with coloured diamonds matching reefs in B classified into outbreak categories in terms of COTS/Manta Tow; No COTS (NC) = 0; No outbreak (NO) < 0.11; Potential outbreak (PO) < 0.22; Established outbreak (EO) < 1; Severe outbreak (SO) > 1. (B) Predicted mean (±50% CI's) COTS/Manta Tow (Colored ribbon) plotted against observation data (black dotted line) for six calibration reefs across the GBR. (C) Predicted mean (± SE) COTS/Manta Tow, aggregated at the GBR level.

particularly in the Capricorn Bunker and inner shelf Pompey sectors. Overall coral cover was predicted to decline 1.3% p.a. across the 22 years of the study.

3.3 Model validation

Predictions were validated against reefs within the AIMS LTMP dataset that were not included in the 46 reefs used for calibration, yet included over 5 years of data ($N = 91$). For this independent dataset, our model captured the impact of multiple disturbances and subsequent coral recovery (mean prediction error = 7.9%; $R^2 = 0.14$) performing similarly as for the calibration data (7.4% $R^2 = 0.1$). For COTS, while our model was able to recreate the general trajectories of COTS outbreaks (Fig. 5), the accuracy

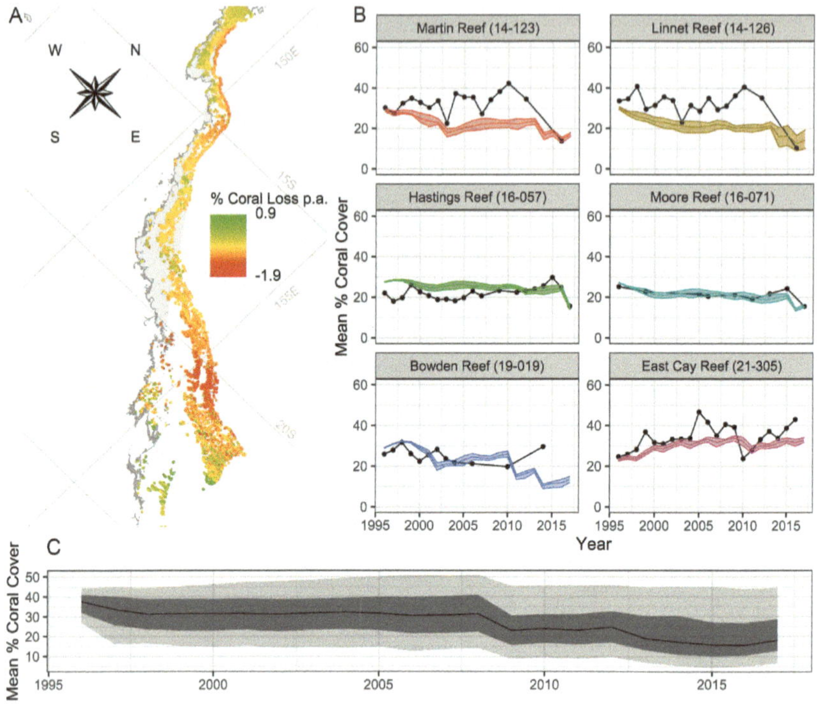

Fig. 6 (A) Predicted mean annual percent coral loss for each GBR reef with coloured diamonds matching reefs in (B) Predicted mean (±50% CI's) percent coral cover (Colored ribbon) plotted against observation data (black dotted line) for six calibration reefs (colour coded diamonds). Predicted mean (±50% and 95% CI's) percent coral cover, aggregated at the GBR level.

of the model for predicting the presence or absence of an outbreak of COTS was non-significant for both calibration (Accuracy = 80.0%, Kappa = 0.193, $P > 0.05$) and validation (78.9%, Kappa = 0.014, $P > 0.05$) datasets (Table 3). Similarly, when classifying for the four outbreak levels model accuracy as non-significant for both calibration (Accuracy = 61.0%, Kappa = 0.176, $P > 0.05$) and validation (52%, Kappa = 0.04, $P > 0.05$) datasets. Additionally, for these models, the mean COTS outbreak density was overestimated in the Whitsundays and Capricorn Bunker sectors while underestimated for Swain, Townsville, Pompeys and Innisfail sectors, indicating a tradeoff between presence/absence accuracy and matching peaks in COTS outbreak densities within the model. Importantly, accuracy for predicting presence/absence of COTS was only marginally better than chance alone, indicating a general tendency to estimate low levels of COTS when none were observed in the data.

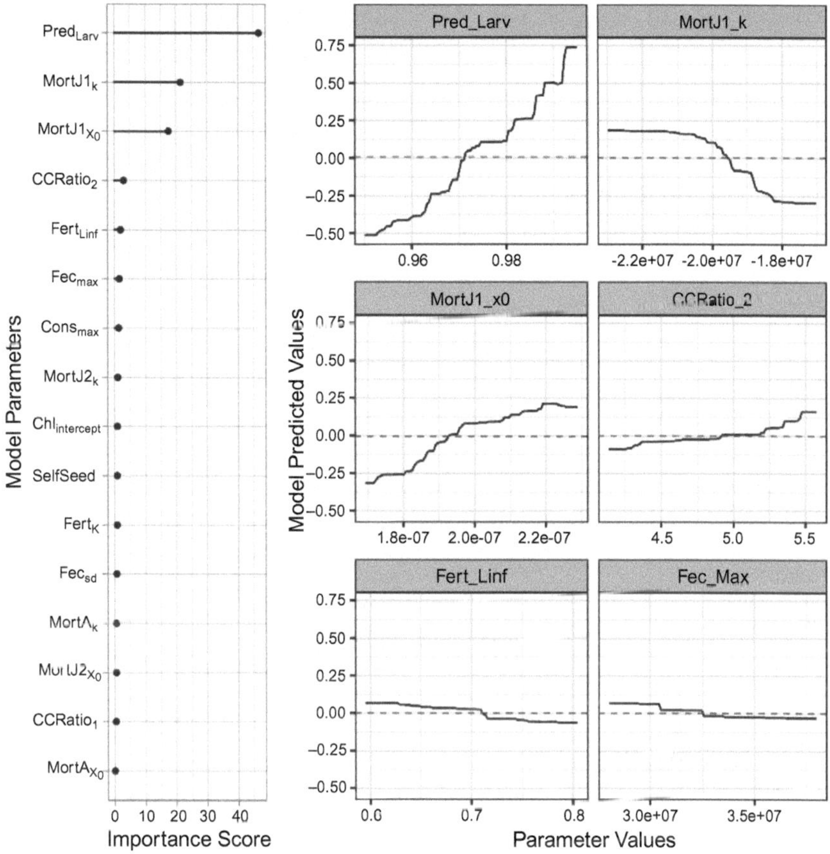

Fig. 7 Relative importance and partial dependency of COTS population model parameters in improving overall accuracy based on the combined V_m validation metric where $Pred_{Larv}$ = larval predation rate; Mort = density dependent mortality parameters (A = Adults, J2 = Subadult; J1 = Juveniles), CCRatio = ratio dependent thresholds for (1) onset of effect and (2) collapse of COTS populations; $Fert_{Linf}$ = max fertilizations rate; $Fert_k$ = rate parameter for fertilization; Fec_{max} = maximum fecundity per female COTS; Fec_{sd} = standard deviation for fecundity; $Chl_{intercept}$ = intercept term for the chlorophyll model, used to adjust larval survival; SelfSeed = scaling parameter to alter the proportion of COTS larvae settling at home reef.

3.4 Sensitivity analysis

Our sensitivity analysis (Fig. 7) using boosted regression trees (BRT) revealed that overall model performance was most sensitive to variation in the constant for larval predation rate (BRT relative importance; $Pred_{Larv}$ = 46.7%), followed by the two parameters controlling the strength of density-dependent mortality for COTS J1 life stage ($MortJ1_k = 21.5\%$, $MortJ1_{k0} = 17.5\%$). The coral cover-COTS ratio below which COTS populations collapsed was less influential in overall prediction accuracy ($CCRatio_2 = 3.0\%$), alongside the maximum rate of fertilization of COTS larvae ($Fert_{Linf} = 2.0\%$) and maximum per capita fecundity ($Fec_{max} = 1.6\%$). Coral cover prediction error was most sensitive to larval predation rate ($Pred_{Larv} = 32.7\%$) and maximum coral consumption rate ($Cons_{max} = 22.6\%$). COTS multilevel classification accuracy was most sensitive to early life stage parameters ($MortJ1_k = 44.8\%$; $MortJ1_{x0} = 34.6\%$; $Pred_{Larv} = 16.3\%$).

4. Discussion

Destructive outbreaks of COTS have been responsible for a significant portion of the loss of coral cover of the last 35 years on Australia's GBR and have thus been the focus of much research and management action (De'ath et al., 2012; Mellin et al., 2019; Pratchett et al., 2017a). This research often focuses on the hypothesized causes of outbreaks (Brodie et al., 2005; Wooldridge and Brodie, 2015), control technologies (Rivera-Posada et al., 2014; Uthicke et al., 2018) or spatial patterns in outbreaks (Matthews et al., 2020, in this volume; Vanhatalo et al., 2017). Moreover, previous modelling studies have generated small scale population models (Chen et al., 2017; Morello et al., 2014; Westcott et al., 2020), larval connectivity networks (Hock et al., 2014), simulated hypothetical reef networks (Condie et al., 2018) or aimed to refine our knowledge of COTS life history characteristics (Rogers et al., 2017). Our work builds on this large foundation of knowledge to provide a framework for spatially and temporally explicit predictions of Coral Cover and COTS abundances and fit them to the extensive observational data that exists for the GBR (Sweatman et al., 2008), while also accounting for the increasing cumulative disturbance in this region (Matthews et al., 2019).

Model predictions were able to match the timing of peaks of outbreak densities in the initiation region and provided reasonable predictions of the presence or absence of COTS outbreaks. The model did not however provide good accuracy at predicting COTS presence or absence in low numbers, and did not provide a satisfactory representation of outbreaks in the southern half of the GBR. Importantly, following the initiation of COTS populations in 1996, the model was able to produce peak densities between 1996 and 1998 for northern reefs in the Cooktown/Lizard Island and Cairns sectors (Fig. 5) while those further south in the Innisfail and

Townsville sectors were predicted to experience peak densities between 1999 and 2001, similar to patterns in the observational data (Figs 1 and 5). This shows that, while COTS densities during an outbreak are difficult to predict accurately, our model captured the spatio-temporal patterns of outbreaks as described by other modelling studies (Vanhatalo et al., 2017). In our model, these spatio-temporal patterns are driven primarily by the variation in chlorophyll concentrations, larval connectivity and prey availability estimates. Our results indicate the utility of these modelled estimates and indicate that larval nutrition (Birkeland, 1982; Brodie et al., 2005) and larval connectivity (Hock et al., 2014; Wooldridge and Brodie, 2015), can be used to reproduce waves of outbreaks in a modelling framework. It must be noted however, that other research has shown that multiple hypotheses are capable of explaining initiation of primary outbreaks (Babcock et al., 2016a), and our model did not include these mechanisms. We are cautious to suggest therefore, that our approach represents a significant improvement compared to other model formulations.

Some limitations on model performance surround the ability to recreate the extreme increases in COTS densities most notably recorded in reefs in the Townsville and southern Swain regions. Using our best-fit model parameters, predictions of mean outbreak density in these regions were underestimated by 80% and 92%, respectively (Table 3). While higher densities were predicted by some model configurations (results not shown), this generally led to overestimating densities in other regions and reducing overall model fit. Additionally, multiple boom-bust cycles were predicted on some Southern reefs that were not supported by empirical data. This is partly due to ratio dependent thresholds in COTS density that promote population declines and alternate parameterizations of ratio and density dependence should be compared explicitly to improve confidence in model performance. This could be countered by forcing the model to only take into account prey-availability but would limit the model's ability to capture population decline when coral resources remain. Previous COTS population models have also had difficulty organically simulating peak COTS outbreak levels (Morello et al., 2014). While this approach produced a more accurate recreation of COTS densities it did require larval input to be artificially introduced into the model to create the second wave of COTS outbreaks beginning 2010–2012. Condie et al. (2018) were more successful in producing realistic cyclic behaviour as an emergent property of a COTS-coral model, which was driven primarily by prey availability and advantageously had a simpler parameterization. Our approach estimated a time series for chlorophyll-a using to drive COTS outbreaks and our model was able to reproduce these peaks in certain areas. This is a novel and useful approach however more development is required to accurately capture dramatic peaks and exact timings of COTS outbreaks.

A number of alterations to the model structure could potentially improve model fit with regards to spatial variability in peak COTS outbreak densities. For example, allowing the threshold that triggers outbreak collapse to vary for each latitudinal sector and cross shelf location combination as estimated by the empirical data, could provide a mechanism for promoting these more severe densities in specific areas. This approach, while not allowing these regional differences to emerge intrinsically from the model, may significantly improve model fit. Allowing these regional differences in mean COTS outbreak density would act as a proxy for calculating spatially explicit carrying capacities that could account for some variation not captured by our current approach. Primarily, as COTS exhibit significant feeding preference for fast growing corals, especially table and branching *Acropora* spp. (Pratchett, 2007), reefs with naturally higher proportions of these species should exhibit higher outbreak densities and thus higher carrying capacities. Increased COTS densities can be explained by both increased food supply supporting more adult COTS and increased maternal nutrition leading to significantly higher rates of larval success (Caballes et al., 2016, 2017). There is also potential to incorporate estimates of coral benthic community type (Mellin et al., 2019) to help define peak outbreak densities, by allowing reefs predicted to be *Acropora* dominated to reach higher thresholds. Additionally, incorporating a "no settlement" or limited settlement period following a collapse, to simulate the attraction of larvae to other nearby reefs with dense adult populations (Cowan et al., 2016) may help capture periods of low density populations. This would lower the influence of parameters that control rapid expansion of COTS and promote more dramatic peaks, without overestimating COTS abundance during the non-outbreak cycle.

While our model incorporates contemporary GBR-wide estimates of COTS larval connectivity, explicitly modelling inter-annual variability in connectivity pathways is an important step to improve model estimates. Larval connectivity pathways are hypothesized to be pivotal in determining COTS spatial distribution during waves of secondary outbreaks (Hock et al., 2014) and, together with primary productivity, inter-annual variability in connectivity has been linked to the initiation of primary outbreaks (Wooldridge and Brodie, 2015). Incorporating more interannual variability in connectivity estimates is integral to increasing the model performance in capturing peak densities at reefs experiencing outbreaks. Due to COTS immense fecundity (Babcock et al., 2016b), any fractional increases in the proportion of larvae arriving at a reef may dramatically increase adult densities. The distribution of arriving larvae in this study was uniform across a reef and did not allow for the attraction of larvae to either coral prey or conspecifics. This would potentially tend to underestimate population establishment through fertilization success and local recruitment and may help to explain the lack of peak densities predicted for some reefs/sectors. Importantly, the connectivity estimates used in this study have received warranted scrutiny in their ability to accurately predict connectivity pathways primarily due to the relatively coarse spatial resolution (4 km) (Bode et al., 2018; Mumby et al., 2018). Further development of the hydrodynamic models of the GBR to incorporate unstructured meshes and finer resolutions (Thomas et al., 2014) should increase the accuracy of these estimates and ultimately the predictions of regional scale ecosystem models. Additionally, recent important advances in satellite imaging have rapidly developed the understanding of potential coral habitat along the GBR (Roelfsema et al., 2018), and incorporating these underlying habitat maps are essential for the refinement of both connectivity models and metacommunity models for the GBR.

Our model makes use of a machine learning approach to sensitivity analyses to identify important parameters when optimizing overall model accuracy, coral cover estimates or COTS outbreak classification (Norton, 2015). This approach highlighted that predictions were particularly sensitive to parameters controlling early life stages such as larval predation rate and density dependent juvenile mortality. Due to the demographic traits of the species (Babcock et al., 2016b), such sensitivity was anticipated, and indeed predatory and density related mortality rates of early life stages represent bottlenecks in COTS population dynamics yet have remained key gaps in the understanding of COTS outbreaks (Pratchett et al., 2017a). While some recent work has highlighted potential predators of COTS larvae (Cowan et al., 2016, , 2017a,b) and investigated juvenile predation rates (Messmer et al., 2017), generating spatially and temporally explicit predictions of predator abundance or a suitable proxy seems beyond the scope of contemporary research. Potentially incorporating the effects of marine park zoning could account for some predatory effect on COTS larvae, however most identified predators of larvae and juveniles are not targeted by fishing activities (Cowan et al., 2017a) although they could indirectly impacted by fishing (e.g., habitat loss due to anchor damage). Moreover, given the links between outbreak probability and zoning are poorly understood (Sweatman, 2008; Vanhatalo et al., 2017), a zoning parameter, while being ecologically significant, seems unlikely to drastically improve model performance. Additionally, incorporating minor spatial and temporal stochasticity to the early life stage parameters would improve the framework.

Our model provides the first framework for predicting coral cover and COTS densities in a spatially and temporally explicit manner, for reef locations across the GBR. While improvements are required to fine-tune parameters, this framework provides a basis for further development, with scope to incorporate proposed management interventions and projections towards the next outbreak cycle. During the course of the most recent outbreak on the GBR there have been vast improvements to both control technologies and strategies (Westcott et al., 2016), with a number of emerging technologies being investigated to help with the early detection of outbreaks (Doyle et al., 2017; Uthicke et al., 2018), use of pheromones for COTS aggregation and/or dispersion (Hall et al., 2017a,b) and remote autonomous surveillance (Llewellyn and Bainbridge, 2015). If deployed effectively (in time and space), some combination of these technologies may prove to be pivotal to reducing the impact of the next wave of COTS outbreaks. Further developing existing modelling frameworks by incorporating a range of intervention strategies could provide an avenue to empirically assess the probability of curtailing the next outbreak cycle and/or protecting high priority areas of the GBR to bolster the resilience GBR (GBRMPA, 2017; Hock et al., 2017). However, with respect to this study, significant improvements to model fit are required before it can be used for management purposes. Additionally, developing probabilistic models for disturbance events would allow these models to be forecast, and the resilience of reefs to be estimated in the context of increasing severity and frequency of disturbance (GBRMPA, 2019; Hughes et al., 2018). To achieve this goal, future projections of chlorophyll concentrations and connectivity patterns could be projected forward based on historical data, climate change projections, ENSO cycles and a variety of water quality management scenarios.

This study aimed to provide the first spatially and temporally explicit COTS-Coral metacommunity model for the GBR, at a 1×1 km resolution. While successfully recreating the trajectories of coral cover and COTS outbreaks, it is currently not able to capture some of the spatial variability with regards to peak outbreak densities. It is important to note that fine-scale models such as these have low generality and thus forecasting skill, and are thus designed to simulate current conditions and management strategies and limited to near future forecasting (i.e., the next COTS outbreak cycle). Furthermore, in its current configuration, the model tends towards overestimating COTS abundances during low density years and does not differentiate between fast and slow growing corals. This study highlights critical knowledge gaps to set priorities for ongoing biological research (e.g., predation rates, natural mortality rates) and also key areas for improvement to this model framework, in order to better capture the spatial and temporal variability in COTS densities. This framework is the first of its kind for the GBR and provides a platform upon which to interrogate the relative effect of important parameters to guide future research and to potentially simulate the relative efficacy of a variety of COTS management scenarios. The results of this study provides strong impetus to explicitly quantify COTS vital rates to allow models to be refined, which will contribute greatly to improved COTS management.

References

Abrams, P.A., Ginzburg, L.R., 2000. The nature of predation: prey dependent, ratio dependent or neither? Trends Ecol. Evol. 15, 337–341.

Arditi, R., Ginzburg, L.R., 1989. Coupling in predator-prey dynamics: ratio-dependence. J. Theor. Biol. 139, 311–326.

Babcock, R.C., Dambacher, J.M., Morello, E.B., Plagányi, É.E., Hayes, K.R., Sweatman, H.P.A., Pratchett, M.S., 2016a. Assessing different causes of crown-of-thorns starfish outbreaks and appropriate responses for management on the Great Barrier Reef. PLoS One 11, e0169048.

Babcock, R.C., Milton, D.A., Pratchett, M.S., 2016b. Relationships between size and reproductive output in the crown-of-thorns starfish. Mar. Biol. 163, 234.

Babcock, R.C., Mundy, C.N., 1992. Reproductive biology, spawning and field fertilization rates of *Acanthaster planci*. Aust. J. Mar. Freshwat. Res. 43, 525–534.

Babcock, R.C., Mundy, C.N., Whitehead, D., 1994. Sperm diffusion models and in situ confirmation of long-distance fertilization in the free-spawning asteroid *Acanthaster planci*. Biol. Bull. 186, 17–28.

Babcock, R., Plaganyi, E., Morello, B., Hoey, J., Pratchett, M., 2014. What Are the Important Ecological Thresholds and Relationships to Inform the Management of COTS? Draft Report, CSIRO.

Birkeland, C., 1982. Terrestrial runoff as a cause of outbreaks of *Acanthaster planci* (Echinodermata: Asteroidea). Mar. Biol. 69, 175–185.

Birkeland, C., Lucas, J.S., 1990. *Acanthaster planci*: Major Management Problem of Coral Reefs. CRC Press, Boca Raton, Florida.

Bode, M., Bode, L., Choukroun, S., James, M.K., Mason, L.B., 2018. Resilient reefs may exist, but can larval dispersal models find them? PLoS Biol. 16, e2005964.

Brodie, J., Devlin, M., Lewis, S., 2017. Potential enhanced survivorship of crown of thorns starfish larvae due to near-annual nutrient enrichment during secondary outbreaks on the central mid-shelf of the great barrier reef, Australia. Diversity 9, 17.

Brodie, J., Fabricius, K., De'ath, G., Okaji, K., 2005. Are increased nutrient inputs responsible for more outbreaks of crown-of-thorns starfish? An appraisal of the evidence. Mar. Pollut. Bull. 51, 266–278.

Caballes, C.F., Pratchett, M.S., 2014. Reproductive biology and early life history of the crown-of-thorns starfish. In: Whitmore, E. (Ed.), Echinoderms: Ecology, Habitats and Reproductive Biology. Nova Science, New York, USA, pp. 101–146.

Caballes, C.F., Pratchett, M.S., Buck, A.C.E., 2017. Interactive effects of endogenous and exogenous nutrition on larval development for crown-of-thorns starfish. Diversity 9, 15.

Caballes, C.F., Pratchett, M.S., Kerr, A.M., Rivera-Posada, J.A., 2016. The role of maternal nutrition on oocyte size and quality, with respect to early larval development in the coral-eating starfish, *Acanthaster planci*. PLoS One 11, e0158007.

Carnell, R., 2009. Latin Hypercube Samples. R CRAN Project, 1–13.

Caswell, H., 2006. Matrix Population Models. Encyclopedia of Environmetrics. Wiley Online Library.

Chen, C.C.M., Drovandi, C.C., Keith, J.M., Anthony, K., Caley, M.J., Mengersen, K.L., 2017. Bayesian semi-individual based model with approximate Bayesian computation for parameters calibration: Modelling Crown-of-Thorns populations on the Great Barrier Reef. Ecol. Model. 364, 113–123.

Chen, Y., Minchin, S.A., Seaton, S., Joehnk, K.D., Robson, B.J., Bai, Q., 2011. eReefs-a new perspective on the Great Barrier Reef. In: 19th International Congress on Modelling and Simulation.

Chesher, R.H., 1969. Destruction of the Pacific corals by the sea star *Acanthaster planci*. Science 165, 280–283.

Cohen, J., 1960. A coefficient of agreement for nominal scales. Educ. Psychol. Meas. 20, 37–46.

Condie, S., Hepburn, M., Mansbridge, J., 2012. Modelling and visualisation of connectivity of the Great Barrier Reef. In: Proceedings of the 12th International Coral Reef Symposium, Cairns, Australia 9–13 July.

Condie, S.A., Plagányi, É.E., Morello, E.B., Hock, K., Beeden, R., 2018. Great Barrier Reef recovery through multiple interventions. Conserv. Biol. 32, 1356–1367.

Cowan, Z.-L., Dworjanyn, S., Caballes, C., Pratchett, M., 2016. Benthic predators influence microhabitat preferences and settlement success of crown-of-thorns starfish (*Acanthaster* cf. *solaris*). Diversity 8, 27.

Cowan, Z.-L., Pratchett, M., Messmer, V., Ling, S., Cowan, Z.-L., Pratchett, M., Messmer, V., Ling, S., 2017a. Known predators of crown-of-thorns starfish (*Acanthaster* spp.) and their role in mitigating, if not preventing, population outbreaks. Diversity 9, 7.

Cowan, Z.L., Ling, S.D., Dworjanyn, S.A., Caballes, C.F., Pratchett, M.S., 2017b. Interspecific variation in potential importance of planktivorous damselfishes as predators of *Acanthaster* spp. eggs. Coral Reefs 36, 653–661.

CSIRO, 2019. eReefs Research. https://research.csiro.au/ereefs/. accessed on 20/4/2019.

De'ath, G., 2003. Analyses of Crown-of-Thorns Starfish Data From the Fine-Scale Surveys and The Long-Term Monitoring Program Manta Tow Surveys. CRC Reef Research Centre Technical Report, CRC Reef Research Centre, Townsville.

De'ath, G., Fabricius, K.E., Sweatman, H., Puotinen, M., 2012. The 27-year decline of coral cover on the Great Barrier Reef and its causes. Proc. Natl. Acad. Sci. U. S. A. 109, 17995–17999.

R Development Core Team, 2017. R: A Language and Environment for Statistical Computing. Austria, Vienna.

Devlin, M., Waterhouse, J., Taylor, J., Brodie, J., 2001. Flood plumes in the Great Barrier Reef: spatial and temporal patterns in composition and distribution. GBRMPA Research Publication 68.

Doyle, J.R., McKinnon, A.D., Uthicke, S., 2017. Quantifying larvae of the coralivorous seastar *Acanthaster* cf. *solaris* on the Great Barrier Reef using qPCR. Mar. Biol. (164), 176.

Elith, J., Leathwick, J.R., Hastie, T., 2008. A working guide to boosted regression trees. J. Anim. Ecol. 77, 802–813.

Endean, R., 1969. Report on investigations made into aspects of the current *Acanthaster planci* (crown-of-thorns) infestations of certain reefs of the Great Barrier Reef. In: Queensland Department of Primary Industries (Fisheries branch). Australia, Brisbane.

Endean, R., 1974. *Acanthaster planci* on the Great Barrier Reef. Proceedings of the Second International Symposium on Coral Reefs 1, 563–576.

Fabricius, K.E., Okaji, K., De'ath, G., 2010. Three lines of evidence to link outbreaks of the crown-of-thorns seastar *Acanthaster planci* to the release of larval food limitation. Coral Reefs 29, 593–605.

GBRMPA, 2017. Great Barrier Reef: Blueprint for Resilience. Townsville, Australia.

GBRMPA, 2019. Great Barrier Reef Outlook Report 2019. Great Barrier Reef Marine Park Authority. Townsville, Australia.

Hall, M.R., Kocot, K.M., Baughman, K.W., Fernandez-Valverde, S.L., Gauthier, M.E.A., Hatleberg, W.L., Krishnan, A., McDougall, C., Motti, C.A., Shoguchi, E., Wang, T., Xiang, X., Zhao, M., Bose, U., Shinzato, C., Hisata, K., Fujie, M., Kanda, M., Cummins, S.F., Satoh, N., Degnan, S.M., Degnan, B.M., 2017a. The crown-of-thorns starfish genome as a guide for biocontrol of this coral reef pest. Nature 544, 231–234.

Hall, M.R., Motti, C.A., Kroon, F., 2017b. The Potential Role of the Giant Triton Snail, Charonia tritonis (Gastropoda: Ranellidae) in Mitigating Populations of the Crown-of-thorns Starfish. Report to the National Environmental Science Programme, Cairns.

Haque, M., 2009. Ratio-dependent predator-prey models of interacting populations. Bull. Math. Biol. 71, 430–452.

Hijmans, R.J., Phillips, S., Leathwick, J.R., Elith, J., 2017. Dismo: Species Distribution Modeling. R CRAN Project: 55.

Hock, K., Wolff, N.H., Condie, S.A., Anthony, K.R.N., Mumby, P.J., 2014. Connectivity networks reveal the risks of crown-of-thorns starfish outbreaks on the Great Barrier Reef. J. Appl. Ecol. 51, 1188–1196.

Hock, K., Wolff, N.H., Ortiz, J.C., Condie, S.A., Anthony, K.R.N., Blackwell, P.G., Mumby, P.J., 2017. Connectivity and systemic resilience of the Great Barrier Reef. PLoS Biol. 15, e2003355.

Hossin, M., Sulaiman, M.N., 2015. A review on evaluation metrics for data classification evaluations. Int. J. Data Mining Knowl. Manage. Process 5, 1.

Hughes, T.P., Kerry, J.T., Baird, A.H., Connolly, S.R., Dietzel, A., Eakin, C.M., Heron, S.F., Hoey, A.S., Hoogenboom, M.O., Liu, G., McWilliam, M.J., Pears, R.J., Pratchett, M.S., Skirving, W.J., Stella, J.S., Torda, G., 2018. Global warming transforms coral reef assemblages. Nature 556, 492–496.

Johnson, C., 1992. Settlement and recruitment of *Acanthaster planci* on the Great Barrier Reef: questions of process and scale. Aust. J. Mar. Freshwat. Res. 43, 611–627.

Kayal, M., Vercelloni, J., Lison de Loma, T., Bosserelle, P., Chancerelle, Y., Geoffroy, S., Stievenart, C., Michonneau, F., Penin, L., Planes, S., Adjeroud, M., 2012. Predator crown-of-thorns starfish (*Acanthaster planci*) outbreak, mass mortality of corals, and cascading effects on reef fish and benthic communities. PLoS One 7, e47363.

Kettle, B.T., Lucas, J.S., 1987. Biometric relationships between organ indices, fecundity, oxygen consumption and body size in *Acanthaster planci*. Bull. Mar. Sci. 41, 541–551.

Llewellyn, L.E., Bainbridge, S.J., 2015. Getting Up Close and Personal: The Need to Immerse Autonomous Vehicles in Coral Reefs. OCEANS 2015—MTS/IEEE Washington.

Lotka, A.J., 1910. Contribution to the theory of periodic reactions. J. Phys. Chem. 14, 271–274.

Lough, J.M., 2001. Climate variability and change on the Great Barrier Reef. In: Wolanski, E. (Ed.), Oceanographic Processes of Coral Reefs: Physical and Biological Links in the Great Barrier Reef. CRC Press, Bocan Raton, FL, pp. 269–300.

MacNeil, M.A., Mellin, C., Matthews, S., Wolff, N.H., McClanahan, T.R., Devlin, M., Drovandi, C., Mengersen, K., Graham, N.A.J., 2019. Water quality mediates resilience on the Great Barrier Reef. Nat. Ecol. Evol. 3, 620–627.
Matérn, B., 1960. Spatial variation: Meddelanden fran statens skogsforskningsinstitut. Lecture Notes in Statist. 36, 21.
Matthews, S.A., Mellin, C., MacNeil, A., Heron, S.F., Skirving, W., Puotinen, M., Devlin, M.J., Pratchett, M., 2019. High-resolution characterization of the abiotic environment and disturbance regimes on the Great Barrier Reef, 1985–2017. Ecology 100, e02574.
Matthews, S., Mellin, C., Pratchett, M., 2020. Larval connectivity and water quality explain spatial distribution of crown-of-thorns starfish outbreaks across the Great Barrier Reef. In: Riegl, B. (Ed.), Population Dynamics of the Reef Crisis, vol. 87. Advances in Marine Biology. Elsevier, pp. 223–258 (in this volume).
McCallum, H., 1993. Are crown-of-thorns starfish populations chaotic? In: The Possible Causes and Consequences of Outbreaks of the Crown-of-Thorns Starfish. Proceedings of A Workshop Held in Townsville, Queensland, Australia, 10 June 1992 at the Sheraton Breakwater Casino Hotel, pp. 83–93.
McKay, M.D., Beckman, R.J., Conover, W.J., 1978. A comparsion of three methods for selecting values of input variables in the analysis of the output from a computer code. Technometrics 21 (2), 239–245.
Mellin, C., Lurgi, M., Matthews, S., MacNeil, M.A., Caley, M.J., Bax, N., Przeslawski, R., Fordham, D.A., 2016. Forecasting marine invasions under climate change: Biotic interactions and demographic processes matter. Biol. Conserv. 204, 459–467.
Mellin, C., Matthews, S., Anthony, K.R.N., Brown, S.C., Caley, M.J., Johns, K.A., Osborne, K., Puotinen, M., Thompson, A., Wolff, N.H., Fordham, D.A., MacNeil, M.A., 2019. Spatial resilience of the Great Barrier Reef under cumulative disturbance impacts. Glob. Chang. Biol. 25, 2431–2445.
Messmer, V., Pratchett, M., Chong-Seng, K., 2017. Variation in incidence and severity of injuries among crown-of-thorns starfish (*Acanthaster* cf. *solaris*) on Australia's great barrier reef. Diversity 9, 1–9.
Mills, S., 2012. Density-dependent prophylaxis in the coral-eating crown-of-thorns sea star, *Acanthaster planci*. Coral Reefs 31, 603–612.
Morello, E.B., Plagányi, É.E., Babcock, R.C., Sweatman, H., Hillary, R., Punt, A.E., 2014. Model to manage and reduce crown-of-thorns starfish outbreaks. Mar. Ecol. Prog. Ser. 512, 167–183.
Mumby, P.J., Hock, K., Condie, S.A., Ortiz, J.C., Wolff, N.H., Anthony, K.R.N., Blackwell, P.G., 2018. Response to Bode and colleagues: "Resilient reefs may exist, but can larval dispersal models find them?". PLoS Biol. 16, e2007047.
Norton, J., 2015. An introduction to sensitivity assessment of simulation models. Environ. Model. Software 69, 166–174.
Pratchett, M.S., 1999. An infectious disease in crown-of-thorns starfish on the Great Barrier Reef. Coral Reefs 18, 272.
Pratchett, M.S., 2005. Dynamics of an outbreak population of *Acanthaster planci* at Lizard Island, northern Great Barrier Reef (1995–1999). Coral Reefs 24, 453–462.
Pratchett, M.S., 2007. Feeding preferences of *Acanthaster planci* (Echinodermata: Asteroidea) under controlled conditions of food availability. Pac. Sci. 61, 113–120.
Pratchett, M.S., 2010. Changes in coral assemblages during an outbreak of *Acanthaster planci* at Lizard Island, northern Great Barrier Reef (1995–1999). Coral Reefs 29, 717–725.
Pratchett, M.S., Caballes, C.F., Rivera-Posada, J.A., Sweatman, H.P.A., 2014. Limits to understanding and managing outbreaks of Crown-of-thorns Starfish (*Ancathanster* spp.). Oceanogr. Mar. Biol. Annu. Rev. 52, 133–200.

Pratchett, M., Caballes, C., Wilmes, J., Matthews, S., Mellin, C., Sweatman, H., Nadler, L., Brodie, J., Thompson, C., Hoey, J., Bos, A., Byrne, M., Messmer, V., Fortunato, S., Chen, C., Buck, A., Babcock, R., Uthicke, S., 2017a. Thirty Years of Research on Crown-of-Thorns Starfish (1986–2016): Scientific Advances and Emerging Opportunities. Diversity 9, 41.

Pratchett, M., Dworjanyn, S., Mos, B., Caballes, C., Thompson, C., Blowes, S., Pratchett, M.S., Dworjanyn, S., Mos, B., Caballes, C.F., Thompson, C.A., Blowes, S., 2017b. Larval survivorship and settlement of crown-of-thorns starfish (*Acanthaster* cf. *solaris*) at varying algal cell densities. Diversity 9, 2.

Rayner, N.A., Parker, D.E., Horton, E.B., Folland, C.K., Alexander, L.V., Rowell, D.P., Kent, E.C., Kaplan, A., 2003. Global analyses of sea surface temperature, sea ice, and night marine air temperature since the late nineteenth century. J. Geophys. Res. D: Atmos. 108.

Rivera-Posada, J., Pratchett, M.S., Aguilar, C., Grand, A., Caballes, C.F., 2014. Bile salts and the single-shot lethal injection method for killing crown-of-thorns sea stars (*Acanthaster planci*). Ocean Coast. Manag. 102, 383–390.

Roelfsema, C., Kovacs, E., Ortiz, J.C., Wolff, N.H., Callaghan, D., Wettle, M., Ronan, M., Hamylton, S.M., Mumby, P.J., Phinn, S., 2018. Coral reef habitat mapping: A combination of object-based image analysis and ecological modelling. Remote Sens. Environ. 208, 27–41.

Rogers, J.G.D., Pláganyi, É.E., Babcock, R.C., 2017. Aggregation, Allee effects and critical thresholds for the management of the Crown-of-thorns starfish *Acanthaster planci*. Mar. Ecol. Prog. Ser. 578, 99–114.

Rossi, R.E., Mulla, D.J., Journel, A.G., Franz, E.H., 1992. Geostatistical tools for modeling and interpreting ecological spatial dependence. Ecol. Monogr. 62, 277–314.

Scandol, J.P., 1999. CotSim—an interactive *Acanthaster planci* metapopulation model for the central Great Barrier Reef. Mar. Model 1, 39–81.

Steinberg, C., 2011. Chapter 3: Impacts of climate change on the physical oceanography of the Great Barrier Reef. Page. In: Johnson, J., Marshall, P. (Eds.), Climate Change and the Great Barrier Reef: A Vulnerability Assessment. Great Barrier Reef Marine Park Authority and Australian Greenhouse Office, Townsville, Australia.

Stubben, C., Milligan, B., 2007. Estimating and analyzing demographic models using the popbio package in R. J. Stat. Softw. 22, 1–23.

Stump, R., 1996. An Investigation to Describe the Population Dynamics of *Acanthaster planci* (L.) around Lizard Island, Cairns section, Great Barrier Reef Marine Park. CRC Reef Research Centre Technical Report No. 10, Townsville, Australia.

Sweatman, H., 2008. No-take reserves protect coral reefs from predatory starfish. Curr. Biol. 18, 598–599.

Sweatman, H.P.A., Cheal, A., Coleman, G., Emslie, M.J., Johns, K.A., Jonker, M., Miller, I., Osborne, K., 2008. Long-term Monitoring of the Great Barrier Reef, Status Report Number 8. AIMS, Townsville, Australia.

Thomas, C.J., Lambrechts, J., Wolanski, E., Traag, V.A., Blondel, V.D., Deleersnijder, E., Hanert, E., 2014. Numerical modelling and graph theory tools to study ecological connectivity in the Great Barrier Reef. Ecol. Model. 272, 160–174.

Uthicke, S., Lamare, M., Doyle, J.R., 2018. eDNA detection of corallivorous seastar (*Acanthaster* cf. *solaris*) outbreaks on the Great Barrier Reef using digital droplet PCR. Coral Reefs 37, 1229–1239.

Uthicke, S., Logan, M., Liddy, M., Francis, D., Hardy, N., Lamare, M., 2015. Climate change as an unexpected co-factor promoting coral eating seastar (*Acanthaster planci*) outbreaks. Sci. Rep. 5, 8402.

Uthicke, S., Schaffelke, B., Byrne, M., 2009. A boom–bust phylum? Ecological and evolutionary consequences of density variations in echinoderms. Ecol. Monogr. 79, 3–24.

Vanhatalo, J., Hosack, G.R., Sweatman, H., 2017. Spatiotemporal modelling of crown-of-thorns starfish outbreaks on the Great Barrier Reef to inform control strategies. J. Appl. Ecol. 54, 188–197.

Volterra, V., 1928. Variations and fluctuations of the number of individuals in animal species living together. ICES J. Mar. Sci. 3, 3–51.

Westcott, D., Fletcher, C.S., Babcock, R., Plaganyi-Lloyd, E., 2016. A Strategy to Link Research and Management of Crown-of-Thorns Starfish on the Great Barrier Reef: An Integrated Pest Management Approach. Report to the National Environmental Science Programme. Reef and Rainforest Research Centre Limited, Cairns.

Westcott, D.A., Fletcher, C.S., Kroon, F.J., Babcock, R.C., Plagányi, E.E., Pratchett, M.S., Bonin, M.C., 2020. Relative efficacy of three approaches to mitigate Crown-of-Thorns Starfish outbreaks on Australia's Great Barrier Reef. Sci. Rep. 10, 12594.

Wolfe, K., Graba-Landry, A., Dworjanyn, S.A., Byrne, M., 2015. Larval starvation to satiation: Influence of nutrient regime on the success of *Acanthaster planci*. PLoS One 10, 1–17.

Wolfe, K., Graba-Landry, A., Dworjanyn, S.A., Byrne, M., 2017. Superstars: Assessing nutrient thresholds for enhanced larval success of *Acanthaster planci*, a review of the evidence. Mar. Pollut. Bull. 116, 307–314.

Wooldridge, S.A., Brodie, J.E., 2015. Environmental triggers for primary outbreaks of crown-of-thorns starfish on the Great Barrier Reef, Australia. Mar. Pollut. Bull. 101, 805–815.

Zann, L., Brodie, J., Berryman, C., Naqasima, M., 1987. Recruitment, ecology, growth and behavior of juvenile *Acanthaster planci* (L.) (Echinodermata: Asteroidea). Bull. Mar. Sci. 41, 561–575.

Zann, L., Brodie, J., Vuki, V., 1990. History and dynamics of the crown-of-thorns starfish *Acanthaster planci* (L.) in the Suva area, Fiji. Coral Reefs 9, 135–144.

CHAPTER ELEVEN

Coral community life histories and population dynamics driven by seascape bathymetry and temperature variability

Tim R. McClanahan*
Wildlife Conservation Society, Marine Programs, Bronx, NY, United States
*Corresponding author: e-mail address: tmcclanahan@wcs.org

Contents

Abstract

Temperature variability, habitat, coral communities, and fishing intensity are important factors influencing coral responses to climate change. Consequently, chronic and acute sea-surface temperatures (SSTs) and their interactions with habitat and fishing were studied along the East African coast (~400 km) by evaluating changes over a ~25-year period in two major reef habitats—island and fringing reefs. These habitats had similar mean and standard deviation temperature measurements but differed in that islands

Advances in Marine Biology, Volume 87
ISSN 0065-2881
https://doi.org/10.1016/bs.amb.2020.08.003

had lower ocean heights and flatter and less right-skewed temperature distributions than fringing reefs. These patterns arise because islands are exposed to deep offshore water passing through deep channels while being protected from the open ocean storms and the strong inter-annual current temperature variability. Within these two seascapes, coral communities are shaped by population responses to the variable temperature distributions as determined by the taxa's associations with the competitive-stress-ruderal (CSR) life history groups. For example, competitive taxa were more abundant where temperature distributions were flat and lacked frequent warm water anomalies. In contrast, ruderal, weedy, and generalist taxa were more common where temperature distributions were centralized, standard deviations high, and warm water anomalies more frequent. Finally, stress-resistant taxa were more common in reefs with high temperature skew but flatter temperature distributions. The rare 1998 thermal anomaly impacted and disturbed the ruderal and stressed reef more than the competitive communities. Ruderal became more similar to stressed communities while the stressed community moved further from the mean before recovering towards the competitive community. Competitive taxa were more common on islands and the deeper fringing reef sites while ruderal were dominant in shallow fringing reef lagoons. Over time, islands were less disturbed than fringing reefs and maintained the highest coral cover, numbers of taxa, and most competitive or space-occupying taxa. However, some island reefs with a history of dynamite fishing aligned with the stress-resistant communities over the full study period. Compared to the in situ SST gauges at the study site, temperature proxies with global coverage were often good at estimating mean and standard deviations of the SSTs but much poorer at estimating the shape of the temperature distributions that reflect chronic and acute stress, as reflected by kurtosis and skewness metrics. Given that these stress variables were critical for understanding the impacts of rare climate disturbances, global climate models that use mean conditions are likely to be poor predictors of future impacts on corals, particularly their species and life history composition. Better predictions should be possible if appropriate chronic and acute stress metrics and their proxies are identified and used.

1. Introduction

Responses of coral reef communities to chronic and acute disturbances will reflect the complex interaction between coral life histories and environmental conditions (Baumann et al., 2016; Gove et al., 2013; Graham et al., 2014; Pandolfi et al., 2011; Wilson et al., 2006). Nevertheless, a limited number of ecological factors, such as temperature variation, the life histories of dominant taxa, and herbivory, are frequently cited as major influences on coral reef resistance to and recovery from climate disturbances (Darling et al., 2019; McClanahan et al., 2012). There is strong evidence for coral resistance associated with background or chronic sea surface temperature (SST) variation and the tolerance of disturbance-resistant coral and symbiont taxa (Baker et al., 2013; Safaie et al., 2018). Recovery is, however, less well understood and evidence for strong effects of herbivory, in general, is lacking but some types of herbivores may play important roles in specific habitats and regions (McClanahan and Muthiga, 2016). Outcomes of acute stress are likely to be influenced by the interactions among many environmental and ecological forces and their variation with geography, habitat, and coral communities. Better understanding of the interplay between climate stress and the coral community organization should help evaluate the response of reef corals to increasingly frequent climate-change impacts (Ban et al., 2012; Hughes et al., 2018; Maina et al., 2008, 2011; McClanahan et al., 2007a, 2019a, 2020a,b; McLeod et al., 2010, 2019; Selig et al., 2010; Teneva et al., 2012).

Background temperature variability or chronic stress interacts with rarer warm water anomalies or acute stress to either attenuate or further stress and potentially kill corals (Donner, 2011; McClanahan et al., 2005, 2007a, 2020a; Oliver and Palumbi, 2011; Selig et al., 2012; Zinke et al., 2018). The interactions may promote acclimation, selection for tolerant taxa, epigenetic, and genetic change (Edmunds and Gates, 2008; Palumbi et al., 2014). Yet, stress that is repeated intensely and frequently beyond an undetermined threshold can deplete coral energy reserves, increase their susceptibility to subsequent thermal anomalies, and increase mortality (Ainsworth et al., 2016; Ateweberhan and McClanahan, 2010; Grottoli et al., 2014). Consequently, further research is needed to know how the background temperature profiles that corals experience either acclimate or stress them further in the face of acute stress. And, how these temperature profiles are influenced by geologic and oceanographic environments, their global distributions, and their values as proxies for coral sanctuaries.

Specific coral taxa differ in their resistance to and recovery from disturbances and thus influence the whole community's tolerance to climate change and increasing thermal stress (Golbuu et al., 2007, Graham et al., 2014; McClanahan et al., 2020a; Selig et al., 2010). Coral taxa will therefore succeed or fail in different thermal environments and the subsequent reassembly of the coral community and associated reef processes after climate disturbances (Darling et al., 2013; Done, 1999; Johns et al., 2014; McClanahan et al., 2015; van Woesik et al., 2011, 2012). Ocean currents and dispersion of thermal anomalies interact with coastal geomorphology, habitats, and reef type to modify open-ocean temperature patterns (Baumann et al., 2016; McClanahan et al., 2007a). For example, windward and leeward sides of islands have different thermal environments and corals have been observed to bleach more in the generally more thermally stable windward than leeward environments when impacted by episodic acute thermal stress (McClanahan et al., 2005). This environmental complexity creates a potential spatial mosaic of taxa, populations and communities with different susceptibility, reassembly, extinction, and adaptation potentials (Darling et al., 2019).

Multivariate analysis of sea surface temperature (SST) variables in the western Indian Ocean (WIO) distinguished three broad temperature environments (Ateweberhan and McClanahan, 2010). These were: (1) the Arabian and Persian Gulfs with high SST variance and strong seasonal bimodality, (2) the northern Indian Ocean represented by lower SST variance and unimodal distributions, and (3) the southern Indian Ocean represented by higher variance and moderate bimodality. The northern Indian Ocean SST environment also has lower rates of temperature rise than the southern Indian Ocean and this information was used in a simple time-to-threshold model to predict the time to extinction for Indian Ocean corals (Sheppard, 2003). These differences in the rate of temperature rise predicted that extinction would be sooner in the southern (East Africa and the Mascarene and Seychelles Islands) than the northern and central atolls of the Indian Ocean (Maldives, Lakshadweep, and Chagos). However, an indication that these predictions were inaccurate was evident shortly thereafter, as a compilation of coral cover data uncovered greater mortality after the 1998 around the equator and to the north and central parts of the Indian Ocean (Ateweberhan et al., 2011; McClanahan et al., 2007a). Moreover, a temperature-rise-to-threshold model, which was the basis of the prediction, had little predictive ability for relationships with coral cover in the Indian Ocean shortly after 1998 (McClanahan et al., 2015). Rather, a multivariate thermal stress model that included various temperature metrics, light intensity and penetration, and currents performed considerably better (Maina et al., 2011).

The boundary between the northern and southern Indian Ocean temperature environments occurs near the national borders of Kenya and Tanzania. Here, there are differences in the morphology of the ocean depths and shoreline that are likely to influence and possibly change the broader water temperature variability (Figs. 1, 2) (Masalu, 2008). These patterns are attributable to sea level change and ancient reef formations driven by glacial cycles and the fluctuating river drainages off the eastern and southern slopes of Meru, Kilamanjaro, and Usambara mountains. The Pangani River is the current drainage and enters the Indian Ocean aligned latitudinally

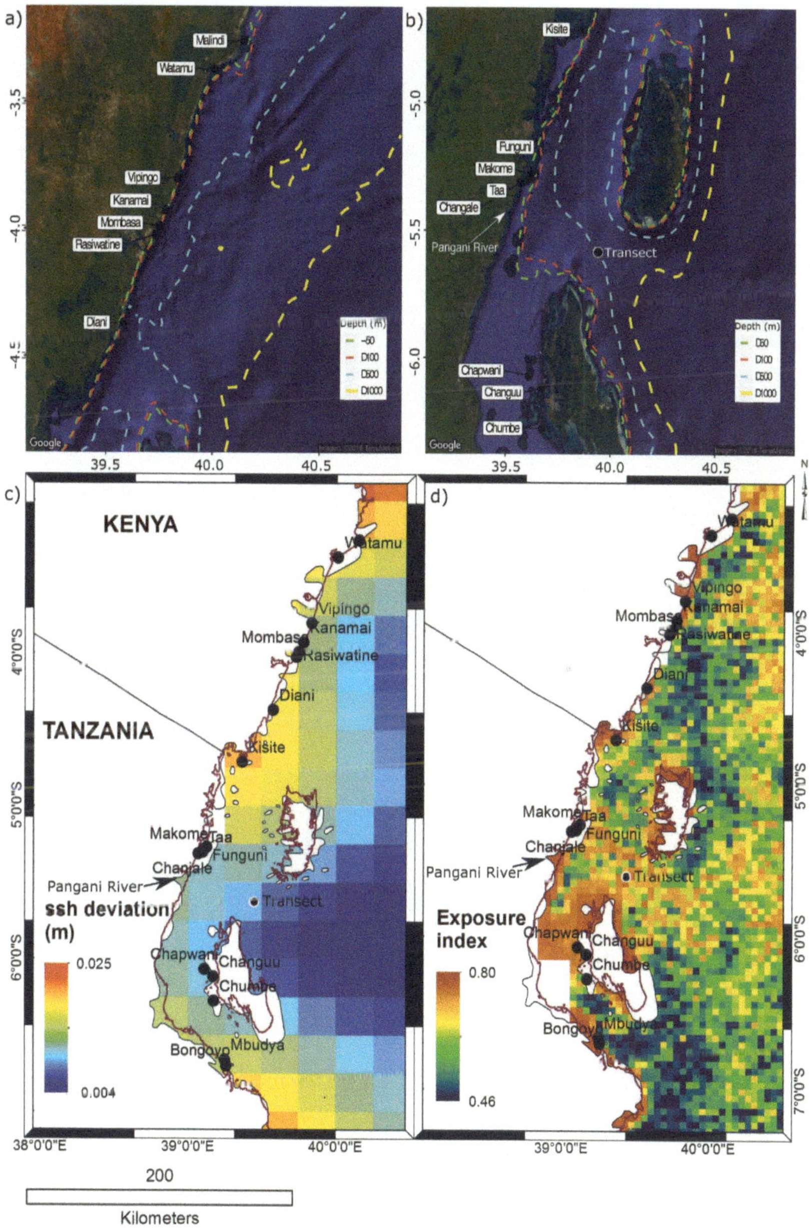

Fig. 1 Map of the study regions showing the distribution of water depths and study sites in (A) fringing reefs and (B) island reefs. Also presented are (C) ocean height variables during the warm season (January to March) and (D) the radiation and reducing portions (exposure) of a multivariate climate stress model. Pangani is the current river outlet for river runoff from the Meru, Kilamanjaro, and Usambara mountains. *Panel (D): Data extracted from Maina, J., McClanahan, T.R., Venus, V., Ateweberhan, M., Madin, J. 2011. Global gradients of coral exposure to environmental stresses and implications for local management. PLoS One 6, e23064.*

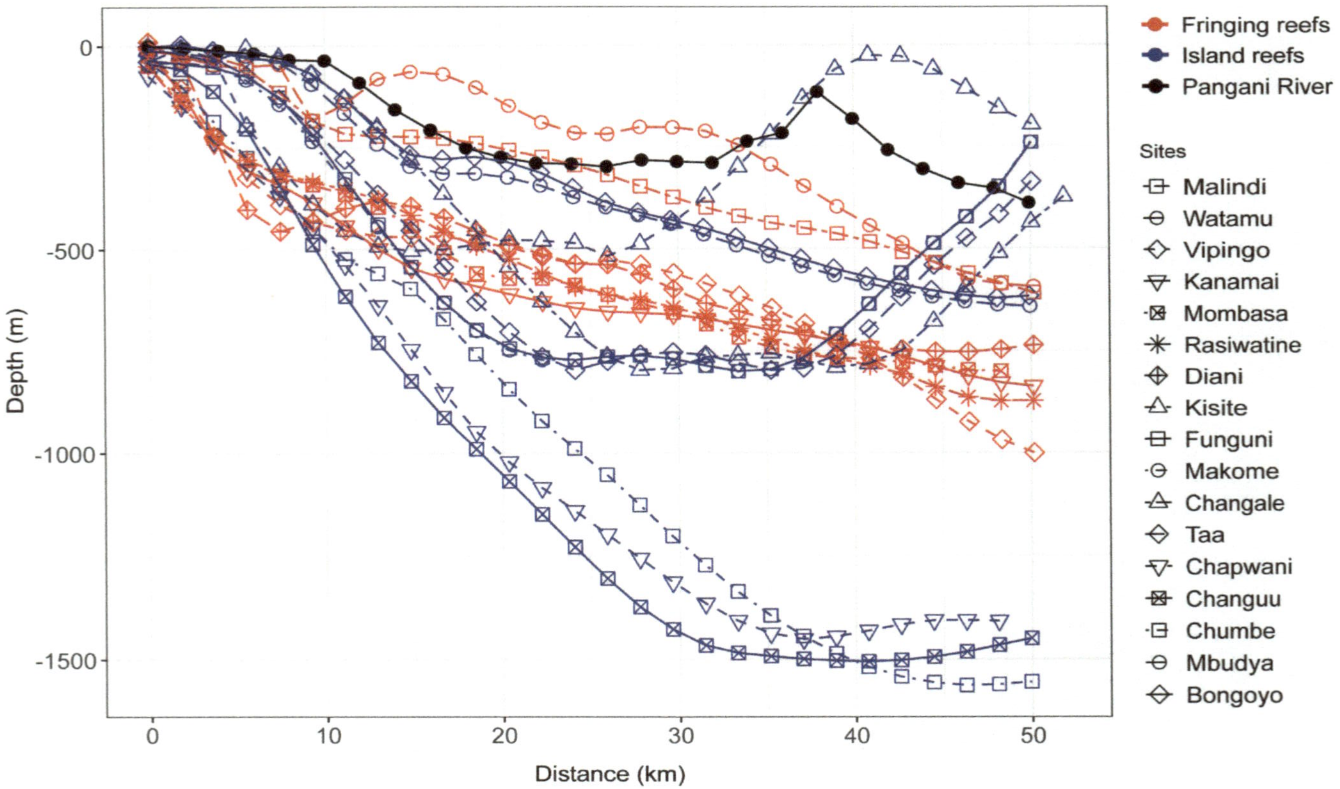

Fig. 2 Changes in ocean depth with distance to shore among the fringing and island reef study sites and off the east coast of Zanzibar aligned with the study sites on the west coast sites. The Pangani River outlet where Meru, Kilamanjaro and Usambara mountain rivers drain and also the southern border of the deep water channel leeward of Pemba Island.

with the deep oceanic channel between Zanzibar and Pemba Islands. The result is that island reefs exist in a more estuarine and dissected shoreline and bathymetry with deep-water channels. Fringing reefs exist in a more continuously inclined coastal bathymetry with fewer large breaks or deep-water, ancient, river estuaries.

The biological communities on reefs in the East African ecoregion (~5000 km of coastline) fall into two subregions (i.e., <500 km, McClanahan and Arthur, 2001) with different larger-scale ocean-temperature patterns. As a consequence of shoreline bathymetry, island and fringing reef environments are expected to differ in their exposures to background and anomalous temperatures, which should influence the coral communities by the variable success of taxa. Specifically, differences should be observable due to interactions between background variability and acute temperature stress. Consequently, the prediction is that the coral community responses to acute ocean-heating events, such as in the summer of 1998, will differ between these two reef environments, because of differential ocean and bathymetry variability.

Fishing is an additional anthropogenic disturbance that is likely to play a key role in influencing long-term impacts and is amenable to local management (McClanahan and Muthiga, 2016). Maintaining herbivorous fish and grazing may give corals a competitive advantage over erect algae and, therefore, unfished and remote reefs are sometimes found to recover more rapidly than fished reefs after thermal impacts (Ateweberhan et al., 2011; Mellin et al., 2016; Mumby et al., 2007). However, protected areas are often selectively placed in stable environments that promote high coral diversity, but may also lack the background environmental variability that creates resistance to thermal extremes (Selig et al., 2012). Thermal anomalies may therefore produce higher bleaching and mortality levels in these more stable environments (McClanahan and Maina, 2003; McClanahan et al., 2001). Despite examples of rapid coral recovery after disturbances, rare thermal anomalies may threaten otherwise stable offshore and protected thermal environments (Gilmour et al., 2013; Sheppard et al., 2008; Zinke et al., 2018). Reefs offshore and protected from fishing are generally not supporting greater coral cover (Bruno and Valdivia, 2016; Graham et al., 2008; McClanahan et al., 2019b; Smith et al., 2016). Resistance and recovery metrics, the two components of resilience, respond independently and each are influenced by many ecological factors (McClanahan et al., 2012).

A full understanding of interactions between thermal histories, coral taxa, reef habitat, and fishing on corals requires measuring changes across periods of acute thermal stress. Since the initiation of East African coral surveys in Kenya and Tanzania in 1991, thermal anomalies occurred in 1998, 2010, 2013, and 2016 (McClanahan, 2017, 2020). The 1997/98 temperature anomaly was among the strongest and best studied and resulted in ~50% mortality averaged across the Indian Ocean (Ateweberhan et al., 2011; McClanahan, 2017, 2020). Acute thermal disturbances provide the opportunity to study responses of benthic communities in the context of their background temperature variations, as described above. The distinct temperature patterns provided by fringing and island reefs also provided a large-scale natural experiment. The outcomes should improve understanding coral responses to climate impacts and possibly provide proxies for identifying coral sanctuaries. Therefore, responses to acute disturbances in these two thermal environments, in the context of fishing, were the focus of this of study.

2. Materials and methods

2.1 Study sites

Southern Kenya is characterized by a nearly continuous fringing reef close (<2 km) to shore and directly impacted by the East African equatorial current (Fig. 1). In contrast, reefs in the southernmost part of Kenya to central Tanzania are located on submerged and emergent islands of mainly Pleistocene limestone (i.e., Wasini, Pemba, Zanzibar, and Mafia Islands) (Arthurton et al., 1999) that are frequently separated by deep-water channels likely created by historical runoff from Meru, Kilamanjaro and Usambara mountains and sea level changes. East African reefs experience a large tidal range (~4 m) and study sites varied in depth at extreme low tides from 0.3 to 3 m. Investigations of water quality and physico-chemical parameters have found variable environmental conditions that indicate land and localized human influences (Carreiro-Silva and McClanahan, 2012; Heikoop et al., 2000; Moynihan et al., 2012).

Previous ecological studies indicate that these two reef environments and fishing impacts influence the composition of coral, fish, and sea urchin communities (McClanahan and Arthur, 2001; McClanahan and Muthiga, 2016; McClanahan et al., 1999). Kenya's shallow fringing reef sites were distributed along ~150 km, located in back reef lagoons, composed of scleractinian and soft (mostly alcyconian cnidarians) corals, various algae, and sponges growing on carbonate surfaces interspersed with sand and seagrass. Kenya's fringing reef ends in the south where two small local rivers (Gazi and Ramisi) enter that coastline. South of these river, the coastline is influenced by tidally-emergent reefs and various islands including Wasini and Pemba Islands. Island and platform reefs are scattered and isolated and vary in diameter from 100 m to 3 km.

Twelve reef sites in 7 fringing and 14 in 10 island reefs were examined (Fig. 1). Three of the fringing reefs were in well-managed national fisheries closures (Malindi, Watamu, and Mombasa) and four reefs were heavily fished (Vipingo, Kanamai, Ras Iwatine, and Diani). Kenyan parks were selectively chosen for their high coral and fish diversity compared to some of the fished sites. Two of the island reefs had effective fisheries closures (Kisite and Chumbe Island Parks), 8 were heavily fished (Mbudya, Bongoyo, Changu, Chapwani) and some had a history of dynamite fishing (Taa, Makome, Funguni, Chanjale) (Wells et al., 2007, 2010). All selected fisheries closures had effective management that had started several years prior to the surveys (McClanahan et al., 2009). Some Tanzanian sites (Bongoyo and Mbuya) were legally gazetted as closures but were heavily fished and therefore included in the fished reefs. Reefs on the islands lacked enclosed lagoons and were therefore expected to exhibit less water temperature variability than Kenya's fringing reef lagoons. Study sites in both island and fringing reefs locations were the shallow leeward sides of reefs that form along the edges of mainly Pleistocene platforms. Shallow leeward sites were chosen for study as they experienced the least wave energy and were most exposed to thermal variability and fishing.

2.2 Environmental factors

Seawater temperature data from in situ automated temperature gauges (OnSet Co. HoboTemp) placed in a variety of shallow-water sites (<2 m at low tide) were used to estimate temperature parameters. Gauges were often placed in shaded locations beneath corals and set to collect temperature measurements every 1–3 h. Gauges were sometimes lost and this led to incomplete coverage of time and place. Nevertheless, the full sampling resulted in ~324,000 measurements in fringing and ~95,000 in island reefs. In the fringing reefs, seven reef lagoons (Malindi, Watamu, Vipingo, Kanamai, Mombasa, Ras Iwatine, and Diani) had gauges placed beginning in 1996. Island reefs had gauges in 5 sites from 1997 (Kisite, Makome, Mbudya, Bongoyo, and Chumbe Island) until 2016. Temperature sampling represented three of the four benthic-sampling study periods.

To evaluate the accuracy of satellite proxy temperature data, in situ data were compared with (1) United States National Oceanographic and

Atmosphere Association (NOAA) - satellite-derived AVHRR sea surface temperature (SST) data, (2) European Space Agency (ESA) Operational SST and Sea Ice Analysis (OSTIA) system, and (3) Global Circulation Model (GCM) hindcast estimates (1984–2005). The NOAA-AVHRR data were weekly 4 × 4 km resolution collected between October 1981 and December 2010 and accessed from the NOAA Coral Reef Temperature Anomaly Database (*CoRTAD v4*) site (http://www.nodc.noaa.gov/sog/cortad/) (Selig et al., 2010). The ESA OSTIA data were SST with global coverage collected on a daily basis (Donlon et al., 2012). OSTIA uses satellite SST data provided by international agencies via the Group for High Resolution SST (GHRSST) Regional/Global Task Sharing (R/GTS) framework. GHRSST products include data from microwave and infrared satellite instruments, in situ SST data available over the Global Telecommunications System (GTS) and a sea-ice concentration product from the EUMETSAT Ocean and Sea Ice Satellite Applications Facility, Global Circulation Model SST time series for the historical experiment and calculated as ensemble median for all the historical runs in the latest suite of Coupled Model Inter-comparison Project (CMIP5) (http://cmip-pcmdi.llnl.gov/cmip5/data_portal.html). The GCM model had a similar integration of physical modelling methods and multiple sources of data used to predict future climate. These models were chosen for comparison with in situ gauges because they are commonly used to predict global patterns in climate change.

The GPS location of each ecological site was used to extract time-series values for the above global data. Temperature distributions from global data source were compared with those from in situ gauges for the metrics of mean, standard deviation (SD), kurtosis, and skewness. A multivariate metric of exposure to thermal stress was based on a fuzzy logic weighting of temperature and light variables known to influence coral bleaching (Maina et al., 2011). Means and variances at sites in each of the two reef geomorphological settings were calculated and tested for differences using *t*-tests.

Sea surface height (SSH) measures current and temperature influences on coastlines. Data were calculated from the multi mission satellite altimetry observations and served at the Archiving, Validation and Interpretation of Satellite Oceanographic data program (AVISO) data server (http://www.aviso.oceanobs.com/en/) (Ducet et al., 2000). Sea Surface Height Deviation (SSHD, also called sea level anomaly) is the deviation from the expected mean SSH. Data were downloaded for October 1992 to January 2010 and monthly mean SSH and SSHD time series data were extracted for the warm months of January to April when warm thermal anomalies occur. Site averages were calculated and mapped for the region at 0.25×0.25 degrees of longitude and latitude. The nearshore bathymetry of the region was also plotted on maps using the MarMap R program as well as depth versus distance to shore profiles for each of the studied coral reef sites (Pante and Simon-Bouhet, 2013).

2.3 Benthic cover field studies

Benthic cover was estimated by laying continuous 10-m transect lines and measuring the intercepts of studied organisms to the nearest centimetre. At 20 sites, between 9 and 18 replicate transects were collected (McClanahan et al., 2001). Sampling was undertaken during calm periods in the warm seasons (November to March). Benthic organisms >3 cm were measured to the nearest centimetre and identified and pooled into gross functional categories of turf, red coralline algae, green calcareous algae, erect fleshy algae, sponge, soft and hard coral, and seagrass. Hard scleractinian corals were further identified to the genus with the exceptions that *Porites* was separated into branching and massive forms and *Synarea* (=*Porites rus*), while *Galaxea* was separated into *Galaxea fascicularis* and *G. astreata*. Such taxonomic distinctions were made because these species or forms are associated with different environments and life histories that are not easily distinguished by genus-level classification (Darling et al., 2012). This study focuses on the response of hard corals and readers interested in acute impacts of the 1998 thermal anomaly on benthic functional groups are referred to McClanahan (2008).

The continuous coverage of each genus and all hard coral taxa combined was calculated as a fraction of the sum of all measurements under the line per transect and for the site. The metrics evaluated in this study were total hard coral cover (cm/m), number of coral taxa per site (9×10 m transects), coral dominance/diversity ($D = 1 - \Sigma n_i/N$, Simpson's Index, where n_i is number of individuals in i-th species, N is total number of individuals), and the coral community bleaching susceptibility index. The bleaching susceptibility index is based on a large database of observations of corals during the warm season and the degree to which they pale, partially or fully bleach, or die (McClanahan et al., 2007b). The relative abundance of each taxon is multiplied by this metric of bleaching and averaged to obtain the index where higher numbers represent communities that are likely to bleach more during warm water anomalies.

2.4 Data analyses

The analyses undertaken and presented were performed to test for the effects of management (fishing and no fishing), reef type (fringing vs island), time periods (pre and post strong and milder coral bleaching and mortality events) on benthic cover functional groups. Therefore, for some comparisons, sites were pooled by reef type to compare the broad-scale habitat-specific distribution patterns. Sites samples were pooled into four periods: period 1 (1991–1997) was the pre-1998 disturbance, period 2 (1999–2005) was the immediate post-1998 disturbance, period 3 (2006–2012) experienced milder bleaching events, and period 4 (2013–2016) had a strong temperature anomaly and bleaching event towards the end of the sampling period (McClanahan, 2017). A generalized linear mixed-effects models (GLMMs) was used to test for the effects of management, time period and reef type on coral community metrics; as well as to evaluate coral responses to the effects of the temperature variation metrics. Reef site was included as a random effect to account for repeated measurements at the same site (Bolker et al., 2009; Zuur et al., 2009). Coral cover and site susceptibility were bounded in the [0,1] interval thus were modelled using a beta distribution with a logit link function in the GLMM, we also logit transformed the variables and applied a Gaussian with error structure and an identity link in the GLMM (Douma and Weedon, 2019). The logit transformed model residuals behaved better than the beta distributed GLMMs and were therefore retained for the analyses. The number of genera was modelled using a Poisson distribution with a log link and included the log-transformed transect length as an offset variable (Zuur et al., 2009).

The temperature distributions during the previous year were used to calculate means, standard deviations (SDs), skewness, and kurtosis. These were then used as independent variables to evaluate relationships with coral community metrics. Temperature variables were moderately correlated, skewness and mean temperature were the most negatively correlated variables (−0.18) while the most positive correlation (0.32) was between kurtosis and skewness. This level of correlations did not present multicollinearity challenges in our analysis. Temperature variables were standardized by dividing each covariate by 2 times its SD prior to the analysis (Gelman, 2008). The resultant GLMMs of temperature evaluations were further subjected to a forward stepwise regressions to sequentially select variables to retain in the model using AIC criteria, that is variables whose inclusion lowered AIC values were retained. We conducted type 3 ANOVA (significant interactions) tests on GLMMs with interaction effects between management, time period and reef type to generate the ANOVA table. Model diagnostics were conducted using the DHARMa R package to check the validity of assumptions (Hartig, 2020) and the performance package to check collinearity of model predictors (Lüdecke et al., 2020). GLMMs were fit using the Template Model Builder (glmmTMB) R package (Brooks et al., 2017).

Coral community structure of the dominant 15 taxa were evaluated using Canonical Correspondence Analysis (CCA) and canonical Principal Components Analysis (PCA, also commonly referred to as RDA-Redundancy Analysis) separately with temperature and the benthic metric variables. CCA assumes unimodal relationships along an environmental gradient common for species—environment relationships and differs from the canonical Principal Components Analysis, which assumes linear relationships along gradients (ter Braak, 1986). Unimodal relationships are common for species distributions. However, along this section of the African coastline the empirical data suggests that the relationships of taxa are linear, possibly due to the gradient in environmental stress being evaluated here (McClanahan et al., 2007a, 2014). Thus, the PCA was given priority in the presentation and interpretation of results. The temperature variables included in these multivariate analysis were the last three sampling periods while the first sampling period was included in the analysis without temperature gauge metrics. These analyses were conducted using the vegan R package (Oksanen et al., 2019).

3. Results

3.1 Bathymetry and environmental variables

The fringing reef depth profiles were largely similar and displayed steady increases in depth to ~500 m at 20 km from shore and declining further to ~800 m at 50-km from shore (Fig. 2). The two exceptions were in the northern sites of Malindi and Watamu where water depths were half the depth of the other more southerly Kenyan fringing reef sites. Water depths seaward of Zanzibar indicate the considerable depths of these waters reaching ~1000 m at 20-km and ~1400 m at 35-km from shore where the depths levelled. The sites leeward or west of Zanzibar and Pemba Islands were more variable but north of the Pangani River, the depth profiles indicate the basin-shaped nature of the bathymetry reaching maximum depths of ~750 m at 25-km from shore before rising on the leeward side of Pemba Island. Kisite in the north was a shallower channel where the East African current exits this basin after entering the gap between Zanzibar and Pemba islands. South of Pangani River the depths were shallower, at ~250 m at 20-km, until the Mbudya and Bongoyo reefs in the north portion of the Tanzanian capital of Dar es Salaam.

3.1.1 Fringing and island reef comparisons

All environmental and temperature variables were statistically significantly different for comparisons of fringing and island reefs including the ocean height and multivariate stress model metrics (*t*-tests; Fig. 3; Table 1a). Altimeter-based sea surface height estimates indicated ~33% higher ocean height on fringing than island reefs. An exception to this overall pattern was the higher ocean height near the islands of Kisite (Wasini and Mpunguti) where the currents leaving the Pemba channel were forced through a shallower gap. Thermal exposure from the multivariate model indicated higher thermal exposure in fringing than the island reefs (Table 1a).

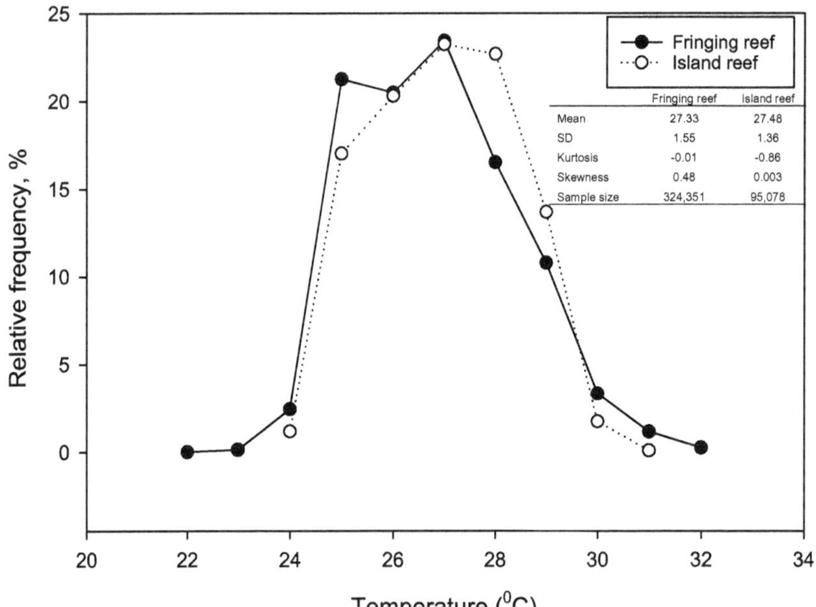

Fig. 3 Summary of in situ temperature gauge data collected in the fringing and island reefs between 1996 and 2016. Satellite and model summary data by site are presented for comparison in Table 1.

Table 1 Summary of (a) ocean heights and climate stress, and average temperatures based on measurements from (b) in situ temperature gauges, (c) satellite radiation measurements (*CoRTAD v4*), (d) European Space Agency blended satellite data, and (e) Global Circulation Model estimates for the locations of the studied fringing and island reefs.

	Fringing reef		Island reefs	
	Mean	**Difference, %**	**Mean**	**Difference, %**
(a) *Ocean stress and sea surface height*				
Ocean height, m	0.012		0.008	
Climate stress model	0.71		0.64	
(b) *Temperature gauges, in situ*				
Mean, °C	27.31		27.06	
Standard Deviation	1.60		1.29	
Kurtosis	−0.12		−0.97	
Skewness	0.44		0.14	
(c) *National Ocean and Atmospheric Association—AVHRR*				
Mean, °C	27.33	0.07	27.52	1.70
Standard Deviation	1.71	6.87	1.61	24.81
Kurtosis	−0.20	66.67	−0.40	−58.76
Skewness	0.22	−50.00	0.04	−71.43
(d) *European Space Agency—OSTIA*				
Mean, °C	27.14	−0.62	27.47	1.52
Standard Deviation	1.32	6.87	1.34	24.81
Kurtosis	−0.81	575.00	−0.98	1.03
Skewness	0.14	−68.18	0.00	−100.00
(e) *Global Circulation Model—CMIP5-historical*				
Mean, °C	26.87	−1.61	26.89	−0.63
Standard Deviation	1.09	−31.88	1.27	−1.55
Kurtosis	−0.80	566.67	−0.87	−10.31
Skewness	0.36	−18.18	−0.05	−135.71

The difference between the in situ and other metrics are compared as percentage differences from the in situ data. The climate stress model (a) is based on weighted and radiation and reducing factors scaled to bleaching responses to a maximum value of 1 as described in Maina et al. (2011). All variables have low standard error terms (~0.01) and are statistically different for reef type by *t*-test comparisons ($P < 0.05$).

Temperature metrics from in situ gauges indicate that mean temperatures, while statistically different, were similar (27.3 in fringing vs 27.1 in islands) but islands had flatter and less skewed temperature distributions than the fringing reefs (Table 1b, Fig. 3). Temperature SDs were also 24% higher on the fringing than island reefs. Overall, the island reefs had slightly lower mean temperatures and lower variance, and less centralized and skewed distributions. In situ measurements indicated that on fringing reefs water temperatures exceeded 31°C in several instances, but rarely so in island reefs. Kurtosis and skewness were the two variables that most distinguished these two reef environments, with kurtosis being ~7 times higher and skewness being ~3 times lower on island than fringing reefs.

3.1.2 Comparison between in situ gauges and global model estimates

The three systems to obtain global proxies for estimating SSTs resulted in estimates within 2% of the in situ gauge data (Table 1c, d, e). Moreover, SD of satellite estimates were similar to fringing reefs (<7%) but less accurate for island reefs—overestimating by ~25%. The Global Circulation Model showed the opposite pattern—underestimating the fringing reefs' SD by ~32% and island reefs by ~2% (Table 1d). However, larger differences were observed between satellite/model and in situ gauge measurements for kurtosis and skewness. For example, the European Space Agency and Global Circulation Model overestimated kurtosis compared to gauge data by >550% in the fringing reefs. In the island reefs, CoRTAD underestimated kurtosis by ~60% and the Global Circulation model was different by 11%. Skewness was positive or right tailed for the in situ and satellite estimates. Nevertheless, the absolute values of the satellite-derived estimates of skewness were consistently underestimated in both habitats. Nevertheless, the relative differences were maintained with the fringing reef having higher skewness than the island reefs. The Global Circulation Model data underestimated skewness by ~135% compared to in situ gauge data.

3.2 Coral variables

Hard coral cover was higher in island (41.8% ± 1.8% ± SEM) than fringing reefs (21.6 ± 0.9%) and in fisheries closures (35.0% ± 1.6%) than fished reefs (28.5% ± 1.2%) (Table 2a, Fig 4). Stepwise regression results indicate that coral cover declined weakly ($r^2 = 0.28$) as the standard deviation of temperature increased (Table 3a). Number of coral taxa was also higher on island

Table 2 Generalized linear mixed model (GLMM) results of coral cover metrics and effects of time period, management, reef type, and their interactions.

	Estimates	SE	*P*	Estimates	SE	*P*	Estimates	SE	*P*
Predictors	(a) Hard coral cover			(b) No of genera			(c) Coral community susceptibility		
Intercept	−1.61	0.35	**<0.001**	−2.42	0.16	**<0.001**	−1.27	0.13	**<0.001**
Time period [Period 1]	*Reference*			*Reference*			*Reference*		
Time period [Period 2]	−0.24	0.28	NS	0.14	0.22	NS	−0.25	0.06	**<0.001**
Time period [Period 3]	0.11	0.28	NS	0.13	0.22	NS	−0.22	0.06	**<0.001**
Time period [Period 4]	0.30	0.28	NS	0.25	0.22	NS	−0.17	0.06	**0.005**
Management [Fished]	*Reference*			*Reference*			*Reference*		
Management [Unfished]	1.08	0.54	**0.04**	0.31	0.22	NS	−0.03	0.21	NS
Reef type [Fringing]	*Reference*			*Reference*			*Reference*		
Reef type [Island]	1.13	0.42	**0.006**	0.42	0.18	**0.02**	−0.26	0.16	NS
Management [Unfished] * Reef type [Island]	−0.63	0.71	NS	−0.21	0.28	NS	0.29	0.27	NS
Time period [Period 2] * Management [Unfished]	−0.85	0.43	**0.04**	−0.10	0.31	NS	−0.12	0.09	NS
Time period [Period 3] * Management [Unfished]	−0.86	0.43	**0.04**	0.00	0.30	NS	−0.14	0.09	NS
Time period [Period 4] * Management [Unfished]	−1.13	0.43	**0.009**	−0.15	0.31	NS	−0.20	0.09	**0.02**
Time period [Period 2] * Reef type [Island]	0.63	0.34	0.06	−0.11	0.25	NS	0.21	0.07	**0.002**

Continued

Table 2 Generalized linear mixed model (GLMM) results of coral cover metrics and effects of time period, management, reef type, and their interactions.—cont'd

	Estimates	SE	*P*	Estimates	SE	*P*	Estimates	SE	*P*
Predictors	(a) Hard coral cover			(b) No of genera			(c) Coral community susceptibility		
Time period [Period 3] * Reef type [Island]	−-0.31	0.34	NS	−0.03	0.25	NS	0.10	0.07	NS
Time period [Period 4] * Reef type [Island]	−0.09	0.34	NS	−0.15	0.26	NS	0.00	0.07	NS
Time period [Period 2] * Management [Unfished * Reef type [Island]	0.72	0.57	NS	0.22	0.39	NS	-0.03	0.12	NS
Time period [Period 3] * Management [Unfished * Reef type [Island]	1.83	0.57	**0.001**	−0.01	0.39	NS	0.06	0.12	NS
Time period [Period 4] * Management [Unfished * Reef type [Island]	1.65	0.57	**0.004**	0.13	0.39	NS	0.22	0.12	0.07
R^2	0.51			0.27			0.19		

Coral cover and susceptibility variables were logit transformed before the analyses, number of genera was modelled with a Poisson distribution and log link and included an offset term of the logarithm of the transect length. All models included reef sites as a random effect to account for repeated measurements. NS = not significant or $P > 0.10$.

Bold values indicate statistically significant variables in the model.

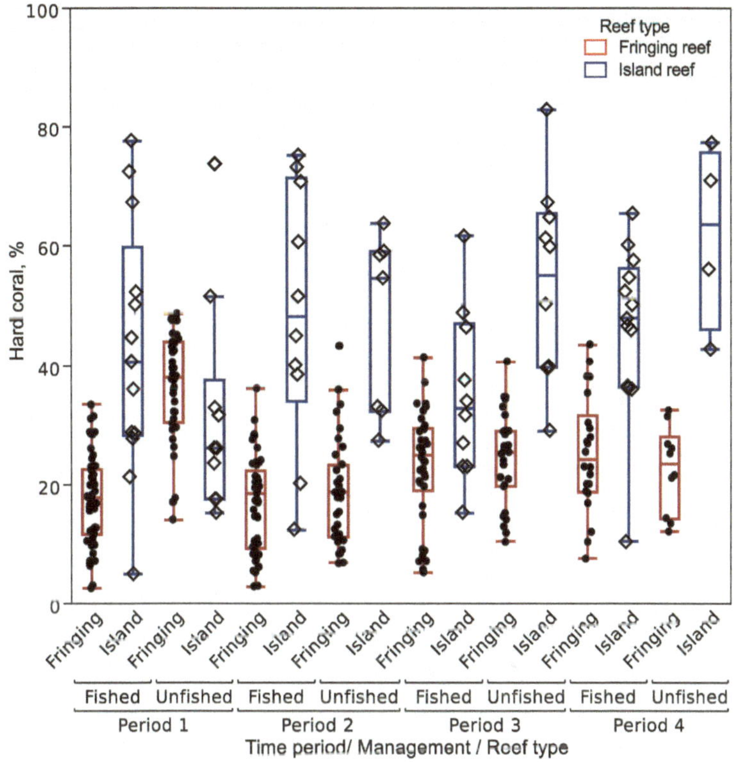

Fig. 4 Changes in hard coral cover over the four sampling periods in fished and unfished reefs of the island and fringing reef environments of East Africa. See Table 2 for statistical results.

(14.8 $\pm$0.7 per 100 m of transect) than fringing (11.6 $\pm$0.3) reefs (Tables 2b and 3b). Number of coral genera was higher but not statistically significant in unfished (14.4 $\pm$0.6) than fished (12.0 $\pm$0.4) reefs and did not change over time by both evaluations (Table 2b). Number of coral taxa declined ($r^2 = 0.31$) with increasing SD, skewness, and mean temperature in that order of influence (Table 3b). Coral community susceptibility to bleaching was not influenced by reef type but was higher in unfished than fished reefs. Susceptibility declined after the 1998 thermal stress or the middle time periods and also interacted with reef type in time period 2 and reef management in time period 4 (Tables 2c; 4c). Coral community susceptibility relationships with temperature metrics were weak ($r^2 = 0.15$) but declined with the mean and SD of the temperature (Table 3c).

Table 3 Forward stepwise regression results on coral metrics and community composition axes 1 and 2 against temperature variables.

Predictors	Estimates	SE	*z* value	*P*	R^2
a) *Hard coral cover*					
Intercept	−0.66	0.23	**−2.75**	**0.006**	**0.28**
Standard deviation	−0.85	0.21	**−3.96**	**<0.001**	
Skewness	−0.38	0.23	−1.69	0.09	
Kurtosis	−0.27	0.19	−1.39	NS	
Mean temperature	-0.00	0.19	−0.03	NS	
b) *No of genera*					
Intercept	−2.03	0.04	**−45.56**	**<0.001**	**0.31**
Standard deviation	−0.30	0.10	**−3.10**	**0.002**	
Skewness	−0.23	0.10	**−2.34**	**0.02**	
Mean temperature	−0.19	0.10	−1.88	0.06	
Kurtosis	−0.05	0.11	−0.47	NS	
c) *Coral community susceptibility*					
Intercept	−1.63	0.05	**-32.66**	**<0.001**	**0.15**
Mean temperature	−0.15	0.05	**−2.90**	**0.004**	
Standard deviation	−0.15	0.06	**−2.54**	**0.01**	
Kurtosis	0.05	0.05	0.96	NS	
Skewness	0.05	0.07	0.78	NS	
d) *PCA, axis 1*					
Intercept	−0.04	0.12	−0.31	NS	0.01
Kurtosis	0.01	0.10	0.12	NS	
Mean temperature	−0.06	0.09	−0.74	NS	
Standard deviation	0.06	0.11	0.53	NS	
Skewness	−0.03	0.12	−0.28	NS	
e) *PCA, axis 2*					
Intercept	−0.04	0.08	−0.06	NS	0.17
Skewness	−0.21	0.08	−2.57	0.01	

Table 3 Forward stepwise regression results on coral metrics and community composition axes 1 and 2 against temperature variables.—cont'd

Predictors	Estimates	SE	*z* value	*P*	R^2
Kurtosis	−0.15	0.08	−1.92	0.05	
Standard deviation	−0.15	0.08	**−1.91**	0.05	
Mean temperature	−0.11	0.06	**−1.61**	NS	

Coral cover and susceptibility predictors were logit transformed prior to the analysis. Number of genera was modelled with a Poisson distribution with transect length included as an offset term. All models included reef sites as a random effect to account for repeated measurements. NS = not significant or $P > 0.10$. SE = standard error of the estimate.
Bold values indicate statistically significant variables in the model.

3.3 Coral communities, change, and temperature variable associations

3.3.1 Coral taxonomic composition

The PCAs differentiated sites and sampling times by the composition of the dominant 15 taxa (Fig. 5). The first and second axis gradients were moderately explained with the GLMM variables of reef type, management, and sampling periods explaining 51% of the first and 60% of the second axes' variability (Table 2). There was, however, high variation in the taxonomic composition of coral communities on the studied reefs with the first and second axis of the PCAs explaining only 22.3% and 14.2% of the total coral community variance. Nevertheless, when evaluating communities at sites and times, some clear patterns emerged based on reef morphology, temperature metrics, and fishing impacts (Tables 3d, e; 4d, e; 5). The first PCA axis separated taxa along reef type whereas the second axis separated taxa by all factors of reef type, management, and time (Tables 4, 5). By the GLMM analysis, there were significant interaction between reef type and time periods 3 and 4. Coral communities in unfished fringing reef sites (Malindi and Watamu) changed after 1998 to align with the island sites damaged by dynamite fishing (Taa, Funguni, Changale) on the right side of axis 1 (Supplementary Fig. 1 in the online version at https://doi.org/10.1016/bs.amb.2020.08.003). Fished and unfished fringing reefs were largely separated by the second axis. Several fished fringing reefs reverted closer to the original state after 1998, whereas recovery was less evident for the unfished fringing reefs or the dynamite-damaged island reefs.

Taxonomically, the first PCA axis separated the *Acropora*, *Montipora*, *Galaxea astreata*, *G. fascicularis*, and *Synarea* (*P. rus*) dominated island reefs from the massive *Porites*, faviids *Pocillopora*, branching *Porites*, *Pavona*, and

Table 4 Type 3 ANOVA (significant interactions) results from fitted GLMM model on (a) hard coral cover, (b) number of coral genera, (c) coral community susceptibility, (d) coral community composition PCA axis 1 and (e) PCA axis 2 with reef type specified as a random variable.

Term	Statistic	df	*P* value
(a) *Hard coral cover*			
Intercept	21.07	1	**<0.001**
Time period	3.86	3	NS
Management	4.03	1	**0.04**
Reef type	7.46	1	**0.006**
Time period * Management	7.69	3	0.05
Time period * Reef type	8.75	3	0.03
Management * Reef type	0.8	1	NS
Time period * Management * Reef type	13.4	3	**0.004**
(b) *Number of genera*			
Intercept	228.5	1	**<0.001**
Time period	1.2	3	NS
Management	2.1	1	NS
Reef type	5.3	1	**0.02**
Time period * Management	0.3	3	NS
Time period * Reef type	0.5	3	NS
Management * Reef type	0.6	1	NS
Time period * Management * Reef type	0.5	3	NS
(c) *Susceptibility*			
Intercept	89.2	1	**<0.001**
Time period	21.4	3	**<0.001**
Management	0.02	1	NS
Reef type	2.6	1	NS
Time period * Management	5.2	3	NS
Time period * Reef type	12.2	3	**0.006**
Management * Reef type	1.1	1	NS

Table 4 Type 3 ANOVA (significant interactions) results from fitted GLMM model on (a) hard coral cover, (b) number of coral genera, (c) coral community susceptibility, (d) coral community composition PCA axis 1 and (e) PCA axis 2 with reef type specified as a random variable.—cont'd

Term	Statistic	df	*P* value
Time period * Management * Reef type	5.1	3	NS
(d) *PCA, axis 1*			
Intercept	2.9	1	0.09
Time period	4.4	3	NS
Management	0.6	1	NS
Reef type	7.9	1	**0.004**
Time period * Management	2.5	3	NS
Time period * Reef type	4.6	3	NS
Management * Reef type	1.1	1	NS
Time period * Management * Reef type	0.1	3	NS
(e) *PCA, axis 2*			
Intercept	27.6	1	<0.001
Time period	8.3	3	0.03
Management	22.4	1	**<0.001**
Reef type	28.5	1	**<0.001**
Time period * Management	1.6	3	NS
Time period * Reef type	5.3	3	NS
Management * Reef type	16.7	1	**<0.001**
Time period * Management * Reef type	3.3	3	NS

Coral cover and susceptibility variables were logit transformed before the analyses, number of genera was modelled with a Poisson distribution and log link and included an offset term of the logarithm of the transect length. Reef types are fringing and island reefs, time periods are 1991/1997, 1998/2005, 20-06/2012, and 2013/2016 and management is either fishing or long-term closures to fishing. NS = not significant or $P > 0.10$. A star between two variables (i.e., Time period * Reef Type) indicates an interaction in the statistical model.
Bold values indicate statistically significant variables in the model.

Stylophora dominated fringing reefs. The second axis split between the fished and unfished fringing reef sites indicated that there were three dominant clusters of coral communities. In many respects, these three divisions largely reflect taxa associations with weedy (ruderal) common to the fished fringing

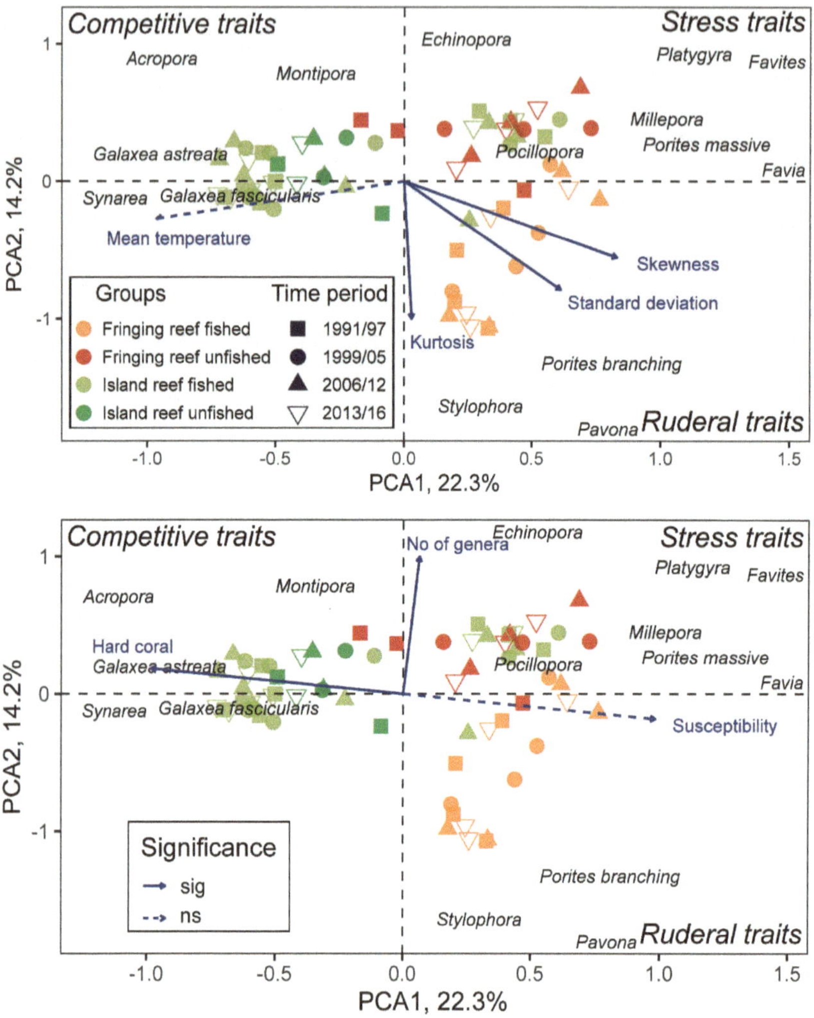

Fig. 5 Principal Component Analysis (PCA) of the sites ordinated by the dominant coral taxa in the island and fringing reefs for the four time periods (A) with temperature metric variables as vectors collected with in situ temperature gauges data for each time period presented and (B) during all periods combined as the statistically selected benthic cover vectors. The coral traits that dominant the coral community in each quadrat are presented.

reefs, stress resistant common to dynamited and unfished fringing reefs after 1998, and competitive life histories common to the island reefs and unfished fringing reefs prior to 1998. Some of the island reefs, particularly those that had been fished with dynamite, were more aligned with the stressed communities. The ruderal community dominated by *Pavona*, branching *Porites*,

Table 5 Generalized linear mixed model (GLMM) results of coral community PCA axes 1 and 2, and effects of time period, management, reef type and their interactions.

	PCA1			PCA2		
Predictors	**Estimates**	**SE**	***P***	**Estimates**	**SE**	***P***
Intercept	0.28	0.17	0.091	−0.66	0.13	**<0.001**
Time period [Period 1]	*Reference*			*Reference*		
Time period [Period 2]	0.15	0.10	NS	0.24	0.09	**0.005**
Time period [Period 3]	0.19	0.10	**0.049**	0.13	0.09	NS
Time period [Period 4]	0.09	0.10	NS	0.08	0.09	NS
Management [Fished]	*Reference*			*Reference*		
Management [Unfished]	−0.19	0.25	0.456	0.91	0.19	**<0.001**
Reef type [Fringing]	*Reference*			*Reference*		
Reef type [Island]	−0.55	0.20	**0.005**	0.79	0.15	**<0.001**
Management [Unfished] * Reef type [Island]	0.10	0.34	NS	−1.04	0.25	**<0.001**
Time period [Period 2] * Management [Unfished]	0.21	0.15	NS	−0.11	0.13	NS
Time period [Period 3] * Management [Unfished]	0.17	0.15	NS	0.05	0.13	NS
Time period [Period 4] * Management [Unfished]	0.19	0.15	NS	0.01	0.13	NS
Time period [Period 2] * Reef type [Island]	−0.18	0.12	NS	−0.23	0.10	**0.023**
Time period [Period 3] * Reef type [Island]	−0.24	0.12	**0.04**	−0.14	0.10	NS
Time period [Period 4] * Reef type [Island]	−0.16	0.12	NS	−0.10	0.10	NS
Time period [Period 2] * Management [Unfished * Reef type [Island]	−0.08	0.20	NS	0.31	0.17	0.07
Time period [Period 3] * Management [Unfished * Reef type [Island]	−0.11	0.20	NS	0.17	0.17	NS
Time period [Period 4] * Management [Unfished * Reef type [Island]	−0.18	0.20	NS	0.18	0.17	NS
R^2	0.51			0.60		

All models included reef sites as a random effect to account for repeated measurements. NS = not significant or $P > 0.10$.

and *Stylophora* were associated with higher SD, skewness, and kurtosis of temperature. The stress-community was associated with lower coral cover and more strongly with skewness than the SD of temperature compared to the ruderal community.

3.3.2 Changes in the coral community and temperature associations

Unfished island reefs changed the least and maintained high dominance of *Montipora*, *Acropora*, *Synarea*, *Galaxea astreata* and *G. fascicularis* over the full study period. Larger changes in the hard coral taxonomic composition was most clearly observed in fringing reefs after the 1998 thermal disturbance (Table 4e, f, Fig. 5b). Fished island reefs were more variable depending on the prior history of dynamite fishing but some of the undamaged reefs (Mbudya, Bongoyo, Chapwani, and Changu) and more damaged reefs (Taa, Funguni, Changale) were fairly stable across the 1998 event. Makome (a community closure) and Chumbe (unfished) showed some movement towards the stressed reef community composition after 1998 followed by recovery closer to the original states.

Change in the fringing reefs was more pronounced in the unfished sites than fished sites, particularly Malindi and Watamu. Taxonomic composition of unfished Malindi and Watamu sites was initially more similar to the island reefs but diverged away from *Acropora* and *Montipora* towards *Platygyra, Favites, Pocillopora, Echinopora* and massive *Porites* dominance after 1998. After 1998, there was little evidence that these unfished fringing reefs returned to their original state due to an increase in *Pocillopora* and *Echinopora* relative to the other taxa. The third and most recently protected unfished reef (Mombasa) always grouped with the stressed community but changed to reflect greater stress after 1998, with some recovery thereafter. Fished fringing reefs harboured a unique ruderal community and, after 1998, the sites' communities moved towards the stress region of the plot but with different movement distances and recovery rates.

4. Discussion

Temperature anomalies, such as the severe one in 1997/98, created an acute disturbance with long-term impacts on the distribution of some but not all coral taxa, their populations, and resultant communities (Ateweberhan et al., 2011; Darling et al., 2019; McClanahan, 2020). In this study location, taxonomic composition was more sensitive than coral cover in exhibiting responses to temperature variables and episodic disturbances. Nevertheless, there was a significant but modest decline in coral cover along the gradient from the competitive to the stress and ruderal coral community types. These different reef and community types associated with bathymetry, SST metrics, and other unmeasured variables harboured different dominant taxa. Specifically, SST-SD, skewness, and kurtosis reflected ocean-land processes that reflected the variable chronic and acute stress conditions experienced by these reefs. Chronic and acute stress interactions influenced coral cover, the dominant taxa, and associated life histories, as was also found in western Australian coral communities (Zinke et al., 2018). SD and kurtosis of the temperature distributions at a site are good metrics for background or chronic variation while skewness better reflects the frequencies of acute stress. Temperature variation metrics also coincided with coastal bathymetry and reef morphology, due to interactions with large-scale and inter-annual ocean current and heating processes. The southern island sub-region had lower background temperature variation than the northern fringing reef sub-region. This is the opposite of larger scale temperature variation patterns in the western Indian Ocean (Ateweberhan and McClanahan, 2016). It also indicates important ecological variation nested within the ecoregion designation. Bathymetry and reef morphologies produced their own unique thermal environments that modified larger scale ocean variability. In the case of the island reefs, this morphology had the potential to reduce or buffer the impacts of large-scale climate impacts.

4.1 Temperature distributions and coral community distributions

The three moments of the temperature distributions (SD, kurtosis, and skewness) helped explain coral community states, particularly in distinguishing the Acroporidae (*Acropora* and *Montipora*), which occurred in temperature regimes of lower variability and fewer and smaller extremes than the massive *Porites*, faviids, *Pavona*, branching *Porites,* and *Stylophora* dominated communities. While the moments of the temperature distributions are related, their relationships also exhibited non-linearity and high scatter (McClanahan et al., 2007a). Collinearity was weak and largely accounted for in the step-wise procedure that identified one or more of these variables as significantly associated with the community structure of the corals. Thus, they represent different aspects of thermal stress and are associated with variable impacts on coral taxa (McClanahan et al., 2019a).

Kurtosis and skewness have a U-shaped relationship in the region—sites with right-skew having strongly centralized or peaky distributions. Whereas locations with low and negatively skewed temperatures have flatter distributions. When the skew is negative or to the left, these locations experience rare cold water anomalies. Most importantly, these distributions are associated with cumulative thermal stress metrics that interact with background conditions to provoke bleaching and coral mortality depending on how the taxa respond to both background and acute stress (McClanahan et al., 2020a,b). During 1998, for example, the Degree-Heating Months (DHM) were strongly negatively associated with SST-SD and positively with skewness and kurtosis (McClanahan et al., 2007a). Here, we see these variables were associated with the state of coral communities, indicating how knowing background temperature distributions assist understanding thermal disturbance outcomes. These findings further identify the likely causes of coral mortality and community change when acute thermal stress impacts corals (Baumann et al., 2016; Donner et al., 2005; Eakin et al., 2009; McClanahan et al., 2007b, 2020a).

The island and fringing reefs reef environments were previously distinguished in terms of the ocean heights and differentiated as having high (fringing) and low (island) thermal impacts in 1998 (McClanahan et al., 2007a). Ocean height is influenced by thermal expansion and retention or drag as ocean currents interact with the coastal bathymetry. Because temperature differences were small, the measured differences in ocean height were due largely to the bathymetry, where islands and associated deep-water channels resist oceanic currents more than fringing reefs growing on an inclined bathymetry. The emergent Pleistocene islands and submerged deep-water channels are likely to produce the water retention that promotes cool and stable temperatures when originating from deep water. The deep water offshore from the island is expected to deliver this stable and cool water through gaps between the islands. In contrast, the small lagoons behind the fringing reefs create some short-term diurnal and tidal variability that will interact with the longer inter-annual variability of ocean-scale variability, such as the Indian Ocean Dipole (IOD) and El Niño Southern Oscillation (ENSO) (Abram et al., 2020; Nguli et al., 2006). The reefs with the shallowest offshore waters in Malindi and Watamu experienced large changes and poor recovery after 1998 despite their long-term closure to fishing. In contrast, the shallower island reefs to the south of Pangani River experienced less disturbance after 1998, suggesting some mediating role of the offshore deep water or the island nature of this environment. Bathymetry and reef habitat patterns were sufficiently influential to override some of the large-scale western Indian Ocean patterns, such as lower SST variation in the north than the south (Ateweberhan and McClanahan, 2010). This variation may have played some role in reversing the predicted impacts of the frequently used time-to-threshold model for coral demise (Hoegh-Guldberg, 1999; Hughes et al., 2017; Sheppard, 2003).

Riverine estuaries were formed during deglaciation by runoff from Meru, Kilimanjaro and the Usambara mountains, and sea-level changes are likely to have assisted in the creation of the dissected nature of this nearshore environment and some of deep water channels close to shore (Arthurton et al., 1999). The deep water seaward of the islands of Zanzibar and Pemba and high seawater retention behind the islands in the Pemba Channel is likely to have had consequences for exposure and penetration of interannual IOD and ENSO ocean variability into nearshore coastal systems. This thermal stability close to shore should provide a buffer to the inter-annual oceanographic disturbances and provide corals with some sanctuary. Given the increasing ocean-scale climate oscillation variability emerging with anthropogenic climate change (Abram et al., 2020; DiNezio et al., 2020), this could be a refuge of increasing importance to the maintenance of the East African ecoregion's unique fauna.

The metrics of SD, kurtosis, and skewness were shown to differentiate the coral communities between the island and fringing reefs along this 400-km of coastline. Kurtosis and skewness appeared to be the most influential metrics, along with biomodality (Ateweberhan et al., 2018; McClanahan et al., 2019b, 2020a,b; Zinke et al., 2018). The step-wise regression evaluations of the coral community axes, indicated that kurtosis, and to a much lesser extent skewness, differentiated coral communities along the two PCA community axes. In contrast, skewness primarily differentiated the Acroporidae from the other two life history trait communities—*Acropora* and *Montipora* being uncommon where skewness was high. Thus temperature variability helps explains the three coral communities and the three-lobed distribution of taxa and their associated traits (Darling et al., 2012, 2019). This pattern in life histories is analogous with the Competitive-Stress-Ruderal (CSR) theory of plant trait ecology (Grime, 2006). Further research will be needed to determine how the various traits, resources, disturbances, and stresses produce these patterns and their consequences for future change (Darling et al., 2012, 2019; van Woesik et al., 2012). East African coral communities are a mixture of life histories differentially associated with reefs and land-sea interactions that influence chronic and acute temperature stress patterns.

4.2 Fisheries management impacts

Coral community responses to 1998 depended on fisheries management. Ecological responses to fishing restriction are variable and may reflect various environment—food web configurations that challenge generalizations about fishing impacts (Graham et al., 2008; McClanahan and Muthiga, 2016; Olds et al., 2014; Selig et al., 2010; Toth et al., 2014). The selection process of protected areas frequently favours high diversity and therefore challenges efforts to scientifically evaluate climate-fishing disturbances (Selig et al., 2010). Many fisheries closures have been selectively placed in stable temperature locations with low temperature variability that support high coral diversity but low tolerance to anomalies (Maina et al., 2008; Selig et al., 2012). Protected areas may also be located in areas where climate stress and future impacts may be greatest (Maina et al., 2008; McClanahan, 2020). As shown here, coral diversity benefits from stable temperatures but a number of key space-occupying taxa are susceptible to SST anomalies (McClanahan and Maina, 2003). When episodic heat stress occurs in stable temperature environments, it causes high mortality among sensitive coral taxa. Stable conditions are, however, expected to decline as climate change intensifies (Hughes et al., 2017, 2018). Here, we see that this oceanic interannual variability had larger effects on the exposed fringing than the island reefs. Evaluation of metrics of variability provides important information for predicting future management needs created by climate change impacts (Ateweberhan and McClanahan, 2016; Beyer et al., 2018).

4.3 Persistence and change in corals

Changes in the coral communities over time suggested more persistence in island than fringing reefs, including those in fisheries closures (Allen et al., 2017). While the higher temperature variance in fringing reefs might provide the capacity for resistance, the skewness was higher and temperature distribution less flat, resulting in larger negative impacts after 1998. Examination of coral mortality in the Western Indian Ocean (WIO) after 1998 found that coral mortality declined with SST-SD until it exceeded 2.5°C SDs, when mortality rose rapidly (Ateweberhan and McClanahan,

2010). Zinke et al. (2018) also showed a general decline in bleaching susceptible corals above this same 2.5°C SST-SD level in western Australia. The SSTs-SDs reported here were below 2.5°C and the moderate differences between habitats suggest that temperature shape parameters of kurtosis and skewness were the more informative metrics of thermal acclimation and stress. Greater persistence in island than fringing reefs was associated with moderate SST-SD but more importantly flatter temperature distributions and low warm-water skew. Some of the weak responses in coral cover, especially across 1998, could be due to pooling taxa and data between years, which can hide short-term changes in coral taxa abundance (Ateweberhan et al., 2011).

Changes in coral communities did not indicate disturbances and rapid recovery to an initial state, as found in some remote *Acropora*-dominated Indian Ocean reefs (Gilmour et al., 2013). The most change in the coral community was after 1998 and, while there was some recovery, it was seldom to the same baseline community. The thermal disturbance in 1998 produced a community change towards and then away from stress-resistant taxa of massive *Porites*, some faviids, and temporary colonization by *Pocillopora* (McClanahan, 2014). While there was some return towards the baseline, this return was frequently arrested depending on the coral community type. The island reefs, regardless of fishing, generally moved towards greater relative dominance of *G. astreata* after 1998. *G. astreata* is a long lived reef foundational taxon more resilient to temperature anomalies than taxa that typically grow on their foundation (Hamilton and Brakel, 1984). Massive *Porites* plays a similar foundational role in the fringing reef environment (Crame, 1981). Consequently, after 1998, there was also a community movement towards and then away from these two massive and resistant taxa. Both the stressed and ruderal communities were differentiated from the competitive communities by temperature skewness and between themselves by SST-SD. The ruderal community having higher chronic but lower acute variation than the stressed community.

Community change differed based on the fisheries management and confounding effects of reef depth, such that slightly deeper unfished reefs maintained more similarity to island reefs until stressed by a rare thermal anomaly in 1998. Thereafter, the relative abundance of the weedy and generalist branching taxa increased (Darling et al., 2013). The causes for the cover increase in unfished island reefs is unclear but could be associated with good environmental conditions for recovery or increased compliance with fisheries management. Thus, competitive coral taxa recover faster and occupy more space, which increases total cover coverage. In contrast, those reefs that had a history of dynamite fishing exhibited the same community structure as thermally stressed reefs and were resistant to change and rapid recovery despite the decline in dynamite fishing over this study period (Wells et al., 2007, 2010). Future trends in coral cover in these reefs were evaluated using a Bayesian autoregressive state-space model and island reefs were projected to be more stable than fringing reefs due to their higher among site variability in coral cover (Allen et al., 2017).

4.4 Climate and community change

Present findings can contribute to evaluating the consequences of increased warming and climatic variability. Clearly, coral responses to increasing frequency and intensity of disturbances depend on background temperature variability. Island environments with deep water offshore and deep channels nearshore have flat and low-skew temperature distributions that reflect resistance or buffering to the intrusion of inter-annual thermal variability and strong waves and currents. Reefs directly exposed to major currents and large storms and waves should lack this buffering capacity and exhibit higher inter-annual temperature skewness. Therefore, in reefs adjacent to deep water and not influenced by waves and currents, temperatures and other physical disturbances should remain below stressful limits, increasing the chances for coral persistence (Kleypas et al., 2008). A summary analysis of coral taxonomic richness—environmental interactions in the Indian Ocean found kurtosis a useful predictor and that richness increased as temperature distributions became flatter (Ateweberhan et al., 2018).

Strong correlations between variance, kurtosis, skewness, and temperature anomalies can challenge the ability to distinguish their various contributions to acclimation and adaptation (Ateweberhan et al., 2018; Zinke et al., 2018). Acclimation, genetic adaptation, and taxonomic or community change, are more likely to arise where variation occurs and produces acclimation or organismic "memory", selective birth, mortality, and evolutionary processes (Palumbi et al., 2014; van Oppen et al., 2015). Shifts in the relative abundance of taxa should be a primary response to increased temperature extremes and variability and therefore important in evaluating the future of corals (Loya et al., 2001; McClanahan and Maina, 2003; Riegl, 1999; Zinke et al., 2018). Here, temperature SD and kurtosis were good descriptors of chronic variability that influenced coral community composition. But, acute temperature anomalies were most influential and catalysed change and recovery that differentiated the traits of the three main communities. Therefore, when corals experience temperature above tolerable thresholds, qualitative changes in the community are expected and rapid declines in cover are associated with the preferential mortality of space-occupying competitive taxa in the Acroporidae. These qualitative changes may be predictable based on chronic-acute temperature interactions and a knowledge of the coral life histories.

Temperatures stresses are predicted to impact competitive space-occupying corals and promote corals with weedy and stress-resistant life histories (Darling et al., 2012; Riegl et al., 2017; van Woesik et al., 2012). Yet, some argue that positive temperature skewness might potentially create experiences and memories that promote adaptation to warm extremes (Ruff and Neelin, 2012). Rare events are likely to be too infrequent to promote acclimation and adaptive responses. Episodic events may promote tolerant taxa that can easily be confused with adaptive taxa (Donner et al., 2005; Riegl, 1999). A study of western Australian corals found that the cover of stress-resistant taxa declined with SST-skewness but the decline was slower than for overall coral cover that was most closely associated with temperature kurtosis (Zinke et al., 2018). Consequently, episodic stress events affected stress-resistant corals less than other coral taxa but there were still limits to their capacity to tolerate stress within the ranges studied.

At the Indian Ocean basin scale, flatter temperature distributions were associated with higher taxonomic richness and lower coral mortality across the 1998 event (Ateweberhan and McClanahan, 2010; Ateweberhan et al., 2018). Flatter temperature distributions are more frequently associated with low skewness but the relationship may be unimodal, concave, and variable (Ateweberhan and McClanahan, 2010). Some influences, such as large tidal ranges may promote variable but flat distributions while interactions with extreme inter-annual temperature oscillations can produce skewed distributions. Very few taxa should be able to tolerate increasingly strong and rare temperature anomalies and, as they increase with climate change, the result should eventually be the impoverishment of the coral community, ecological functions, and ecosystem services (D'agata et al., 2014; Hooper et al., 2005; Riegl and Purkis, 2015). High chronic and acute temperatures combined are likely to favour non-hard coral taxa, such as algae, seagrass, and other stress-tolerant invertebrates.

4.5 Weaknesses and caveats

The large-scale experimental design helped uncover the influence of reef geomorphology, oceanography, temperature histories, habitat, management, and their interactions in influencing coral reef community responses to temperature anomalies. Typical of most large-scale geographic studies, the factors of reef type and the environmental variables of currents, water flow, and temperature interacted, were driven by geography, and were not always statistically independent. It is not possible to strictly control and randomly assign any reef to these environmental and management factors. The statistical methods used in this study should account for some of this dependency. Future efforts to test these temperature factors on key coral taxa in controlled, smaller scale, random-assignment, and balanced design experiments are likely to be insightful. Nevertheless, controlling experimental influences often produces experimental artefacts that make it difficult to fully replicate large spatial and temporal natural experiments (McClanahan et al., 2011).

4.6 Future directions and predictions

Future research is needed to understand whether temperature buffering capacity is widespread and associated with low ocean height, deep water offshore, ancient estuaries, or island and complex reef environments. Reef archipelagos, such as the Indo-Malaysian and Pacific region, may be locations where seawater retention and thermal stress attenuation are present (McClanahan et al., 2020b; Peñaflor et al., 2009). If so, there is the possibility that some island environments have temperature buffering capacity that can reduce impacts of episodic warm temperatures, making them candidates for climate refugia (McClanahan et al., 2019a, 2020a; McLeod et al., 2010). Additionally, given the importance of temperature metrics in reflecting the stress that corals experience, the global distribution of stress-reducing conditions requires further investigation (Hoegh-Guldberg et al., 2018). Clearly, more investigation is required to better understand the causes of the problem, appropriate measurements and metrics, proxies for them, and their interactions with nearshore conditions. Otherwise, predictions and the consequences of future climate impacts will be poorly understood and possibly undermined by a poor connection between causative factors and proxies of stress used in models.

Acknowledgements

This work was supported by the Wildlife Conservation Society through grants from the Western Indian Ocean Marine Science Association (WIOMSA) from their Marine Science for Management Program (MASMA), Pew Charitable Trust, Tiffany Foundation, and the John D. and Catherine T. MacArthur Foundations. Clearance to do research in Kenya was provided by Kenya's Office of Science and Technology and in the parks by

Kenya Wildlife Services. The University of Dar es Salaam and The Board of Trustees of the Marine Parks and Reserves of Tanzania provided permission to study in the parks and reserves of Tanzania. Many people assisted with the benthic monitoring including R. Arthur, E. Darling, C. Hicks, R. Kiambo, A.T. Kamukuru, B. Kaunda-Arara, J. Kawaka, R. Machaku, H. Machano Ali, S. Mangi, J. Maina, J. Mariara, R. Moothien-Pillay, J. Mutere, N.A. Muthiga, S. Mwacheriya, E. Mwakalapa, J. Ndagala, J. Omukoto, and M.J. Rodrigues. R. Machaku and M. Azali assisted with producing the graphs and statistics and J. Maina assisted with the map and model and satellite-derived information. C. Muhando collected and shared his long-term in situ temperature data from Chumbe Island, Zanzibar. M. Ateweberhan, E. Darling, and M. Spencer provided useful comments on an earlier draft.

References

Abram, N.J., Wright, N.M., Ellis, B., Dixon, B.C., Wurtzel, J.B., et al., 2020. Coupling of Indo-Pacific climate variability over the last millennium. Nature 579, 385–392.

Ainsworth, T.D.A., Heron, S.F., Ortiz, J.C., Mumby, P.J., Grech, A., et al., 2016. Climate change disables coral bleaching protection on the Great Barrier Reef. Science 352, 338–342.

Allen, K.A., Bruno, J.F., Chong, F., Clancy, D., McClanahan, T.R., et al., 2017. Among-site variability in the stochastic dynamics of East African coral reefs. PeerJ 5, e3290.

Arthurton, R., Brampton, A., Kaaya, C.Z., Mohamed, S., 1999. Late Quaternary coastal stratigraphy on a platform-fringed tropical coast: a case study from Zanzibar, Tanzania. J. Coast. Res. 15, 635–644.

Ateweberhan, M., McClanahan, T.R., 2010. Relationship between historical sea-surface temperature variability and climate change-induced coral mortality in the western Indian Ocean. Mar. Pollut. Bull. 60, 961–970.

Ateweberhan, M., McClanahan, T.R., 2016. Partitioning scleractinian coral diversity across reef sites and regions in the Western Indian Ocean. Ecosphere 7, e01243.

Ateweberhan, M., McClanahan, T.R., Graham, N.A.J., Sheppard, C.R.C., 2011. Episodic heterogeneous decline and recovery of coral cover in the Indian Ocean. Coral Reefs 30, 739–752.

Ateweberhan, M., McClanahan, T.R., Maina, J., Sheppard, C., 2018. Thermal energy and stress properties as the main drivers of regional distribution of coral species richness in the Indian Ocean. J. Biogeogr. 45, 1355–1366.

Baker, A.C., McClanahan, T.R., Starger, C.J., Boonstra, R.K., 2013. Long-term monitoring of algal symbiont communities in corals reveals stability is taxon dependent and driven by site-specific thermal regime. Mar. Ecol. Prog. Ser. 479, 85–97.

Ban, N.C., Pressey, R.L., Weeks, S., 2012. Conservation objectives and sea-surface temperature anomalies in the Great Barrier Reef. Conserv. Biol. 26, 799–809.

Baumann, J.H., Townsend, J.E., Courtney, T.A., Aichelman, H.E., Davies, S.W., et al., 2016. Temperature regimes impact coral assemblages along environmental gradients on lagoonal reefs in Belize. PLoS One 11, e0162098.

Beyer, H.L., Kennedy, E.V., Beger, M., Chen, C.A., Cinner, J.E., et al., 2018. Risk-sensitive planning for conserving coral reefs under rapid climate change. Conserv. Lett. 11, e12587.

Bolker, B.M., Brooks, M.E., Clark, C.J., Geange, S.W., Poulsen, J.R., et al., 2009. Generalized linear mixed models: a practical guide for ecology and evolution. Trends Ecol. Evol. 24, 127–135.

Brooks, M.E., Kristensen, K., van Benthem, K.J., Magnusson, A., Berg, C.W., et al., 2017. glmmTMB balances speed and flexibility among packages for zero-inflated generalized linear mixed modeling. The R Journal 9, 378–400.

Bruno, J.F., Valdivia, A., 2016. Coral reef degradation is not correlated with local human population density. Sci. Rep. 6, 29778.
Carreiro-Silva, M., McClanahan, T.R., 2012. Macrobioerosion of dead branching Porites, 4 and 6 years after coral mass mortality. Mar. Ecol. Prog. Ser. 458, 103–122.
Crame, J.A., 1981. Ecological stratification in the Pleistocene coral reefs of the Kenya Coast. Palaeontology 24, 609–646.
D'agata, S., Mouillot, D., Kulbicki, M., Andrefouet, S., Bellwood, D.R., et al., 2014. Human-mediated loss of phylogenetic and functional diversity in coral reef fishes. Curr. Biol. 24, 555–560.
Darling, E.S., Alvarez-Filip, L., Oliver, T.A., McClanahan, T.R., Côté, I.M., 2012. Evaluating life-history strategies of reef corals from species traits. Ecol. Lett. 15, 1378–1386.
Darling, E.S., McClanahan, T.R., Côté, I.M., 2013. Life histories predict coral community disassembly under multiple stressors. Glob. Chang. Biol. 19, 1930–1940.
Darling, E.S., McClanahan, T.R., Maina, J., Gurney, G.G., Graham, N.A., et al., 2019. Social–environmental drivers inform strategic management of coral reefs in the Anthropocene. Nat. Ecol. Evol. 3, 1341–1350.
DiNezio, P.N., Puy, M., Thirumalai, K., Jin, F.-F., Tierney, J.E., 2020. Emergence of an equatorial mode of climate variability in the Indian Ocean. Sci. Adv. 6, eaay7684.
Done, T.J., 1999. Coral community adaptability to environmental change at the scales of regions, reefs and reef zones. Am. Zool. 39, 66–79.
Donlon, C.J., Martin, M., Stark, J., Roberts-Jones, J., Fiedler, E., et al., 2012. The operational sea surface temperature and sea ice analysis (OSTIA) system. Remote Sens. Environ. 116, 140–158.
Donner, S.D., 2011. An evaluation of the effect of recent temperature variability on the prediction of coral bleaching events. Ecol. Appl. 21, 1718–1730.
Donner, S.D., Skirving, W.J., Little, C.M., Oppenheimer, M., Hoegh-Guldberg, O., 2005. Global assessment of coral bleaching and required rates of adaptation under climate change. Glob. Chang. Biol. 11, 2251–2265.
Douma, J.C., Weedon, J.T., 2019. Analysing continuous proportions in ecology and evolution: a practical introduction to beta and Dirichlet regression. Methods Ecol. Evol. 10, 1412–1430.
Ducet, N., Le Traon, P.Y., Reverdin, G., 2000. Global high resolution mapping of ocean circulation from TOPEX/Poseidon and ERS-1 and -2. J. Geophys. Res. 105, 19477–19498.
Eakin, C.M., Lough, J.M., Heron, S.F., 2009. Climate variability and change: Monitoring data and evidence for increased coral bleaching stress. In: van Oppen, M.J.H., Lough, J.M. (Eds.), Coral Bleaching. Springer, pp. 41–67.
Edmunds, P.J., Gates, R.D., 2008. Acclimatization in tropical reef corals. Mar. Ecol. Prog. Ser. 361, 307–310.
Gelman, A., 2008. Scaling regression inputs by dividing by two standard deviations. Stat. Med. 27, 2865–2873.
Gilmour, J.P., Smith, L.D., Heyward, A.J., Baird, A.H., Pratchett, M.S., 2013. Recovery of an isolated coral reef system following severe disturbance. Science 340, 69–71.
Golbuu, Y., Victor, S., Penland, L., Idip Jr., D., Emaurois, C., et al., 2007. Palau's coral reefs show differential habitat recovery following the 1998-bleaching event. Coral Reefs 26, 319–332.
Gove, J.M., Williams, G.J., McManus, M.A., Heron, S.F., Sandin, S.A., et al., 2013. Quantifying climatological ranges and anomalies for Pacific coral reef ecosystems. PLoS One 8, e61974.
Graham, N.A., McClanahan, T.R., MacNeil, M.A., Wilson, S.K., Polunin, N.V., et al., 2008. Climate warming, marine protected areas and the ocean-scale integrity of coral reef ecosystems. PLoS One 3. https://doi.org/10.1371/journal.pone.0003039.

Graham, N.A., Cinner, J.E., Norström, A.V., Nyström, M., 2014. Coral reefs as novel ecosystems: embracing new futures. Curr. Opin. Environ. Sustain. 7, 9–14.
Grime, J.P., 2006. Plant Strategies, Vegetation Processes, and Ecosystem Properties. John Wiley & Sons, New York.
Grottoli, A.G., Warner, M.E., Levas, S.J., Aschaffenburg, M.D., Schoepf, V., et al., 2014. The cumulative impact of annual coral bleaching can turn some coral species winners into losers. Glob. Chang. Biol. 20, 3823–3833.
Hamilton, H.G.H., Brakel, W.H., 1984. Structure and coral fauna of East African reefs. Bull. Mar. Sci. 34, 248–266.
Hartig, F., 2020. DHARMa: Residual Diagnostics for Hierarchical (Multi-Level/Mixed) Regression Models. R package version 0.3.2.0.
Heikoop, J.M., Risk, M.J., Lazier, A.V., Edinger, E.N., Jompa, J., et al., 2000. Nitrogen-15 signals of anthropogenic nutrient loading in reef corals. Mar. Pollut. Bull. 40, 628–636.
Hoegh-Guldberg, O., 1999. Climate change, coral bleaching and the future of the world's coral reefs. Mar. Freshw. Res. 50, 839–866.
Hoegh-Guldberg, O., Kennedy, E.V., Beyer, H.L., McClennen, C., Possingham, H.P., 2018. Securing a long-term future for coral reefs. Trends Ecol. Evol. 33, 936–944.
Hooper, D.U., Chapin, F.S., Ewel, J.J., Hector, A., Inchausti, P., et al., 2005. Effects of biodiversity on ecosystem functioning: a consensus of current knowledge. Ecol. Monogr. 75, 3–35.
Hughes, T.P., Kerry, J.T., Álvarez-Noriega, M., Álvarez-Romero, J.G., Anderson, K.D., et al., 2017. Global warming and recurrent mass bleaching of corals. Nature 543, 373–377.
Hughes, T.P., Anderson, K.D., Connolly, S.R., Heron, S.F., Kerry, J.T., et al., 2018. Spatial and temporal patterns of mass bleaching of corals in the Anthropocene. Science 359, 80–83.
Johns, K.A., Osborne, K.O., Logan, M., 2014. Contrasting rates of coral recovery and reassembly in coral communities on the Great Barrier Reef. Coral Reefs 33, 553–563.
Kleypas, J.A., Danabasoglu, G., Lough, J.M., 2008. Potential role of the ocean thermostat in determining regional differences in coral reef bleaching events. Geophys. Res. Lett. 35, L03613.
Loya, Y., Sakai, K., Yamazato, K., Nakano, Y., Sambali, H., et al., 2001. Coral bleaching: the winners and losers. Ecol. Lett. 4, 122–131.
Lüdecke, D., Makowski, D., Waggoner, P., Patil, I., 2020. Assessment of Regression Models Performance. R package version 0.4.7.
Maina, J., Venus, V., McClanahan, T.R., Ateweberhan, M., 2008. Modelling susceptibility of coral reefs to environmental stress using remote sensing data and GIS models in the western Indian Ocean. Ecol. Model. 212, 180–199.
Maina, J., McClanahan, T.R., Venus, V., Ateweberhan, M., Madin, J., 2011. Global gradients of coral exposure to environmental stresses and implications for local management. PLoS One 6, e23064.
Masalu, D.C.P., 2008. An overview of the bathymetry and geomorphology of the Tanzania EEZ. Open Oceanogr. J. 2.
McClanahan, T.R., 2008. Response of the coral reef benthos and herbivory to fishery closure management and the 1998 ENSO disturbance. Oecologia 155, 169–177.
McClanahan, T., 2014. Decadal coral community reassembly on an African fringing reef. Coral Reefs 33, 939–950.
McClanahan, T.R., 2017. Changes in coral sensitivity to thermal anomalies. Mar. Ecol. Prog. Ser. 570, 71–85.
McClanahan, T.R., 2020. Taxonomic turnover of thermally-stressed coral communities. Coral Reefs. https://doi.org/10.1007/s00338-020-01984-w.
McClanahan, T.R., Arthur, R., 2001. The effect of marine reserves and habitat on populations of East African coral reef fishes. Ecol. Appl. 11, 559–569.

McClanahan, T.R., Maina, J., 2003. Response of coral assemblages to the interaction between natural temperature variation and rare warm-water events. Ecosystems 6, 551–563.

McClanahan, T.R., Muthiga, N.A., 2016. Similar impacts of fishing and environmental stress on calcifying organisms in Indian Ocean coral reefs. Mar. Ecol. Prog. Ser. 560, 87–103.

McClanahan, T.R., Muthiga, N.A., Kamukuru, A.T., Machano, H., Kiambo, R.W., 1999. The effects of marine parks and fishing on coral reefs of northern Tanzania. Biol. Conserv. 89, 161–182.

McClanahan, T.R., Muthiga, N.A., Mangi, S., 2001. Coral and algal changes after the 1998 coral bleaching: interaction with reef management and herbivores on Kenyan reefs. Coral Reefs 19, 380–391.

McClanahan, T.R., Maina, J., Moothien-Pillay, R., Baker, A.C., 2005. Effects of geography, taxa, water flow, and temperature variation on coral bleaching intensity in Mauritius. Mar. Ecol. Prog. Ser. 298, 131–142.

McClanahan, T.R., Ateweberhan, M., Graham, N.A.J., Wilson, S.K., Sebastián, C.R., et al., 2007a. Western Indian Ocean coral communities: bleaching responses and susceptibility to extinction. Mar. Ecol. Prog. Ser. 337, 1–13.

McClanahan, T.R., Ateweberhan, M., Muhando, C.A., Maina, J., Mohammed, S.M., 2007b. Effects of climate and seawater temperature variation on coral bleaching and mortality. Ecol. Monogr. 77, 503–525.

McClanahan, T.R., Cinner, J.E., Graham, N.A.J., Daw, T.M., Maina, J., et al., 2009. Identifying reefs of hope and hopeful actions: contextualizing environmental, ecological, and social parameters to respond effectively to climate change. Conserv. Biol. 23, 662–671.

McClanahan, T.R., Huntington, B.E., Cokos, B., 2011. Coral responses to macroalgal reduction and fishery closure on Caribbean patch reefs. Mar. Ecol. Prog. Ser. 437, 89–102.

McClanahan, T.R., Donner, S.D., Maynard, J.A., MacNeil, M.A., Graham, N.A., et al., 2012. Prioritizing key resilience indicators to support coral reef management in a changing climate. PLoS One 7, e42884.

McClanahan, T.R., Ateweberhan, M., Darling, E.S., Graham, N.A.J., Muthiga, N.A., 2014. Biogeography and change among regional coral communities across the Western Indian Ocean. PLoS One 9, e93385.

McClanahan, T.R., Maina, J., Ateweberhan, M., 2015. Regional coral responses to climate disturbances and warming is predicted by multivariate stress model and not temperature threshold metrics. Clim. Change 131, 607–620.

McClanahan, T.R., Darling, E.S., Maina, J.M., Muthiga, N.A., D'agata, S., et al., 2019a. Temperature patterns and mechanisms influencing coral bleaching during the 2016 El Niño. Nat. Clim. Change 9, 845–851.

McClanahan, T.R., Schroeder, R.E., Friedlander, A.M., Vigliola, L., Wantiez, L., et al., 2019b. Global baselines and benchmarks for fish biomass: comparing remote reefs and fisheries closures. Mar. Ecol. Prog. Ser. 612, 167–192.

McClanahan, T., Darling, E., Maina, J., Muthiga, N., D'agata, S., et al., 2020a. Highly variable taxa-specific coral bleaching responses to thermal stresses. Mar. Ecol. Prog. Ser., M13402.

McClanahan, T.R., Maina, J.M., Darling, E.S., Muthiga, N.A., D'agata, S., et al., 2020b. Resistance of corals to thermal exposure higher in global biodiversity hotspot. Glob. Ecol. Biogeogr. https://doi.org/10.1111/geb.13191.

McLeod, E., Moffitt, R., Timmermann, A., Salm, R., Menviel, L., et al., 2010. Warming seas in the coral triangle: Coral reef vulnerability and management implications. Coast. Manage. 38, 518–539.

McLeod, E., Anthony, K.R.N., Mumby, P.J., Maynard, J., Beeden, R., et al., 2019. The future of resilience-based management in coral reef ecosystems. J. Environ. Manage. 233, 291–301.
Mellin, C., Mouillot, D., Kulbicki, M., McClanahan, T.R., Vigliola, L., et al., 2016. Humans and seasonal climate variability threaten large-bodied coral reef fish with small ranges. Nat. Commun. 7, 10491.
Moynihan, M.A., Baker, D.M., Mmochi, A.J., 2012. Isotopic and microbial indicators of sewage pollution from Stone Town, Zanzibar, Tanzania. Mar. Pollut. Bull. 64, 1348–1355.
Mumby, P.J., Harborne, A.R., Williams, J., Kappel, C.V., Brumbaugh, D.R., et al., 2007. Trophic cascade facilitates coral recruitment in a marine reserve. Proc. Natl. Acad. Sci. U. S. A. 104, 8362–8367.
Nguli, M.M., Ryderg, L., Francis, J., 2006. Estimate of water residence times in Tudor Creek, Kenya based on sea surface heat fluxes and observations of the horizontal temperature gradient during different seasons. West. Indian Ocean J. Mar. Sci. 5, 163–178.
Oksanen, J., Blanchet, F.B., Friendly, M., Kindt, R., Legendre, P., et al., 2019. Package 'vegan'. Accessed.
Olds, A.D., Pitt, K.A., Maxwell, P.S., Babcock, R.C., Rissik, D., et al., 2014. Marine reserves help coastal ecosystems cope with extreme weather. Glob. Chang. Biol. 20, 3050–3058.
Oliver, T.A., Palumbi, S.R., 2011. Many corals host thermally resistant symbionts in high-temperature habitat. Coral Reefs 30, 241–250.
Palumbi, S.R., Barshis, D.J., Traylor-Knowles, N., Bay, R.A., 2014. Mechanisms of reef coral resistance to future climate change. Science 344, 895–898.
Pandolfi, J.M., Connolly, S.R., Marshall, D.J., Cohen, A.L., 2011. Projecting coral reef futures under global warming and ocean acidification. Science 333, 418–422.
Pante, E., Simon-Bouhet, B., 2013. Marmap: A package for importing, plotting and analyzing bathymetric and topographic data in R. PLoS One 8, e73051.
Peñaflor, E.L., Skirving, W.J., Strong, A.E., Heron, S.F., David, L.T., 2009. Sea-surface temperature and thermal stress in the Coral Triangle over the past two decades. Coral Reefs 28, 841–850.
Riegl, B., 1999. Corals in a non-reef setting in the southern Arabian Gulf (Dubai, UAE): fauna and community structure in response to recurring mass mortality. Coral Reefs 18, 63–73.
Riegl, B., Purkis, S., 2015. Coral population dynamics across consecutive mass mortality events. Glob. Chang. Biol. 21, 3995–4005.
Riegl, B., Cavalcante, G., Bauman, A.G., Feary, D.A., Steiner, S., et al., 2017. Demographic mechanisms of reef coral species winnowing from communities under increased environmental stress. Front. Mar. Sci. 4, 344.
Ruff, T.W., Neelin, J.D., 2012. Long tails in regional surface temperature probability distributions with implications for extremes under global warming. Geophys. Res. Lett. 39. https://doi.org/10.1029/2011GL050610.
Safaie, A., Silbiger, N.J., McClanahan, T.R., Pawlak, G., Barshis, D.J., et al., 2018. High frequency temperature variability reduces the risk of coral bleaching. Nat. Commun. 9, 1–12.
Selig, E.R., Casey, K.S., Bruno, J.F., 2010. New insights into global patterns of ocean temperature anomalies: Implications for coral reef health and management. Glob. Ecol. Biogeogr. 19, 397–411.
Selig, E.R., Casey, K.S., Bruno, J.F., 2012. Temperature-driven coral decline: the role of marine protected areas. Glob. Chang. Biol. 18, 1561–1570.
Sheppard, C.R.C., 2003. Predicted recurrences of mass coral mortality in the Indian Ocean. Nature 425, 294–297.

Sheppard, C.R.C., Harris, A., Sheppard, A.L.S., 2008. Archipelago-wide coral recovery patterns since 1998 in the Chagos Archipelago, central Indian Ocean. Mar. Ecol. Prog. Ser. 362, 109–117.

Smith, J.E., Brainard, R., Carter, A., Grillo, S., Edwards, C., et al., 2016. Re-evaluating the health of coral reef communities: Baselines and evidence for human impacts across the central Pacific. Proc. R. Soc. B 283, 20151985.

Teneva, L., Karnauskas, M., Logan, C.A., Bianucci, L., Currie, J.C., et al., 2012. Predicting coral bleaching hotspots: the role of regional variability in thermal stress and potential adaptation rates. Coral Reefs 31, 1–12.

ter Braak, C.J.F., 1986. Canonical correspondence analysis: a new eigenvector technique for multivariate direct gradient analysis. Ecology 67, 1167–1179.

Toth, L.T., van Woesik, R., Murdoch, T.J.T., Smith, S.R., Ogden, J.C., et al., 2014. Do no-take reserves benefit Florida's corals? 14 years of change and stasis in the Florida keys national marine sanctuary. Coral Reefs 33, 565–577.

van Oppen, M.J.H., Oliver, J.K., Putnam, H.M., Gates, R.D., 2015. Building coral reef resilience through assisted evolution. Proc. Natl. Acad. Sci. U. S. A. 112, 2307–2313.

van Woesik, R., Sakai, K., Ganase, A., Loya, Y., 2011. Revisiting the winners and loser a decade after coral bleaching. Mar. Ecol. Prog. Ser. 434, 67–76.

van Woesik, R., Franklin, E.C., O'Leary, J.K., McClanahan, T.R., Klaus, J.S., et al., 2012. Hosts of the Plio-Pleistocene past reflect modern-day coral vulnerability. Proc. R. Soc. B Biol. Sci. 279, 2448–2456.

Wells, S., Melita, S., Anderson, J., Kalombo, H., Makoloweka, S., 2007. Collaborative fisheries management in Tanga, Northern Tanzania. In: McClanahan, T.R., Castilla, J.C. (Eds.), Fisheries Management: Progress Towards Sustainability. Blackwell Publishing Ltd. pp. 139–165.

Wells, S., Samoilys, M., Makoloweka, S., Kalombo, H., 2010. Lessons learnt from a collaborative management programme in coastal Tanzania. Ocean Coast. Manag. 53, 161–168.

Wilson, S.K., Graham, N.A., Pratchett, M.S., Jones, G.P., Polunin, N.V., 2006. Multiple disturbances and the global degradation of coral reefs: Are reef fishes at risk or resilient? Glob. Chang. Biol. 12, 2220–2234.

Zinke, J., Gilmour, J.P., Fisher, R., Puotinen, M., Maina, J., et al., 2018. Gradients of disturbance and environmental conditions shape coral community structure for south-eastern Indian Ocean reefs. Divers. Distrib. 24, 605–620.

Zuur, A., Ieno, E.N., Walker, N., Saveliev, A.A., Smith, G.M., 2009. Mixed Effects Models and Extensions in Ecology With R. Springer Science & Business Media.

CHAPTER TWELVE

The timing and causality of ecological shifts on Caribbean reefs

William F. Precht[a,*], Richard B. Aronson[b], Toby A. Gardner[c], Jennifer A. Gill[d], Julie P. Hawkins[e], Edwin A. Hernández-Delgado[f], Walter C. Jaap[g], Tim R. McClanahan[h], Melanie D. McField[i], Thaddeus J.T. Murdoch[j], Maggy M. Nugues[k], Callum M. Roberts[e], Christiane K. Schelten[l], Andrew R. Watkinson[m], and Isabelle M. Côté[n]

[a]Marine and Coastal Programs, Dial Cordy and Associates, Miami, FL, United States
[b]Department of Ocean Engineering and Marine Sciences, Florida Institute of Technology, Melbourne, FL, United States
[c]Stockholm Environment Institute, Stockholm, Sweden
[d]Centre for Ecology, Evolution and Conservation, School of Biological Sciences, University of East Anglia, Norwich, United Kingdom
[e]Environment Department, University of York, York, United Kingdom
[f]Department of Environmental Sciences and Center for Applied Tropical Ecology and Conservation, Applied Marine Ecology Laboratory, University of Puerto Rico, San Juan, Puerto Rico
[g]Lithophyte Research LLC, Saint Petersburg, FL, United States
[h]Wildlife Conservation Society, Marine Programs, Bronx, NY, United States
[i]Smithsonian Marine Station, Ft Pierce, FL, United States
[j]Murdoch Marine Ltd., Warwick, WKBX, Bermuda
[k]EPHE, Laboratoire d'Excellence "CORAIL", PSL Research University, UPVD, CNRS, USR, Perpignan, France
[l]GEOMAR, Helmholtz Centre for Ocean Research Kiel, Kiel, Germany
[m]Living with Environmental Change, School of Environmental Sciences, University of East Anglia, Norwich, United Kingdom
[n]Department of Biological Sciences, Simon Fraser University, Burnaby, BC, Canada
*Corresponding author: e-mail address: william.precht@gmail.com

Contents

Advances in Marine Biology, Volume 87
ISSN 0065-2881
https://doi.org/10.1016/bs.amb.2020.08.008

Abstract

Caribbean reefs have experienced unprecedented changes in the past four decades. Of great concern is the perceived widespread shift from coral to macroalgal dominance and the question of whether it represents a new, stable equilibrium for coral-reef communities. The primary causes of the shift—grazing pressure (top-down), nutrient loading (bottom-up) or direct coral mortality (side-in)—still remain somewhat controversial in the coral-reef literature. We have attempted to tease out the relative importance of each of these causes. Four insights emerge from our analysis of an early regional dataset of information on the benthic composition of Caribbean reefs spanning the years 1977–2001. First, although three-quarters of reef sites have experienced coral declines concomitant with macroalgal increases, fewer than 10% of the more than 200 sites studied were dominated by macroalgae in 2001, by even the most conservative definition of dominance. Using relative dominance as the threshold, a total of 49 coral-to-macroalgae shifts were detected. This total represents ~35% of all sites that were dominated by coral at the start of their monitoring periods. Four shifts (8.2%) occurred because of coral loss with no change in macroalgal cover, 15 (30.6%) occurred because of macroalgal gain without coral loss, and 30 (61.2%) occurred owing to concomitant coral decline and macroalgal increase. Second, the timing of shifts at the regional scale is most consistent with the side-in model of reef degradation, which invokes coral mortality as a precursor to macroalgal takeover, because more shifts occurred after regional coral-mortality events than expected by chance. Third, instantaneous observations taken at the start and end of the time-series for individual sites showed these reefs existed along a continuum of coral and macroalgal cover. The continuous, broadly negative relationship between coral and macroalgal cover suggests that in some cases coral-to-macroalgae phase shifts may be reversed by removing sources of perturbation or restoring critical components such as the herbivorous sea urchin *Diadema antillarum* to the system. The five instances in which macroalgal dominance was reversed corroborate the conclusion that macroalgal dominance is not a stable, alternative community state as has been commonly assumed. Fourth, the fact that the loss in regional coral cover and concomitant changes to the benthic community are related to punctuated, discrete events with known causes (i.e. coral disease and bleaching), lends credence to the hypothesis that coral reefs of the Caribbean have been under assault from climate-change-related maladies since the 1970s.

1. Introduction

In their natural state, coral reefs are non-equilibrial, dynamic, disturbance-dominated ecosystems. However, in the past four decades coral reefs around the world have experienced unprecedented ecological changes (Bruno and Selig, 2007; Gardner et al., 2003; Wilkinson, 2002). Most notable have been a drastic decline in living cover of hard corals and an increase in macroalgal abundance (Aronson and Precht, 2001a,b; McManus and

Polsenberg, 2004). Increases in macroalgae have been interpreted by some authors as the single best indicator of reef degradation (Mumby et al., 2006, 2007; Steneck and Sala, 2005).

The first regional-scale, quantitative assessment of the extent of coral loss showed that, in aggregate, Caribbean reefs experienced a decline in coral cover from ~50% to ~10%, from 1977 to 2001 (Gardner et al., 2003). Using a larger data set, Schutte et al. (2010) calculated that the average coral cover was around 16% in 2006. More recently, Jackson et al. (2014) noted "that the average coral cover for the wider Caribbean was around 17% in 2012." On some reefs, the decline of coral cover has been accompanied by an increase in macroalgal cover, leading to a community phase shift sensu Done (1992). The prevalence of shifts from the coral-dominated state to a macroalgae-dominated state was reviewed by Bruno et al. (2009). They argued that most reefs around the world are not dominated by macroalgae (with dominance defined as ≥50% cover); rather, most reefs are characterized by macroalgal cover and highly variable coral cover that averages <50%. Although on many reefs there are more macroalgae today than compared with any recorded historical baseline (see Bruno et al., 2014), the causes of the increase, particularly on Caribbean reefs, are still debated in the literature. Three non-exclusive scenarios have been proposed: the top-down (overfishing), bottom-up (eutrophication and pollution) and side-in (perturbations causing direct coral mortality and habitat loss) models (Precht and Aronson, 2006).

The top-down model of reef decline, which suggests that macroalgal dominance resulted primarily from declines in grazing intensity and was derived mainly from observations of Jamaican reefs, holds that phase shifts are primarily a consequence of overfishing (Bellwood et al., 2004; Hughes, 1994; Hughes et al., 1999; Jackson et al., 2001; Knowlton, 1992). In this model, the depletion of large predators, which began centuries ago (Jackson, 2001; Jackson et al., 2001), the subsequent switch to overfishing of herbivores (Pauly et al., 1998), and finally the mass mortality in 1983–1984 of the herbivorous sea urchin *Diadema antillarum* (Lessios, 1988), the grazing activity of which had concealed the loss of herbivorous fishes (Jackson, 2001), triggered a proliferation of macroalgae and concomitant coral mortality (Hughes, 1994; Jackson et al., 2001). Although the link between herbivory and macroalgal abundance is clear (Adam et al., 2015; Aronson and Precht, 2000; Idjadi et al., 2010; Lessios, 1988; Williams and Polunin, 2001), the role of macroalgal growth in causing significant coral mortality is ambiguous (Aronson and Precht, 2006; Nugues and Bak, 2006; see also Alevizon and Porter, 2014).

The bottom-up scenario maintains that declining water quality, particularly as a result of nutrient enrichment, causes shifts from coral to macroalgal dominance, as was observed in the classic example from Kaneohe Bay, Hawaii (Bahr et al., 2015; Banner, 1974; Hunter and Evans, 1995). Lapointe (1997) argued that nutrient enrichment is the main cause of shifts to algae-dominated states on reefs throughout the Caribbean, and others have argued that "the collapse of many Caribbean coral reefs was long preceded by... increased nutrient and sediment runoff from land" (Bellwood et al., 2004). A number of reviews have assessed the role of nutrient enrichment in the decline of corals (Lirman, 2001; McCook, 1999; Precht and Miller, 2007; Szmant, 2002) and the relationships among nutrients, algae, and algae–coral competition (Furman and Heck, 2008; Hughes et al., 1999; Littler et al., 2006; McClanahan et al., 2004; McManus and Polsenberg, 2004; Miller et al., 1999; Mumby and Steneck, 2008; Nugues and Bak, 2006; Sotka and Hay, 2009). In the Florida Keys, Lirman and Fong (2007) showed that patterns of coral cover, population size-structure, growth, and mortality are not related to water-quality gradients. These authors have all essentially concluded that the regional losses of coral cover and increases in macroalgal abundance are not due to nutrient enrichment alone but to a variety of causes.

The side-in scenario invokes perturbations such as infectious diseases, outbreaks of predators, bleaching events, tsunamis, earthquakes, hurricane impacts, and hypothermic stress as important causes of ecological change on coral reefs. These types of disturbance are well-documented, direct causes of the loss of coral cover and habitat on Caribbean reefs (Aronson et al., 2012; Aronson and Precht, 2001a, 2001b; Gardner et al., 2005; Kaufman, 1983; Knowlton et al., 1981; Lirman et al., 2011; McWilliams et al., 2005; Scheffers et al., 2006; Stoddart, 1985; Toth et al., 2014; Woodley et al., 1981). Many authors have suggested that coral mortality itself was the crucial precondition for macroalgal dominance (Alevizon and Porter, 2014; Aronson and Precht, 2001a, 2001b, 2006; Lirman, 2001; McManus and Polsenberg, 2004; Nugues and Bak, 2008; Ostrander et al., 2000). Macroalgae are also perceived as harmful because in high abundance they can reduce coral recruitment (Birrell et al., 2008; Box and Mumby, 2007; Idjadi et al., 2010; Kuffner et al., 2006; Ritson-Williams et al., 2009), potentially slowing the recovery of coral populations from natural and anthropogenic disturbances.

Identifying the causes of ecological change on Caribbean reefs is a complex task because they are likely to vary among reefs and include a combination of localized and regional stressors that have acted separately or in concert at different times on different reefs. We explicitly subscribe to the pluralistic view of Quinn and Dunham (1983) that ecology seeks to evaluate the relative importance of the many causes underlying an observed pattern. There is evidence that on some reefs a combination of factors, including top-down, bottom-up, and side-in causes, may act in concert leading to an extreme situation of dominance by macroalgae (Côté et al., 2013; McClanahan et al., 2002; Paine et al., 1998; Vieira et al., 2016). It should, however, be possible to gain more precise insight into the regional cause(s) of algal dominance by examining the temporal pattern of coral-to-macroalgae phase shifts in relation to large-scale, punctuated disturbances, such as the catastrophic loss of the keystone herbivore *D. antillarum* to a pandemic disease in 1983–1984 and to episodes of coral disease and mass coral bleaching.

In this paper, we quantify the prevalence of coral-to-macroalgae and macroalgae-to-coral shifts across the Caribbean region over three decades. By examining the timing of the phase shifts, we evaluate the relative likelihood of the causes proposed for these transitions. We also comment on the idea that reefs of the Caribbean have become vast carpets of seaweed (Jackson et al., 2001; Ogden, 1995; Pandolfi et al., 2005), and we attempt to discern whether coral- and macroalgae-dominated communities are stable endpoints, each resistant to conversion to the other. Clarifying these hypothetical and empirical aspects of reef ecology is critical to developing pragmatic approaches to coral-reef conservation and management.

2. Methods

2.1 Data acquisition

Time-series of data on hard-coral and macroalgal cover for reef sites spanning the wider Caribbean basin (Fig. 1 and Table S1 in the online version at https://doi.org/10.1016/bs.amb.2020.08.008) were obtained through electronic and manual literature searches, as well as direct personal communications with reef scientists, site managers and institutional librarians. Electronic literature searches were conducted using the Science Citation Index (SCI) and Aquatic Sciences and Fisheries Abstracts (ASFA) from 1981 to 2001 and 1988 to 2001, respectively, covering a period of extensive changes in benthic composition on coral reefs (Côté et al., 2005;

Gardner et al., 2003). All relevant references cited in these publications were also checked. The only selection criterion employed was that the study reported percent cover of either or both target benthic components, with replicated measurements, from a site within the region. Sites were deemed separate if they were defined as such in the study. When a site crossed a steep depth gradient, parts of the replicate transects were pooled into smaller depth-ranges (e.g. Dustan and Halas, 1987).

2.2 Coral–macroalgae phase shifts and reversals

We defined macroalgae following Steneck (1988) as all erect and anatomically complex algal forms with canopy heights usually >10 mm. Crustose coralline algae and algal turfs are also functionally important members of the coral-reef algal assemblage and changes in their relative abundance over time and space may also be critical metrics in diagnosing reef health (e.g. Birrell et al., 2005; Jorissen et al., 2016). However, we did not consider these benthic components in our analysis because many reef-monitoring programs do not differentiate between turfs and other non-erect algal forms (Aronson and Precht, 2006; Bruno et al., 2014). Crustose coralline algae, fine algal turfs, and bare space (CTB) are essentially indistinguishable from one

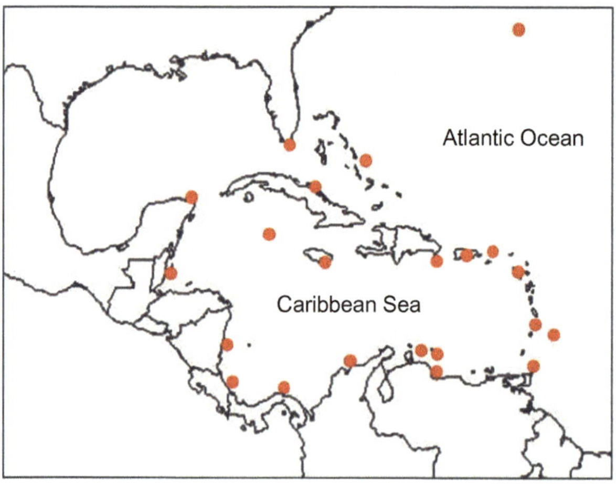

Fig. 1 Regional distribution of study sites in the wider Caribbean basin. The study sites from which monitoring data were sourced are shown as solid-red circles (see Table S1 in the online version at https://doi.org/10.1016/bs.amb.2020.08.008). *Figure modified from original in Gardner, T.A., Côté, ,I.M., Gill, J.A., Grant, A., Watkinson, A.R., 2003. Long-term region-wide declines in Caribbean corals. Science 301, 958–960.*

another using rapid-assessment monitoring protocols (Aronson et al., 1994; Aronson and Precht, 2000).

There is no generally accepted definition of what constitutes a phase shift from corals to macroalgae (Rogers and Miller, 2006). The term "dominance" itself is ill-defined in the ecological literature (Alevizon and Porter, 2014; Bruno et al., 2009). For example, the definition from Biology Online (2013) is "the species that predominates in an ecological community, particularly when they are most numerous..." But by what comparative threshold do we evaluate the component that is "most numerous?" In quantitative terms, dominance can be a simple majority, which is to say a component (e.g. hard corals or macroalgae) that is more than half of the entire group (i.e. all benthic components combined). By contrast, plurality is the largest subset in a multi-part system. A plurality is not necessarily a majority, as the largest subset may be less than half of the entire group. Finally, relative majority (or relative dominance) is also a subset in a multi-part system but it applies only to one component (e.g. coral) relative to another component (e.g. macroalgae) in the system. At present, there is no consensus in the ecological literature as to whether a simple majority, a plurality, or a relative majority constitutes ecological dominance.

We examined the prevalence of macroalgal dominance in three ways. First, we considered absolute dominance (or majority) of macroalgae. For each year of the dataset, we calculated the proportion of sites with more than 50% macroalgal cover (see also Bruno et al., 2009). Hughes et al. (2010) noted, however, that "dominance" may not be an appropriate metric, because an arbitrary abundance threshold, in this case 50% absolute cover, may be overly conservative. Because macroalgal cover is measured in terms of proportion of the total area rather than the area suitable for colonization (i.e. total area minus the area covered by sand), the usual threshold of 50% may underestimate dominance. We therefore also calculated the proportion of sites with more than 40% macroalgal cover, which would be equivalent to 50% cover where only 80% of the total area surveyed was available for colonization (i.e. average sand cover was 20%, a figure at the high end of our collective personal experience of benthic surveys).

Second, we examined relative dominance (or relative majority) by calculating the proportion of sites in each year with more macroalgae than coral. Both methods estimate the prevalence of macroalgal dominance, but they have four limitations. (1) They do not account for the fact that sites may have lost coral cover and/or gained macroalgal cover without having become dominated (in either absolute or relative terms) by the latter while they were monitored. (2) They do not effectively capture changes occurring at sites with naturally low coral cover and high macroalgal cover (see Vroom et al., 2006 for a Pacific example), which may have been that way historically due to regular perturbation. (3) They also do not capture seasonal or local variations in macroalgal cover (Lirman and Biber, 2000). (4) Finally, they do not account for other classes of benthic organisms (octocorals, sponges, zoanthids, turfs, crustose coralline algae, etc.) that may have higher absolute cover than either corals or macroalgae or that may ascend to dominance on some temporal scale (e.g. Aronson et al., 2012; Norström et al., 2009; Ruzicka et al., 2013; Toth et al., 2014; Tsounis and Edmunds, 2017).

Third, due to a paucity of data, we are unsure what the natural cover of coral and macroalgae was on undisturbed Caribbean coral reefs prior to the 1970s (Bruno et al., 2014; Eddy et al., 2018). Knowing this baseline metric would be helpful in interpreting change on reefs through time and developing quantitative targets for management and restoration. Some authors, however, have argued that coral and macroalgal cover by themselves are not reliable metrics of reef health and resilience and that cover only becomes a definitive indicator of phase shifts if the same site is monitored for multiple years (Hughes et al., 2010). To overcome this possible limitation, we examined site-specific ratios of coral to macroalgae at the start and end of each time-series to generate a more dynamic metric of dominance. We recorded the proportion of sites that had moved towards having proportionally less coral than macroalgae, as well as the specific year in which any change of relative dominance was observed at a site. We counted the number of sites in each year that had two attributes: (1) more coral than macroalgae (so they could have shifted); and (2) data for the previous year (so a shift could have been detected). We expressed that number of sites for each year as a proportion of all such sites, and then multiplied by the number of observed shifts (which was 49). This procedure distributed the shifts in proportion to the availability of sites that could shift.

The year of shift could only be identified at sites for which we had concurrent yearly estimates of both coral and macroalgal cover that were separated by no more than 1 year. Of the 208 sites in the database, 193 met this criterion. To evaluate whether more shifts were observed in each year than expected by chance alone, we used *G*-tests to compare the number of shifts recorded in any given year to the number expected given the total number of shifts observed across all years and the number of sites available to assess a shift in that year (expressed as the proportion of all sites across all years for which we had both coral and macroalgal cover data in that year and the previous year). To account for multiple tests, we adjusted the critical α using the simple Bonferroni correction. The adjusted significance threshold was $\alpha_{adj}=0.002$. Finally, we recorded "phase-shift reversals" when a site that had exhibited greater cover of macroalgae than of coral, either at the start of the study or for at least two consecutive years during the study, shifted to having a higher cover of coral than of macroalgae at the end of its monitoring period.

3. Results

A total of 193 reef sites from 71 separate studies across the Caribbean reported concurrent cover of hard corals and macroalgae. The sites were distributed evenly around the Greater Caribbean (similar to the distribution shown in Gardner et al., 2003, with the addition of sites in St. Lucia and Saba; Fig. 1). Overall, the time-series of coral and macroalgal cover spanned the years 1977 to 2001, although individual time-series were on average 6 ($\pm$4.7 SD) years long.

When coral-to-macroalgae phase shifts are defined by absolute (or majority) dominance—that is, shifts to $\geq$50% cover of macroalgae—the prevalence of phase shifts appears to have varied since 1977. Prior to 1981, all sites that were surveyed at the time ($n=19$) had less than 50% cover of macroalgae. The proportion of sites dominated by macroalgae peaked in

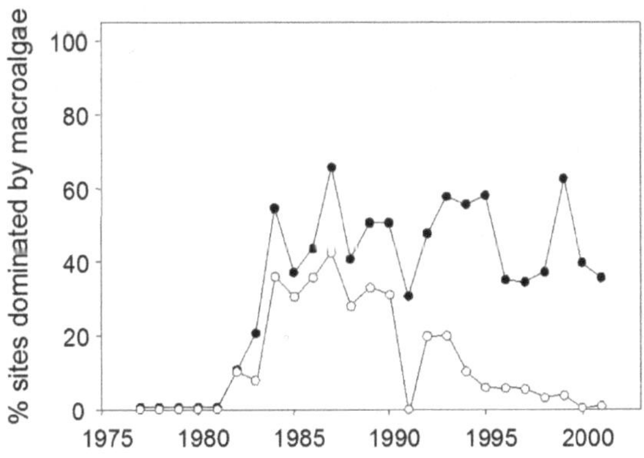

Fig. 2 Prevalence of macroalgal dominance in Caribbean reef sites in each year between 1977 and 2001. Macroalgal dominance is defined either in absolute terms (>50% cover; open circles) or relative terms (higher macroalgal than coral cover; solid circles).

1987 and then decreased to fewer than 10% of sites since 1994 (Fig. 2). This change through time is related in part to the diminishing proportion in the dataset of Jamaican sites, which experienced a unique series of disturbances and ensuing ecological changes in the early 1980s (Côté et al., 2013; Liddell and Ohlhorst, 1992). Setting the threshold for absolute dominance of macroalgae to 40% cover does not change the temporal pattern or proportion of sites dominated by macroalgal cover.

The prevalence of phase shifts is higher if phase shifts are defined as occurring when macroalgal cover exceeds coral cover (i.e., relative dominance of macroalgae). In the late 1970s, all sites exhibited higher cover of hard corals than of macroalgae (Fig. 2). Using this more liberal definition, by 1987 approximately half the sites had more macroalgae than coral, a proportion that has since remained relatively constant. The site-specific analysis reveals an even greater prevalence of phase shifts. Overall, 72% of individual sites moved towards proportionally less coral than macroalgae during their monitoring periods. The difference between coral and macroalgal cover declined until the early 1990s and has remained relatively stable since then at the majority of sites (Fig. 2; see similar analysis in Jackson et al., 2014).

A total of 49 coral-to-macroalgae shifts in relative dominance were detected. This total represents ~35% of all sites that were dominated by coral at the start of their monitoring periods. Four shifts (8.2%) occurred because of coral loss with no change in macroalgal cover, 15 (30.6%) occurred because of macroalgal gain without coral loss, and 30 (61.2%) occurred owing to concomitant coral decline and macroalgal increase. The shifts occurred in 14 of the 25 years of the study period (Fig. 3). The years posting significantly more shifts to macroalgae than expected by chance were 1983, 1984, 1990 and 1998 (Fig. 3). Note that adjustment of α to maintain a familywise error rate of 0.05 yielded the same results using the conservative simple-Bonferroni procedure and the less conservative sequential-Bonferroni and false-detection-rate procedures (Benjamini and Hochberg, 1995). In 1983 and 1984, two-thirds of shifts were caused by macroalgal increase with no change in coral cover, whereas in the 1990 and 1998, 81% of shifts involved coral loss, either with or without macroalgal increases ($G=4.62$, df$=1$, $P=0.03$).

Of 57 sites that had more macroalgae than coral at the start of their monitoring periods, 52 (91%) had more macroalgae for the duration of monitoring; however, five reefs showed a clear reversal of the usual coral-to-macroalgae sequence of relative dominance. These five reefs represent ~5% of all reefs that could have shifted to coral dominance:

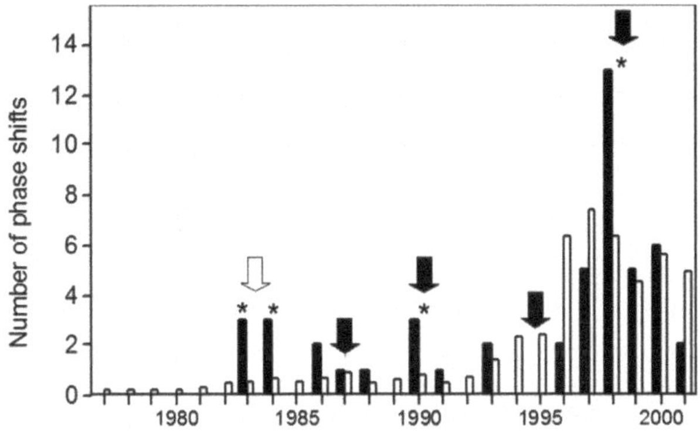

Fig. 3 Numbers of coral-to-macroalgae shifts, expressed in relative terms, on Caribbean coral reefs (black bars) each year from 1977 to 2001. Also shown (white bars) are numbers of coral-to-macroalgae shifts expected to occur if these shifts simply occurred in proportion to the number of sites available to shift towards macroalgae dominance each year in the database. * indicates a significantly greater number of observed shifts (G-tests, Bonferroni-corrected $\alpha = 0.002$). The white arrow denotes the year of mass mortality of the sea urchin *Diadema antillarum*. The black arrows indicate years of mass coral bleaching, the most severe of which was 1998.

the 57 sites that were initially macroalgae-dominated, plus 49 sites that had shifted to more macroalgae during the time they were monitored. The reversals all occurred in the 1990s (1994, 1995, 1998 [2 sites], and 1999) and were geographically widespread (Mexico, Colombia, Saba [2 sites], and Puerto Rico). In three cases, the reversals were the result of decreased macroalgal cover only, whereas in two cases, macroalgae decreased and corals simultaneously increased.

4. Discussion

Coral-reef degradation and phase shifts are highly variable processes that involve different combinations of coral loss and macroalgal gain in time and space (Bruno et al., 2009). Although we found that nearly three-quarters of reef sites across the Caribbean region changed in the general direction of having less coral and more macroalgae, our analysis also reveals that most Caribbean reefs were not overrun by macroalgae. After 1995, fewer than 10% of sites were dominated by macroalgae in absolute-majority terms. In this respect, our results are remarkably similar to the global analysis of

Bruno et al. (2009), who came to a similar conclusion about the extent of macroalgal dominance in the Caribbean using different data.

The timing of shifts from coral to macroalgae sheds light on the causes of these ecological changes. The top-down model of reef decline, which suggests that macroalgal dominance resulted primarily from declines in grazing intensity, implies that the mass mortality of *Diadema antillarum* in 1983–1984 should have had a detectable effect on the timeline of coral–macroalgae shifts in the Caribbean. In 1983, a water-borne pathogen decimated populations of this important herbivore, causing the loss of ~93% of the *Diadema* throughout the region in less than 1 year (Lessios, 1988, 2016). The mass mortality triggered cascading effects on the reef benthos, as well as on herbivorous and predatory fishes (Hughes et al., 1987; McClanahan, 1999; Robertson, 1991). We found that in 1983 and 1984 more reefs exhibited shifts towards macroalgae than expected by chance alone, suggesting a regional link between decreased herbivory and increased macroalgal cover. During this two-year period, however, observed phase shifts were caused mainly by increases in macroalgae and not decreases in coral cover. This result calls into question the claim of Jackson et al. (2001) that the *Diadema* epizootic was the principal cause of coral mortality throughout the Caribbean.

The early years of our time-series are heavily influenced by Jamaica. The independent contributions of reduced herbivory and coral mortality are, unfortunately, difficult to partition because other events immediately preceded and overlapped the loss of *Diadema* in Jamaica. Most notable were Hurricane Allen in 1980, which caused extensive coral mortality (Kaufman, 1983; Woodley et al., 1981), and subsequent losses of coral due to predation and white-band disease (Knowlton et al., 1981). Even in Jamaica, there appears to have been an interaction between coral mortality and reduced herbivory in the shift from coral to macroalgal dominance (Andres and Witman, 1995; Hughes, 1994; Precht and Aronson, 2006). Jamaican reefs had much more coral than macroalgae until 1980, after which the relative difference in cover between these two benthic components began to decline markedly. By 1984, macroalgal cover exceeded coral cover on Jamaican reefs. The difference between macroalgal and coral cover peaked in 1992, with reefs having had on average 70% higher absolute cover of macroalgae than living coral. Corals recovered and macroalgae declined between 1992 and 1999, the last year of our dataset for Jamaica (Fig. 4). The pattern for other Caribbean reefs was superficially similar but quantitatively different. On non-Jamaican reefs, the difference between

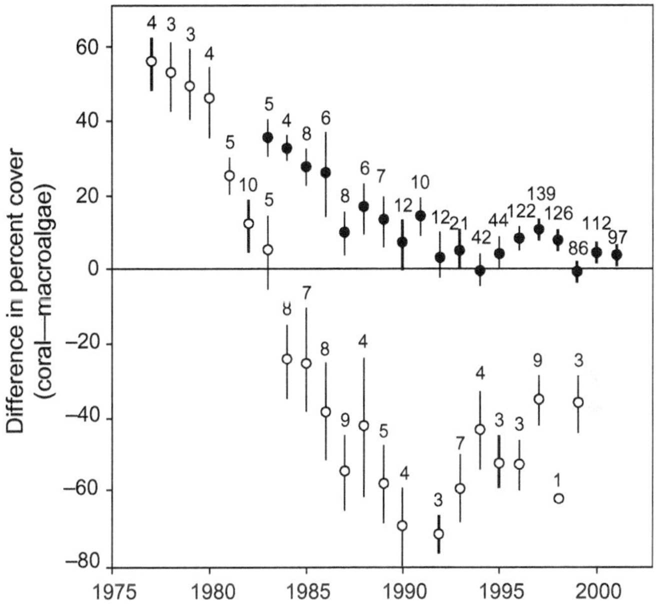

Fig. 4 Magnitudes of the difference between percent cover of live coral and macroalgae on Caribbean reefs in each year between 1977 and 2001. Means (±1 SE) are shown separately for Jamaican sites (open circles) and all other sites (filled circles). Positive values indicate that the cover of coral was higher than that of macroalgae. Numbers above the error bars show the number of sites contributing to each mean. *Figure modified from original in Côté, I.M., Precht, W.F., Aronson, R.B., Gardner, T.A., 2013. Is Jamaica a good model for understanding Caribbean coral reef dynamics? Mar. Pollut. Bull. 76, 28–31.*

coral and macroalgal dominance also declined from 1984 to 1994 but less-so, and they remained relatively stable for the remainder of the time-series (Fig. 4).

At the regional scale, as in Jamaica, significant losses of acroporid corals from white-band disease preceded and overlapped the loss of *Diadema* (Aronson and Precht, 2001a, 2006, 2016; Gladfelter, 1982). The years 1977–1983 showed significant losses of coral, with up to 40% decline in coral cover (Fig. 5; Gardner et al., 2003). The timing of coral loss is consistent with white-band disease having been the primary cause of coral mortality in the Caribbean (Alvarez-Filip et al., 2009; Aronson and Precht, 2006, 2016; Gardner et al., 2003; Jaap et al., 1988; Kuffner and Toth, 2016; Schutte et al., 2010). The loss of coral cover per se during this period was not coincident with widespread coral-to-macroalgae phase shifts, highlighting the importance of *Diadema* in ameliorating the impacts of macroalgae in

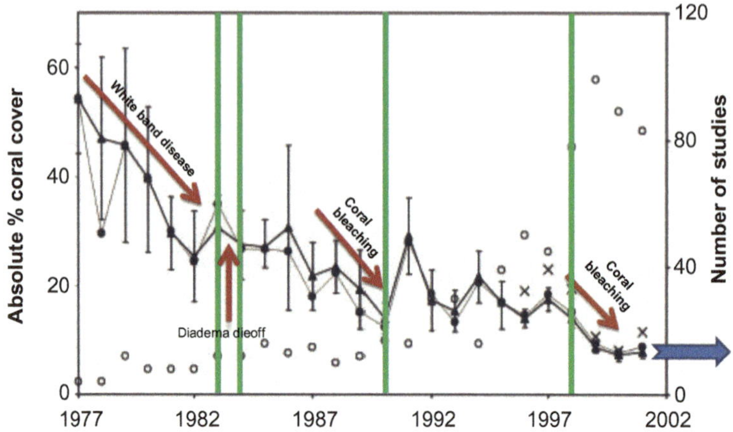

Fig. 5 Long-term trajectory of coral cover on Caribbean reefs, based on a meta-analysis of ecological studies. Solid circles are means, and error bars are 95% bootstrapped confidence intervals. The black, solid line connects the mean coral cover by year from 1977 to 2001. The red arrows are major punctuated disturbance events that were recorded region-wide. The initial decline in 1977–1982 represents a decrease in absolute coral cover from approximately 50–25% and corresponds to the regional loss of acroporid corals caused by white-band disease (Aronson and Precht, 2006). The declines in 1987–1990 and 1998–1999 are associated with major, El Niño-induced bleaching events (Aronson and Precht, 2006). Green horizontal bars represent the years in which there was a significantly greater number of observed shifts from coral to macroalgal dominance. Coral cover has continued to decline in the Caribbean in a stepwise fashion since 2001, associated with impacts from coral diseases, hurricanes and warm-water bleaching events. *Figure modified from Gardner, T.A., Côté, ,I.M., Gill, J.A., Grant, A., Watkinson, A.R., 2003. Long-term region-wide declines in Caribbean corals. Science 301, 958–960. Aronson, R.B., Precht, W.F., 2006. Conservation, precaution, and Caribbean reefs. Coral Reefs 25, 441–450.*

the face of large-scale disturbances. Once *Diadema* were removed from the system in 1983–1984, however, much of the space made available by catastrophic mortality of the acroporids was rapidly colonized by macroalgae. The two other years, 1990 and 1998, in which there were significantly more coral-to-macroalgae shifts than expected—virtually all of which involved declines in coral cover—were years of mass coral-bleaching events in the Caribbean (Fig. 5; Gill et al., 2006; McWilliams et al., 2005). Coral mortality was followed by macroalgal overgrowth on the dead-coral surfaces. The major coral-bleaching event in 1997–1998 resulted in widespread loss of coral, not only in the Caribbean (Aronson et al., 2002a) but also in other regions of the world (Lough, 2000; Wilkinson, 2002). The trend in coral cover followed a step-type function driven by coral disease and bleaching events as the proximal drivers of coral mortality (Fig. 5; Aronson and Precht, 2006). The ecological changes occurring in these years are, therefore, most consistent with the side-in model of reef degradation, which invokes coral mortality as the primary factor initiating phase shifts (Precht and Aronson, 2006). Mass coral-bleaching events in 1987 and 1995 resulted in only localized areas of coral mortality (Lang et al., 1992; McField, 1999; Ostrander et al., 2000). The latter two events were undetectable at the regional scale in our analysis (Fig. 3).

Our results suggest a limited role for changes in grazing pressure in driving phase shifts after 1990. Although we do not have site-specific estimates of herbivory, a regional analysis of Caribbean reef-fish abundance showed declines in herbivorous-fish density over the duration of our study (Paddack et al., 2009). Since 1984 there has been no significant regional-scale recovery of *Diadema* (Hughes et al., 2010; Levitan et al., 2014; Rogers and Lorenzen, 2016), suggesting that the intensity of herbivory has been declining on most Caribbean reefs. The temporal patterns in the prevalence of phase shifts—roughly stable or declining since the mid-1980s, depending on the definition of dominance (Fig. 2)—are not consistent with a regional reduction in herbivory. It is possible that reduced herbivory, caused by historical overfishing, set the stage for the extreme impact of coral mortality, as suggested by Jackson (2001, 2014). In a similar vein, Hughes et al. (2012) argued that the regime shifts observed across the Caribbean unfolded so slowly and gradually as to be imperceptible until a tipping point was reached. However, the non-random timing of phase-shifts across a broad gradient of fishing pressure and human influences (Kramer, 2003; Newman et al., 2006), and the combined effects of two lethal, essentially concurrent epizootics with regional consequences unprecedented in at least the last several thousand years (Aronson et al., 2002b; Knowlton, 2001; Sheppard, 1993), suggest that the observed recent drivers contributed far more strongly to the pattern than inferred historical drivers.

Coral-to-macroalgae phase shifts on Caribbean reefs have been postulated to represent transitions from one alternative stable state to another (Knowlton, 1992). Whether the coral- and macroalgae-dominated states are stable alternatives, each of which resists conversion to the other, or whether the coral-to-macroalgae transition is a reversible phase shift (Petraitis and Dudgeon, 2004) is critical to determining the resilience of reef communities in the Caribbean (Bellwood et al., 2004; Mumby, 2009;

Scheffer and Carpenter, 2003). Degraded reefs initially showed no evidence of recovery from a macroalgae-dominated state (e.g., Rogers and Miller, 2006). Later literature, however, described a number of phase-shift reversals in the Caribbean including at highly degraded sites in Jamaica (Bechtel et al., 2006; Idjadi et al., 2006). We found evidence of shifts away from macroalgal dominance at five sites during the 1990s, where large changes in herbivory by fish were not reported. Since 2001, reduced macroalgal cover and then increased recruitment of juvenile corals, both associated with the local reappearance of *Diadema*, have occurred, first on Jamaican reefs (Bechtel et al., 2006; Cho and Woodley, 2002; Edmunds and Carpenter, 2001; Idjadi et al., 2010; Precht and Aronson, 2006) and then on reefs elsewhere across the Caribbean (Carpenter and Edmunds, 2006; Macintyre et al., 2005; Myhre and Acevedo-Gutiérrez, 2007). A full reversal to the coral-dominated state was noted on one Jamaican reef (Crabbe, 2009; Idjadi et al., 2006; Precht and Aronson, 2006), where *Diadema* had recovered to half their pre-mortality densities (Edmunds and Carpenter, 2001) without noticeable changes in fish populations (Hardt, 2008; Woodley and Sary, 2002) or improvements in water quality (Cho and Woodley, 2002; Greenaway and Gordon-Smith, 2006; Webber et al., 2005).

There is little evidence to support the notion of discrete, low coral/high macroalgae and high coral/low macroalgae alternative states. In our study, the abundances of hard corals and macroalgae exhibited a continuous, broadly negative relationship (see also Bruno et al., 2009; Lowe et al., 2011; Lester et al., 2020). Reefs with little coral had highly variable amounts of macroalgae, just as reefs with little macroalgae varied greatly in coral cover (Fig. 6). It was more common for Caribbean reefs to shift from coral to macroalgal dominance, with 35% of sites that could do so having done so during the study period, than to shift in the reverse direction, with only 5% of eligible sites having done so. This asymmetry in the direction of shifts is more parsimoniously explained by (1) the relatively short time-series of our data set, (2) the catastrophic and escalating nature of recent disturbances, many of which have caused (and continue to cause) extensive coral mortality (Eakin et al., 2010; Miller et al., 2009; Precht et al., 2016), (3) the rapid colonization and turnover times of the replacement algal species (McCook et al., 2001), and (4) the slow natural succession of long-lived reef fauna following regional, catastrophic mortality of the major foundation species (*Acropora palmata* and *A. cervicornis*) and the keystone herbivore (*Diadema antillarum*)—than by resistance to recovery from the macroalgae-dominated state (Dudgeon et al., 2010). Clearly, if coral-mortality events become more frequent and more severe, the likelihood of complete recovery between disturbance events will diminish (see Palumbi et al., 2008).

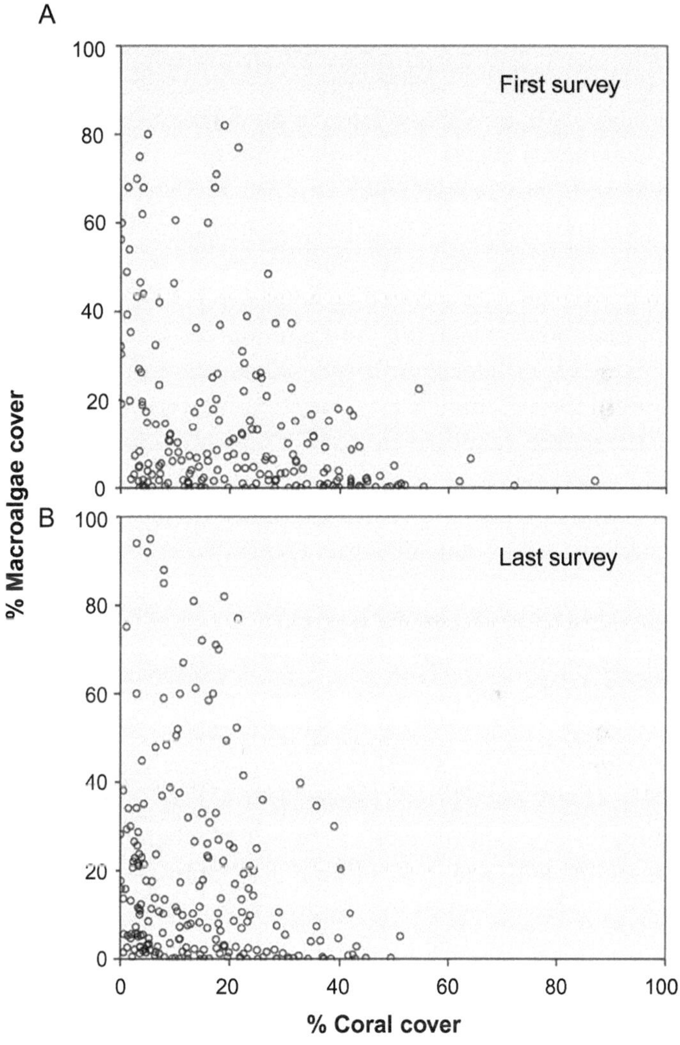

Fig. 6 Relationship between macroalgal and coral cover on the reefs included in the study. Data are from the first (A) and last (B) survey of each reef only. The relationship between macroalgal and coral cover is virtually identical. Note the absence of two distinct clusters of reefs, one with high coral and low macroalgal cover and the other with low coral and high macroalgal cover, which would have been predicted by the hypothesis of alternative stable states.

It is worth considering another interpretation of the apparently continuous range of combinations of coral and macroalgal cover exhibited in Fig. 6. If each reef exists in alternative, quasi-stable states but the configurations of those states are slightly different on different reefs because of differing environmental conditions, and if some reefs at some times are in transition away from unstable equilibria, then surveys across sites and the same or at different times might yield what appears to be a continuum of combinations of coral and macroalgal cover. Individual reefs in the Caribbean show a similar pattern when coral and macroalgal cover are plotted against each other, recording the transition between coral dominance and other states. One could argue that an individual site showing such a continuous pattern through time was in transition between basins of attraction for part of the time it was observed, just as a snapshot across sites at one time might yield observations of some reefs that happened to be in transition. A transition between states might appear continuous on a decadal time scale but discontinuous on a centennial or longer scale.

Two lines of evidence argue against the interpretation that Fig. 6 indicates latent alternative states. First, numerous studies from around the Caribbean region have found that coral and algal dynamics have been decoupled at both low and high values of coral cover on a multidecadal time scale. Coral cover has shown little temporal variation, whereas macroalgae and the category known as CTB—crustose coralline algae, fine turfs, and bare space—have varied in a complementary fashion (Aronson et al., 2012; Johnston et al., 2016; Toth et al., 2014; Tsounis and Edmunds, 2017). Second, Bruno et al. (2019), using data from Jackson et al. (2014), showed that, aggregated across the Caribbean and through time, there has been no negative relationship between coral cover and macroalgal cover, nor has there been a relationship between change in coral cover and change in algal cover. Taken together, these results corroborate our interpretation of Fig. 6 and do not support the hypothesis of alternative community states on Caribbean reefs (see also Dudgeon et al., 2010).

We anticipate that if *Diadema* populations continue to increase across the region, so will the number of phase-shift reversals (Precht and Precht, 2016; but see Levitan et al., 2014). Early evaluations suggest, however, that reversing phase shifts may depend on successfully reducing the causative stressors (Dudgeon et al., 2010) and having sufficient rates of activity by functional groups that play key roles in reversing the changes (Francis et al., 2019; McClanahan et al., 2011b). Given enough time and in the absence of additional, significant coral mortality, recovery of Caribbean coral populations seems plausible. Devising restoration and management programs that enhance or restore critical components of the ecosystem, reduce macroalgae, and promote the success of corals should foster resilience and accelerate recovery, at least locally (Adam et al., 2015; Edwards et al., 2011; Furman and Heck, 2009; Idjadi et al., 2010; Jackson, 2014; Mumby, 2006; Newman et al., 2006; Precht and Aronson, 2006; Sandin and McNamara, 2012; Smith et al., 2010).

Evidence suggests, however, that overfishing may not be the major threat to benthic assemblages on coral reefs, nor may the recovery of parrotfish be the only solution (Aronson and Precht, 2006; Bruno et al., 2019; Kuffner and Toth, 2016; Precht and Precht, 2016; Rotjan et al., 2006; Russ et al., 2015; Suchley et al., 2016). For example, reefs in the Florida Keys, which have maintained relatively high populations of herbivorous fishes (Alevizon and Porter, 2014; Bohnsack et al., 1994), should have exhibited considerably less coral loss than reported for the rest of the Caribbean region. Coral loss in Florida, however, was significantly greater than the Caribbean average throughout this period (Schutte et al., 2010), while the survival of coral recruits and reef recovery were limited (Toth et al., 2014; van Woesik et al., 2014). Protecting fish stocks does not necessarily reduce the cover of macroalgae, increase coral populations, or preserve or increase the topographic complexity that is critical to maintaining and increasing those fish stocks (Alvarez-Filip et al., 2009, 2011; Bates et al., 2019; Bood, 2006; Coelho and Manfrino, 2007; Cox et al., 2017; Huntington et al., 2011; Idjadi et al., 2006; Kramer and Heck, 2007; Ledlie et al., 2007; Lowe et al., 2011; McClanahan et al., 2011a,b; Mora, 2008; Myers and Ambrose, 2009; Reyes-Bonilla et al., 2014; Selig and Bruno, 2010; Stockwell et al., 2009; Toth et al., 2014; Żychaluk et al., 2012).

5. Summary

The timing of macroalgal phase-shifts at the regional scale spanning the years 1977–2001 is most consistent with side-in model of reef degradation, which invokes coral mortality as a precursor to macroalgal takeover, because more shifts occurred after regional coral-mortality events than expected by chance. The implication of this conclusion is that regional causes of coral mortality far outweighed local causes. Why is this important? Reversing coral decline will depend on simultaneously addressing the proximal causes of regional mortality of reef-building corals: the local-scale manifestations of global climate change and other local causes (Aronson and Precht, 2006, 2016; Birkeland et al., 2013; Bruno et al., 2019; Bruno and Valdivia, 2016; Buddemeier et al., 2004; Côté and Darling, 2010; Hoegh-Guldberg and Bruno, 2010; Hughes et al., 2003; McClanahan, 2002; Rogers, 2008; Selig et al., 2012). If the side-in impacts of diseases, hurricanes and bleaching continue unabated (Eakin et al., 2010; Hoegh-Guldberg et al., 2011; Manzello, 2015; Miller et al., 2009; Precht et al., 2016; Randall et al., 2014; Veron et al., 2009), regional-scale coral mortality will outpace local recovery, resulting in reefs that will look and function very differently from the way they do now (Aronson and Precht, 2016; Burman et al., 2012; Carpenter et al., 2008; Kuffner et al., 2014; Kuffner and Toth, 2016; Precht et al., 2016; Smith et al., 2013; Toth et al., 2019). Macroalgae-dominated reefs could then truly become the norm.

Acknowledgements

We thank the many people who contributed unpublished data or manuscripts in press. John Bruno, Lauren Toth and Maggie Toscano commented on early versions of this manuscript. Special thanks to Bernhard Riegl for editing this volume and for his insightful comments. Portions of this work were carried out under permits from the Florida Keys National Marine Sanctuary (FKNMS) and the Belize Department of Fisheries. This research was supported by the UK Natural Environment Research Council (T.A.G.); the Tyndall Centre for Climate Change Research (I.M.C., J.A.G., A.R.W.); the US National Science Foundation (OCE-1535007 to R.B.A.); the National Geographic Society; NOAA's National Undersea Research Program,Sanctuaries and Reserves Division, Coastal Ocean Program, and FKNMS (R.B.A. and W.F.P.); and the Natural Science and Engineering Research Council of Canada (I.M.C.). This is contribution number 232 from the Institute for Global Ecology at the Florida Institute of Technology.

References

Adam, T.C., Burkepile, D.E., Ruttenberg, B.I., Paddack, M.J., 2015. Herbivory and the resilience of Caribbean coral reefs: knowledge gaps and implications for management. Mar. Ecol. Prog. Ser. 520, 1–20.

Alevizon, W.S., Porter, J.W., 2014. Coral loss and fish guild stability on a Caribbean coral reef: 1974–2000. Environ. Biol. Fishes 98, 1035–1045.

Alvarez-Filip, L., Dulvy, N.K., Gill, J.A., Côté, I.M., Watkinson, A.R., 2009. Flattening of Caribbean coral reefs: region-wide declines in architectural complexity. Proc. R. Soc. Lond. B Biol. Sci. 276, 3019–3025.

Alvarez-Filip, L.O., Côté, I.M., Gill, J.A., Watkinson, A.R., Dulvy, N.K., 2011. Region-wide temporal and spatial variation in Caribbean reef architecture: is coral cover the whole story? Glob. Chang. Biol. 17, 2470–2477.

Andres, N.G., Witman, J.D., 1995. Trends in community structure on a Jamaican reef. Mar. Ecol. Prog. Ser. 118, 305–310.

Aronson, R.B., Precht, W.F., 2000. Herbivory and algal dynamics on the coral reef at Discovery Bay, Jamaica. Limnol. Oceanogr. 45, 251–255.

Aronson, R.B., Precht, W.F., 2001a. White-band disease and the changing face of Caribbean coral reefs. Hydrobiologia 460, 25–38.

Aronson, R.B., Precht, W.F., 2001b. Evolutionary Paleoecology of Caribbean Coral Reefs. *In*: Allmon W.D., and D.J. Bottjer, Editors. Evolutionary Paleoecology: The Ecological Context of Macroevolutionary Change. Columbia University Press, New York, pp. 171–233.

Aronson, R.B., Precht, W.F., 2006. Conservation, precaution, and Caribbean reefs. Coral Reefs 25, 441–450.

Aronson, R.B., Precht, W.F., Lipps, J.H., 2016. Physical and biological drivers of coral-reef dynamics. In: Hubbard, D.K., Rogers, C.S., Stanley Jr., G.D. (Eds.), Coral Reefs at the Crossroads. Coral Reefs of the World. Vol. 6. Springer-Verlag, New York, pp. 435–461.

Aronson, R.B., Edmunds, P.J., Precht, W.F., Swanson, D.W., Levitan, D.R., 1994. Large scale, long-term monitoring of Caribbean coral reefs: simple, quick, inexpensive techniques. Atoll Res. Bull. 421, 1–19.

Aronson, R.B., Macintyre, I.G., Precht, W.F., Murdoch, T.J.T., Wapnick, C.M., 2002a. The expanding scale of species turnover events on coral reefs. Ecol. Monogr. 72, 233–249.

Aronson, R.B., Precht, W.F., Toscano, M.A., Koltes, K.H., 2002b. The 1998 bleaching event and its aftermath on a coral reef in Belize. Mar. Biol. 141, 435–437.

Aronson, R.B., Precht, W.F., Macintyre, I.G., Toth, L.T., 2012. Catastrophe and the life span of coral reefs. Ecology 93, 303–313.

Bahr, K.D., Jokiel, P.L., Toonen, R.J., 2015. The unnatural history of Kāne'ohe bay: coral reef resilience in the face of centuries of anthropogenic impacts. Peer J. 3, e950. https://doi.org/10.7717/peerj.950.

Banner, A.H., 1974. Kaneohe Bay, Hawaii: urban pollution and a coral reef ecosystem. In: Proceedings of the 2nd International Coral Reef Symposium Brisbane. 2, pp. 685–702.

Bates, A.E., Cooke, R.S., Duncan, M.I., Edgar, G.J., Bruno, J.F., Benedetti-Cecchi, L., Côté, I.M., Lefcheck, J.S., Costello, M.J., Barrett, N., Bird, T.J., 2019. Climate resilience in marine protected areas and the 'protection paradox'. Biol. Conserv. 236, 305–314.

Bechtel, J.D., Gayle, P., Kaufman, L., 2006. The return of *Diadema antillarum* to Discovery Bay: patterns of distribution and abundance. In: Proceedings of the 10th International Coral Reef Symposium Okinawa. 1, pp. 367–375.

Bellwood, D.R., Hughes, T.P., Folke, C., Nyström, M., 2004. Confronting the coral reef crisis. Nature 429, 827–833.

Benjamini, Y., Hochberg, Y., 1995. Controlling the false discovery rate: a practical and powerful approach to multiple testing. J. R. Stat. Soc. B (Methodology) 57, 289–300.

Birkeland, C., Miller, M.W., Piniak, G.A., Eakin, C.M., Weijerman, M., McElhany, P., Dunlap, M., Brainard, R.E., 2013. Safety in numbers? Abundance may not safeguard corals from increasing carbon dioxide. Bioscience 63, 967–974

Birrell, C.L., McCook, L.J., Willis, B.L., 2005. Effects of algal turfs and sediment on coral settlement. Mar. Pollut. Bull. 51, 408–414.

Birrell, C.L., McCook, L.J., Willis, B.L., Diaz-Pulido, G.A., 2008. Effects of benthic algae on the replenishment of corals and the implications for the resilience of coral reefs. Oceanogr. Marine Biol. An Annual Rev. 46, 25–64.

Bohnsack, J.A., Harper, D.E., McClellan, D.B., 1994. Fisheries trends from Monroe County, Florida. Bull. Mar. Sci. 54, 982–1018.

Bood, N.D., 2006. Recovery and Resilience of Coral Assemblages on Managed and Unmanaged Reefs in Belize: A Long-Term Study. MS Thesis. University of South Alabama, Mobile.

Box, S.J., Mumby, P.J., 2007. Effect of macroalgal competition on growth and survival of juvenile Caribbean corals. Mar. Ecol. Prog. Ser. 342, 139–149.

Bruno, J.F., Selig, E.R., 2007. Regional decline of coral cover in the indo-Pacific: timing, extent, and subregional comparisons. PLoS ONE 2, e711. doi.org/10.1371/journal.pone.0000711.

Bruno, J.F., Valdivia, A., 2016. Coral reef degradation is not correlated with local human population density. Sci. Rep. 6, 29778. https://doi.org/10.1038/srep29778.

Bruno, J.F., Sweatman, H., Precht, W.F., Selig, E.R., Schutte, V.G.W., 2009. Assessing evidence of phase shifts from coral to macroalgal dominance on coral reefs. Ecology 90, 1478–1484.

Bruno, J.F., Precht, W.F., Vroom, P.S., Aronson, R.B., 2014. Coral reef baselines: how much macroalgae is natural? Mar. Pollut. Bull. 80, 24–29.

Bruno, J.F., Côté, I.M., Toth, L.T., 2019. Climate change, coral loss, and the curious case of the parrotfish paradigm: why don't marine protected areas improve reef resilience? Ann. Rev. Mar. Sci. 11, 307–334.

Buddemeier, R.W., Kleypas, J.A., Aronson, R.B., 2004. Coral Reefs and Global Climate Change: Potential Contributions of Climate Change to Stresses on Coral Reef Ecosystems. Pew Center on Global Climate Change, Arlington, Virginia. http://www.c2es.org/docUploads/Coral_Reefs.pdf.

Burman, S.G., Aronson, R.B., van Woesik, R., 2012. Biotic homogenization of coral assemblages along the Florida reef tract. Mar. Ecol. Prog. Ser. 467, 89–96.

Carpenter, R.C., Edmunds, P.J., 2006. Local and regional scale recovery of *Diadema* promotes recruitment of scleractinian corals. Ecol. Lett. 9, 268–277.

Carpenter, K.E., Abra, M., Aeby, G., Aronson, R.B., Banks, S., Bruckner, A., Chiriboga, A., Cortes, J., Delbeek, J.C., DeVantier, L., Edgar, G.J., Edwards, A.J., Fenner, D., Guzman, H.M., Hoeksema, B.W., Hodgson, G., Johan, O., Licuanan, W.Y., Livingstone, S.R., Lovell, E.R., Moore, J.A., Obura, D.O., Ochavillo, D., Polidoro, B.A., Precht, W.F., Quibilan, M.C., Reboton, C., Richards, Z.T., Rogers, A.D., Sanciangco, J., Sheppard, A., Sheppard, C., Smith, J., Stuart, S., Turak, E., Veron, J.E.N., Wallace, C., Weil, E., Wood, E., 2008. One-third of reef-building corals face elevated extinction risk from climate change and local impacts. Science 321, 560–563.

Cho, L.L., Woodley, J.D., 2002. Recovery of reefs at Discovery Bay, Jamaica and the role of *Diadema antillarum*. In: Proceedings of the 9th International Coral Reef Symposium, Bali. 1, pp. 331–338.

Coelho, V.R., Manfrino, C., 2007. Coral community decline at a remote Caribbean island: marine no-take reserves are not enough. Aquat. Conserv. Mar. Freshwat. Ecosyst. 17, 666–685.

Côté, I.M., Darling, E.S., 2010. Rethinking ecosystem resilience in the face of climate change. PLoS Biol. 8. https://doi.org/10.1371/journal.pbio.1000438.

Côté, I.M., Gill, J.A., Gardner, T.A., Watkinson, A.R., 2005. Measuring coral reef decline through meta-analyses. Philos. Trans. R. Soc. Lond. B 360, 385–395.

Côté, I.M., Precht, W.F., Aronson, R.B., Gardner, T.A., 2013. Is Jamaica a good model for understanding Caribbean coral reef dynamics? Mar. Pollut. Bull. 76, 28–31.

Cox, C., Valdivia, A., McField, M., Castillo, K., Bruno, J.F., 2017. Establishment of marine protected areas alone does not restore coral reef communities in Belize. Mar. Ecol. Prog. Ser. 563, 65–79.

Crabbe, M.J.C., 2009. Scleractinian coral population size structures and growth rates indicate coral resilience on the fringing reefs of North Jamaica. Mar. Environ. Res. 67, 189–198.

Done, T.J., 1992. Phase shifts in coral reef communities and their ecological significance. Hydrobiologia 247, 121–132.

Dudgeon, S.R., Aronson, R.B., Bruno, J.F., Precht, W.F., 2010. Phase shifts and stable states on coral reefs. Mar. Ecol. Prog. Ser. 413, 201–216.

Dustan, P., Halas, J.C., 1987. Changes in the reef-coral community of Carysfort reef, key largo, Florida: 1974 to 1982. Coral Reefs 6, 91–106.

Eakin, C.M., Morgan, J.A., Heron, S.F., Smith, S., Liu, G., Alvarez-Filip, L., Baca, B., Bartels, E., Bastidas, C., Bouchon, C., Brandt, M., Bruckner, A.W., Bunkley-Williams, L., Cameron, A., Causey, B.D., Chiappone, M., Christensen, T.R.L., Crabbe, M.J.C., Day, O., de la Guardia, E., Díaz-Pulido, G., DiResta, D., Gil-Agudelo, D.L., Gilliam, D.S., Ginsburg, R.N., Gore, S., Guzmán, H.M., Hendee, J.C., Hernández-Delgado, E.A., Husain, E., Jeffrey, C.F.G., Jones, R.J., Jordán-Dahlgren, E., Kaufman, L.S., Kline, D.I., Kramer, P.A., Lang, J.C., Lirman, D., Mallela, J., Manfrino, C., Maréchal, J.-P., Marks, K., Mihaly, J., Miller, W.J., Mueller, E.M., Muller, E.M., Orozco Toro, C.A., Oxenford, H.A., Ponce-Taylor, D., Quinn, N., Ritchie, K.B., Rodríguez, S., Rodríguez Ramírez, A., Romano, S., Samhouri, J.F., Sánchez, J.A., Schmahl, G.P., Shank, B.V., Skirving, W.J., Steiner, S.C.C., Villamizar, E., Walsh, S.M., Walter, C., Weil, E., Williams, E.H., Woody Roberson, K., Yusuf, Y., 2010. Caribbean corals in crisis: record thermal stress, bleaching, and mortality in 2005. PLoS ONE 5, e13969. doi.org/10.1371/journal.pone.0013969.

Eddy, T.D., Cheung, W.W., Bruno, J.F., 2018. Historical baselines of coral cover on tropical reefs as estimated by expert opinion. Peer J. 6, e4308. https://doi.org/10.7717/peerj.4308.

Edmunds, P.J., Carpenter, R.C., 2001. Recovery of *Diadema antillarum* reduces macroalgal cover and increases abundance of juvenile corals on a Caribbean reef. Proce. Natl. Acad. Sci. USA 98, 5067–5071.

Edwards, H.J., Elliott, I.A., Eakin, C.M., Irikawa, A., Madin, J.S., McField, M., Morgan, J.A., van Woesik, R., Mumby, P.J., 2011. How much time can herbivore protection buy for coral reefs under realistic regimes of hurricanes and coral bleaching? Glob. Chang. Biol. 17, 2033–2048.

Francis, F.T., Filbee-Dexter, K., Yan, H.F., Côté, I.M., 2019. Invertebrate herbivores: overlooked allies in the recovery of degraded coral reefs? Global Ecol. Conserv., e00593. https://doi.org/10.1016/j.gecco.2019.e00593.

Furman, B.T., Heck, K.L., 2008. Effects of nutrient enrichment and grazers on coral reefs: an experimental assessment. Mar. Ecol. Prog. Ser. 363, 89–101.

Furman, B., Heck, K.L., 2009. Differential impacts of echinoid grazers on coral recruitment. Bull. Mar. Sci. 85, 121–132.

Gardner, T.A., Côté, I.M., Gill, J.A., Grant, A., Watkinson, A.R., 2003. Long-term region-wide declines in Caribbean corals. Science 301, 958–960.

Gardner, T.A., Côté, I.M., Gill, J.A., Grant, A., Watkinson, A.R., 2005. Hurricanes and Caribbean coral reefs: immediate impacts, recovery trajectories, and contribution to long-term coral decline. Ecology 86, 174–184.

Gill, J.A., Watkinson, A.R., McWilliams, J.P., Côté, I.M., 2006. Opposing forces of aerosol cooling and El Niño drive coral bleaching on Caribbean reefs. Proce. Natl. Acad. Sci. USA 103, 18870–18873.

Gladfelter, W.B., 1982. White band disease in *Acropora palmata*: implications for the structure and growth of shallow reefs. Bull. Mar. Sci. 32, 639–643.

Greenaway, A.M., Gordon-Smith, D., 2006. The effects of rainfall on the distribution of inorganic nitrogen and phosphorus in Discovery Bay, Jamaica. Limnol. Oceanogr. 51, 2206–2220.

Hardt, M.J., 2008. Lessons from the past: the collapse of Jamaican coral reefs. Fish Fish. 10, 1–16.

Hoegh-Guldberg, O., Bruno, J.F., 2010. The impact of climate change on the world's marine ecosystems. Science 328, 1523–1528.

Hoegh-Guldberg, O., Ortiz, J.C., Dove, S., 2011. The future of coral reefs. Science 334, 1494–1495.
Hughes, T.P., 1994. Catastrophes, phase-shifts, and large-scale degradation of a Caribbean coral reef. Science 265, 1547–1551.
Hughes, T.P., Reed, D.C., Boyle, M.J., 1987. Herbivory on coral reefs: community structure following mass mortalities of sea urchins. J. Exp. Mar. Biol. Ecol. 113, 39–59.
Hughes, T.P., Szmant, A.M., Steneck, R., Carpenter, R., Miller, S., 1999. Algal blooms on coral reefs: what are the causes? Limnol. Oceanogr. 44, 1583–1586.
Hughes, T.P., Baird, A.H., Bellwood, D.R., Card, M., Connolly, S.R., Folke, C., Grosberg, R., Hoegh-Guldberg, O., Jackson, J.B.J., Kleypas, J., Lough, J.M., Marshall, P., Nystrom, M., Palumbi, S.R., Pandolfi, J.M., Rosen, B., Roughgarden, J., 2003. Climate change, human impacts, and the resilience of coral reefs. Science 301, 929–933.
Hughes, T.P., Graham, N.A.J., Jackson, J.B.C., Mumby, P.J., Steneck, R.S., 2010. Rising to the challenge of sustaining coral reef resilience. Trends Ecol. Evol. 25, 633–642.
Hughes, T.P., Linares, C., Dakos, V., van de Leemput, I.A., van Nes, E.H., 2012. Living dangerously on borrowed time during slow, unrecognized regime shifts. Trends Ecol. Evol. 28, 149–155.
Hunter, C.L., Evans, C.W., 1995. Coral reefs in Kaneohe Bay, Hawaii: two centuries of western influence and two decades of data. Bull. Mar. Sci. 57, 501–515.
Huntington, B.E., Karnauskas, M., Lirman, D., 2011. Corals fail to recover at a Caribbean marine reserve despite ten years of reserve designation. Coral Reefs 30, 1077–1085.
Idjadi, J.A., Lee, S.C., Bruno, J.F., Precht, W.F., Allen-Requa, L., Edmunds, P.J., 2006. Rapid phase-shift reversal on a Jamaican coral reef. Coral Reefs 25, 209–211.
Idjadi, J.A., Haring, R.N., Precht, W.F., 2010. Recovery of the sea urchin *Diadema antillarum* promotes scleractinian coral growth and survivorship on shallow Jamaican reefs. Mar. Ecol. Prog. Ser. 403, 91–100.
Jaap, W.C., Halas, J.C., Muller, R.G., 1988. Community dynamics of stony corals (Milleporina and Scleractinia) at Key Largo National Marine Sanctuary Florida, during 1981–1986. Proc. 6th Int. Coral Reef Symp. 2, 237–243. Townsville.
Jackson, J.B.C., 2001. What was natural in the coastal oceans? Proce. Natl. Acad. Sci. USA 98, 5411–5418.
Jackson, J.B.C., 2014. Executive summary. In: Jackson, J.B.C., Donovan, M.K., Cramer, K.L., Lam, V.V. (Eds.), Status and Trends of Caribbean Coral Reefs: 1970–2012. Global Coral Reef Monitoring Network, IUCN, Gland, Switzerland.
Jackson, J.B.C., Kirby, M.X., Berger, W.H., Bjorndal, K.A., Botsford, L.A., Bourque, B.J., Bradbury, R.H., Cooke, R., Erlandson, J., Estes, J.A., Cooke, T., Hughes, T.P., Kidwell, S., Lange, C.B., Lenihan, H.S., Pandolfi, J.M., Peterson, C.H., Steneck, R.S., Tegner, M.J., Warner, R.R., 2001. Historical overfishing and the recent collapse of coastal ecosystems. Science 293, 629–638.
Jackson, J.B.C., Donovan, M.K., Cramer, K.L., Lam, V.V. (Eds.), 2014. Status and Trends of Caribbean Coral Reefs: 1970–2012. Global Coral Reef Monitoring Network, IUCN, Gland, Switzerland.
Johnston, M.A., Embesi, J.A., Eckert, R.J., Nuttall, M.F., Hickerson, E.L., Schmahl, G.P., 2016. Persistence of coral assemblages at east and west flower garden Banks, Gulf of Mexico. Coral Reefs 25, 821–826.
Jorissen, H., Skinner, C., Osinga, R., de Beer, D., Nugues, M.M., 2016. Evidence for water-mediated mechanisms in coral–algal interactions. Proc. R. Soc. London B 283, 20161137. doi.org/10.1098/rspb.2016.1137.
Kaufman, L.S., 1983. Effects of hurricane Allen on reef fish assemblages near Discovery Bay. Jamaica, Coral Reefs 2, 43–47.

Knowlton, N., 1992. Thresholds and multiple stable states in coral reef community dynamics. Am. Zool. 32, 674–682.

Knowlton, N., 2001. Sea urchin recovery from mass mortality: new hope for Caribbean coral reefs? Proce. Natl. Acad. Sci. USA 98, 4822–4824.

Knowlton, N., Lang, J.C., Rooney, M.C., Clifford, P.A., 1981. Evidence for delayed mortality in hurricane-damaged Jamaican staghorn corals. Nature 294, 251–252.

Kramer, P.A., 2003. Synthesis of coral reef health indicators for the western Atlantic: results of the AGRRA program (1997-2000). Atoll Res. Bull. 496, 1–57.

Kramer, K.L., Heck Jr., K.L., 2007. Top-down trophic shifts in Florida keys patch reef marine protected areas. Mar. Ecol. Prog. Ser. 349, 111–123.

Kuffner, I.B., Toth, L.T., 2016. A geological perspective on the degradation and conservation of western Atlantic coral reefs. Conserv. Biol. 30, 706–715.

Kuffner, I.B., Walters, L.J., Becerro, M.A., Paul, V.J., Ritson-Williams, R., Beach, K.S., 2006. Inhibition of coral recruitment by macroalgae and cyanobacteria. Mar. Ecol. Prog. Ser. 323, 107–117.

Kuffner, I.B., Lidz, B.H., Hudson, J.H., Anderson, J.S., 2014. A century of ocean warming on Florida keys coral reefs: historic in situ observations. Estuaries Coast 38, 1085–1096.

Lang, J.C., Lasker, H.R., Gladfelter, E.H., Hallock, P., Jaap, W.C., Losada, F.J., Muller, R.G., 1992. Spatial and temporal variability during periods of "recovery" after mass bleaching on western Atlantic coral reefs. Am. Zool. 32, 696–706.

Lapointe, B.E., 1997. Nutrient thresholds for bottom up control of macroalgal blooms on coral reefs in Jamaica and Southeast Florida. Limnol. Oceanogr. 42, 1119–1131.

Ledlie, M.H., Graham, N.A.J., Bythell, J.C., Wilson, S.K., Jennings, S., Polunin, N.V.C., Hardcastle, J., 2007. Phase shifts and the role of herbivory in the resilience of coral reefs. Coral Reefs 26, 641–653.

Lessios, H.A., 1988. Mass mortality of *Diadema antillarum* in the Caribbean: what have we learned? Annu. Rev. Ecol. Syst. 19, 371–393.

Lessios, H.A., 2016. The great *Diadema antillarum* die-off: 30 years later. Ann. Rev. Mar. Sci. 8, 267–283.

Lester, S.E., Rassweiler, A., McCoy, S.J., Dubel, A.K., Donovan, M.K., Miller, M.W., Miller, S.D., Ruttenberg, B.I., Samhouri, J.F., Hay, M.E., 2020. Caribbean Reefs of the Anthropocene: Variance in ecosystem metrics indicates bright spots on coral depauperate reefs. Global Change Biol. 26, 4785–4799. https://doi.org/10.1111/gcb.15253.

Levitan, D.R., Edmunds, P.J., Levitan, K.E., 2014. What makes a species common? No evidence of density-dependent recruitment or mortality of the sea urchin *Diadema antillarum* after the 1983–1984 mass mortality. Oecologia 175, 117–128.

Liddell, W.D., Ohlhorst, S.L., 1992. Ten years of disturbance and change on a Jamaican fringing reef. In: Proceedings of the 7th International Coral Reef Symposium, Guam, 1, pp. 144–150.

Lirman, D., 2001. Competition between macroalgae and corals: effects of herbivore exclusion and increased algal biomass on coral survivorship and growth. Coral Reefs 19, 392–399.

Lirman, D., Biber, P., 2000. Seasonal dynamics of macroalgal communities of the northern Florida reef tract. Bot. Mar. 43, 305–314.

Lirman, D., Fong, P., 2007. Is proximity to land-based sources of coral stressors an appropriate measure of risk to coral reefs? An example from the Florida reef tract. Mar. Pollut. Bull. 54, 779–791.

Lirman, D., Schopmeyer, S., Manzello, D., Gramer, L.J., Precht, W.F., Muller-Karger, F., Banks, K., Barnes, B., Bartels, E., Bourque, A., Byrne, J., Donahue, S., Duquesnel, J., Fisher, L., Gilliam, D., Hendee, J., Johnson, M., Maxwell, K., McDevitt, E., Monty, J., Rueda, D., Ruzicka, R., Thanner, S., 2011. Severe 2010 cold-water event caused

unprecedented mortality to corals of the Florida reef tract and reversed previous survivorship patterns. PLoS One 6. https://doi.org/10.1371/journal.pone.0023047.
Littler, M.M., Littler, D.S., Brooks, B.L., 2006. Harmful algae on tropical coral reefs: bottom-up eutrophication and top-down herbivory. Harmful Algae 5, 565–585.
Lough, J.M., 2000. 1997–98: unprecedented thermal stress to coral reefs? Geophys. Res. Lett. 27, 3901–3904.
Lowe, P.K., Bruno, J.F., Selig, E.R., Spencer, M., 2011. Empirical models of transitions between coral reef states: effects of region, protection, and environmental change. PLoS ONE 6, e26339. doi.org/10.1371/journal.pone.0026339.
Macintyre, I.G., Glynn, P.W., Hinds, F., 2005. Evidence of the role of *Diadema antillarum* in the promotion of coral settlement and survivorship. Coral Reefs 24, 273.
Manzello, D.P., 2015. Rapid recent warming of coral reefs in the Florida keys. Sci. Rep. 5, 16762. https://doi.org/10.1038/srep16762.
McClanahan, T.R., 1999. Predation and the control of the sea urchin Echinometra viridis and fleshy algae in the patch reefs of glovers reef. Belize, Ecosystems 2, 511–523.
McClanahan, T.R., 2002. The near future of coral reefs. Environ. Conserv. 29, 460–483.
McClanahan, T., Polunin, N., Done, T., 2002. Ecological states and the resilience of coral reefs. Conserv. Ecol. 6, 18. http://www.consecol.org/vol6/iss2/art18/.
McClanahan, T.R., Sala, E., Mumby, P.J., Jones, S., 2004. Phosphorus and nitrogen enrichment do not enhance brown frondose "macroalgae.". Mar. Pollut. Bull. 48, 196–199.
McClanahan, T.R., Huntington, B.E., Cokos, B., 2011a. Coral responses to macroalgal reduction and fisheries closure on Caribbean patch reefs. Mar. Ecol. Prog. Ser. 437, 89–102.
McClanahan, T.R., Muthiga, N.A., Coleman, R.A., 2011b. Testing for top-down control: can post-disturbance fisheries closures reverse algal dominance? Aquat. Conserv. Mar. Freshwat. Ecosyst. 21, 658–675.
McCook, L.J., 1999. Macroalgae, nutrients and phase shifts on coral reefs: scientific issues and management consequences for the great barrier reef. Coral Reefs 18, 357–367.
McCook, L., Jompa, J., Diaz-Pulido, G., 2001. Competition between corals and algae on coral reefs: a review of evidence and mechanisms. Coral Reefs 19, 400–417.
McField, M.D., 1999. Coral response during and after mass bleaching in Belize. Bull. Mar. Sci. 64, 155–172.
McManus, J.W., Polsenberg, J.F., 2004. Coral-algal phase shifts on coral reefs: ecological and environmental aspects. Prog. Oceanogr. 60, 263–279.
McWilliams, J.P., Côté, I.M., Gill, J.A., Sutherland, W.J., Watkinson, A.R., 2005. Accelerating impacts of temperature-induced coral bleaching in the Caribbean. Ecology 86, 2055–2060.
Miller, M.W., Hay, M.E., Miller, S.L., Malone, D., Sotka, E.E., Szmant, A.M., 1999. Effects of nutrients versus herbivores on reef algae: a new method for manipulating nutrients on coral reefs. Limnol. Oceanogr. 44, 1847–1861.
Miller, J., Muller, E., Rogers, C., Waara, R., Atkinson, A., Patterson, M., Witcher, B., 2009. Coral disease following massive bleaching in 2005 causes 60% decline in coral cover on reefs in the US Virgin Islands. Coral Reefs 28, 925–937.
Mora, C., 2008. A clear human footprint in the coral reefs of the Caribbean. Aquat. Conserv. Mar. Freshwat. Ecosyst. 275, 767–773.
Mumby, P.J., 2006. The impact of exploiting grazers (Scaridae) on the dynamics of Caribbean coral reefs. Ecol. Appl. 16, 747–769.
Mumby, P.J., 2009. Phase shifts and the stability of macroalgal communities on Caribbean coral reefs. Coral Reefs 28, 761–777.
Mumby, P.J., Steneck, R.S., 2008. Coral reef management and conservation in light of rapidly emerging ecological paradigms. Trends Ecol. Evol. 23, 555–563.

Mumby, P.J., Hedley, J.D., Żychaluk, K., Harborne, A.B., Blackwell, P.G., 2006. Revisiting the catastrophic die-off of the urchin *Diadema antillarum* on Caribbean coral reefs: fresh insights on resilience from a simulation model. Ecol. Model. 196, 131–148.

Mumby, P.J., Hastings, A., Edwards, H.J., 2007. Thresholds and the resilience of Caribbean coral reefs. Nature 450, 98–101.

Myers, M.R., Ambrose, R.F., 2009. Differences in benthic cover inside and outside marine protected areas on the great barrier reef: influence of protection or disturbance history? Aquat. Conserv. Mar. Freshwat. Ecosyst. 19, 736–747.

Myhre, S., Acevedo-Gutiérrez, A., 2007. Recovery of sea urchin *Diadema antillarum* populations is correlated to increased coral and reduced macroalgal cover. Mar. Ecol. Prog. Ser. 329, 205–210.

Newman, M.J.H., Paredes, G.A., Sala, E., Jackson, J.B.C., 2006. Structure of Caribbean coral reef communities across a large gradient of fish biomass. Ecol. Lett. 9, 1216–1227.

Norström, A.V., Nyström, M., Lokrantz, J., Folke, C., 2009. Alternative states on coral reefs: beyond coral–macroalgal phase shifts. Mar. Ecol. Prog. Ser. 376, 295–306.

Nugues, M.M., Bak, R.P.M., 2006. Differential competitive abilities between Caribbean coral species and a brown alga: a year of experiments and a long-term perspective. Mar. Ecol. Prog. Ser. 315, 75–86.

Nugues, M.M., Bak, R.P.M., 2008. Long-term dynamics of the brown macroalga *Lobophora variegata* on deep reefs in curacao. Coral Reefs 27, 389–393.

Ogden, J.C., 1995. Coral reef decline in the Caribbean region. In: Proceedings of the 2nd European Regional Meeting of the International Society for Reef Studies, Publications du Service Geologique du Luxembourg XXIX, p. 224.

Biology Online. 2013. Definition of Ecological Dominance. http://www.biology online.org/dictionary/Dominant_species.

Ostrander, G.K., Armstrong, K.M., Knobbe, E.T., Gerace, D., Scully, E.P., 2000. Rapid transition in the structure of a coral reef community: the effects of coral bleaching and physical disturbance. Proce. Natl. Acad. Sci. USA 97, 5297–5302.

Paddack, M., Reynolds, J., Aguilar, C., Appledorn, R., Beets, J., Burkett, E.W., Chittaro, P.M., Clarke, K., Esteves, R., Fonseca, A.C., Forrester, G.E., Friedlander, A.M., García-Sais, J., González-Sansón, G., Jordan, L.K.B., McClellan, D.B., Miller, M.W., Molloy, P.P., Mumby, P.J., Nagelkerken, I., Nemeth, M., Navas-Camacho, R., Pitt, J., Polunin, N.V.C., Reyes-Nivia, M.C., Robertson, D.R., Rodríguez-Ramírez, A., Salas, E., Smith, S.R., Spieler, R.E., Steele, M.A., Williams, I.D., Wormald, C.L., A.R. Watkinson AR, and I.M. Côté., 2009. Recent region-wide declines in Caribbean reef fish abundance. Curr. Biol. 19, 590–595.

Paine, R.T., Tegner, M.J., Johnson, E.A., 1998. Compounded perturbations yield ecological surprises. Ecosystems 1, 535–545.

Palumbi, S.R., McLeod, K.L., K.L. and D. Grünbaum, D., 2008. Ecosystems in action: lessons from marine ecology about recovery, resistance, and reversibility. Bioscience 58, 33–42.

Pandolfi, J.M., Bradbury, E., Sala, T.P., Hughes, K.A., Bjorndal, R.G., Cooke, D., McArdle, L., McClenachan, M.J.H., Newman, G., Paredes, R.R.W., Jackson, J.B.C., 2005. Are U.S. coral reefs on the slippery slope to slime? Science 307, 1725–1726.

Pauly, D., Christiansen, V., Dalsgaard, J., Froese, R., Torres Jr., F., 1998. Fishing down marine food webs. Science 279, 860–863.

Petraitis, P.S., Dudgeon, S.R., 2004. Detection of alternative states in marine communities. J. Exp. Mar. Biol. Ecol. 300, 343–371.

Precht, W.F., Aronson, R.B., 2006. Death and resurrection of Caribbean coral reefs: a palaeoecological approach. In: Côté, I.M., Reynolds, J.D. (Eds.), Coral Reef Conservation. Cambridge University Press, Cambridge, UK, pp. 40–77.

Precht, W.F., Miller, S.L., 2007. Ecological shifts along the Florida Reef Tract: the past is a key to the future. In: Aronson, R.B. (Ed.), Geological Approaches to Coral Reef Ecology. Springer, New York, USA, pp. 237–312.
Precht, L.L., Precht, W.F., 2016. The sea urchin *Diadema antillarum*—keystone herbivore or redundant species? Peer J PrePrints 3, e1951. doi.org/10.7287/peerj.preprints.1565v2.
Precht, W.F., Gintert, B.E., Robbart, M.L., Fura, R., van Woesik, R., 2016. Unprecedented disease-related coral mortality in southeastern Florida. Sci. Rep. 6, 31374. https://doi.org/10.1038/srep31374.
Quinn, J.F., Dunham, A.E., 1983. On hypothesis testing in ecology and evolution. Am. Nat. 122, 602–617.
Randall, C.J., Jordán-Garza, A.G., Muller, E.M., van Woesik, R., 2014. Relationships between the history of thermal stress and the relative risk of diseases of Caribbean corals. Ecology 95, 1981–1994.
Reyes-Bonilla, H., Millet-Encalada, M., Álvarez-Filip, L., 2014. Community structure of scleractinian corals outside protected areas in Cozumel Island, Mexico. Atoll Res. Bull. 601, 1–13.
Ritson-Williams, R., Arnold, S.N., Fogarty, N.D., Steneck, R.S., Vermeij, M.J., Paul, V.J., 2009. New perspectives on ecological mechanisms affecting coral recruitment on reefs. Smithson. Contrib. Mar. Sci. 38, 437–457.
Robertson, D.R., 1991. Increases in surgeonfish populations after mass mortality of the sea urchin *Diadema antillarum* in Panama indicate food limitation. Mar. Biol. 111, 437–444.
Rogers, C.S., 2008. Coral bleaching and disease should not be underestimated as causes of Caribbean coral reef decline. Proc. R. Soc. London B 276, 197–198.
Rogers, A., Lorenzen, K., 2016. Does slow and variable recovery of *Diadema antillarum* on Caribbean fore-reefs reflect density-dependent habitat selection? Front. Mar. Sci. 3, 63. https://doi.org/10.3389/fmars.2016.00063.
Rogers, C.S., Miller, J., 2006. Permanent 'phase shifts' or reversible declines in coral cover? Lack of recovery of two coral reefs in St. John, US Virgin Islands. Mar. Ecol. Prog. Ser. 306, 103–114.
Rotjan, R.D., Dimond, J.L., Thornhill, D.J., Leichter, J.J., Helmuth, B., Kemp, D.W., Lewis, S.M., 2006. Chronic parrotfish grazing impedes coral recovery after bleaching. Coral Reefs 25, 361–368.
Russ, G.R., Questel, S.L.A., Rizzari, J.R., Alcala, A.C., 2015. The parrotfish–coral relationship: refuting the ubiquity of a prevailing paradigm. Mar. Biol. 162, 2029–2045.
Ruzicka, R.R., Colella, M.A., Porter, J.W., Morrison, J.M., Kidney, J.A., Brinkhuis, V., Lunz, K.S., Macaulay, K.A., Bartlett, L.A., Meyers, M.A., Colee, J., 2013. Temporal changes in benthic assemblages on Florida keys reefs 11 years after the 1997/1998 El Niño. Mar. Ecol. Prog. Ser. 489, 125–141.
Sandin, S., McNamara, D., 2012. Spatial dynamics of benthic competition on coral reefs. Oecologia 168, 1079–1090.
Scheffer, M., Carpenter, S.R., 2003. Catastrophic regime shifts in ecosystems: linking theory to observation. Trends Ecol. Evol. 18, 648–656.
Scheffers, S.R., Scheffers, A., Radtke, U., Kelletat, D., Staben, K., Bak, R.P.M., 2006. Tsunami triggers long-lasting phase-shifts in coral reef ecosystem. In: Scheffers, A., Kelletat, D. (Eds.), Tsunamis, Hurricanes and Neotectonics as Driving Mechanisms in Coastal Evolution. *Proceedings Bonaire Field Symposium, March 2–6, 2006. IGCP Contribution #495*, Zeitschrift fur Geomorphologie Supplementband. 146, pp. 59–79.
Schutte, V.G.W., Selig, E.R., Bruno, J.F., 2010. Regional spatio-temporal trends in Caribbean coral reef benthic communities. Mar. Ecol. Prog. Ser. 402, 115–122.
Selig, E.R., Bruno, J.F., 2010. A global analysis of the effectiveness of marine protected areas in preventing coral loss. PLoS ONE 5. https://doi.org/10.1371/journal.pone.0009278.

Selig, E.R., Casey, K.S., Bruno, J.F., 2012. Temperature-driven coral decline: the role of marine protected areas. Glob. Chang. Biol. 18, 1561–1570.
Sheppard, C., 1993. Coral reef environmental science: dichotomies, not the Cassandras, are false. Reef Encounter 14, 12–13.
Smith, J.E., Hunter, C.L., Smith, C.M., 2010. The effects of top–down versus bottom–up control on benthic coral reef community structure. Oecologia 163, 497–507.
Smith, T.B., Brandt, M.E., Calnan, J.M., Nemeth, R.S., Blondeau, J., Kadison, E., Taylor, M., Rothenberger, P., 2013. Convergent mortality responses of Caribbean coral species to seawater warming. Ecosphere 4 (7), 87.
Sotka, E.E., Hay, M.E., 2009. Effects of herbivores, nutrient enrichment, and their interactions on macroalgal proliferation and coral growth. Coral Reefs 28, 555–568.
Steneck, R.S., 1988. Herbivory on coral reefs: a synthesis. In: Proceedings of the 6th International Coral Reef Symposium, Townsville. 1, pp. 37–49.
Steneck, R.S., Sala, E.A., 2005. Large marine carnivores: trophic cascades and top-down controls in coastal ecosystems past and present. In: Ray, J.C., Redford, K.H., Steneck, R., Berger, J. (Eds.), Large Carnivores and the Conservation of Biodiversity. Island Press, Washington, D.C., USA, pp. 110–137.
Stockwell, B., Jadloc, C.R.L., Abesamis, R.A., Alcala, A.C., Russ, G.R., 2009. Trophic and benthic responses to no-take marine reserve protection in the Philippines. Mar. Ecol. Prog. Ser. 389, 1–15.
Stoddart, D.R., 1985. Hurricane effects on coral reefs. In: Proceedings of the 5th International Coral Reef Congress, Tahiti. 3, pp. 349–350.
Suchley, A., McField, M.D., Alvarez-Filip, L., 2016. Rapidly increasing macroalgal cover not related to herbivorous fishes on Mesoamerican reefs. Peer J. 4, e2084. https://doi.org/10.7717/peerj.2084.
Szmant, A.M., 2002. Nutrient enrichment on coral reefs: Is it a major cause of coral reef decline? Estuaries 25, 743–766.
Toth, L.T., Van Woesik, R., Murdoch, T.J.T., Smith, S.R., Ogden, J.C., Precht, W.F., Aronson, R.B., 2014. Do no-take reserves benefit Florida's corals? 14 years of change and stasis in the Florida keys National Marine Sanctuary. Coral Reefs 33, 565–577.
Toth, L.T., Stathakopoulos, A., Kuffner, I.B., Ruzicka, R.R., Colella, M.A., Shinn, E.A., 2019. The unprecedented loss of Florida's reef-building corals and the emergence of a novel coral-reef assemblage. Ecology, e02781. https://doi.org/10.1002/ecy.2781
Tsounis, G., Edmunds, P.J., 2017. Three decades of coral reef community dynamics in St. John, USVI: a contrast of scleractinians and octocorals. Ecosphere 8, e01646.
van Woesik, R., Scott IV, W.J., Aronson, R.B., 2014. Lost opportunities: coral recruitment does not translate to reef recovery in the Florida keys. Mar. Pollut. Bull. 88, 110–117.
Veron, J.E.N., Hoegh-Guldberg, O., Lenton, T.M., Lough, J.M., Obura, D.O., Pearce-Kelly, P., Sheppard, C.R.C., Spalding, M., Stafford-Smith, M.G., Rogers, A.D., 2009. The coral reef crisis: the critical importance of <350 ppm CO_2. Mar. Pollut. Bull. 58, 1428–1436.
Vieira, C., Payri, C., De Clerck, O., 2016. A fresh look at macroalgal-coral interactions: are macroalgae a threat to corals? Perspect. Phycol. 3, 129–140.
Vroom, P.S., Page, K.N., Kenyon, J.C., Brainard, R.E., 2006. Algae-dominated reefs. Am. Sci. 94, 430–437.
Webber, M., Edwards-Myers, E., Campbell, C., Webber, D., 2005. Phytoplankton and zooplankton as indicators of water quality in Discovery Bay, Jamaica. Hydrobiologia 545, 177–193.
Wilkinson, C.R., 2002. Status of Coral Reefs of the World: 2002. GCRMN, Townsville, Australia.
Williams, I.D., Polunin, N.V.C., 2001. Large-scale associations between macroalgal cover and grazer biomass on mid-depth reefs in the Caribbean. Coral Reefs 19, 358–366.

Woodley, J.D., Chornesky, E.A., Clifford, P.A., Jackson, J.B.C., Kaufman, L.S., Knowlton, N., Lang, J.C., Pearson, M.P., Porter, J.W., Rooney, M.C., Rylaarsdam, K.W., Wahle, C.M., Wulff, J.L., Curtis, A.S.G., Dallmeyer, M.D., Jupp, B.P., Koehl, M.A.R., Neigel, J., Sides, E.M., 1981. Hurricane Allen's impact on Jamaican coral reefs. Science 214, 749–755.

Woodley, J.D., Sary, Z., 2002. Development of a locally managed fisheries reserve at Discovery Bay, Jamaica. Proc. 9th Intl. Coral Reef Symp. 2, 627–633.

Żychaluk, K., Bruno, J.F., Clancy, D., McClanahan, T.R., Spencer, M., 2012. Data-driven models for regional coral-reef dynamics. Ecol. Lett. 15, 151–158.

CHAPTER THIRTEEN

The rise of octocoral forests on Caribbean reefs

Howard R. Lasker[a,*], Lorenzo Bramanti[b], Georgios Tsounis[c], and Peter J. Edmunds[c]

[a]Department of Environment and Sustainability and Department of Geology, University at Buffalo, Buffalo, NY, United States
[b]CNRS-Sorbonne Université, Laboratoire d'Ecogéochimie des Environnements Benthiques, LECOB, Observatoire Océanologique de Banyuls sur Mer, Banyuls sur Mer, France
[c]Department of Biology, California State University, Northridge, CA, United States
*Corresponding author: e-mail address: hlasker@buffalo.edu

Contents

Abstract

Coral reefs throughout the tropics have experienced large declines in the abundance of scleractinian corals over the last few decades, and some reefs are becoming functionally dominated by animal taxa other than scleractinians. This phenomenon is striking on many shallow reefs in the tropical western Atlantic, where arborescent octocorals

Advances in Marine Biology, Volume 87
ISSN 0065-2881
https://doi.org/10.1016/bs.amb.2020.08.009

now are numerically and functionally dominant. Octocorals are one of several taxa that have been overlooked for decades in analyses of coral reef community dynamics, and our understanding of why octocorals are favoured (whereas scleractinians are not) on some modern reefs, and how they will affect the function of future reef communities, is not commensurate with the task of scientifically responding to the coral reef crisis. We summarize the biological and ecological features predisposing octocorals for success under contemporary conditions, and focus on those features that could have generated resistance and resilience of octocoral populations to environmental change on modern reefs. There is a rich set of opportunities for rapid advancement in understanding the factors driving the success of octocorals on modern reefs, but we underscore three lines of inquiry: (1) the functional implications of strongly mixotrophic, polytrophic, and plastic nutrition, (2) the capacity to recruit at high densities and maintain rapid initial rates of vertical growth, and (3) the emergent properties associated with dense animal forests at high colony densities.

1. Introduction

Most tropical coral reefs have dramatically changed from those first experienced by Europeans when they explored the tropics (Jackson, 1997), and many reefs bear little resemblance to those of the 1950s and 1960s (Goreau, 1959; Loya, 1972). These changes have been prominently featured in empirical and theoretical analyses (Dudgeon et al., 2010; Hughes et al., 2018), and have deeply penetrated into mainstream media. Yet the trends most frequently discussed—declining abundances of stony coral and fishes (Bellwood et al., 2004; Jackson et al., 2001)—are narrow in taxonomic breadth given the high diversity of tropical reef communities (Connell, 1978; Huston, 1985; Plaisance et al., 2011), and do not capture the full nature of the changes taking place on tropical reefs.

Studies of the "usual suspects" on coral reefs capture the implications of changing community structure as defined by the taxa that have been the focus of previous studies. Such studies have limited ability to evaluate the functional and ecological implications of the state into which modern reefs are transitioning. At one extreme, changes to coral reef communities have been characterized as a descent to "slime" (Pandolfi et al., 2005). Yet at the other extreme there are cases in which tropical reefs appear to be transitioning to domination by animal taxa other than scleractinian corals (Norstrom et al., 2009).

A variety of taxa have replaced scleractinians on tropical reefs, including sponges, corallimorpharians, tunicates, and octocorals (Norstrom et al., 2009). Sponges have been suggested as a dominant member of future reef communities (Pawlik and McMurray, 2020), although evidence of this trend remains sparse (Wulff, 2016). Octocorals, the focus of this review, are functionally important on Caribbean and Indo-Pacific reefs (Bayer, 1961; Benayahu and Loya, 1977; Cary, 1914; Goldberg, 1973; Reichelt et al., 1986), yet their biology has been overlooked for decades. Unanswered questions regarding the transitions in community structure taking place on reefs focus on the underlying mechanisms: do the species assemblages represent stable communities, or placeholders on the road to slime? What are the implications of these changes to reef ecosystem services? In the Caribbean, evidence that octocorals are becoming more abundant and adopting greater functional roles than in the recent past (Lenz et al., 2015; Ruzicka et al., 2013), suggests they may represent a "new normal" for these shallow reefs (Lasker et al., 2020).

This review focuses on the biology and ecology of octocorals on present-day coral reefs in the tropical western Atlantic. Rather than presenting a comprehensive review of this topic [(which has been covered (Kupfner Johnson and Hallock, 2020)], we interpret the task through the prism of the ecological changes underway on reefs in the tropical western Atlantic to favour octocorals (Lenz et al., 2015; Ruzicka et al., 2013; Tsounis and Edmunds, 2017). We address the features of octocorals that promote their ecological success in habitats where they are exposed to the same conditions driving large declines in the abundance of scleractinians (Hughes, 1994; Jackson et al., 2014), and we explore how octocorals alter their habitat. We focus on the tropical western Atlantic because the demise of scleractinian-dominated reefs in this region is well advanced (Bruno and Selig, 2007; Jackson et al., 2014; Roff and Mumby, 2012). The ecological future of coral reefs remains uncertain, but it is reasonable to hypothesize that present-day reefs in the tropical western Atlantic provide a glimpse of the broader future of "coral" reefs in the Anthropocene (Hughes et al., 2017).

Much of the scientific history of modern coral reefs has addressed declines in population sizes of fishes and large vertebrates that began in the 1500s (Jackson, 1997). Large-scale reductions in the abundance of scleractinians are more recent, and virtually were absent from the literature before Indo-Pacific reefs were devastated by crown of thorns sea stars in the 1960s (Weber and Woodhead, 1970). The last 50 years has seen widespread mortality of scleractinians through natural and anthropogenic disturbances (e.g. Birkeland, 2015; Jackson et al., 2014), but these losses have been modest compared to the mortalities resulting from thermal bleaching driven by global warming, effects that will be amplified by ocean acidification (Hoegh-Guldberg et al., 2007). These events have transitioned many reefs into a phase functionally dominated by taxa other than scleractinians, usually macroalgae (Roff and Mumby, 2012), and some examples may represent alternative stable states (Dudgeon et al., 2010; Schmitt et al., 2019).

The functional attributes of reefs depleted of scleractinians are different from those of reefs dominated by this taxon. As ecosystem engineers (Jones et al., 1994), reduced abundances of scleractinians deprive coral reefs of their capacity to produce massive, wave resistant platforms (Stoddart, 1969), within which habitats are created for other taxa (Alvarez-Filip et al., 2009). The loss of scleractinians suggests the reef framework will succumb to net dissolution through ocean acidification (Eyre et al., 2018), thereby threatening the goods and services delivered by scleractinian corals (Birkeland, 2015). Yet the critical question is not whether coral reefs will disappear, but in what form will they persist? Against this backdrop, we evaluate the factors favouring increased abundances of octocorals in the tropical western Atlantic (Figs 1 and 2). Neglect of the taxa (e.g. octocorals) that now are dominating coral reefs deprives ecologists of the capacity to understand the attributes of present-day reef communities. This also means that there is a limited understanding of the potential for a return to a scleractinian-dominated state should the environmental conditions favouring scleractinian mortality reverse.

Fig. 1 A typical octocoral forest at ~8-m depth at Booby Rock, St. John (US Virgin Islands), in 2013. Such forests are dominated by a high density of octocorals, with some reaching >1 m height. These communities are not "new" per se (cf. Kinzie, 1973), but in the last few decades they have increased in spatial extent. The canopy created by octocoral forests forms an understory habitat that contributes to the emergent functional properties of the forest.

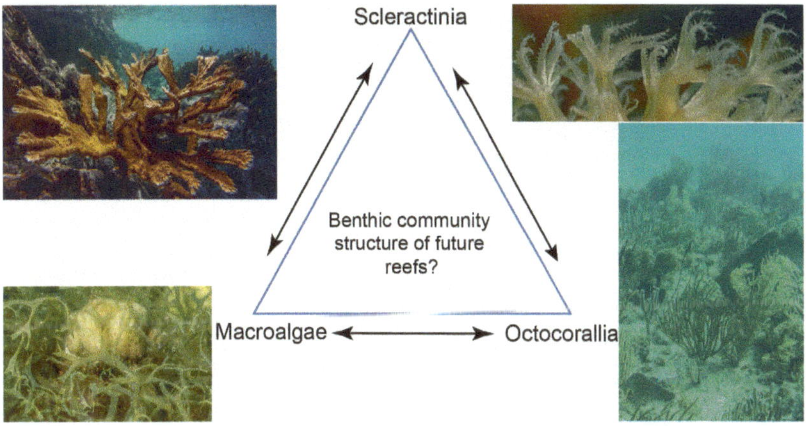

Fig. 2 Schematic illustrating the community structure of present-day reefs in the tropical western Atlantic in which scleractinians, macroalgae, and octocorals compete for dominance. Sponges can also occupy substantial portions of benthic space, but time series data for this taxon are sparse, and in St. John, they do not appear to have played a dominant role in occupying space for over 26 years. Environmental conditions and biotic processes over the last few decades have favoured rising abundances of arborescent octocorals relative to the other benthic space holders.

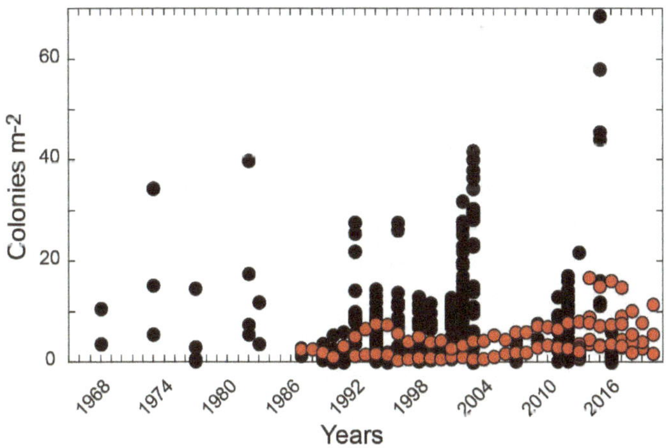

Fig. 3 Octocoral density from 1968 to 2019 in the tropical western Atlantic, as compiled from published and unpublished sources. Red shows data from St. John, black shows all other data. *Updated from Lenz, E.A., Bramanti, L., Lasker, H., Edmunds, P.J., 2015. Long-term variation of octocoral populations in St. John, US Virgin Islands. Coral Reefs 34, 1–11.*

The context for this review is created by evidence that octocorals have become more abundant in the tropical western Atlantic (Fig. 3). Ruzicka et al. (2013) were the first to describe changes in the community structure of Caribbean reefs in favour of octocorals versus scleractinians.

Although ecological analyses of octocorals have remained scarce, a data compilation for 1968–2013 supports the hypothesis that octocorals have increased in abundance (Lenz et al., 2015; Tsounis and Edmunds, 2017). Analyses of reefs in St. John from 2014 to present also support this hypothesis, as well as the notion that octocoral domination is the "new normal" for shallow Caribbean reefs (Lasker et al., 2020). Updating Lenz et al. (2015) with results published since 2013, including our analyses from St. John reveals that the trends reported in Lenz et al. (2015) have continued (Fig. 3). The data suggest that the region-wide abundance of octocorals has risen 84% since 1968–90 (versus 2015–19), and now represents a mean density of 9.2 ± 2.2 colonies m^{-2} (averaged over 2015–19, $n = 48$).

2. Historical context

Octocorals have been ecologically meaningful members of reef communities in the tropical western Atlantic for as long as researchers have described reefs. In 1915, Lewis R. Cary wrote, "On many reef areas in the Florida-Antillean region the areas occupied by stony corals and calcareous algae are relatively small while flexible alcyonaria or the family Gorgonaceae make up the most characteristic feature of the fauna" (Cary, 1915). While that characterization lacks the quantitative data of modern ecology, it is clear that flexible, arborescent octocorals, which are commonly referred to as gorgonians (Gorgonaceae is no longer a recognized taxon), have long been recognized as important on reefs throughout the tropical western Atlantic.

Tracing the origins of the Caribbean octocoral fauna is difficult. There is not much of a fossil record of octocorals because their largely proteinaceous skeletons reduce the likelihood of preservation, and their sclerites are only visible as microfossils, and being composed of high Mg calcite, they are unlikely to be preserved. In some cases, evidence of Octocorals in the fossil record is indirect, such as the presence of *Cyphoma* sp., a specialist octocoral grazer, in fossil Pleistocene reef deposits (Fig. 4), indicating octocorals have been members of Caribbean reef communities for at least 120,000 years. Direct evidence includes gorgoniid and plexaurid sclerites, which have been found in the Lower Oligocene (27–34 million years ago) of Mississippi (Giammona and Stanton, 1980). Some species have secondary calcification in their central axis, and gorgoniid holdfasts and branch fragments have been found preserved in deposits from Texas, dating from the Middle Eocene (47–41 million years ago) (Kocurko and Kocurko, 1992). Both reports of

Fig. 4 *Cyphoma* sp. found in reef deposits of a Pleistocene Reef, San Salvador, The Bahamas. The reef has been dated to 120,000 to 132,000 BP (Chen et al., 1991).

fossilized octocorals come from the Gulf of Mexico in depositional environments reflecting "hard ground" habitats that are subject to bouts of clastic deposition.

The lack of a robust fossil record relegates much of the evolutionary history of the Caribbean octocoral fauna to conjecture. It is likely that it has paralleled that of the scleractinians, with one noteworthy difference, the diversification of plexaurids. At the start of the Tertiary, the ancestral Caribbean fauna probably was part of the circumtropical Tethys Sea fauna, and was steadily isolated from the Eurasian fauna as the Atlantic grew and westward currents weakened (von der Heydt and Dijkstra, 2006). Connectivity with reefs in the Indo-Pacific, to the west, would already have been low due to both the distances and the generally westward flowing circumtropical surface currents (von der Heydt and Dijkstra, 2006), much as the Eastern Pacific is currently isolated. Over that span of time, the scleractinians were subject to several episodes of faunal turnover (Budd, 2000). If changes affecting the octocorals were similar to those affecting the scleractinians, those events would have marked the start of the development of the endemic octocoral fauna of the tropical western Atlantic. Two plexaurid genera, *Plexaura* and *Eunicea* are endemic to the Caribbean/western Atlantic and the *Eunicea* underwent extensive diversification generating the 16 currently recognized species (Cordeiro et al., 2020).

A relatively recent diversification in these genera is suggested by the low levels of sequence divergence among species in those genera compared to patterns in the gorgoniids, which have broader geographic distributions. The ecological and evolutionary consequences of diversification are complex and undoubtedly case specific (c.f., Tanentzap et al., 2020). This diversification may have laid the foundations of increased abundance of octocorals on modern reefs, but it is also possible that the specialization implied by diversification will put octocorals at greater risk of extinction as the environment continues to change.

3. Octocorals: The right taxon at the right moment

The increasing abundance of octocorals (Lenz et al., 2015; Tsounis and Edmunds, 2017; Fig. 3) has been driven by multiple factors, but the change can be characterized as resilience driven by recruitment following disturbance events in concert with resistance driven by the survival of adults. This is summarized in Fig. 5, which illustrates how mortality of scleractinians has made substratum available for octocorals. The relevant question becomes, why, unlike the scleractinians, are octocorals surviving

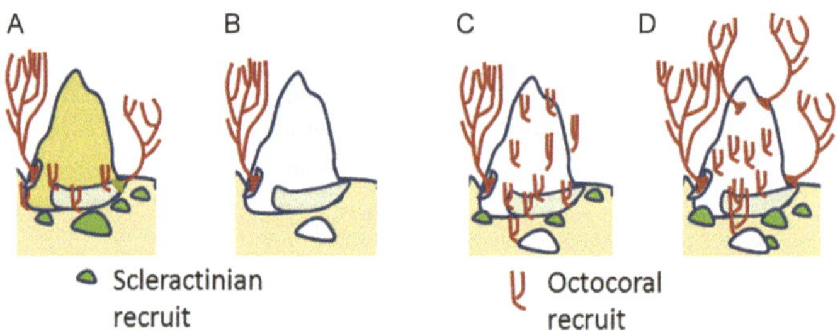

Fig. 5 Schematic illustrating hypotheses by which octocorals attain enhanced resilience following disturbance, and promoting an octocoral-dominated community ((A–D) in a chronology). (A) A coral reef before a major disturbance with a few large colonies of scleractinians and octocorals, as well as their recruits. Differences in recruit positioning reflect dissimilar settlement niches, and octocorals are constrained by their inability to settle on living scleractinians. (B) Immediately after a disturbance that has killed many scleractinians and some octocorals. (C) Octocorals and scleractinians recruit. The altered abundance and distribution of octocorals relative to scleractinians reflects differences in larval supply, and the increase in substrata matching octocoral settlement niches. (D) With time, mature octocorals dominate the benthos. Canopy effects on recruitment enhance self-seeding, which differentially benefits octocorals due to their greater abundance and high reproductive output. Macroalgae, which are not shown, are common in (B–D) and have less effect on octocorals due to the octocorals' escape in size (height).

on present-day reefs, both in the context of adult survival and ecological resilience following major disturbances? Our knowledge of the basic biology of octocorals has lagged behind that of scleractinians, and we have too often assumed that our knowledge of scleractinians carries over to the octocorals. Octocoral biology is not identical to that of scleractinians, but only a few studies have focused on the ways in which they differ. Studies of octocorals also have been limited because the species are difficult to delineate, and taxonomic identification at species level is difficult. Acknowledging these limits, we focus our review on those aspects of octocoral autecology that may predispose them for high survival on tropical western Atlantic reefs in the Anthropocene.

3.1 Trophic status

Almost all octocorals on shallow Western Atlantic reefs host symbiodinacean dinoflagellate algae, which suggests that, as for the scleractinians, photosynthate from their symbionts plays a large role in their diet. As light declines with depth, azooxanthellate species (lacking algal symbionts), become common, and when the entire tropical western Atlantic is considered, there are as many azooxanthellate as zooxanthellate species (Schubert et al., 2016). If the number of species can be interpreted as a measure of ecological success, then a heterotrophic feeding strategy has been as successful as the mixotrophic feeding of shallow water octocorals. However, in the shallow waters of Caribbean reefs, where the two modes of feeding could co-occur, purely heterotrophic reef octocorals are almost entirely absent. Within families, such as the Gorgoniidae, that are present in both shallow and deeper water, shallow water species are more commonly zooxanthellate and the Caribbean reef gorgoniids are all zooxanthellate.

While a mixotrophic carbon budget dominates the octocoral reef fauna, the dominance of heterotrophic octocorals in communities such as Mediterranean hard bottoms (Gili and Coma, 1998) and the apparent resistance of octocorals to environmental degradation in the tropical western Atlantic, has led to increasing scrutiny of the role of heterotrophy in the success of reef octocorals. Heterotrophy among octocorals ranges from the ability to take up dissolved organic matter, to the filtration of fine particulates ranging in size from bacteria to detritus, and finally, to the active capture of zooplankton (Cocito et al., 2013). The hypothesized importance of heterotrophy to the nutrition of tropical octocorals is attractive, but the evidence that heterotrophy is the key to reef octocoral success is scant.

If the capacity for heterotrophy is one of the factors facilitating the recent success of octocorals in the tropical western Atlantic, it appears unlikely to be due to the capture of zooplankton. Zooplankton feeding has been the most commonly quantified mode of heterotrophy among octocorals, primarily because ingested prey can be searched for in dissections of polyps (Lasker, 1981). Surveys of Caribbean species have found low numbers plankters within polyps (Lasker, 1981; Ribes et al., 1998; Rossi et al., 2020), supporting the hypothesis of limited feeding on zooplankton. Leversee (1976) found that *Leptogorgia virgulata* from North Carolina, which is not a typical member of the Caribbean reef fauna, captured *Artemia* nauplii in an experimental setting. Lasker (1981) made similar observations for some octocorals from reef communities, but also reports that three species of Caribbean octocoral were incapable of capturing the highly motile zooplankton that were attracted to lights at night. Capture and ingestion of artificial non-motile particulates has been demonstrated for several Caribbean species (Lasker, 1981; Lasker et al., 1983). The inability to capture larger, faster zooplankton and the ability to capture non-motile particles, and plankters that are poor swimmers, is probably related to the relatively simple cnidom present in octocorals (Fautin, 2009; Mariscal, 1974). Differences in the cnidom between octocorals and other anthozoans have not been quantified with respect to either the abundance of nematocysts or their effectiveness in securing prey items.

Quantification of in situ zooplankton capture by tropical octocorals report low capture rates. Lasker (1981) found one prey item in 100 polyps in *Briareum asbestinum* and *Pseudoplexaura porosa* branches that were collected at night. Ribes et al. (1998) examined gut content of *Pseudoplexaura porosa* and *Eunicea flexuosa* over a 24h cycle and reported 0.03–0.1 prey polyp^{-1}, which they extrapolated to a feeding rate of 0.086–0.227 prey polyp^{-1} day^{-1}. This rate corresponds to between 1 of 12 and 1 of 4 polyps capturing a prey item each day. Ribes et al. (1998) estimated those capture rates accounted for <1% of the carbon required for a colony's basal metabolism. Rossi et al. (2020) report capture rates for *Plexaurella nutans* and *Pterogorgia anceps* of 0.26 and 0.31 prey polyp^{-1} day^{-1}, respectively, and estimated heterotrophy could only account for 0.5–2.0% of colonies' metabolic needs.

The feeding rates observed among octocorals in the tropical western Atlantic are an interesting contrast to those observed in two azooxanthellate species from the Mediterranean. Rossi et al. (2004) report capture rates for *Leptogorgia sarmentosa* that can be extrapolated to 1.50 to 3.80 prey polyp^{-1} day^{-1} depending on the season. That feeding rate corresponds to 66% of a colony's metabolic demand, based on estimates of carbon content of plankters, octocoral assimilation efficiency, and octocoral respiration rates (Ribes et al., 2003; Rossi et al., 2004). Feeding rates for the azooxanthellate Mediterranean octocoral *Paramuricea clavata* are 0.12–0.80 prey polyp^{-1} day^{-1} (Coma et al., 1994). Although the capture rate is not as great as for *L. sarmentosa*, zooplankton feeding by *P. clavata* accounts for ~50% of a colony's metabolic needs. As in the Caribbean studies, the most commonly captured prey items were invertebrate larvae. Rossi et al. (2004) suggest that prey species are captured proportionate to their abundance in the water column. This suggests the differences between octocorals from the Mediterranean and the tropical western Atlantic may be more a function of the availability of plankton than inherent differences in the species that have been studied. A zooxanthellate species in the Mediterranean, *Eunicella singularis,* has capture rates of 0.33 prey polyp^{-1} day^{-1} which was estimated to be 6–30% of the colonies metabolic needs (Coma et al., 2015). That feeding rate is on the lower end of the rates observed in the azooxanthellate species and the upper end of feeding rates observed for Caribbean zooxanthellate gorgonians.

A relative inability to capture zooplankton does not preclude a large role for heterotrophy among octocorals. Uptake of dissolved organic matter, and the successful capture/filtration of fine particulate organic matter, nanoplankton, and phytoplankton may contribute to the species' trophic strategies. Octocorals would seemingly be well adapted for the uptake of small particulate matter. The presence of pinnules on the tentacles (Fig. 6) decreases the "sieve size" of the tentacular crown and increases surface area both of which can be important components of particulate capture abilities (Rubenstein and Koehl, 1977). Uptake of phytoplankton has been observed in experiments with octocorals other than Caribbean reef species. Roushdy and Hansen (1961) reported uptake of phytoplankton in laboratory experiments with the temperate alcyonacean *Alcyonium digitatum.* Fabricius et al. (1998) found that in the Red Sea the azooxanthellate alcyonacean *Dendronepthya hemprichi* was able to capture 3–20 μm phytoplankton at rates that provided carbon well beyond that required for basal metabolism (zooplankton feeding contributed <5% to the colony's metabolic needs). Tsounis et al. (2006) found that particulate organic matter was the main source of dietary carbon in *Corallium rubrum.*

Direct measures of nutritional modes for Caribbean octocorals are uncommon. Ribes et al. (1998) measured predation on pico-, nano- and

Fig. 6 Expanded polyps of *Eunicea clavigera* illustrating how the tentacles and pinnules create a complex surface for capturing particles. *Photo, H.R. Lasker.*

microplankton by *Eunicea flexuosa* within incubation chambers that were placed in situ on the reef and were supplied with reef water. Using flow cytometry to characterize changes in prey density they report uptake of dinoflagellates, pennate diatoms and ciliates, among other particles, which supplied half as much carbon as feeding on zooplankton. In sum the two modes of heterotrophy accounted for <1% of the basal metabolic rate of *E. flexuosa*. Grazing on plankters <5 μm has not been detected. Although the role of heterotrophy in the nutrient budget of *E. flexuosa* was small, Ribes et al. (1998) suggested that *E. flexuosa* alone was capable of removing ~10% of the available diatoms, dinoflagellates, and ciliates in the water column. That again suggests that heterotrophy may be limited by the availability of prey, which would not be an unexpected result for what is functionally passive suspension feeding.

Given the difficulties of quantifying uptake rates of food items such as phytoplankton, ciliates and bacteria, there has been increasing use of stable isotopes to identify the role of heterotrophy in octocoral nutrition. Ward-Paige et al. (2005) suggest that *Plexaura homomalla* and *Eunicea flexuosa* from sites with higher levels of eutrophication had $\delta^{15}N$ and $\delta^{13}C$ values indicative of greater reliance on heterotrophy. Rossi et al. (2018) present $\delta^{15}N$ of 4–8‰ and $\delta^{13}C$ of −14 to −18‰ from nine species from the Yucatan. Those values are consistent with a dominantly autotrophic system but do not preclude heterotrophy. Baker et al. (2015, their supplemental material) present $\delta^{13}C$ of host and symbiont from 11 species from coastal Panamanian reefs. Seven of those species have mean tissue $\delta^{13}C$ values that are lower than the mean symbiont $\delta^{13}C$ values, which is consistent with a signature of heterotrophy. However, the differences between host and symbiont $\delta^{13}C$ are <1‰. While indicative of multiple food sources quantification of the relative roles of heterotrophy and autotrophy from stable isotope data is difficult. The relationships between trophic status and stable isotope concentrations are complex in symbiotic systems where compounds are recycled within the holobiont (Ferrier-Pages and Leal, 2019).

An interesting speculation is that the relative ineffectiveness of heterotrophic feeding among Caribbean octocorals is related to the low productivity of tropical seawater. However, *D. hemprichi*, which is azooxanthellate, effectively captured phytoplankton on a Red Sea coral reef (Fabricius et al., 1998). An inverse association between heterotrophic capabilities and symbiosis with Symbiodiniaceae is suggested from the more productive waters of the Mediterranean. Zooplankton capture by *Eunicella singularis*, the one zooxanthellate species in the fauna, accounts for less of its metabolic demands than the two azooxanthellate species that have been studied. In a study of scleractinians, Grottoli et al. (2006) reported increased feeding among bleached colonies of *Montipora capitata* compared to unbleached colonies. Among the octocorals it would be interesting to determine the extent to which differences in behaviour, or perhaps the cnidom, can explain different feeding capabilities and to determine the extent to which those characteristics exhibit plasticity.

Direct measures of the contribution of autotrophy to metabolic needs of Caribbean octocorals are more available than measures of heterotrophy. Rossi et al. (2018) measured P:R ratios that were >1.0 for all 10 species investigated, and while Rossi et al. (2020) found seasonal differences in both metabolic rate and gross photosynthetic rates, P:R ratios remained ≥ 1 throughout the year. Baker et al. (2015) report P:R ratios ranging from of 0.7–1.6. These studies acknowledge some likely heterotrophic inputs to the carbon budget of these organisms, but based on P:R ratios determined in oxygen units, much of the daily nutrition of the octocorals could have been met through the carbon fixed and translocated from algal symbionts.

A hypothesized relationship between the ecological success of octocorals in the tropical western Atlantic and heterotrophy remains to be experimentally tested. A complete evaluation will require assessments of heterotrophy that go beyond measures of zooplankton capture to consider the input of resources obtained from the capture of phytoplankton. Ribes et al. (1998), for example, report that the capture of phytoplankton contributed <1% of the carbon necessary to meet the respiratory demand of *Eunicea flexuosa*. Studies are needed that evaluate carbon (or energy) sources and sinks comprehensively, in compatible units, and under ecologically relevant conditions that capture the complex flow regimes under which arborescent octocorals function. Stable isotope analyses have promise in assessing trophic strategies of octocorals, and have the benefit over other methods of being suitable for community-scale surveys and integrating the effects of nutritional strategies over relevant time-scales of weeks to months. However, the signature of consuming phytoplankton may be very similar to that of photosynthate translocated from Symbiondiniacaea, and it will be necessary to develop tools that can discriminate the carbon isotopic signatures of carbon translocated from symbiodiniacaea versus the capture of phytoplankton.

3.2 Growth form and growth rate

The signature feature of all but two of the octocorals found on Caribbean coral reefs is their arborescent colony form. All but *Erythropodium caribbeaorum* and the encrusting form of *Briareum asbestinum* (historically reported as *B. polyanthes)* exploit the water column from a diminutive site of basal attachment (encrusting *B. asbestinum* also can take on an upright aspect as it often grows on exposed axes of other octocoral species). A vertical morphological strategy removes the colony from a variety of the vicissitudes of life on the substratum, elevating colonies above processes such as scour by sediments, shading by macroalgae, and overgrowth by other benthos.

Examination of the substratum on a coral reef quickly reveals that there is very little "open" space, and bare carbonate rock is quickly colonized by an array of algae and encrusting invertebrates. A variety of sessile invertebrates settle on crustose coralline algae, but in general, potential competitors settle onto the substratum among, but not on, already established benthos, and then interact with their neighbours. The outcomes of those interactions are determined by a variety of both direct and indirect processes (Chadwick and Morrow, 2011). Those potentially harmful interactions include the direct effects of overgrowth and killing of the competitor's tissue and indirect effects such as reducing access to resources (i.e. shading or altering the boundary layer from which nutrients are extracted) or altering the microenvironment [c.f., anoxic conditions under dense algal mats (Wangpraseurt et al., 2012)].

Although coral reefs are complex 3-D structures, overgrowth is primarily a 2-D process, and elevating most of the gorgonian colony above the substratum minimizes the interactions with competitors. Many of the potential competitors of gorgonians can grow over their holdfasts (e.g. Edmunds et al., 2019), but they tend not to grow up the length of the primary branch, and overgrowth of the base on its own will not necessarily weaken a colony. Building a rigid structure on a flexible object also may be a limiting factor for some competitors. Those structural constraints would seemingly be less important for potential sponge competitors. Species such as *Desmapsamma anchorata* can be found on Caribbean gorgonians, but probably settle on damaged areas on the colony where the axis is exposed (McLean and Yoshioka, 2008). Massive sponges such as *Xestospongia* spp. will "envelope" gorgonians as they expand (McLean and Yoshioka, 2008), but Caribbean sponges typically do not spread along the substratum and up the base of octocorals. Some Caribbean reef sponges bore into the carbonate substratum, and the gorgonin core of gorgonians would not be a suitable habitat for these species. Finally, peyssonellid algal crusts (PAC, calcifying red algae), notably those including *Ramicrusta* spp., recently have undergone rapid growth on some Caribbean reefs (Bramanti et al., 2017; Eckrich and Engel, 2013). PAC is capable of overgrowing a variety of benthic taxa, but it does not seem to grow up beyond the bases of octocorals (Edmunds et al., 2019). The importance of vertical growth by octocorals may be accentuated on present-day reefs, where the proliferation of turf and macroalgae over the past four decades (Hughes, 1994; Roff and Mumby, 2012) has, at the least, altered the nature of boundary layer over the substratum (c.f., Stocking et al., 2016; Wangpraseurt et al., 2012), thus placing strong selective value on the ability to project above this layer. In some cases, however, vertical branching growth could be counterproductive, such as in the case of the *Lyngbya* bloom in SE Florida, elements of which were transported by seawater current and killed gorgonians by entanglement and subsequent suffocation (Paul et al., 2005).

An arborescent morphology is advantageous for colonies that are decimeters tall, but all sexually produced colonies begin their benthic life as a single polyp. Per capita risk of mortality decreases with increasing size among Caribbean octocorals (Gotelli, 1991; Lasker, 1990; Yoshioka, 1994) and octocorals in general (Bramanti et al., 2009; Garrabou and Harmelin, 2002; Linares et al., 2007; Santangelo et al., 2007). Data on survival of recently settled recruits are limited. Lasker et al. (1998) settled *Plexaura kuna* larvae onto acrylic plates, which were then returned to the reef and survival tracked. Overall mortality was 0.03 day^{-1}, but survival was substantially greater on the most protected area of the plate. Evans et al. (2013) found survival of *B. asbestinum* on similar plates to be 8% after 10 days on the reef. Following polyps that had naturally recruited to ceramic tiles, Wells et al. (2020) observed mortality of 0.024 day^{-1}. Extrapolating those data to an annual rate yields survival rates that range from 0.000018 to 0.048. A number of researchers have characterized survival among small octocoral colonies (Table 1). Those values are all higher than for survival of recently settled polyps, which suggests that recruits that rapidly increase in size can lower their mortality rate, with rapid growth to ~10 cm height further lowering the risk of mortality.

Growth rates of octocorals are markedly variable both within and between species, but all illustrate the high degree to which octocorals exploit vertical growth to escape the risks of mortality through increases in size. Species-specific growth rates among Caribbean gorgonians range from 23.1 cm $year^{-1}$ for *Antillogorgia acerosa* colonies from Colombia (Cadena and Sanchez, 2010) to 0.8 cm $year^{-1}$ among *Plexaurella dichotoma* colonies

Table 1 Annual survival of octocoral colonies.

Species	Locations	Size class (cm)	Survival	Source
Leptogorgia virgulata	Gulf of Mexico	0.5–4.0 4.5–8.0 8.5–12.0	0.780 0.880 0.964	Gotelli (1991)
Plexaura kuna	Caribbean—Panama	<10 10–19.9	0.49 0.82	Lasker (1990)
Paramuricea clavata	Mediterranean—Spain, two sites	0.3–3.0 3.0–10.0 11–20	0.667–0.769 0.636–0.768 0.770–0.962	Linares et al. (2007)
Antillogorgia spp.	Caribbean—Puerto Rico	<6 6–10 11–15	0.768 0.893 0.939	Yoshika and Yoshioka (1991)

on the southwest coast of Puerto Rico (Yoshika and Yoshioka, 1991). The range is even greater as most reported values are averages. There also are location/habitat differences as suggested by the fivefold difference in observed growth rates of *A. acerosa* from Colombia and Puerto Rico. All of the studies that have characterized growth also note many cases of negative growth (i.e. partial mortality) presumably caused by some combination of predation, disease, and physical disturbance. For instance, average growth of *Plexaura kuna* is 2–4 cm year^{-1} lower than the potential growth suggested by growth among colonies that had net positive growth (Table 2). The highest growth rates are found among species with thin branches such as the *Antillogorgia* spp. (Borgstein et al., 2020). Among those species, recruits can escape the highest mortality section of the life table in 1–2 years and reach 10 cm in 3–4 years. Given the variance associated with the averages and the possibility of negative growth, many colonies will take longer to reach those milestones, while some will do so in substantially less time. In contrast to the octocorals, scleractinians have lower growth rates. Edmunds (2018) reported growth in colony diameter among small, <4 cm diameter scleractinians on St. John, Virgin Islands, ranging from −3.45 to 5.55 cm year^{-1}, but average growth was 0.29 cm year^{-1}. Edmunds (2018) noted that the data for scleractinians did not support an expectation of rapid growth that could provide a refuge in size. Growth at the colony edge does not elevate colonies above the substratum, and most scleractinians expand their base before substantial vertical growth occurs.

3.3 Reproductive biology

Reproductive strategies among Caribbean octocorals are not dramatically different from those observed in Caribbean scleractinians (as described in Carlon, 1999; Szmant, 1986). The one difference is the occurrence of surface brooding, in which eggs are released but remain on the branch surface until they develop (Benayahu and Loya, 1983). With respect to fertilization and then subsequent dispersal, the strategy is akin to internal brooding. The four combinations of sexuality and mode of spawning are almost equally common among the scleractinians (Baird et al., 2011), but all of the Caribbean reef gorgonians are gonochoric (Kahng et al., 2011) and the majority are broadcast spawners. The mode of reproduction appears to correlate with biogeographic patterns in the Western Atlantic. Species richness is relatively uniform throughout the Caribbean with differences between areas reflecting combinations of the habitats studied and the intensity of surveys. However, on the reefs of Bermuda, the most isolated of Western

Table 2 Growth rates of Caribbean gorgonians.

Species	Location(s)	Colony size	Growth (cm/year)	Reference
Antillogorgia acerosa	Colombia, Puerto Rico		23.05, 2.12, 4.48, 5.17	Cadena and Sanchez (2010), Yoshika and Yoshioka (1991) and Borgstein et al. (2020)
Antillogorgia americana	Puerto Rico (three sites)		3.44, 4.48, 5.08	Yoshika and Yoshioka (1991) and Borgstein et al. (2020)
Antillogorgia elisabethae	The Bahamas	<10 cm	3.73	Lasker et al. (2003)
		10–19.9 cm	4.54	
		20–29.9 cm	3.87	
Briareum asbestinum	Panama, Puerto Rico		5.93, 3.01	Brazeau and Lasker (1992) and Borgstein et al. (2020)
Eunicea calyculata	Puerto Rico		1.98	Borgstein et al. (2020)
Eunicea flexuosa	Puerto Rico		2.15, 1.77, 2.98	Yoshika and Yoshioka (1991) and Borgstein et al. (2020)
Eunicea laciniata	Puerto Rico		3.01	Borgstein et al. (2020)
Eunicea laxispica	Puerto Rico		2.21, 3.49	Yoshika and Yoshioka (1991) and Borgstein et al. (2020)
Eunicea succinea	Puerto Rico		1.36	Borgstein et al. (2020)
Eunicea tourneforti	Puerto Rico		2.06, 2.76	Yoshika and Yoshioka (1991) and Borgstein et al. (2020)
Gorgonia ventalina	Puerto Rico		3.74	Borgstein et al. (2020)
Muricea elongata	Puerto Rico		3.71	Borgstein et al. (2020)
Muriceopsis flavida	Puerto Rico		1.85, 4.52	Yoshika and Yoshioka (1991) and Borgstein et al. (2020)

P. wagenaari	Puerto Rico		2.57, 2.13	Yoshika and Yoshioka (1991)
Plexaura homomalla	Puerto Rico		1.99, 3.71	Yoshika and Yoshioka (1991) and Borgstein et al. (2020)
Plexaura kukenthali	Puerto Rico		1.18	Yoshika and Yoshioka (1991)
Plexaura kuna	Panama	<10	3.9	Lasker (1990)
		10–19.9	3.49	
		>20 cm	3.77	
Plexaurella dichotoma	Puerto Rico		0.8	Yoshika and Yoshioka (1991)
Plexaurella spp.	Puerto Rico		2.63	Borgstein et al. (2020)
Pseudoplexaura porosa	Puerto Rico		1.98, 2.22	Yoshioka and Yoshioka (1991)
Pseudoplexaura spp.	Puerto Rico		3.58	Borgstein et al. (2020)

Growth rates from Borgstein et al. (2020) were taken from their fig. 1 and are median values. All other rates are averages. Calculations excluded cases of negative growth. Unless indicated growth rates are for colonies >10 cm height.

Table 3 Number of octocoral species found in Bermuda compared to the Caribbean and Florida, with respect to reproductive mode.

Reproductive mode	Caribbean	Bermuda
Brooder and external brooder	11	2
Spawner	12	11

Totals are limited by the number of species for which reproductive mode and sexuality are known.

Atlantic reef complexes, brooding and surface brooding octocorals make up a smaller proportion of the fauna than in the Caribbean and Florida (Table 3, Test of independence, $G=4.09$, $P=0.04$).

Egg size might also affect the recruitment of octocorals and scleractinians. Large egg size implies greater parental provisioning, which can then affect some combination of longevity in the water column (i.e. greater dispersal), size at metamorphosis and recruit survival (Hartmann et al., 2013; Vance, 1973; but see Bernardo, 1996). Our knowledge of gamete traits for octocorals is incomplete, but Caribbean octocorals for which there are data (six species, Kahng et al., 2011) have mean oocyte diameters of 722 μm, and produce 2.6 mature oocytes polyp^{-1}. In contrast, the eggs of Caribbean scleractinians have a mean diameter of 317 μm and produce on average 130 oocytes polyp^{-1} (data from Szmant, 1986). Genetic analysis of *A. elisabethae,* a surface brooder, suggests modal dispersal distances of 100 m to 1 km in The Bahamas (Lasker and Porto-Hannes, 2015). In biophysical model simulations of the dispersal of *A. americana*, which is a broadcast spawner, Coelho (2018) reports a median dispersal distance of 39 km with more settlement within 1 km than any other distance. If these congeners are representative of the Western Atlantic reef fauna, octocorals generally have modest dispersal distances, even the widely distributed *A. americana*.

We have relatively limited knowledge on how reproduction and dispersal of octocorals translates into settlement, but there are data on octocoral recruitment at several Caribbean sites. Recruitment, the first appearance of an individual in a population (Caley et al., 1996), is operationally defined in manners that often are unique to individual studies. Recruitment in annual surveys of octocorals has been quantified as the appearance of small colonies whose height suggests they are a year or less old. Heights from ≤3 cm (Lasker, 1990) to ≤6 cm (Yoshioka, 1997) have been used. As discussed above, the height of a 1-year-old recruit can vary considerably both within and especially among species. Thus, any height-based definition of a recruit will err by excluding fast growing recruits and by including colonies that are over a year old. As importantly, because recruits that have died do not leave a skeleton behind, recruitment based on annual or even monthly surveys will miss the settlement of individuals that subsequently died.

We compiled published, octocoral recruitment data to address two questions: (1) is there evidence of changing rates of octocoral recruitment? and (2) are there differences in rates of recruitment between octocorals and scleractinians? The former addresses the issue of whether changing conditions on reefs have altered any of the processes controlling successful recruitment, and the latter provides insight into whether the current differences in the demographics of octocorals and scleractinians might occur in the early life history stages that control recruitment. Our data compilation yielded 18 studies that together provided 198 records of the density of octocoral recruits. Most data were for individuals 3–6 cm tall, which probably are about 1 year in age, although a handful of records of polyps settling on tiles were obtained. Most records were for octocorals pooled among taxa, but substantial records also were obtained for *Antillogorgia* spp. ($n = 63$, mostly *A. elisabethae, A. acerosa, A. americana* and *A. rigida*), plexaurids ($n = 40$, mostly *P. kuna*), and *Pseudoplexaura* spp. ($n = 11$). To provide a comparative context to these data, we included 13 years of scleractinian recruitment to settlement tiles deployed at ~5-m depth in St. John, and a decade of density data for juvenile scleractinians on natural reef surfaces (Edmunds, 2018, and unpublished data). Densities of octocoral recruits in these annual surveys ranged from 0 to 84 colonies m^{-2}, and the highest recorded density was from 1985 on the south shore of Puerto Rico (Yoshioka, 1996). These data substantially underestimate settlement. Surveys that were conducted 1–2 months after spawning over 4 years found recruitment at sites in the Bahamas for *Antillogorgia* spp. that ranged from 3 to 231 recruits m^{-2} (Lasker, unpublished data). Annual versus monthly surveys for *Plexaura kuna* recruits in the San Blas Islands, Panama, generated recruitment rates of 0.37 individuals m^{-2} (annual) and 1.0 individuals m^{-2} (monthly). Finally, surveys of tiles deployed on the south shore of St John, Virgin Islands, that were conducted at 3-day intervals found recruitment over a 19-day period of 443 individuals m^{-2} (Wells et al., 2020). Although they are an incomplete characterization of the processes controlling recruitment, the data address the presence of temporal patterns. Overall, octocoral recruitment was unrelated to time from 1962 to 2019 ($r = -0.06$, df$=196$, $P = 0.40$).

From 1962 to 2019, octocoral recruitment has remained lower than the median recruitment of scleractinians in St. John from 2007 to 2019 (126 corals m^{-2}) (Figs 7 and 8). The scleractinian recruitment data probably include many individuals that settled and did not survive until detection at the annual sampling. The few data on octocorals that provide more

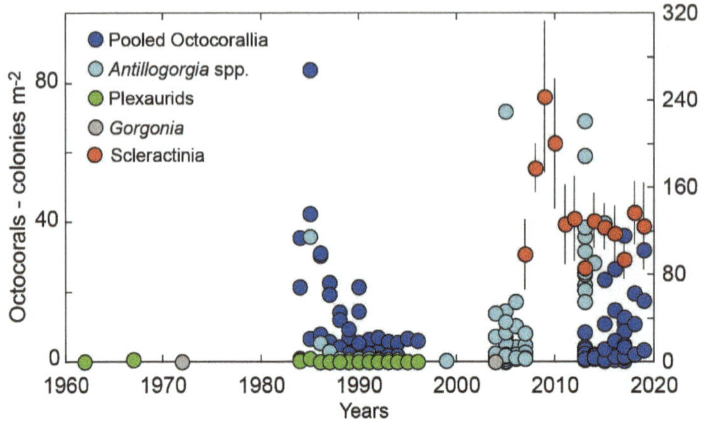

Fig. 7 Density of recruits of octocorals (colonies 3–6 cm tall) and scleractinians (recruits on tiles at 5-m depth in St. John, mean ± SE) from 1962 to 2019. Over this period, octocoral recruitment statistically has not changed.

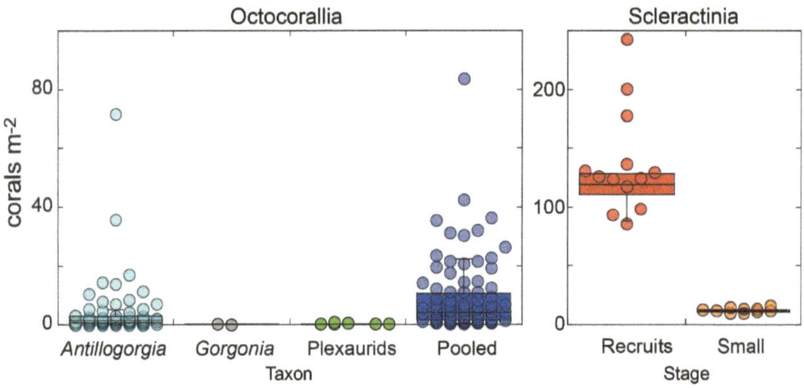

Fig. 8 Box plots displaying a comparison of density of octocoral recruits (colonies 3–6 cm tall) and recruits (corals on settlement tiles) and small colonies (≤4-cm diameter) of scleractinians. Scleractinian data come from St. John, and refer to the use of settlement tiles at ~5-m depth, or surveys of natural reef surfaces (at 5- to 9-m depth).

effective summaries of settlement than the annual surveys, suggest that the disparity between octocoral and scleractinian recruitment may not be as great as suggested in Figs 7 and 8. The octocoral annual recruitment data are comparable to numbers of living ≤4-cm diameter scleractinian data. These data suggest that scleractinians have greater, or perhaps similar recruitment as the octocorals, and thus the failure of scleractinian populations to recover may be more related to their post-settlement survival than to settlement.

3.4 Bleaching

Over the last four decades bleaching has been responsible for widespread scleractinian mortality (Hoegh-Guldberg et al., 2007; Hughes et al., 2017; Oliver et al., 2018), but the effects on octocorals may be less severe. Bleaching, the paling of tissue as chlorophyll and algal cells are lost from the holobiont (Lesser, 2011) is elicited by a variety of environmental conditions (Gates, 1990; Glynn, 1993; van Oppen and Lough, 2018), the best known of which is elevated seawater temperature (Brown, 1997; Lough and van Oppen, 2018). Although bleached octocorals were present in some of the first reports of Caribbean bleaching (Lasker et al., 1984), the notion has persisted that octocorals are more resistant than scleractinians to warm water bleaching (Goulet et al., 2017; Prada et al., 2010). However, an explicit test of this hypothesis has yet to be conducted. It is unclear whether octocorals are less susceptible than scleractinians, or whether bleaching has been more conspicuous on scleractinian-dominated reefs and thus more commonly reported.

Lasker et al. (1984) reported limited bleaching of octocorals during the 1983 Caribbean bleaching event, but Hannes et al. (2009) and Prada et al. (2010) report more extensive bleaching of octocorals in the 2005 bleaching events in the Florida Keys and southwest coast of Puerto Rico. In those studies, *Plexaurella, Pterogorgia, Muricea and Briareum* were the most common species to bleach. Mortality has been observed among *Plexaurella* (M.A. Coffroth, Personal communication) and significant mortality has been reported for *Muricea* (Prada et al., 2010). Reliance on autotrophy does not appear to explain sensitivity to bleaching. For instance, two *Plexaurella* spp. and *Briareum asbestinum* are among the small number of taxa that bleach and they have the lowest P:R ratios reported in the Rossi et al. (2018) study.

The relative success of octocorals and the variance among octocoral species during scleractinian coral bleaching events may be associated with the Symbiodiniaceae species that they host. *B. asbestinum*, which harbours a *Breviolium* sp. as an adult (B178 in the nomenclature of Lewis and Coffroth, 2004), is one of the few octocorals that visibly bleaches (Goulet and Coffroth, 2004; Lasker et al., 1984). Poland (2010) observed a loss of B178 from *B. asbestinum* under heat-induced stress bleaching, and Lewis and Coffroth (2004) reported a loss of B178 from *B. asbestinum* under dark-induced stress bleaching. Poland and Coffroth (2019) suggested that colonies with this symbiont type are more susceptible to bleaching in a perturbed environment. Hosting thermal tolerant symbionts, which may be the more general case among octocorals, may be a key aspect of their continued success on Caribbean reefs.

3.5 Disease

Octocorals are affected by a diversity of diseases (Kim, 2016; Weil et al., 2016), many similar in aetiology to the numerous diseases affecting scleractinians (Weil and Rogers, 2011). Like most aspects of octocoral biology, diseases of this taxon have been less well studied than in scleractinians, although *Aspergillus* and its effects on *Gorgonia* is an exception (Geiser et al., 1998; Smith et al., 1996). While population sizes of *G. ventalina* throughout the Caribbean were depressed by this disease, there were signs of acquired resistance which ultimately attenuated further population decline due to Aspergillosis (Bruno et al., 2011). The analysis of this disease illustrates the potential threats to octocorals of emerging diseases (Weil and Rogers, 2011; Weil et al., 2016). However, while numerous diseases of octocorals in the tropical western Atlantic have been documented, with the exception of Aspergillosis, they have not had the impacts on populations and communities that diseases such as white band disease (Aronson and Precht, 2001) and most recently stony coral tissue loss disease (Muller et al., 2020) have had on scleractinians. Octocorals have robust immune systems (van de Water et al., 2018) and it is tempting to think they may be less susceptible to epizootics than scleractinians. However, octocorals have undergone catastrophic losses due to disease in the Mediterranean (Coma et al., 2004, 2006; Linares et al., 2005). Furthermore, disease incidence in octocorals as for scleractinians appears to increase with temperature (van de Water et al., 2018). In a world with increasing likelihood of disease outbreaks (Maynard et al., 2015) there is no certainty that octocorals will not also be subject to more significant disease outbreaks.

3.6 Ocean acidification

Coral reef biology has been transformed by an appreciation of the threats posed to calcareous taxa by ocean acidification (OA) (Chan and Connolly, 2012; Kroeker et al., 2013), which is driven by rising atmospheric CO_2 and its effects on seawater chemistry. The effects of these trends on scleractinian corals have been extensively studied to reveal inhibitory effects on net calcification (Chan and Connolly, 2012; Comeau et al., 2014; Kroeker et al., 2013). The effects of OA, particularly the consequences of exposure to low pH and elevated pCO_2, have diverse implications for scleractinians, including gamete mobility, fertilization success (Albright, 2011), recruitment (Albright et al., 2010; Doropoulos et al., 2012), and abundance (Kroeker et al., 2013). Since octocorals do not produce a massive calcareous skeleton, it is reasonable to hypothesize that they will be less affected by OA than scleractinians. However, while octocorals are "soft" they contain calcified sclerites that have important functional implications for skeletal strength and flexibility (Bayer, 1961). There are signs that the capacity to deposit these structures is impaired under OA (Bramanti et al., 2013; Gomez et al., 2015), although octocoral tissue appears to reduce or prevent dissolution of sclerites under low pH (7.3 and 7.8) (Gabay et al., 2014).

The trend for increasing abundance of octocorals in the tropical western Atlantic could reflect the appearance of communities composed of new combinations of taxa reflecting prevailing physical and chemical conditions, but comparison of present-day octocoral communities in St. John (Tsounis and Edmunds, 2017) with older studies from Jamaica (Kinzie, 1973; Wahle, 1980) and Puerto Rico (Yoshika and Yoshioka, 1991; Yoshioka and Yoshioka, 1987, and others) reveal broad similarities between historic and extant communities. Current evidence therefore favours a "habitat expansion" of historic octocoral communities rather than the development of a "novel community" per se.

4. Resilience and resistance of modern octocoral communities

In Section 3 we highlighted aspects of the biology of octocorals that have served them well in mediating recent increases in abundance. In this section we address features of their ecology that may be critical to determining whether they will persist or are simply a placeholder community in an ecological succession driven by the continuing anthropogenic alteration of tropical reef environments.

4.1 Disturbance, resilience and the octocoral forest

To the extent that octocorals have increased in abundance on Caribbean reefs (Lasker et al., 2020; Lenz et al., 2015; Ruzicka et al., 2013), it is reasonable to now examine their resilience. The lack of analyses of octocoral communities with high taxonomic resolution and decadal resolution impedes the capacity to test for community resilience, but in those cases where appropriate information is available, the results suggest octocorals, unlike scleractinians, have exhibited resilience. St. John octocorals communities, which were analysed with genus resolution, quickly recovered from hurricanes and bleaching from 1987 to 2013 (Tsounis and Edmunds, 2017)

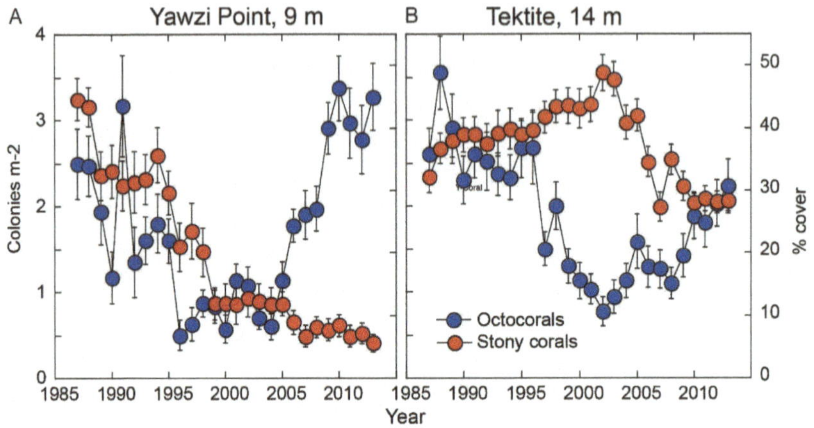

Fig. 9 Abundance (mean ± SE) of octocorals and cover of stony corals on two reefs in St. John, US Virgin Islands. Population sizes shown from 1987 to 2013 and the plots contrast resilience for octocorals (high resilience, blue) versus stony corals (low resilience, red). *After Tsounis, G., Edmunds, P.J., 2017. Three decades of coral reef community dynamics in St. John, USVI: a contrast of scleractinians and octocorals. Ecosphere 8.*

(Fig. 9). Where species resolution has been applied, results from 2014 to 2019 reveal resilience in the face of the unprecedented Category 5 hurricanes in 2017 (Lasker et al., 2020).

The origins of that resilience are largely the traits that have enabled their survival under steady-state conditions. First, as described above, the morphological strategy and rapid linear growth of arborescent octocorals provides a means by which recruits can rapidly grow to adult size and avoid mortality on the substratum where they can be smothered by macroalgae. Second, and in contrast to scleractinians, octocoral recruitment has enabled recovery from events such as hurricanes (Lasker et al., 2020). Third, while octocorals are not immune to bleaching and disease, they have been less affected by these disturbances than scleractinians. Fourth, historic and recent surveys of octocoral communities reveal high species diversity. Recent research is beginning to reveal the extent to which octocoral species differ from one another with respect to their nutritional strategies, structure, and symbiont performance (Rossi et al., 2018, 2020). It is interesting to speculate that the co-occurrence of morphologically similar species results in a high degree of functional redundancy (Rosenfeld, 2002), which could contribute to high community resilience.

Finally, we need to consider the degree to which octocoral communities and octocoral forests (sensu Rossi et al., 2017) modify their environment, and thus their interactions with the physical environment and with each other. It is this last component that we focus on in this section. A terrestrial forest biome classically is defined as a complex ecosystem consisting mainly of trees, which are the ecosystem engineers of the community (sensu Jones et al., 1994). The central characteristic of a forest is the emergent property of the three-dimensional structure, which supports biodiversity and provides shelter and food for numerous species of plants and animals (Sasaki and Putz, 2009). Physically, a forest can be divided into several zones, among them the distal canopy composed of leaves and branches, and the understory community that operates in physical conditions that differ from those prevailing outside the forest. On tropical reefs, scleractinians, octocorals, and hydrocorals, are the "trees", the ecosystem engineers of the community (Rossi et al., 2017) (Figs 10 and 11). Like terrestrial forests, octocoral forests generate 3-D frameworks with high structural complexity. While lacking basal roots and the capacity for basal-distal internal transport, dense populations of arborescent octocorals share many functional similarities with terrestrial forests. Viewing octocoral communities in this manner identifies theoretical and empirical approaches that may be important in understanding the processes responsible for structuring and maintaining octocoral forests.

Fig. 10 Long-lived organisms with three-dimensional structures that function as ecosystem engineers. (A) Trees and (B) arborescent octocorals on the reefs of St. John.

Fig. 11 Branching morphology. (A) *Muriceopsis* sp. (B) *Pinus* sp.

5. Canopies formed by dense stands of octocorals

Structure-forming marine organisms modify, create, and maintain habitat and their presence can define entire ecosystems, including, for example, kelp forests, mangroves and coral reefs (Bruno and Bertness, 2001). Arborescent octocorals that form dense canopies modify the physical and chemical environmental conditions and can also be considered ecosystem engineers.

5.1 Effects on seawater currents

Habitat modifications by erect canopies include, for example, alterations of the mean turbulent flow of seawater, affecting sedimentation and propagule retention (Folkard, 2019; Reidenbach et al., 2007). When the density of octocorals is high enough, colonies produce wakes that affect the flow of seawater itself, with strong spatial variations of the flow (i.e. turbulence) over a region known as the roughness sublayer (RSL) (Guizien and Ghisalberti, 2016). In a spatially homogeneous canopy, two sections can be distinguished within the RSL: a lower part (i.e. within the canopy) and an upper part, which extends ~two–three times the height of the canopy above the seafloor (Macdonald, 2000).

Changes in flow over and through an octocoral canopy should also affect the physico-chemical environment beneath the canopy and conditions within the canopy are likely to differ from those outside of the canopy (Umanzor et al., 2018), thereby creating a unique in-canopy microclimate. Inside a dense octocoral canopy, the modification of the flow is associated with changes in the sedimentation regime. Both deposition and resuspension are expected to be affected by higher turbulence in the upper RSL, and the lower flow velocity beneath the canopy (Reidenbach et al., 2007). In fact, in the water column directly above the canopy the downward vector of flow (upper RSL; Fig. 12), should generate the transport of dissolved nutrients and particulate material into the canopy, thus increasing deposition. This mass transfer could also increase capture rates by octocoral polyps. The slow passage of water in the lower RSL (beneath canopy; Fig. 12), can lead to an accumulation of the products released by understory species (e.g. DOC) and to a depletion of the nutrients consumed by those species (e.g. oxygen). For instance, Hauri et al. (2010) showed that the reduction in oxygen and the increase in DOC concentration below dense canopies of macroalgae, negatively affected the understory corals.

The effect of the canopy on the water flow within octocoral forests on St. John has been examined by measuring the current speed across the bottom boundary layer at different octocoral densities and over a vertical profile from the seafloor to 3 m above the seafloor. Measures with an Acoustic Doppler Current Profiler (ADCP) provide high spatial resolution to distinguish current profiles at 3 cm intervals. The octocoral canopy affected the RSL at densities >7 colonies m^{-2}, suggesting a threshold in colony density above which the canopy has a hydrodynamic effect, which is expressed as a decay in the exponential flow velocity in the lower portion of the RSL

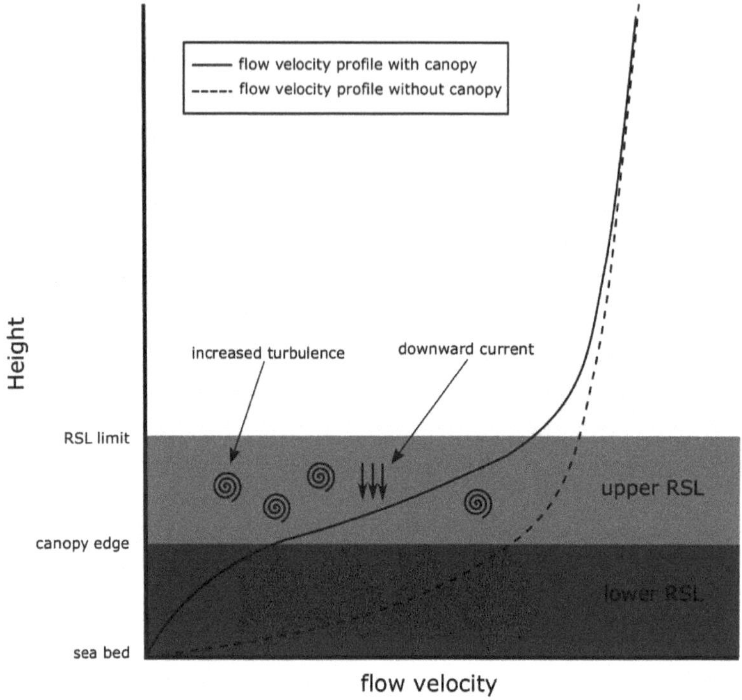

Fig. 12 Schematic showing the development of the roughness sublayer (RSL) as a function of the canopy formed by octocorals.

(i.e. within the canopy) (Fig. 12). In the upper portion of the RSL, enhanced turbulence extends above the canopy to 130% of the total height of the canopy, thus creating a downward vector of flow (in the Z axis) (L. Bramanti, Unpublished data). The turbulence structure within the canopy affects resuspension and sedimentation (Cerpovicz and Lasker, unpublished data), and the increased turbulence over the RSL may modify the supply of waterborne resources (e.g. zooplankton and nutrients) to the octocoral colonies beneath the canopy as well as the understory community. This effect would affect resource competition between community members with an advantage for the species in the higher portion of the canopy which would experience higher supply of waterborne resources generating a trophic shadow (sensu Porter, 1976) over the understory community, which would experience a reduced particle flux. The potential importance of resource competition is suggested by Kim and Lasker (1997) who showed that *Plexaura homomalla* branches in the center of a linear array of branches had lower growth rates than the exterior branches.

5.2 Shading

The shading of light by closely-spaced octocoral colonies may also influence autotrophy among neighbouring colonies as well as members of the understory community. Reduction of photon flux density (PFD) beneath octocoral colonies, which is inversely proportional to the distance from the colony holdfast, has been observed on the shallow (<9 m depth) reefs of St. John (Nelson, 2017). PFD inside and outside gorgonian canopies at a density of 8 colonies m^{-2} was on average ~40% lower than outside the canopy.

Additionally, wave-driven movement of gorgonian colonies can generate sunflecks on the seafloor that affect understory organisms (Iluz et al., 2012). In situ measurements found that in-canopy PFD regimes were more variable than outside the canopy when averaged over 6–10 s intervals, suggesting the presence of sun flecks (J. Girard, Unpublished data). However, on the shallow reef of St. John, the community structure of the understory algae and invertebrates was not associated with the presence of the octocoral canopy, or the perturbed light regime it created (J. Girard, Unpublished data).

5.3 Self-thinning

Competition would seem to be an inevitable consequence of the presence of a canopy in which colonies can restrict the access of adjacent colonies to light and nutrients (Kim and Lasker, 1997), or directly interfere with each other (Gambrel and Lasker, 2016). Sessile organisms cannot escape those effects, and growth will be affected by the density of potential competitors. In plant communities this is manifested as the self-thinning "rule" (Tadaki and Shidei, 1959; Yoda et al., 1963). Self-thinning refers to a reduction in the number of surviving individuals in a cohort as those individuals increase in size, presumably due to competition for limiting resources, such as food or space (Frechette and Lefaivre, 1990). As a result, there is an inverse relationship between the average size of individuals (i.e. biomass) and population density (Westoby, 1981; White et al., 2007). In recent years, support for the self-thinning rule has come from several coral populations. Linares et al. (2008) found that self-thinning applies to the Mediterranean gorgonian *Paramuricea clavata*, but not to *Eunicella singularis*. The applicability of self-thinning to the precious Mediterranean red coral, *Corallium rubrum*, was rejected by Garrabou et al. (2017) based on a study conducted at ≤30-m depth and focusing on colonies ≤9-cm tall. However, when this species was studied over a wider depth range (i.e. ≤155 m) and included colonies as big as 35-cm tall, self-thinning emerged as a potential mechanism driving differences between the shallow- and deep-water populations (Cau et al., 2016), as predicted by modelling (Bramanti et al., 2009).

In tropical octocoral communities, evidence of self-thinning has rarely been sought (Nelson, 2017). Edmunds and Lasker (2019) tested for self-thinning in octocoral communities in shallow water off St. John, U.S. Virgin Islands (Fig. 13). They found weak evidence for such relationships but only at two of six sites. At one site, the density of octocorals was high in 2014, and declined over 4 years with a weak trend of increasing height (as required for self-thinning). At a second site, densities initially categorized as intermediate, changed over time, but they increased while sizes declined (which is inconsistent with self-thinning). Densities and heights of all octocorals did not change at the other four sites (Fig. 13).

Finding self-thinning is consistent with the presence of common competitive interactions. Not finding such effects in communities composed of multiple species with a range of ages does not demonstrate an absence of competition. The self-thinning concept was developed and tested in monocultures of even-aged trees (Weller, 1987; Westoby, 1984). Future research should combine measures of density and size at the population level (i.e. by species) in order to focus on evaluating the extent to which self-thinning might affect octocoral community dynamics, and the conditions under which its effects might be most strongly expressed. Understanding if the self-thinning rule scales up from populations to multispecific communities would improve capacities both to evaluate the stability of communities formed by arborescent octocorals, and to predict the effects of future disturbances (Linares et al., 2008).

5.4 Density-dependent recruitment

Among the biological processes affected by density dependent mechanisms, recruitment is one of the most studied (Hixon et al., 2002). Classically, a decline in per-capita recruitment with rising population density can regulate population density at a finite size (Caley et al., 1996). In the US Virgin Islands, for example, a diverse octocoral community (Tsounis et al., 2018) produces a dense animal forest consisting of an upper canopy and a shaded understory community (Figs 1 and 14). Within the understory, octocoral recruitment is positively associated with the density of adults

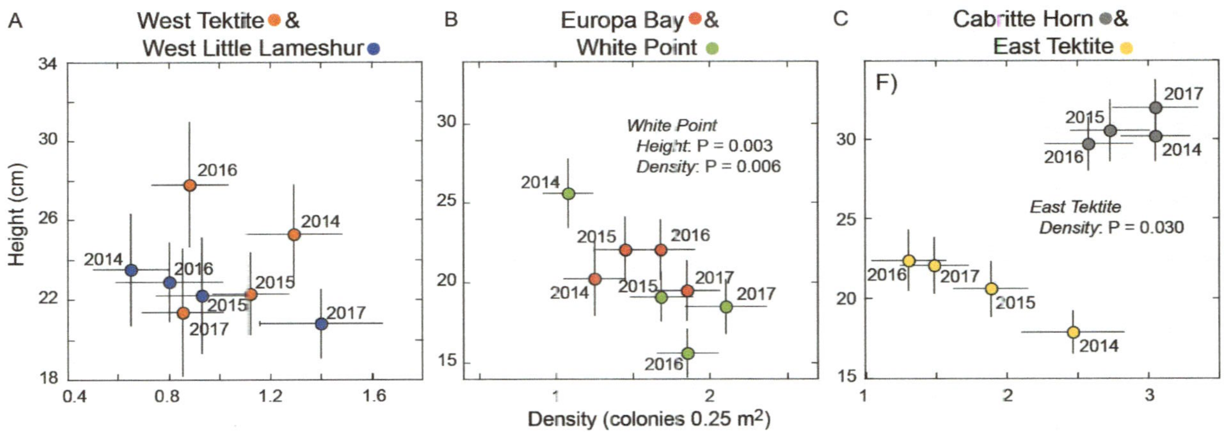

Fig. 13 Relationships between size (height) and density (holdfasts in each quadrat) over time (2014–17) for arborescent octocorals (pooled among taxa) at six sites on shallow reefs in St. John (mean ± SE). Height and density changed over time at White Point, and density changed at East Tektite; variation in height and density was not significant at any other site. These data were used as a test of self-thinning, with only weak evidence in support of the hypothesis at East Tektite. *After Edmunds, P.J., Lasker, H.R., 2019. Regulation of population size of arborescent octocorals on shallow Caribbean reefs. Mar. Ecol. Prog. Ser. 615, 1–14.*

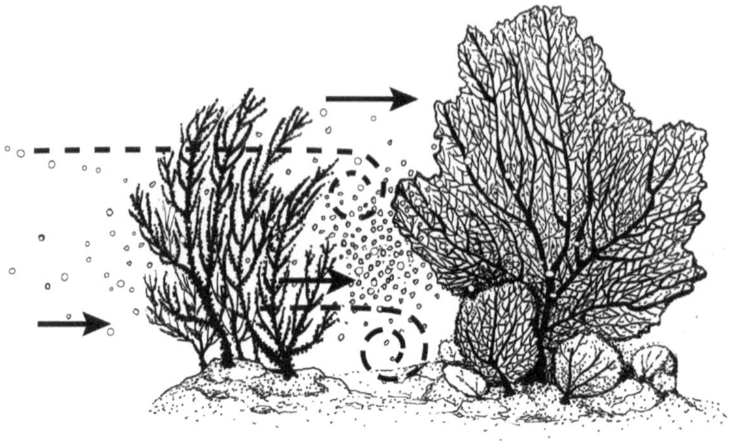

Fig. 14 Schematic displaying the interaction of an octocoral forest with moving seawater and the capacity to entrain, trap, and retain propagules that can support density dependent recruitment. *Credit M. Lenz.*

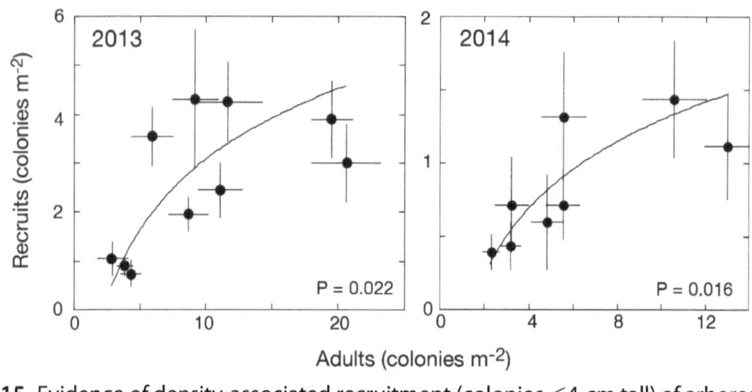

Fig. 15 Evidence of density-associated recruitment (colonies ≤4-cm tall) of arborescent octocorals on the shallow reefs of St. John in 2013 (10 sites) and 2014 (8 sites). *After Privitera-Johnson, K., Lenz, E.A., Edmunds, P.J., 2015. Density-associated recruitment in octocoral communities in St. John, US Virgin Islands. J. Exp. Mar. Biol. Ecol. 473, 103–109.*

(pooled among taxa or for all plexaurids), suggesting that recruitment for this group is density dependent (Privitera-Johnson et al., 2015) (Fig. 15). Density-associated recruitment was not characteristic of all octocoral species, and while it was well developed for *Gorgonia ventalina*, it was absent for *Antillogorgia* spp.

6. Goods and services delivered by octocorals

Ecosystems and their constituent species often provide benefits to human life, and these benefits are characterized as goods and ecosystem services (Daily, 2003). Goods include seafood, biomass fuels (Mooney et al., 2009), and many pharmaceuticals (Rodríguez, 1995), while services can range from carbon sinks (Wild et al., 2011), erosion protection (Sheppard et al., 2005), to attractions for recreation and tourism (Graham and Nash, 2013).

The principal ecosystem service that Caribbean octocoral canopies provide is habitat structure. Physical structure provided by corals generally increases biodiversity of associated organisms, such as invertebrates and fish (Reaka-Kudla, 1997). Similarly, octocoral canopies increase surface cover that acts as living space, and partitions space to provide refuge from predators (Tsounis et al., 2020), and by creating a 3-D structure, the canopy should reduce the visual detection of benthic organisms by fish predators hunting in demersal waters (e.g. Anderson, 2001; Dahlgren and Eggleston, 2000; Pollux et al., 2007; Walter and Haynes, 2006). Caribbean fishes associate with and dwell within octocoral canopies that are efficient at providing visual cover to individual fish that can confuse their visual predators (Tsounis et al., 2020). Little is known about the effects Caribbean of octocoral canopies on the invertebrate fauna, but preliminary data indicate that invertebrate communities at the south shore of St. John, US Virgin Islands, are composed of 66 genera and 82 species (J. Girard, Personal communication). Further research is required to quantify canopy-mediated increase of diversity and productivity, but it appears likely that the invertebrate fauna present provides a base for a food chain that has cascading effects into commercial fisheries of carnivorous fish species.

Octocoral habitat is being increasingly used as commercial fishing grounds in the Caribbean, such as in the US Virgin Islands (Wolff et al., 1999), as a result of degrading structural complexity of reefs (Alvarez-Filip et al., 2009; Wolff et al., 1999). While no data confirm increases of fish biomass under octocoral canopies, the concentration of fish in a specific location due to attraction (Tsounis et al., 2020) is arguably an ecosystem service of octocoral communities. Such effects, once detected, can reduce fishing effort and make fisheries more profitable.

Octocoral communities may influence reef succession and aid the development of stony coral reefs (and associated ecosystem services). Increased feeding rates of demersal fishes (including herbivores) within octocoral canopies (Tsounis et al., 2020) may mediate a top-down control by herbivores on macroalgae (Adam et al., 2015; Hughes et al., 2007; Mumby, 2006). If octocorals act as a pioneer species (sensu Connell and Slatyer, 1977), they may play a key role in regime shifts between coral- and macroalgal-dominated states (given the return of adequate environmental conditions for scleractinians).

There are no estimates that quantify the productivity and carbon budgets of octocoral forests in the tropical western Atlantic, but octocorals are known to contribute to the carbonate budget of coral reefs. Cary (1915) estimated that octocorals occurring at high density (15.9 octocorals m^{-2}) produced up to $13.6 \, kg \, m^{-2}$ of calcium carbonate per year in the form of spicules supplied to the sediment. His estimate was based on an average mortality of 17.2% of the octocoral fauna (a conservative estimate of dead skeletons present on the reef), over 5 years from 1910 to 1915 (Cary, 1915). The proportion of those sclerites that become incorporated into the sediments and ultimately lithified is unknown.

Octocorals are also a source of bioactive metabolites. These compounds have promising pharmaceutical activity, and octocorals have long been systematically screened for secondary compounds with biomedical and cosmetic value (Gerhart and Coll, 1993; Rodríguez, 1995). Direct harvesting of octocorals for extraction of commercially important compounds occurs with *Antillogorgia* (*Pseudopterogorgia*) *elisabethae* on the shallow reefs of the Little Bahama Bank (Goffredo and Lasker, 2008). *A. elisabethae* contains pseudopterosins (Look et al., 1986), that are used by the cosmetic industry for their anti-inflammatory and analgesic properties (Bruckner, 2002). The fishery for *A. elisabethae* is regulated by the Bahamas Department of Marine Resources, and when last reported had an estimated annual value of $ 3–4 million (Bruckner, 2002), meeting an estimated demand of 30–45 tons of *A. elisabethae* per year (Puyana et al., 2004). Finally, octocorals can also be goods themselves, as some species are globally harvested (legally and illegally) for use of their skeleton as raw material in jewelry, or for use of entire small colonies in the aquarium curio trade (Tsounis et al., 2010).

7. The future

Research priorities. The future of coral reefs in the 21st century will be different from the past, particularly in the tropical western Atlantic. Our premise has been that octocorals are well suited for future reefs, but that conclusion is based on observational data, and research that experimentally explores those conclusions is essential. Furthermore, if octocorals are well adapted to current conditions, it is critical that we determine the extent to which that capacity will continue to promote their success as environmental conditions further change. The degree to which octocorals modify environmental conditions and, in turn, affect the stability and resilience of the ecosystem, requires further study. Finally, understanding how octocoral populations interact with each other as well as other reef taxa is critical in evaluating how reef communities might change. Below we highlight critical needs for future research on octocorals in the context of understanding their roles on future reefs on which they will likely dominate.

1. *Discerning and identifying species.* Correctly delineating species and identifying colonies is critical to understanding octocoral communities. On reefs in the tropical western Atlantic, numerous octocoral species share similar gross morphology and can only be distinguished through sclerites sampled in voucher specimens and analysed in the laboratory (Bayer, 1961). Those characterizations are further complicated by genetic data (c.f. Prada and Hellberg, 2013), which suggest many species designations require revision. The inability to quickly and reliably identify octocorals underwater during ecological surveys remains a significant impediment to progress in understanding the biology of this taxon, and the implications of upward trends in their abundance and spatial extent on present-day reefs (Kupfner Johnson and Hallock, 2020).
2. *Bleaching.* Much of the current success of octocorals on western Atlantic reefs may be due to their relative resistance to thermal stress and bleaching. Extensive surveys during bleaching events and experimental studies are needed to identify the mechanistic basis of that resistance, to identify the role of the host and Symbiodiniacae symbionts, and to characterize the variation that occurs between species and habitats.
3. *Recruitment.* The colonization of substratum released through mortality of scleractinians may also be a critical component of the modern success of octocorals. Although little is known about the processes controlling octocoral recruitment, there are some indications for possible mechanisms:
   - **(A)** Octocoral recruitment declined following Hurricanes Irma and Maria on St. John, and then recovered after 1 year (Lasker et al., 2020). We do not know if the decline in recruitment was a consequence of reduced fecundity caused by damage to adult colonies or

if it was due to adverse environmental conditions in the months following the hurricanes.

(B) There is some evidence for positive associations between adult abundances and recruitment, but we need to determine if that reflects habitats that are especially suitable for both adults and settlers or if the octocoral canopy modulates flow in a manner that provides a positive feedback on recruitment.

Most *importantly*, whatever the process regulating octcocoral recruitment might be, we need to ascertain how they will be affected by future changes in the environment.

4. *Mixotrophy*. Complete trophic analyses of multiple octocoral species are needed. The extent to which varying levels of heterotrophy and autotrophy explain the success of the octocorals and their diversity is unknown. Relevant data are beginning to emerge for octocorals (Rossi et al., 2018, 2020), but much more is needed, particularly under ecologically relevant conditions that capture the complex hydrodynamic regimes under which branching octocorals operate.
5. *Octocoral demography*. The upward trend in abundance of octocorals on reefs in the tropical western Atlantic reflects successful recruitment and growth over shrinkage and mortality. While little is known about the biology of octocorals in this region, their demographics are even less understood. Demographic models exist for only a handful of species (e.g. Bramanti et al., 2009, 2014; Bruno et al., 2011; Goffredo and Lasker, 2008; Gotelli, 1991; Lasker, 1991; Linares and Doak, 2010; Santangelo et al., 2007; Yoshioka, 1994). Demographic models can identify life history characteristics that are key factors in controlling population abundance and generate hypotheses regarding the processes controlling population growth. More detailed spatial knowledge on octocoral distribution and demography will be key to identifying threats to octocoral communities as well as the services they provide.
6. *Holistic analyses of coral reef communities*. For decades, coral reef biology has advanced through analyses focused on scleractinians or only a handful of benthic groups, and in the majority of cases the studies have focused on a single stressor (e.g. ocean acidification, temperature, disease, etc.). As scleractinians have declined in abundance, "traditional" functional groups have less influence on the ecological processes affecting benthic communities, and now the future of "coral" reefs arguably is best described by focusing on a wide diversity of taxa. Reefs dominated by different taxa are less tractable to traditional approaches (e.g. benthic

cover), particularly as taxonomic resolution will be challenged by the profusion of morphologically similar, but genetically diverse, taxa including sponges and octocorals. Where these taxa are tall and erect, their planar cover provides a poor proxy for biomass or ecological impact, and their quantification will need to move beyond this simple metric. Finally, holistic analyses that incorporate interactions of octocorals with other taxa will be key to evaluate ecosystem services.

8. Summary

Octocorals on reefs in the tropical western Atlantic have remained understudied for over 50 years. Most studies have described deteriorating reef communities by the abundances of scleractinians and reef fishes, leading to the modern notion of the "coral reef crisis" (Bellwood et al., 2004; Gardner et al., 2003; Hughes, 1994). The impact of those changes has been enormous, but those characterizations do not capture all of the dynamics of 21st century reefs. Octocorals have increased in abundance and expanded their occupancy of reef surfaces (Lenz et al., 2015; Ruzicka et al., 2013; Sanchez et al., 2019), and it has become clear that the dynamics of octocorals differs from that of scleractinians.

With fewer scleractinians to study and community dynamics influenced more strongly by other taxa, biologists increasingly study octocorals (Kupfner Johnson and Hallock, 2020). This review focused on the ecology of octocorals and on the characteristics that could have favoured their increasing abundance. The trends described above point to the ecological significance of a tree-like morphology and the exploitation of the seawater volume above the benthos as reasons for the success of octocorals on modern reefs. The region-wide population decline of branching *Acropora* spp. (Aronson and Precht, 2001) left the functional niche of the water column underexploited by sessile taxa. The capacity for rapid linear growth allows octocorals to quickly evade competition with benthic macroalgae, and transition through the risky small size classes to reduce post-settlement mortality. By growing in dense clusters and forming dense forests (Rossi et al., 2017), octocorals benefit from canopy formation, promoting particle capture, larval retention, and beneficially exploiting hydrodynamic regimes (Guizien and Ghisalberti, 2016).

While there is compelling evidence that octocorals are gaining greater ecological significance on the coral reefs of the tropical western Atlantic, it remains unknown whether they are ecological "placeholders" on the "slippery slope to slime" (Pandolfi et al., 2005), or a "new normal" that will define the structure and function of "coral" reefs for centuries (or longer). The notion that Caribbean reef communities might be dominated for at least a few decades by octocorals, highlights the ecological opportunities remaining. This provides an alternative to focusing solely on the extinction of scleractinians and the loss of coral reef ecosystems (Carpenter et al., 2008; Hoegh-Guldberg et al., 2007). Octocorals can deliver many goods and services similar to those of scleractinians, but their inability to deliver a calcified, sturdy, and wave resistant reef framework with habitats for other taxa, means that "octocoral forests" cannot be considered replacements for what is being lost with living coral reefs. Nevertheless, they likely will characterize a large part of what we have, rather than what we might want, so this is the time to continue and expand studies of their biology and ecology.

Acknowledgements

This review was completed in part with support from the US National Science Foundation (most recently, OCE 13-32915 and 17-56678 to P.J.E. and OCE 17-56381 to H.R.L.) and by Project GORGOSHIFT (PICS-CNRS 2017–2020 n. 279738 to L.B.). The empirical work was completed in St. John, US Virgin Islands under permits from the Virgin Islands National Park (most recently VIIS-2019-SCI-0022).

References

Adam, T.C., Burkepile, D.E., Ruttenberg, B.I., Paddack, M.J., 2015. Herbivory and the resilience of Caribbean coral reefs: knowledge gaps and implications for management. Mar. Ecol. Prog. Ser. 520, 1–20.

Albright, R., 2011. Reviewing the effects of ocean acidification on sexual reproduction and early life history stages of reef-building corals. J. Mar. Biol. 2011, 1–14.

Albright, R., Mason, B., Miller, M., Langdon, C., 2010. Ocean acidification compromises recruitment success of the threatened Caribbean coral Acropora palmata. Proc. Natl. Acad. Sci. U. S. A. 107, 20400–20404.

Alvarez-Filip, L., Dulvy, N.K., Gill, J.A., Cote, I.M., Watkinson, A.R., 2009. Flattening of Caribbean coral reefs: region-wide declines in architectural complexity. Proc. R. Soc. B Biol. Sci. 276, 3019–3025.

Anderson, T.W., 2001. Predator responses, prey refuges, and density-dependent mortality of a marine fish. Ecology 82, 245–257.

Aronson, R.B., Precht, W.F., 2001. White-band disease and the changing face of Caribbean coral reefs. Hydrobiologia 460, 25–38.

Baird, A.H., Blakeway, D.R., Hurley, T.J., Stoddart, J.A., 2011. Seasonality of coral reproduction in the Dampier Archipelago, northern Western Australia. Mar. Biol. 158, 275–285.

Baker, D.M., Freeman, C.J., Knowlton, N., Thacker, R.W., Kim, K., Fogel, M.L., 2015. Productivity links morphology, symbiont specificity and bleaching in the evolution of Caribbean octocoral symbioses. ISME J. 9, 2620–2629.

Bayer, F.M., 1961. The Shallow Water Octocorallia of the West Indian Region. Martinus Nijhoff, The Hague.

Bellwood, D.R., Hughes, T.P., Folke, C., Nystrom, M., 2004. Confronting the coral reef crisis. Nature 429, 827–833.
Benayahu, Y., Loya, Y., 1977. Space partitioning by stony corals soft corals and benthic algae on coral reefs of northern Gulf of Eilat (Red-Sea). Helgoländer Meeresun. 30, 362–382.
Benayahu, Y., Loya, Y., 1983. Surface brooding in the Red-Sea soft coral *Parerythropodium fulvum fulvum* (Forskal, 1775). Biol. Bull. 165, 353–369.
Bernardo, J., 1996. The particular maternal effect of propagule size, especially egg size: patterns, models, quality of evidence and interpretations. Am. Zool. 36, 216–236.
Birkeland, C., 2015. Coral Reefs in the Anthropocene. Springer, Dordrecht.
Borgstein, N., Beltrán, D.M., Prada, C., 2020. Variable growth across species and life stages in Caribbean reef octocorals. Front. Mar. Sci. 7, 483. https://doi.org/10.3389/fmars.2020.00483.
Bramanti, L., Iannelli, M., Santangelo, G., 2009. Mathematical modelling for conservation and management of gorgonians corals: youngs and olds, could they coexist? Ecol. Model. 220, 2851–2856.
Bramanti, L., Lasker, H.R., Edmunds, P.J., 2017. The encrusting alga Peyssonnelia preempts vacant space and overgrows corals in St. John, US Virgin Islands. Reef Encount. 32 (1), 68–70.
Bramanti, L., Movilla, J., Guron, M., Calvo, E., Gori, A., Dominguez-Carrio, C., Grinyo, J., Lopez-Sanz, A., Martinez-Quintana, A., Pelejero, C., Ziveri, P., Rossi, S., 2013. Detrimental effects of ocean acidification on the economically important Mediterranean red coral (*Corallium rubrum*). Glob. Chang. Biol. 19, 1897–1908.
Bramanti, L., Vielmini, I., Rossi, S., Tsounis, G., Iannelli, M., Cattaneo-Vietti, R., Priori, C., Santangelo, G., 2014. Demographic parameters of two populations of red coral (*Corallium rubrum* L. 1758) in the North Western Mediterranean. Mar. Biol. 161, 1015–1026.
Brazeau, D.A., Lasker, H.R., 1992. Growth rates and growth strategy in a clonal marine invertebrate, the Caribbean Octocoral *Briareum-asbestinum*. Biol. Bull. 183, 269–277.
Brown, B.E., 1997. Adaptations of reef corals to physical environmental stress. In: Blaxter, J.H.S., Southward, A.J. (Eds.), Advances in Marine Biology. vol. 31, pp. 221–299.
Bruckner, A.W., 2002. Life-saving products from coral reefs. Issues Sci. Technol. 18, 39–44.
Bruno, J.F., Bertness, M.D., 2001. Habitat modification and facilitation in benthic marine communities. In: Bertness, M.D., Gaines, S.D., Hay, M. (Eds.), Marine Community Ecology. Sinauer Associates, Sunderland, Massachusetts, USA, pp. 201–218.
Bruno, J.F., Selig, E.R., 2007. Regional decline of coral cover in the Indo-Pacific: timing, extent, and subregional comparisons. Plos One 2, e711.
Bruno, J.F., Ellner, S.P., Vu, I., Kim, K., Harvell, C.D., 2011. Impacts of aspergillosis on sea fan coral demography: modeling a moving target. Ecol. Monogr. 81, 123–139.
Budd, A.F., 2000. Diversity and extinction in the Cenozoic history of Caribbean reefs. Coral Reefs 19, 25–35.
Cadena, N.J., Sanchez, J.A., 2010. Colony growth in the harvested octocoral *Pseudopterogorgia acerosa* in a Caribbean coral reef. Mar. Ecol. 31, 566–573.
Caley, M.J., Carr, M.H., Hixon, M.A., Hughes, T.P., Jones, G.P., Menge, B.A., 1996. Recruitment and the local dynamics of open marine populations. Annu. Rev. Ecol. Syst. 27, 477–500.
Carlon, D.B., 1999. The evolution of mating systems in tropical reef corals. Trends Ecol. Evol. 14, 491–495.
Carpenter, K.E., Abrar, M., Aeby, G., Aronson, R.B., Banks, S., Bruckner, A., Chiriboga, A., Cortes, J., Delbeek, J.C., DeVantier, L., Edgar, G.J., Edwards, A.J., Fenner, D., Guzman, H.M., Hoeksema, B.W., Hodgson, G., Johan, O., Licuanan, W.Y., Livingstone, S.R., Lovell, E.R., Moore, J.A., Obura, D.O.,

Ochavillo, D., Polidoro, B.A., Precht, W.F., Quibilan, M.C., Reboton, C., Richards, Z.T., Rogers, A.D., Sanciangco, J., Sheppard, A., Sheppard, C., Smith, J., Stuart, S., Turak, E., Veron, J.E.N., Wallace, C., Weil, E., Wood, E., 2008. One-third of reef-building corals face elevated extinction risk from climate change and local impacts. Science 321, 560–563.

Cary, L.R., 1914. Observations Upon the Growth and Ecology of Gorgonians. Carnegie Inst. Washington Pop. Tortugas Lab, publ.182, vol. 5, pp. 81–90.

Cary, L.R., 1915. The Alcyonaria as a factor in reef limestone formation. Proc. Natl. Acad. Sci. U. S. A. 1, 285–289.

Cau, A., Bramanti, L., Cannas, R., Follesa, M.C., Angiolillo, M., Canese, S., Bo, M., Cuccu, D., Guizien, K., 2016. Habitat constraints and self-thinning shape Mediterranean red coral deep population structure: implications for conservation practice. Sci. Rep. 6, 23322.

Chadwick, N.E., Morrow, K.M., 2011. Competition among sessile organisms on coral reefs. In: Dubinsky, Z., Stambler, N. (Eds.), Coral reefs: An ecosystem in transition. Springer, Dordrecht, pp. 347–371.

Chan, N.C.S., Connolly, S.R., 2012. Sensitivity of coral calcification to ocean acidification: a meta-analysis. Glob. Chang. Biol. 19, 282–290.

Chen, J.H., Curran, H.A., White, B., Wasserburg, G.J., 1991. Precise chronology of the last interglacial period: U^{234} Th^{230} data from fossil coral reefs in the Bahamas. Geol. Soc. Am. Bull. 103, 82–97.

Cocito, S., Ferrier-Pages, C., Cupido, R., Rottier, C., Meier-Augenstein, W., Kemp, H., Reynaud, S., Peirano, A., 2013. Nutrient acquisition in four Mediterranean gorgonian species. Mar. Ecol. Prog. Ser. 473, 179–188.

Coelho, M.A.G., 2018. Life History Traits of Caribbean Octocorals of the Genus Antillogorgia and Their Implications for Larval Dispersal (PhD Dissertation). University at Buffalo, State University of New York, Buffalo, NY.

Coma, R., Gili, J.M., Zabala, M., Riera, T., 1994. Feeding and prey capture in the aposymbiotic gorgonian *Paramuricea clavata*. Mar. Ecol. Prog. Ser. 115, 257–270.

Coma, R., Pola, E., Ribes, M., Zabala, M., 2004. Long-term assessment of temperate octocoral mortality patterns, protected vs. unprotected areas. Ecol. Appl. 14, 1466–1478.

Coma, R., Linares, C., Ribes, M., Diaz, D., Garrabou, J., Ballesteros, E., 2006. Consequences of a mass mortality in populations of *Eunicella singularis* (Cnidaria: Octocorallia) in Menorca (NW Mediterranean). Mar. Ecol. Prog. Ser. 327, 51–60.

Coma, R., Llorente-Llurba, E., Serrano, E., Gili, J.M., Ribes, M., 2015. Natural heterotrophic feeding by a temperate octocoral with symbiotic zooxanthellae: a contribution to understanding the mechanisms of die-off events. Coral Reefs 34, 549–560.

Comeau, S., Edmunds, P.J., Spindel, N.B., Carpenter, R.C., 2014. Fast coral reef calcifiers are more sensitive to ocean acidification in short-term laboratory incubations. Limnol. Oceanogr. 59, 1081–1091.

Connell, J.H., 1978. Diversity in tropical rain forest and coral reefs. Science 1999, 1302–1310.

Connell, J., Slatyer, R.O., 1977. Mechanisms of succession in natural communities and their role in community stablility and organization. Am. Nat. 111, 1119–1144.

Cordeiro, R., McFadden, C., van Ofwegen, L., Williams, G., 2020. World List of Octocorallia. Eunicea Lamouroux, 1816. Accessed through: World Register of Marine Species at: http://www.marinespecies.org/aphia.php?p=taxdetails&id=291322. on 2020-07-01.

Dahlgren, C.P., Eggleston, D.B., 2000. Ecological processes underlying ontogenetic habitat shifts in a coral reef fish. Ecology 81, 2227–2240.

Daily, G., 2003. What are ecosystem services. In: Lorey, D.E. (Ed.), Global Environmental Challenges for the Twenty-First Century: Resources, Consumption and Sustainable Solutions. Rowman & Littlefield Publishers, pp. 227–231.

Doropoulos, C., Ward, S., Diaz-Pulido, G., Hoegh-Guldberg, O., Mumby, P.J., 2012. Ocean acidification reduces coral recruitment by disrupting intimate larval-algal settlement interactions. Ecol. Lett. 15, 338–346.

Dudgeon, S.R., Aronson, R.B., Bruno, J.F., Precht, W.F., 2010. Phase shifts and stable states on coral reefs. Mar. Ecol. Prog. Ser. 413, 201–216.

Eckrich, C.E., Engel, M.S., 2013. Coral overgrowth by an encrusting red alga (Ramicrusta sp.): a threat to Caribbean reefs? Coral Reefs 32, 81–84.

Edmunds, P.J., 2018. The hidden dynamics of low coral cover communities. Hydrobiologia 818, 193–209.

Edmunds, P.J., Lasker, H.R., 2019. Regulation of population size of arborescent octocorals on shallow Caribbean reefs. Mar. Ecol. Prog. Ser. 615, 1–14.

Edmunds, P.J., Zimmermann, S.A., Bramanti, L., 2019. A spatially aggressive peyssonnelid algal crust (PAC) threatens shallow coral reefs in St. John, US Virgin Islands. Coral Reefs 38, 1329–1341.

Evans, M.J., Coffroth, M.A., Lasker, H.R., 2013. Effects of predator exclusion on recruit survivorship in an octocoral (*Briareum asbestinum*) and a scleractinian coral (*Porites astreoides*). Coral Reefs 32, 597–601.

Eyre, B.D., Cyronak, T., Drupp, P., De Carlo, E.H., Sachs, J.P., Andersson, A.J., 2018. Coral reefs will transition to net dissolving before end of century. Science 359, 908–911.

Fabricius, K., Yahel, G., Genin, A., 1998. *In situ* depletion of phytoplankton by an azooxanthellate soft coral. Limnol. Oceanogr. 43, 354–356.

Fautin, D.G., 2009. Structural diversity, systematics, and evolution of cnidae. Toxicon 54, 1054–1064.

Ferrier-Pages, C., Leal, M.C., 2019. Stable isotopes as tracers of trophic interactions in marine mutualistic symbioses. Ecol. Evol. 9, 723–740.

Folkard, A.M., 2019. Biophysical interactions in fragmented marine canopies: fundamental processes, consequences, and upscaling. Front. Mar. Sci. 6, 17.

Frechette, M., Lefaivre, D., 1990. Discriminating between food and space limitation in benthic suspension feeders using self-thinning relationships. Mar. Ecol. Prog. Ser. 65, 15–23.

Gabay, Y., Fine, M., Barkay, Z., Benayahu, Y., 2014. Octocoral tissue provides protection from declining oceanic pH. Plos One 9, e91553.

Gambrel, B., Lasker, H.R., 2016. Interactions in the canopy among Caribbean reef octocorals. Mar. Ecol. Prog. Ser. 546, 85–95.

Gardner, T.A., Côté, I.M., Gill, J.A., Grant, A., Watkinson, A.R., 2003. Long-term region-wide declines in Caribbean corals. Science 301, 958–960.

Garrabou, J., Harmelin, J.G., 2002. A 20-year study on life-history traits of a harvested long-lived temperate coral in the NW Mediterranean: insights into conservation and management needs. J. Anim. Ecol. 71, 966–978.

Garrabou, J., Sala, E., Linares, C., Ledoux, J.B., Montero-Serra, I., Dominici, J.M., Kipson, S., Teixido, N., Cebrian, E., Kersting, D.K., Harmelin, J.G., 2017. Re-shifting the ecological baseline for the overexploited Mediterranean red coral. Sci. Rep. 7, 6.

Gates, R.D., 1990. Seawater temperature and sublethal coral bleaching in Jamaica. Coral Reefs 8, 193–197.

Geiser, D.M., Taylor, J.W., Ritchie, K.B., Smith, G.W., 1998. Cause of sea fan death in the West Indies. Nature 394, 137–138.

Gerhart, D.J., Coll, J.C., 1993. Pukalide, a widely distributed octocoral diterpenoid, induces vomiting in fish. J. Chem. Ecol. 19, 2697–2704.

Giammona, C.P., Stanton, R.J., 1980. Octocorals from the Middle Eocene Stone city formation, Texas. J. Paleontol. 54, 71–80.

Gili, J.M., Coma, R., 1998. Benthic suspension feeders: their paramount role in littoral marine food webs. Trends Ecol. Evol. 13, 316–321.

Glynn, P.W., 1993. Coral-reef bleaching—ecological perspectives. Coral Reefs 12, 1–17.

Goffredo, S., Lasker, H.R., 2008. An adaptive management approach to an octocoral fishery based on the Beverton-Holt model. Coral Reefs 27, 751–761.

Goldberg, W.M., 1973. Ecology of coral-octocoral communities off Southeast Florida coast—geomorphology, species composition, and zonation. Bull. Mar. Sci. 23, 465–488.

Gomez, C.E., Paul, V.J., Ritson-Williams, R., Muehllehner, N., Langdon, C., Sanchez, J.A., 2015. Responses of the tropical gorgonian coral *Eunicea fusca* to ocean acidification conditions. Coral Reefs 34, 451–460.

Goreau, T.F., 1959. The ecology ofJamaican coral reefs I. Species composition and zonation. Ecology 40, 67–89.

Gotelli, N.J., 1991. Demographic models for *Leptogorgia virgulata*, a shallow-water gorgonian. Ecology 72, 457–467.

Goulet, T.L., Coffroth, M.A., 2004. The genetic identity of dinoflagellate symbionts in Caribbean octocorals. Coral Reefs 23, 465–472.

Goulet, T.L., Shirur, K.P., Ramsby, B.D., Iglesias-Prieto, R., 2017. The effects of elevated seawater temperatures on Caribbean gorgonian corals and their algal symbionts, Symbiodinium spp. Plos One 12, e0171032.

Graham, N.A.J., Nash, K.L., 2013. The importance of structural complexity in coral reef ecosystems. Coral Reefs 32, 315–326.

Grottoli, A.G., Rodrigues, L.J., Palardy, J.E., 2006. Heterotrophic plasticity and resilience in bleached corals. Nature 440, 1186–1189.

Guizien, K., Ghisalberti, M., 2016. Living in the canopy of the animal forest: physical and biogeochemical aspects. In: Rossi, S., Bramanti, L., Gori, A., Orejas Saco del Valle, C. (Eds.), Marine Animal Forests: The Ecology of Benthic Biodiversity Hotspots. Springer International Publishing AG, Germany, pp. 507–528.

Hannes, A.R., Barbeitos, M., Coffroth, M.A., 2009. Stability of symbiotic dinoflagellate type in the octocoral *Briareum asbestinum*. Mar. Ecol. Prog. Ser. 391, 65–72.

Hartmann, A.C., Marhaver, K.L., Chamberland, V.F., Sandin, S.A., Vermeij, M.J.A., 2013. Large birth size does not reduce negative latent effects of harsh environments across life stages in two coral species. Ecology 94, 1966–1976.

Hauri, C., Fabricius, K.E., Schaffelke, B., Humphrey, C., 2010. Chemical and physical environmental conditions underneath mat- and canopy-forming macroalgae, and their effects on understorey corals. Plos One 5, e12685.

Hixon, M.A., Pacala, S.W., Sandin, S.A., 2002. Population regulation: historical context and contemporary challenges of open vs. closed systems. Ecology 83, 1490–1508.

Hoegh-Guldberg, O., Mumby, P.J., Hooten, A.J., Steneck, R.S., Greenfield, P., Gomez, E., Harvell, C.D., Sale, P.F., Edwards, A.J., Caldeira, K., Knowlton, N., Eakin, C.M., Iglesias-Prieto, R., Muthiga, N., Bradbury, R.H., Dubi, A., Hatziolos, M.E., 2007. Coral reefs under rapid climate change and ocean acidification. Science 318, 1737–1742.

Hughes, T.P., 1994. Catastrophes, phase shifts, and large-scale degradation of a Caribbean coral reef. Science 265, 1547–1551.

Hughes, T.P., Rodrigues, M.J., Bellwood, D.R., Ceccarelli, D., Hoegh-Guldberg, O., McCook, L., Moltschaniwskyj, N., Pratchett, M.S., Steneck, R.S., Willis, B., 2007. Phase shifts, herbivory, and the resilience of coral reefs to climate change. Curr. Biol. 17, 360–365.

Hughes, T.P., Barnes, M.L., Bellwood, D.R., Cinner, J.E., Cumming, G.S., Jackson, J.B.C., Kleypas, J., van de Leemput, I.A., Lough, J.M., Morrison, T.H., Palumbi, S.R., van Nes, E.H., Scheffer, M., 2017. Coral reefs in the Anthropocene. Nature 546, 82–90.

Hughes, T.P., Kerry, J.T., Baird, A.H., Connolly, S.R., Dietzel, A., Eakin, C.M., Heron, S.F., Hoey, A.S., Hoogenboom, M.O., Liu, G., McWilliam, M.J., Pears, R.J., Pratchett, M.S., Skirving, W.J., Stella, J.S., Torda, G., 2018. Global warming transforms coral reef assemblages. Nature 556, 492–496.

Huston, M.A., 1985. Patterns of species-diversity on coral reefs. Annu. Rev. Ecol. Syst. 16, 149–177.

Iluz, D., Alexandrovich, L., Dubinsky, Z., 2012. The enhancement of photosynthesis by fluctuating light. In: Najafpour, M.M. (Ed.), Artificial Photosynthesis. Intech, Rijeka, Croatia, pp. 115–134.

Jackson, J.B.C., 1997. Reefs since Columbus. Coral Reefs 16, S23–S32.

Jackson, J.B.C., Kirby, M.X., Berger, W.H., Bjorndal, K.A., Botsford, L.W., Bourque, B.J., Bradbury, R.H., Cooke, R., Erlandson, J., Estes, J.A., Hughes, T.P., Kidwell, S., Lange, C.B., Lenihan, H.S., Pandolfi, J.M., Peterson, C.H., Steneck, R.S., Tegner, M.J., Warner, R.R., 2001. Historical overfishing and the recent collapse of coastal ecosystems. Science 293, 629–638.

Jackson, J.B.C., Donovan, M.K., Cramer, K.L., Lam, W. (Eds.), 2014. Status and Trends of Caribbean Coral Reefs: 1970–2012. Glob. Coral Reef Monit, IUCN, Gland. Switz, Network, p. 306.

Jones, C.G., Lawton, J.H., Shachak, M., 1994. Organisms as ecosystem engineers. Oikos 69, 373–386.

Kahng, S.E., Benayahu, Y., Lasker, H.R., 2011. Sexual reproduction in octocorals. Mar. Ecol. Prog. Ser. 443, 265–283.

Kim, K., 2016. Diseases of octocorals. In: Woodley, C.M., Downs, C.A., Bruckner, A.W., Porter, J.W., Galloway, S.B. (Eds.), Diseases of Coral. John Wiley & Sons Inc, Hoboken, pp. 410–415.

Kim, K., Lasker, H.R., 1997. Flow-mediated resource competition in the suspension feeding gorgonian *Plexaura homomalla* (Esper). J. Exp. Mar. Biol. Ecol. 215, 49–64.

Kinzie, R.A., 1973. The zonation of West Indian gorgonians. Bull. Mar. Sci. 23, 93–155.

Kocurko, M.J., Kocurko, D.J., 1992. Fossil octocorallia of the Red Bluff formation, lower oligocene, Mississippi. J. Paleo. 66, 594–602.

Kroeker, K.J., Kordas, R.L., Crim, R., Hendriks, I.E., Ramajo, L., Singh, G.S., Duarte, C.M., Gattuso, J.P., 2013. Impacts of ocean acidification on marine organisms: quantifying sensitivities and interaction with warming. Glob. Chang. Biol. 19, 1884–1896.

Kupfner Johnson, S., Hallock, P., 2020. A review of symbiotic gorgonian research in the western Atlantic and Caribbean with recommendations for future work. Coral Reefs 39, 239–258.

Lasker, H.R., 1981. A comparison of the particulate feeding abilities of three species of gorgonian soft coral. Mar. Ecol. Prog. Ser. 5, 61–67.

Lasker, H.R., 1990. Clonal propagation and population dynamics of a gorgonian coral. Ecology 71, 1578–1589.

Lasker, H.R., 1991. Population-growth of a gorgonian coral—equilibrium and non-equilibrium sensitivity to changes in life-history variables. Oecologia 86, 503–509.

Lasker, H.R., Porto-Hannes, I., 2015. Population structure among octocoral adults and recruits identifies scale dependent patterns of population isolation in the Bahamas. PeerJ 3, e1019.

Lasker, H.R., Gottfried, M.D., Coffroth, M.A., 1983. Effects of depth on feeding capabilities of two octocorals. Mar. Biol. 73, 73–78.

Lasker, H.R., Peters, E.C., Coffroth, M.A., 1984. Bleaching of reef coelenterates in the San-Blas Islands, Panama. Coral Reefs 3, 183–190.

Lasker, H.R., Kim, K., Coffroth, M.A., 1998. Production, settlement, and survival of plexaurid gorgonian recruits. Mar. Ecol. Prog. Ser. 162, 111–123.

Lasker, H.R., Boller, M.L., Castanaro, J., Sanchez, J.A., 2003. Determinate growth and modularity in a gorgonian octocoral. Biol. Bull. 205, 319–330.

Lasker, H.R., Martínez-Quintana, Á., Bramanti, L., Edmunds, P.J., 2020. Resilience of octocoral forests to catastrophic storms. Sci. Rep. 10, 4286.

Lenz, E.A., Bramanti, L., Lasker, H., Edmunds, P.J., 2015. Long-term variation of octocoral populations in St. John, US Virgin Islands. Coral Reefs 34, 1–11.

Lesser, M.P., 2011. Coral bleaching: causes and mechanisms. In: Dubinsky, Z., Stambler, N. (Eds.), Coral Reefs: An Ecosystem in Transition. Springer, Dordrecht, pp. 405–419.

Leversee, G., 1976. Flow and feeding in fan-shaped colonies of the octocoral, Leptogorgia virgulata. Biol. Bull. 151, 344–356.

Lewis, C.L., Coffroth, M.A., 2004. The acquisition of exogenous algal symbionts by an octocoral after bleaching. Science 304, 1490–1492.

Linares, C., Doak, D.F., 2010. Forecasting the combined effects of disparate disturbances on the persistence of long-lived gorgonians: a case study of *Paramuricea clavata*. Mar. Ecol. Prog. Ser. 402, 59–68.

Linares, C., Coma, R., Diaz, D., Zabala, M., Hereu, B., Dantart, L., 2005. Immediate and delayed effects of a mass mortality event on gorgonian population dynamics and benthic community structure in the NW Mediterranean Sea. Mar. Ecol. Prog. Ser. 305, 127–137.

Linares, C., Doak, D.F., Coma, R., Diaz, D., Zabala, M., 2007. Life history and viability of a long-lived marine invertebrate: the octocoral *Paramuricea clavata*. Ecology 88, 918–928.

Linares, C., Coma, R., Garrabou, J., Diaz, D., Zabala, M., 2008. Size distribution, density and disturbance in two Mediterranean gorgonians: *Paramuricea clavata* and *Eunicella singularis*. J. Appl. Ecol. 45, 688–699.

Look, S.A., Fenical, W., Jacobs, R.S., Clardy, J., 1986. The pseudopterosins: anti-inflammatory and analgesic natural products from the sea whip *Pseudopterogorgia elisabethae*. Proc. Natl. Acad. Sci. U. S. A. 83, 6238–6240.

Lough, J.M., van Oppen, M.J.H., 2018. Introduction: coral bleaching-patterns, processes, causes and consequences. In: VanOppen, M.J.H., Lough, J.M. (Eds.), Coral Bleaching: Patterns, Processes, Causes and Consequences, second ed. Springer-Verlag Berlin Heidelberg, pp. 1–8.

Loya, Y., 1972. Community structure and species diversity of hermatypic corals at Eilat, Red Sea. Mar. Biol. 13, 100–123.

Macdonald, R.W., 2000. Modelling the mean velocity profile in the urban canopy layer. Bound. Lay. Meteorol. 97, 25–45.

Mariscal, R.N., 1974. Nematocysts. In: Muscatine, L., Lenhoff, H.M. (Eds.), Coelenterate Biology: Reviews and New Perspectives. Academic Press, New York, pp. 129–178.

Maynard, J., van Hooidonk, R., Eakin, C.M., Puotinen, M., Garren, M., Williams, G., Heron, S.F., Lamb, J., Weil, E., Willis, B., Harvell, C.D., 2015. Projections of climate conditions that increase coral disease susceptibility and pathogen abundance and virulence. Nat. Clim. Chang. 5, 688–694.

McLean, E.L., Yoshioka, P.M., 2008. Substratum effects on the growth and survivorship of the sponge Desmapsamma anchorata. Caribb. J. Sci. 44, 83–89.

Mooney, H., Larigauderie, A., Cesario, M., Elmquist, T., Hoegh-Guldberg, O., Lavorel, S., Mace, G.M., Palmer, M., Scholes, R., Yahara, T., 2009. Biodiversity, climate change, and ecosystem services. Curr. Opin. Environ. Sustain. 1, 46–54.

Muller, E.M., Sartor, C., Alcaraz, N.I., van Woesik, R., 2020. Spatial epidemiology of the stony-coral-tissue-loss disease in Florida. Front. Mar. Sci. 7. https://doi.org/10.3389/fmars.2020.00163.

Mumby, P.J., 2006. The impact of exploiting grazers (scaridae) on the dynamics of Caribbean coral reefs. Ecol. Appl. 16, 747–769.

Nelson, H., 2017. From Trees to Octocorals: The Role of Phenotypic Integration, Shading, and Self-Thinning in Underwater Animal Forests (Master's Thesis). California State University, Northridge.

Norstrom, A.V., Nystrom, M., Lokrantz, J., Folke, C., 2009. Alternative states on coral reefs: beyond coral-macroalgal phase shifts. Mar. Ecol. Prog. Ser. 376, 295–306.

Oliver, E.C.J., Donat, M.G., Burrows, M.T., Moore, P.J., Smale, D.A., Alexander, L.V., Benthuysen, J.A., Feng, M., Sen Gupta, A., Hobday, A.J., Holbrook, N.J., Perkins-Kirkpatrick, S.E., Scannell, H.A., Straub, S.C., Wernberg, T., 2018. Longer and more frequent marine heatwaves over the past century. Nat. Commun. 9, 1–12. https://doi.org/10.1038/s41467-018-03732-9.

Pandolfi, J.M., Jackson, J.B.C., Baron, N., Bradbury, R.H., Guzman, H.M., Hughes, T.P., Kappel, C.V., Micheli, F., Ogden, J.C., Possingham, H.P., Sala, E., 2005. Ecology—are US coral reefs on the slippery slope to slime? Science 307, 1725–1726.

Paul, V.J., Thacker, R.W., Banks, K., Golubic, S., 2005. Benthic cyanobacterial bloom impacts the reefs of South Florida (Broward County, USA). Coral Reefs 24, 693–697.

Pawlik, J.R., McMurray, S.E., 2020. The emerging ecological and biogeochemical importance of sponges on coral reefs. In: Carlson, C.A., Giovannoni, S.J. (Eds.), Annual Review of Marine Science. vol. 12, pp. 315–337.

Plaisance, L., Caley, M.J., Brainard, R.E., Knowlton, N., 2011. The diversity of coral reefs: what are we missing? Plos One 6, e25026.

Poland, D.M., 2010. Specificity Versus Flexibility in Cnidarian Algal Symbioses. PhD Dissertation, State University of New York at Buffalo.

Poland, D.M., Coffroth, M.A., 2019. Host growth and survivorship varies with endosymbiotic algal partner in developing cnidarians. Mar. Ecol. Prog. Ser. 612, 87–100.

Pollux, B.J.A., Verberk, W., Dorenbosch, M., de la Moriniere, E.C., Nagelkerken, I., van der Velde, G., 2007. Habitat selection during settlement of three Caribbean coral reef fishes: indications for directed settlement to seagrass beds and mangroves. Limnol. Oceanogr. 52, 903–907.

Porter, J.W., 1976. Autotrophy, heterotrophy, and resource partitioning in Caribbean reef-building corals. Am. Nat. 110, 731–742.

Prada, C., Hellberg, M.E., 2013. Long prereproductive selection and divergence by depth in a Caribbean candelabrum coral. Proc. Natl. Acad. Sci. U. S. A. 110, 3961–3966.

Prada, C., Weil, E., Yoshioka, P.M., 2010. Octocoral bleaching during unusual thermal stress. Coral Reefs 29, 41–45.

Privitera-Johnson, K., Lenz, E.A., Edmunds, P.J., 2015. Density-associated recruitment in octocoral communities in St. John, US Virgin Islands. J. Exp. Mar. Biol. Ecol. 473, 103–109.

Puyana, M., Narvaez, G., Paz, A., Osorno, O., Duque, C., 2004. Pseudopterosin content variability of the purple sea whip Pseudopterogorgia elisabethae at the Islands of San Andres and Providencia (SW Caribbean). J. Chem. Ecol. 30, 1183–1201.

Reaka-Kudla, M.L., 1997. The global biodiversity of coral reefs: a comparison with rain forests. In: Reaka-Kudla, M.L., Wilson, D.E., Wilson, E.O. (Eds.), Bidiversity II: Understanding and Protecting Our Biological Resources. Joseph Henry Press, Washington, D.C, pp. 83–108.

Reichelt, R.E., Loya, Y., Bradbury, R.H., 1986. Patterns in the use of space by benthic communities on two coral reefs of the great barrier reef. Coral Reefs 5, 73–79.

Reidenbach, M.A., Koseff, J.R., Monismith, S.G., 2007. Laboratory experiments of fine-scale mixing and mass transport within a coral canopy. Phys. Fluids 19, 075107.

Ribes, M., Coma, R., Gili, J.M., 1998. Heterotrophic feeding by gorgonian corals with symbiotic zooxanthella. Limnol. Oceanogr. 43, 1170–1179.

Ribes, M., Coma, R., Rossi, S., 2003. Natural feeding of the temperate asymbiotic octocoral-gorgonian *Leptogorgia sarmentosa* (Cnidaria: Octocorallia). Mar. Ecol. Prog. Ser. 254, 141–150.

Rodríguez, A.D., 1995. The natural products chemistry of West Indian gorgonian octocorals. Tetrahedron 51, 4571–4618.

Roff, G., Mumby, P.J., 2012. Global disparity in the resilience of coral reefs. Trends Ecol. Evol. 27, 404–413.

Rosenfeld, J.S., 2002. Functional redundancy in ecology and conservation. Oikos 98, 156–162.

Rossi, S., Ribes, M., Coma, R., Gili, J.M., 2004. Temporal variability in zooplankton prey capture rate of the passive suspension feeder *Leptogorgia sarmentosa* (Cnidaria: Octocorallia), a case study. Mar. Biol. 144, 89–99.

Rossi, S., Bramanti, L., Gori, A., Orejas, C., 2017. An overview of the animal forests of the world. In: Rossi, S. (Ed.), Marine Animal Forests: The Ecology of Benthic Biodiversity Hotspots. Springer International Publishing AG, Germany, pp. 1–26.

Rossi, S., Schubert, N., Brown, D., Soares, M.D., Grosso, V., Rangel-Huerta, E., Maldonado, E., 2018. Linking host morphology and symbiont performance in octocorals. Sci. Rep. 8, 14.

Rossi, S., Schubert, N., Brown, D., Gonzalez-Posada, A., Soares, M.O., 2020. Trophic ecology of Caribbean octocorals: autotrophic and heterotrophic seasonal trends. Coral Reefs 39, 433–449.

Roushdy, H.M., Hansen, V.K., 1961. Filtration of phytoplankton by the octocoral *Alcyonium digitatum* L. Nature 190, 649–650.

Rubenstein, D.I., Koehl, M.A.R., 1977. Mechanisms of filter feeding—some theoretical considerations. Am. Nat. 111, 981–994.

Ruzicka, R.R., Colella, M.A., Porter, J.W., Morrison, J.M., Kidney, J.A., Brinkhuis, V., Lunz, K.S., Macaulay, K.A., Bartlett, L.A., Meyers, M.K., Colee, J., 2013. Temporal changes in benthic assemblages on Florida keys reefs 11 years after the 1997/1998 El Nino. Mar. Ecol. Prog. Ser. 489, 125–141.

Sanchez, J.A., Gomez-Corrales, M., Gutierrez-Cala, L., Vergara, D.C., Roa, P., Gonzalez-Zapata, F.L., Gnecco, M., Puerto, N., Neira, L., Sarmiento, A., 2019. Steady decline of corals and other benthic organisms in the SeaFlower biosphere reserve (Southwestern Caribbean). Front. Mar. Sci. 6, 73. https://doi.org/10.3389/fmars.2019.00073.

Santangelo, G., Bramanti, L., Iannelli, M., 2007. Population dynamics and conservation biology of the over-exploited Mediterranean red coral. J. Theor. Biol. 244, 416–423.

Sasaki, N., Putz, F.E., 2009. Critical need for new definitions of "forest" and "forest degradation" in global climate change agreements. Conserv. Lett. 2, 226–232.

Schmitt, R.J., Holbrook, S.J., Davis, S.L., Brooks, A.J., Adam, T.C., 2019. Experimental support for alternative attractors on coral reefs. Proc. Natl. Acad. Sci. U. S. A. 116, 4372–4381.

Schubert, N., Brown, D., Rossi, S., 2016. Symbiotic versus non-symbiotic octocorals: physiology and ecological implications. In: Rossi, S., Bramanti, L., Gori, A., Orejas, C. (Eds.), Marine Animal Forests: The Ecology of Benthic Biodiversity Hotspots. Springer International Publishing AG, Germany, pp. 887–918.

Sheppard, C., Dixon, D.J., Gourlay, M., Sheppard, A., Payet, R., 2005. Coral mortality increases wave energy reaching shores protected by reef flats: examples from the Seychelles. Estuar. Coast. Shelf Sci. 64, 223–234.

Smith, G.W., Ives, L.D., Nagelkerken, I.A., Ritchie, K.B., 1996. Caribbean sea-fan mortalities. Nature 363, 487.

Stocking, J.B., Rippe, J.P., Reidenbach, M.A., 2016. Structure and dynamics of turbulent boundary layer flow over healthy and algae-covered corals. Coral Reefs 35, 1047–1059.

Stoddart, D.R., 1969. Ecology and morphology of recent coral reefs. Biol. Rev. Camb. Philos. Soc. 44, 433–498.

Szmant, A.M., 1986. Reproductive ecology of Caribbean reef corals. Coral Reefs 5, 43–53.
Tadaki, Y., Shidei, T., 1959. Studies on the competition of forest trees. II. The thinning experiment on small model stand of sugi (*Cryptomeria japonica*) seedlings. J. Physical Soc. Japan 41, 341–349.
Tanentzap, A.J., Igea, J., Johnston, M.G., Larcombe, M.J., 2020. Does evolutionary history correlate with contemporary extinction risk by influencing range size dynamics? Am. Nat. 195, 569–576.
Tsounis, G., Edmunds, P.J., 2017. Three decades of coral reef community dynamics in St. John, USVI: a contrast of scleractinians and octocorals. Ecosphere 8, 22.
Tsounis, G., Rossi, S., Laudien, J., Bramanti, L., Fernandez, N., Gili, J.M., Arntz, W., 2006. Diet and seasonal prey capture rates in the Mediterranean red coral (*Corallium rubrum* L.). Mar. Biol. 149, 313–325.
Tsounis, G., Rossi, S., Grigg, R., Santangelo, G., Bramanti, L., Gili, J.M., 2010. The exploitation and conservation of precious corals. In: Gibson, R.N., Atkinson, R.J.A., Gordon, J.D.M. (Eds.), Oceanography and Marine Biology: An Annual Review. vol. 48. CRC Press, Taylor & Francis Group, pp. 161–211.
Tsounis, G., Edmunds, P.J., Bramanti, L., Gambrel, B., Lasker, H.R., 2018. Variability of size structure and species composition in Caribbean octocoral communities under contrasting environmental conditions. Mar. Biol. 165, 1–14.
Tsounis, G., Steele, M.A., Edmunds, P.J., 2020. Elevated feeding rates of fishes within octocoral canopies on Caribbean reefs. Coral Reefs 39, 1299–1311.
Umanzor, S., Ladah, L., Zertuche-Gonzalez, J.A., 2018. Intertidal seaweeds modulate a contrasting response in understory seaweed and microphytobenthic early recruitment. Front. Mar. Sci. 5, 114.
van de Water, J., Allemand, D., Ferrier-Pages, C., 2018. Host-microbe interactions in octocoral holobionts—recent advances and perspectives. Microbiome 6, 64.
van Oppen, M.J.H., Lough, J.M., 2018. Coral Bleaching: Patterns, Processes, Causes and Consequences, second ed. Springer, Cham.
Vance, R.R., 1973. More on reproductive strategies in marine benthic invertebrates. Am. Nat. 107, 353–361.
von der Heydt, A., Dijkstra, H.A., 2006. Effect of ocean gateways on the global ocean circulation in the late Oligocene and early Miocene. Paleoceanography 21, 18.
Wahle, C.M., 1980. Detection, pursuit, and overgrowth of tropical gorgonians by milleporid hydrocorals—perseus and medusa revisited. Science 209, 689–691.
Walter, R.P., Haynes, J.M., 2006. Fish and coral community structure are related on shallow water patch reefs near San Salvador, Bahamas. Bull. Mar. Sci. 79, 365–374.
Wangpraseurt, D., Weber, M., Roy, H., Polerecky, L., de Beer, D., Suharsono, Nugues, M.M., 2012. In situ oxygen dynamics in coral-algal interactions. Plos One 7, e31192.
Ward-Paige, C.A., Risk, M.J., Sherwood, O.A., 2005. Reconstruction of nitrogen sources on coral reefs: delta N-15 and delta C 13 in gorgonians from Florida Reef Tract. Mar. Ecol. Prog. Ser. 296, 155–163.
Weber, J.N., Woodhead, P.M., 1970. Ecological studies of coral predator Acanthaster planci in the South-Pacific. Mar. Biol. 6, 12–17.
Weil, E., Rogers, C.S., 2011. Coral reef diseases in the Atlantic-Caribbean. In: Dubinsky, Z., Stambler, N. (Eds.), Coral Reefs: An Ecosystem in Transition. Springer, Dordrecht, pp. 465–491.
Weil, E., Rogers, C.S., Croquer, A., 2016. Octocoral diseases in a changing ocean. In: Rossi, S., Bramanti, L., Gori, A., Orejas Saco del Valle, C. (Eds.), Marine Animal Forests: The Ecology of Benthic Biodiversity Hotspots. Springer International Publishing AG, Germany, pp. 1–55.
Weller, D.E., 1987. A reevaluation of the $-3/2$ power rule of plant self-thinning. Ecol. Monogr. 57, 23–43.

Wells, C.D., Martínez-Quintana, Á., Tonra, K.J., Lasker, H.R., 2020. Welcome to the jungle: algal turf negatively affects recruitment of a Caribbean octocoral. BioRxiv. 2020.2005.2027.119404.
Westoby, M., 1981. The place of the self-thinning rule in population-dynamics. Am. Nat. 118, 581–587.
Westoby, M., 1984. The self-thinning rule. Adv. Ecol. Res. 14, 167–225.
White, E.P., Ernest, S.K.M., Kerkhoff, A.J., Enquist, B.J., 2007. Relationships between body size and abundance in ecology. Trends Ecol. Evol. 22, 323–330.
Wild, C., Hoegh-Guldberg, O., Naumann, M.S., Colombo-Pallotta, M.F., Ateweberhan, M., Fitt, W.K., Iglesias-Prieto, R., Palmer, C., Bythell, J.C., Ortiz, J.C., Loya, Y., van Woesik, R., 2011. Climate change impedes scleractinian corals as primary reef ecosystem engineers. Mar. Freshw. Res. 62, 205–215.
Wolff, N., Grober-Dunsmore, R., Rogers, C.S., Beets, J., 1999. Management implications of fish trap effectiveness in adjacent coral reef and gorgonian habitats. Environ. Biol. Fishes 55, 81–90.
Wulff, J., 2016. Sponge contributions to the geology and biology of reefs: past, present, and future. In: Hubbard, D.K., Rogers, C.S., Lipps, J.H., Stanley, G.D. (Eds.), Coral Reefs at the Crossroads. Springer, Dordrecht, pp. 103–126.
Yoda, K., Kira, T., Ogawa, H., Hozami, K., 1963. Self thinning in overcrowded pure stands under cultivated and natural conditions. J. Biol. 14, 107–129.
Yoshika, P.M., Yoshioka, B.B., 1991. A comparison of the survivorship and growth of shallow-water gorgonian species of Puerto-Rico. Mar. Ecol. Prog. Ser. 69, 253–260.
Yoshioka, P.M., 1994. Size-specific life-history pattern of a shallow-water gorgonian. J. Exp. Mar. Biol. Ecol. 184, 111–122.
Yoshioka, P.M., 1996. Variable recruitment and its effects on the population and community structure of shallow-water gorgonians. Bull. Mar. Sci. 59, 433–443.
Yoshioka, P.M., 1997. In: Are Variations in Gorgonian Recruitment Determined by "Pre-Recruitment" or "Post-Recruitment" Processes? 8th ICRS 2, pp. 1175–1178.
Yoshioka, P.M., Yoshioka, B.B., 1987. Variable effects of Hurricane David on the shallow water gorgonians of Puerto Rico. Bull. Mar. Sci. 40, 132–144.

CHAPTER FOURTEEN

Octocoral populations and connectivity in continental Ecuador and Galápagos, Eastern Pacific

Sascha C.C. Steiner[a,d,*], Priscilla Martínez[b], Fernando Rivera[b], Matthew Johnston[c], and Bernhard M. Riegl[c]

[a]Cooperación Alemana al Desarrollo (GIZ), Quito, Ecuador
[b]Instituto Nazca de Investigaciones Marinas, Salinas, Ecuador
[c]Department of Marine and Environmental Sciences, Halmos College of Arts and Sciences, Nova Southeastern University, Dania Beach, FL, United States
[d]Institute for Tropical Marine Ecology (ITME) Inc., Roseau, Dominica
*Corresponding author: e-mail address: scc.steiner@itme.org

Contents

Abstract

Octocorals are important zoobenthic organisms, contributing to structural heterogeneity and species diversity on hardgrounds. Their persistence amidst global coral reef degradation and ocean acidification, has prompted renewed interest in this taxon. Octocoral assemblages at 52 sites in continental Ecuador and Galápagos (23 species, 3742 colonies) were examined for composition, size distributions within and among populations, and connectivity patterns based on ocean current models. Species richness varied from 1 to 14 species per site, with the richest sites on the continent. Three assemblage clusters were recognised based on species richness and population size, one with a mix of sites from the mainland and Galápagos (defined by *Muricea fruticosa* and *Leptogorgia alba, Muricea plantaginea* and *Pacifigorgia darwinii*), the second from Santa Elena in southern Ecuador (defined by *M. plantaginea* and *L. alba*) and the third from the northernmost sites on the continent, in Esmeraldas (defined by *Muricea fruticosa, Heterogorgia hickmani, Leptogorgia manabiensis*). Based on biophysical larval flow models with 30, 60, 90-day Pelagic Larval Duration, good connectivity existed along the South American mainland, and from the continent to Galápagos. Connectivity between Galápagos, Cocos, Malpelo

Advances in Marine Biology, Volume 87
ISSN 0065-2881
https://doi.org/10.1016/bs.amb.2020.07.002

and the Colombian mainland may explain the wide distribution of *L. alba*. *Muricea plantaginea* had the densest populations with the largest colonies and therewith was an important habitat provider both in continental Ecuador and Galápagos. Continental Ecuador harbours the most speciose populations of octocorals so far recorded in the southern Eastern Tropical Pacific (ETP). Most species were uncommon and possibly vulnerable to local extirpation. The present study may serve as a base line to determine local and regional impacts of future disturbances on ETP octocorals.

1. Introduction

Octocorallia is a subclass within Anthozoa (Ehrenberg), phylum Cnidaria (Hatschek), with a wide geographic, bathymetric and solely marine distribution (Bayer, 1981). In shallow water ecosystems such as temperate or tropical hardgrounds and coral reefs, octocorals contribute to carbonate sediments (Cary, 1918; Konishi, 1981) and are an important component of the zoobenthos (Bayer, 1961; Fabricius and Alderslade, 2001; Sammarco and Coll, 1992; Spalding et al., 2001). Similar to stony reef corals (Scleractinia), the presence of octocoral taxa with arborescent growth forms compartmentalize the benthic boundary layer by increasing structural habitat complexity for many epizoic algae and invertebrates (Cantera et al., 1987; Glynn et al., 2019; Lasker and Coffroth, 1988; Mosher and Watling, 2009; Patton, 1972; Ramos, 1995; Vreeland and Lasker, 1989) as well as fishes (Etnoyer and Warernchuck, 2007; Lasker, 1985; Taylor et al., 2013) and other reef biota. Dense octocoral stands also create a mosaic of physio-chemical gradients in light penetration, hydrodynamics and sedimentation rates (Valisano et al., 2016; Wainwright et al., 1976), as well as gas exchange and material cycles (see Fréchette et al., 1989; González-Ortiz et al., 2014), analogous to seagrass meadows (Irlandi, 1996; Vogel, 1994). They thus augment the heterogeneity of conditions in micro-habitats and interstitial spaces driving biodiversity (Graham and Nash, 2013; Plass-Johnson et al., 2016; Richardson et al., 2020), as well as the survival of rare species (Mouillot et al., 2013).

The impacts of the global reef crisis on octocorals have received less attention than in other taxa (but see Lasker et al., 2020a, in this volume). Unlike scleractinian reef corals, at least in Caribbean settings, population densities of octocorals and the ecosystem benefits of increased structural habitat heterogeneity, showed persistence (Lasker et al., 2020b; Lenz et al., 2015; Tsounis and Edmunds, 2017) and even expansion (Ruzicka et al., 2013;

Sánchez et al., 2019) during recent decades, amidst a general structural flattening of reefs caused by the degradation of stony coral communities (Alvarez-Filip et al., 2009, Glynn et al., 2018). Furthermore, Enochs et al. (2016) presented evidence of a degree of octocoral resistance to ocean acidification. These observations were based on zooxanthellate octocoral species sharing habitat with reef-building Scleractinia. In the wake of global bleaching events and mass mortalities (Hughes et al., 2017), winners and losers among mixotrophic zooxanthellate Scleractinia (Loya et al., 2001) were identified, gauged on population performance following disruption. The concept of winners and losers requires an established regional baseline of populations, in which some species have either a disturbance-resistant population or can be effectively replenished through connectivity. It is unclear whether winners and losers can also be identified among octocorals, which despite demonstrated persistence are nonetheless not exempt from population declines and anthropogenic impacts (Cerrano et al., 2000; Gomez et al., 2015; Sánchez and Ballesteros, 2014; Tsounis et al., 2012). Eastern Pacific octocorals are heterotrophic (Van Oppen et al., 2005) and it is unclear how their populations may fare in future disturbances like thermal anomalies.

Octocoral assemblage structure is a building block in defining its constituent populations for comparisons across wider geographic regions. In the Eastern Tropical Pacific (ETP) such studies have so far been limited to select locations in Mexico (Abeytia et al., 2013), Panama (Gomez et al., 2014), Colombia (Sánchez et al., 2011; Sánchez and Ballesteros, 2014) and part of Ecuador (Steiner et al., 2018a, 2018b). The majority of octocoral studies from the region describe species taxonomy and range extensions (summarised in Steiner et al., 2018a, 2018b), but few studies develop data that could be used as baselines to evaluate impacts on populations and assemblages following large-scale, or even local, perturbances.

To identify species and populations that best lend themselves for monitoring and possibly anticipating the susceptibility of octocoral populations towards future disturbances in the ETP, assemblage composition, population and size distributions were examined at 37 sites along the continental coast of Ecuador and 15 sites in Galápagos. In view of the paucity of reproductive data for ETP species (Gomez et al., 2018; Grigg, 1977; Kahng et al., 2011), surface current models were applied to discern possible connectivity patterns among continental and insular regions and to establish the potential role of Ecuador's octocoral populations as larval sources or recipients in a wider geographic context. Similar larval connectivity models based on hydrodynamics have been successfully applied in shallow and deep communities (Lequeux et al., 2018; Metaxas et al., 2019; Wanatabe et al., 2009; Wood et al., 2016). The emerging picture provides a regional baseline for this habitat-shaping taxon, upon which conservation initiatives and further studies can build.

2. Materials and methods

2.1.1 Octocoral assemblages

Octocorallia occur along the entire Ecuadorian coast, exposed to the convergence of the seasonal southward–flowing Panama Flow and the northward flowing Peru Current, as well the southward–flowing sporadic El Niño current in the months between December and April (Fiedler, 1992; Fiedler and Lavine, 2016; Strub et al., 1998), and throughout the Galápagos archipelago (Fig. 1). Local anthropogenic disturbances include colony abrasion and detachment from extensive damage by bottom-trawling up to the year 2009 (Government of Ecuador, 2009; Rivera and Martínez, pers. comm.) and ongoing fishing gear loss (Figueroa-Pico et al., 2020). Octocorals were encountered in 37 of 40 continental sites and in 15 of the

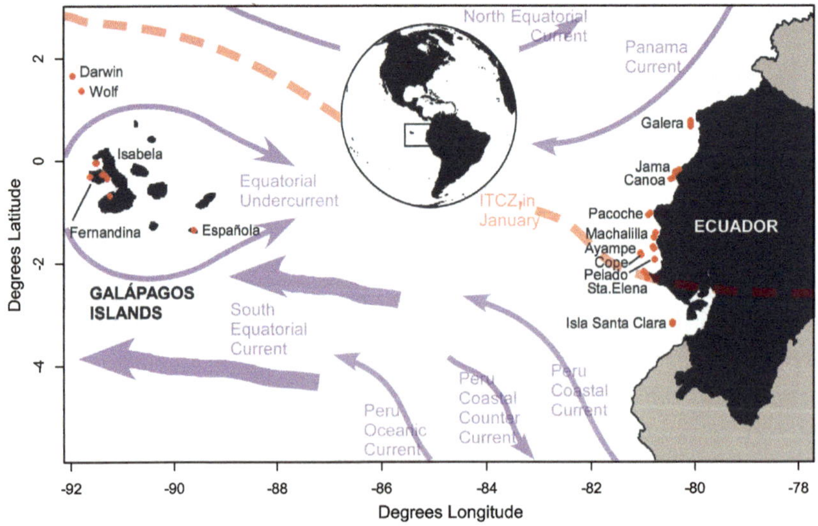

Fig. 1 Sampling sites along the mainland coast and the Galápagos Archipelago in Ecuador, Eastern Tropical Pacific. Cocos and Malpelo Islands, mentioned in the text, are to the north of the mapped area. ITCZ oscillating Intertropical Conversion Zone.

29 Galápagos sites explored in depths between 6 and 20 m (Table 1). Surveys employed belt transects for colony counts and quantification of size classes in evenly distributed subsamples (2015–2018). Sampling effort per site varied among sampling regions and covered 50 m^2 (n = 29, mainland sites), 80 m^2 (n = 8, mainland), and 100 m^2 (n = 15 Galápagos). Colony density per species, per site was determined and assigned to a semi-quantitative, 9-point density scale and ranked along a gradient from rare (lowest density = 1) to common (highest density = 9), to reduce the range of densities and render the data more compatible with the chosen analyses (Borcard et al., 2011; Elzinga et al., 2001; Legendre and Legendre, 2012). Diversity indices (H′ = Shannon Entropy, J = Pielou's Evenness; Legendre and Legendre, 2012) were calculated from untransformed colony frequency per species, per site.

The identification of species was based on a previously built photographic repository of in situ colonies, as well as specimen collections (2015–2017), which had been examined macro- and microscopically, in some cases with the extraction of sclerites, and compared to the latest species descriptions as detailed in Steiner et al. (2018a, 2018b).

2.1.2 Evaluation of assemblage patterns

Assemblage patterns were explored by hierarchical, agglomerative cluster analysis in the R-package vegan (Oksanen et al., 2017; R Development Core Team, 2012) using procedures outlined in Borcard et al. (2011). A Euclidian distance matrix was calculated for the rank-transformed data and several clustering methods were evaluated (single and complete linkage, UPGMA and Ward; Borcard et al., 2011; Legendre and Legendre, 2012). The optimal number of interpretable clusters was investigated by graphing Silhouette widths, which are a measure of membership of an object to its cluster (Borcard et al., 2011). This was checked against the Mantel-optimal number of clusters, which is a comparison between the distance matrix and binary matrices that represent the partitions (Borcard et al., 2011). Upon coincidence of the number of optimal clusters as checked by these methods, the quality of appropriate group-membership (i.e. the incidence of misclassifications) was examined with another silhouette plot. When misclassifications were minimised, the appropriate clustering method and optimal number of clusters had been determined (Borcard et al., 2011). The chosen dendrogram was then combined with a heatmap to bring forth patterns of similarity, and the species content in the clusters was visualised by sorting them according to weighted averages on site scores.

Table 1 Location of study sites.

	Site code	Site name	Depth	Latitude	Longitude	Area	Province/ Island	Region
1	CEGSF1	Piedra d Quingue	7	0.725000	−80.100000	Reserva Marina Galera San Francisco	Esmeraldas	Continental Ecuador
2	CEGSF2	Tortuga 1	12	0.769000	−80.101000			
3	CEGSF3	Punta Alta	8	0.661000	−80.099000			
4	CEGSF4	Frente al Horno	8	0.665000	−80.098000			
5	CMJAM1	Vaca Brava 1	10	−0.240708	−80.394160	Jama	Manabí	
6	CMJAM2	Vaca Brava 2	12	−0.237985	−80.394880			
7	CMJAM3	Punta Venado	12	−0.223220	−80.387960			
8	CMJAM4	Punta Ballena	10	−0.174546	−80.329940			
9	CMCAN1	Cabo Pasado 1	6	−0.357037	−80.488210	Canoa		
10	CMCAN2	Cabo Pasado 2	7	−0.357589	−80.485070			
11	CMCAN3	La Saibita	12	−0.342415	−80.445000			
12	CMPAC1	Roca Ahogada	7	−1.019466	−80.881630	Refugio Pacoche		
13	CMPAC2	Cabo S. Lorenzo	7	−1.060216	−80.911900			
14	CMMAC1	Bajo Frailes	12	−1.493247	−80.806530	Parque Nacional Machalilla		
15	CMMAC2	Horno de Pan	10	−1.498630	−80.809010			
16	CMMAC3	Islote Sombrerito	12	−1.405504	−80.770590			
17	CMAYA1	Rinconada	10	−1.712225	−80.805660	Ayampe Canta Gallo-Machalilla		
18	CMAYA2	Rinconada 1	12	−1.710244	−80.806620			
19	CMAYA3	Ahorcados E	6	−1.677532	−80.835570			

20	CSCOP1	Bajo Mantas	10	−1.812314	−81.063320	Reserva Marina, Bajo Copé	Santa Elena
21	CSCOP2	Bajo Copé 1	12	−1.845219	−81.052770		
22	CSCOP3	Bajo Copé 2	14	−1.816403	−81.061280		
23	CSCOP4	Bajo Fer 3	14	−1.812929	−81.063670		
24	CSPEL1	EPPNW 2	20	−1.933333	−80.792500	Reserva Marina El Pelado[a]	
25	CSPEL2	EPPNW 3	19	−1.931944	−80.790556		
26	CSPEL 3	Islote Pelado 5	13	−1.933611	−80.788889		
27	CSPEL4	EPPNE 6	14	−1.931389	−80.787222		
28	CSPEL5	I Pelado 8 W	15	−1.936111	−80.788889		
29	CSPEL6	I Pelado 8E	15	−1.935833	−80.791111		
30	CSPEL7	EPPSE	15	−1.933889	−80.786667		
31	CSAYA	Ayangue	15	−1.987222	−80.757500	Ayangue	
32	CSSEL1	Guarro	10	−2.2992161	−80.934220	Reserva PFM Santa Elena	
33	CSSEL2	Bajo Ballena	8	−2.1963506	−80.957000		
34	CSSEL3	Casa de Lobos	14	−2.1837304	−81.003650		
35	CSSEL4	Piedras Altas	12	−2.1666222	−81.003300		
36	COSCL1	Sitio Oeste	10	−3.189198	−80.452840	Isla Santa Clara	El Oro
37	COSCL2	Sitio Oeste 1	10	−3.1614192	−80.442280		
38	COSCL3	Sitio NO	10	−3.1623725	−80.448590		

Continued

Table 1 Location of study sites.—cont'd

	Site code	Site name	Depth	Latitude	Longitude	Area	Province/ Island	Region
39	GDDAR1	Arch NE	18	1.673667	−91.989806	Darwin	Darwin I.	Galápagos
40	CDDAR2	Arch E	18	1.673065	−91.989113			
41	GWWOL1	Wolf	15	1.393709	−91.818958	Wolf	Wolf	
42	GEGAR1	Gardener	15	−1.346000	−89.642000	Gardener	Española	
43	GEGAR2	Canones	10	−0.343000	− 91.324000	Canones		
44	GICAS	Caseta	10	−0.295000	−91.357000	Caseta	Isabela	
45	GIPNE	Playa Negra	12	−0.244000	−91.395000	Playa Negra		
46	GIPMO1	Pesca	10	−0.686000	−91.270000	Punta Moreno		
47	GIPMO2	Prot	10	−0.686000	−91.270000			
48	GIPMO3	Turismo	10	−0.686000	−91.270000			
49	GIVRO1	Pared	13	−0.051000	−91.560000	Punta Vincente Rocafuerte		
50	GIVRO2	Pesca	11	−0.686000	−91.270000			
51	GFPES	Punta Espinoza	10	−0.270000	−91.435000	Punta Espinoza	Fernandina	
52	GFDOU	Cabo Douglas	12	−0.301672	−91.653835	Gabo Douglas		

[a]Detailed site descriptions for El Pelado Marine Reserve available in Steiner et al. (2018a).

2.1.3 Evaluation of population

The structure of individual species populations was examined via the distribution of size-classes, which were expressed as an area based on a planar view of the colony in photographic images and videos containing scale bars with 5 or 10 cm units, from which measurements of colony "width" (x-axis) and "height" (y-axis) were taken. Colony sizes were sorted into four size classes (SC1 < 100 cm^2, SC2 100–399 cm^2, SC3 400–899 cm^2 and SC4 ≥ 900 cm^2). The number and dimensions of the size classes were strategically created in order to reflect distinct size classes among the smaller species and the many small colonies of larger species observed during the surveys.

2.1.4 Biophysical model for connectivity

Given the few reproductive strategies available for ETP octocoral species considered, we assumed that the species were gonochoric (Gomez et al., 2018; Kahng et al., 2011; Simpson, 2009) and released actively swimming larvae, with larval durations of up to 90 days (Baird et al., 2009; Coelho and Lasker, 2016; Dahan and Benayahu, 1998), and a spawning period similar to the one reported for stony corals, from January to May (Glynn et al., 2017; Riegl et al., 2019b), which is compatible with Grigg's (1977) observation of spawning during annual temperature peaks in two EP species, but see Gomez et al. (2018). Whether fertilization was external or internal (brooding), as well as the time from spawning to fertilization and completed embryogenesis (planulae) was unknown. Based on this premise, larval dispersal and connectivity among known occurrences of octocorals were examined with a biophysical model (Johnston and Akins, 2016; Riegl et al., 2019b). The applied model simulated the movement of larvae (planulae) across a 2-D, fixed cell-dimension grid using stochastic, Lagrangian particle tracking. The grid incorporated the Galápagos and northern Eastern Tropical Pacific (ETP) regions with a grid pixel size of 1/12th decimal degree (~9 km), using ocean current data from the global Hybrid Current Ocean Model (HYCOM). Daily data from 2005 to 2015 were employed in a 10-year simulation. This time period included several ENSO events (see Glynn et al., 2017, Riegl et al., 2019a, 2019b). Planulae diffusion in ocean surface currents and recruitment were tracked from points of origin (spawning) to settlement in known octocoral sites in the study region over the 30, 60, 90-day PLD (Pelagic Larval Duration). Known octocoral populations were clustered into precincts (southern and northern Galápagos; Cocos, Costa Rica; Malpelo, Chocó and Gorgona, Colombia, sites on the Ecuadorian coast in Esmeraldas,

Manabí, Santa Elena and El Oro provinces). One geographic location every 9 km within a precinct was selected as a site for a reproductively mature octocoral cluster to seed the simulations (density-independent and one cluster per point, total 350 clusters). Simulations selected reproductive clusters within all precincts at the start of each year. Each mature cluster was allowed to produce larvae every 30 days from January to May. Larval release was simultaneous from all colonies and randomly staggered within each model colony over 5 days.

From HYCOM simulations, a daily vectorized flow-field of ocean currents (measured in m/s) was obtained over which planulae dispersed following a Lagrangian path during the pelagic larval duration (PLD, 30, 60, and 90 days). The u (east-west) and v (north-south) positions of the larvae (P_i = east-west, north-south position at time i) and their hourly transition times were tracked. The vector ($\vec{v}_i$) traversed by any planula per Δt was calculated by bilinearly interpolating the nearest four ocean current vectors to the planula's x/y position. A timer tracked the hours elapsed over each planula's PLD. After a pre-competency period of two lapsed days, known coordinates of coral areas within 1 km of the planula's position at each Δt were determined. If a planula was within 1 km of a reef, it was allowed to recruit to that position and then remain non-reproductive for the duration of the simulation (10 years), to conservatively account for timing until sexual maturity. Planulae were considered lost if they did not come within 1 km of a suitable recruitment position during their PLD.

Using this protocol, the path from spawning to recruitment was chronicled (as were founder colonies and subsequent recruitment positions) in order to produce a connectivity matrix. To model conservatively, since larval survival rates are unknown, only one larva per colony was incorporated in the simulation per monthly reproductive cycle. To account for possible differences in larval duration, we constructed 30, 60 and 90-day models and connectivity matrices that cross-plotted the founder clusters on the x-axis with the recruitment positions on the y-axis. Each recruitment position was colour coded via a cool (blue—low) to warm (red—high) scale to represent relative densities of recruits.

3. Results

A total of 3742 octocoral colonies were encountered, of which 63 and 8 colonies from continental and Galápagos sites, respectively, could not be assigned to a species and were predominantly *Muricea* spp. The 23 identified species belonged to the families Clavulariidae (1), Gorgoniidae (13) and Plexauridae (9), (Table 2). From here on, species are referred to as "common" if >500 colonies were encountered in this survey, "uncommon" with 100–500 colonies and "rare" with <100 colonies. Only three species reached locally high densities of >2–3 colonies m^{-2}, *Muricea plantaginea, Muricea fruticosa* and *Leptogorgia alba*, while most species were widely distributed but at low density and rare. Co-occurrence on the mainland and Galápagos was observed in *M. plantaginea, Heterogorgia hickmani* and *L. alba.*

3.1 Distribution and structure of octocoral assemblages

Cluster analysis to define assemblage groups expected three major groups of sites. These were based on silhouette width and the correlation between the original distance matrix of rescaled octocoral densities per site and binary matrices from several possible numbers of partitions. Silhouette plotting of partition accuracy in three clusters was satisfactory (Fig. 2). From this point forward, we worked on the assumption that the Ecuadorian octocoral communities differentiated into three broad groups.

The largest cluster (henceforth referred to as cluster C1 and shown in green in Figs. 3, 5, 6, and 7) included a mixture of sites from the mainland and Galápagos (Fig. 3A). The cluster itself partitioned into two further groups, one consisting of sites mainly in Galápagos, the other mainly in continental Ecuador. The species with highest weight in the Galápagos subcluster were *M. plantaginea* and *Pacifigorgia darwinii*, which was encountered exclusively in Galápagos. The second sub-cluster, which contained all continental sites plus Darwin (Galápagos), was characterized by the widespread *L. alba* and *M. fruticosa*.

Cluster C2 (shown in red) was made up exclusively by samples from the Santa Elena region (Fig. 3) and was characterized by high species richness and by weight of *M. plantaginea* and the less abundant *L. alba* at Pelado. Almost unique to this cluster were the overall uncommon *Muricea purpurea, Leptogorgia taboguillae*, and a colour morph of *M. plantaginea* (Fig. 3B).

A third well-defined cluster (C3, shown in blue) consisted of samples from the northern province of Esmeraldas, with one mis-grouped sample (Fig. 2C) from Santa Elena. This was a species-rich group of samples with high weights of widespread species (*M. fruticosa, H. hickmani, Leptogorgia manabiensis*) and species largely confined to this region (*Pacifigorgia irene, Pacifigorgia firma, Pacifigorgia rubicunda*).

Table 2 Species and colonies (out of 3742) recorded in the surveys.

Species	Mainland	Galápagos
Suborder Holaxonia Studer, 1887		
Family Plexauridae Gray, 1859		
1) *Heterogorgia hickmani* Breedy and Guzman, 2004[b]	36	17
2) *Heterogorgia verrucosa* Verrill, 1868	2	–
3) *Muricea austera* Verrill, 1869	78	–
4) *Muricea crassa* Verrill, 1868	16	–
5) *Muricea fruticosa* Verrill, 1869[b]	528	–
6) *Muricea plantaginea* (Valenciennes, 1846)[b]	1011	525
7) *Muricea purpurea* Verrill, 1864	247	–
8) *Muricea squarrosa* Verrill, 1869	63	–
9) *Psammogorgia* cf. *arbuscula* Verrill, 1866	84	–
Family Gorgoniidae Lamoroux, 1812		
10) *Leptogorgia alba* (Duchassaing and Michelotti, 1860)[b]	104	533
11) *Leptogorgia cuspidata* Verrill, 1865	4	–
12) *Leptogorgia laxa* Hickson, 1928	5	–
13) *Leptogorgia manabiensis* Soler-Hurtado, Megina, Machordom, López-González, 2017	163	–
14) *Leptogorgia pumila* (Verrill, 1868)	18	–
15) *Leptogorgia taboguillae* (Hickson, 1928)	38	–
16) *Leptogorgia* cf. *rigida* Verrill, 1864	1	–
17) *Pacifigorgia adamsii* (Verrill, 1868)	30	–
18) *Pacifigorgia darwinii* (Hickson, 1928)[a]	–	97
19) *Pacifigorgia firma* Breedy and Guzman, 2003	16	–
20) *Pacifigorgia irene* Bayer, 1951	12	–
21) *Pacifigorgia rubicunda* Breedy and Guzman, 2003	16	–
22) *Pacifigorgia stenobrochis* (Valenciennes, 1846)	5	–
Suborder Stolonifera Thompson and Simpson, 1809		
Family Clavulariidae Hickson, 1894		
23) *Carijoa riisei* (Duchassaing and Michelotti, 1860)	20	–

[a]Reported for Galápagos.
[b]Reported for Galápagos and continental Ecuador.
Species are referred to as common with over 500 colonies, uncommon with 100–500 colonies and rare with less than 100 colonies.

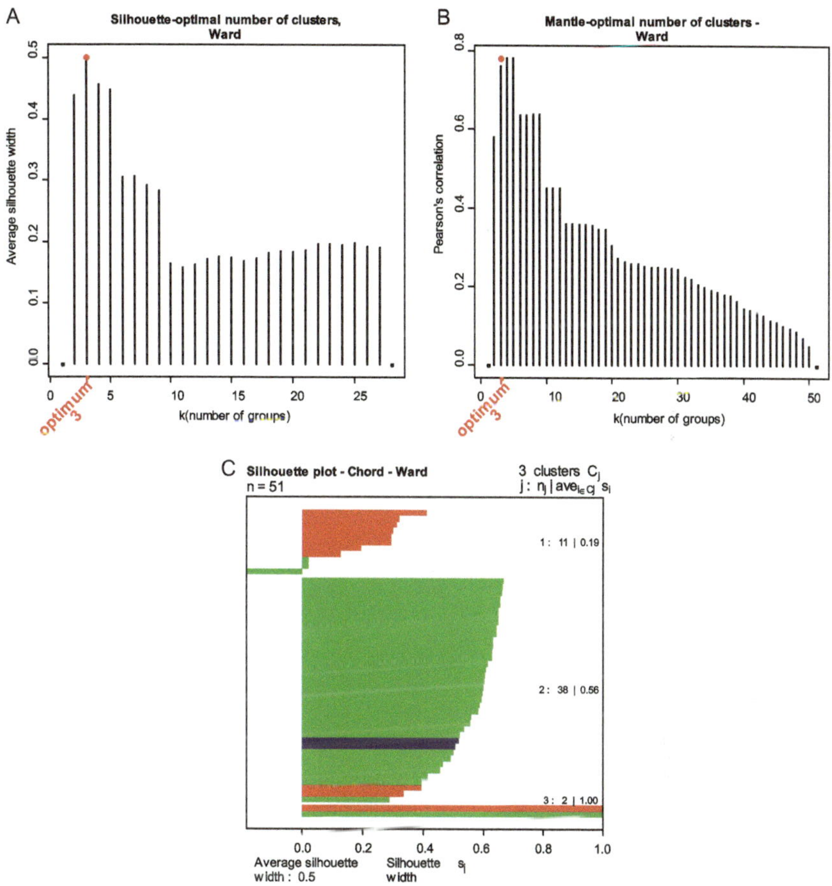

Fig. 2 (A) Bar plot showing the average silhouette widths for 27 groups. The best partition (shown by red text "optimum") is at the largest silhouette width. (B) Bar plot of correlations between original distance matrix and binary matrices computed from the dendrogram at the level of each vertical bar. Again, the highest correlation is shown at a level of 3 clusters. (C) Silhouette plot of the finally chosen three-group partition from the Ward clustering. Most bars towards the right are properly classified, only one bar to left indicates a misclassification. Relatively poor differentiation between the blue and green (clusters 3 and 1 in Fig. 3) is evident. Overall, the result gives high confidence in the observed clustering pattern (see Fig. 3).

3.2 Distribution and structure of octocoral populations

Species richness was variable across the sampling area, ranging from 1 to 14 species, with the richest sites found on small islands and rocks off the mainland, both in the drier and cooler South (Pelado ($H'=1.5$, $J=0.6$—$H'=2.0$, $J=0.8$) and Santa Elena peninsula ($H'=1.3$, $J=0.5$—$H'=2.3$, $J=0{,}6$)) and

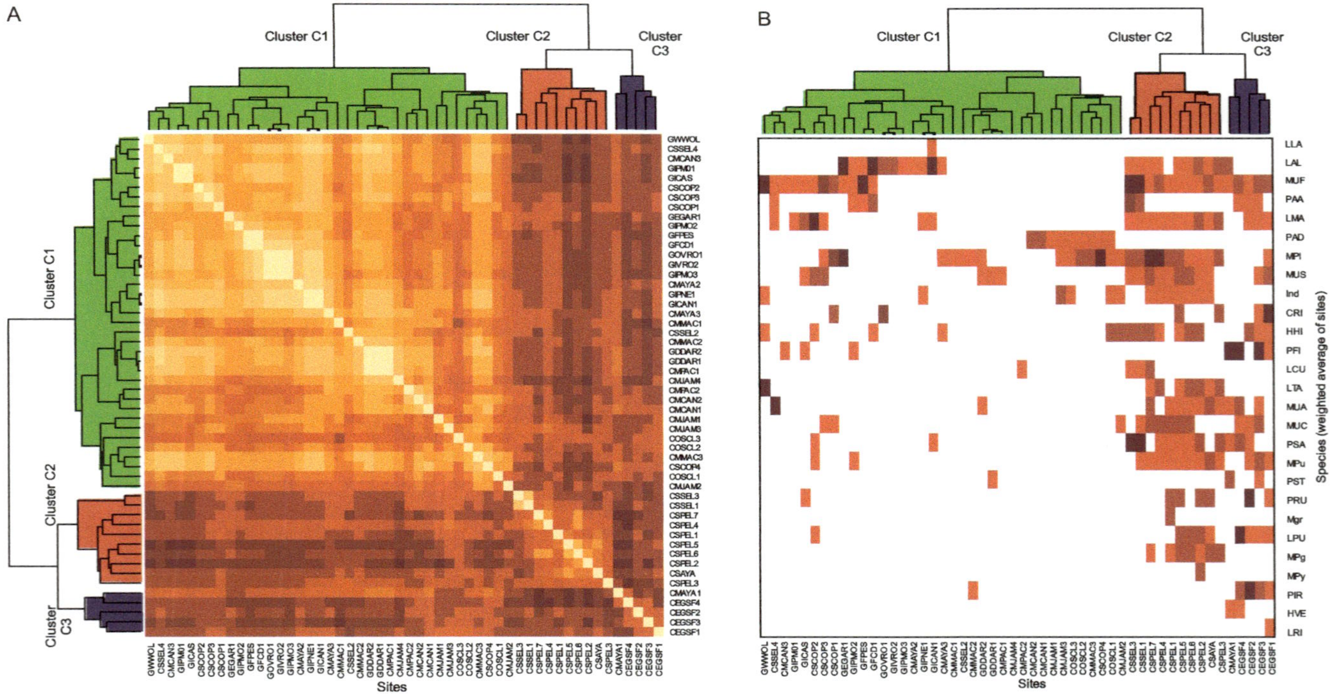

Fig. 3 Heatmap of Euclidian distances among samples and dendrogram. (A) The dendrograms to the left and above the heatmap are identical and the information contained within the heatmap is equivalent to that in the dendrograms. Colour scheme of clusters follows Fig. 2. The combination of these two graphical methods allows easy detection in patterns. (B) The same dendrogram above a heatmap of a community table ordered by cluster grouping and weighted averages on site scores (values between 0 = white, and brown = 9; the redder, the higher the value). This depicts which species can be considered characteristic for specific sites.

the wetter and warmer North (sites in Esmeraldas (H′=2.2, J=0.9—H′=2.4, J=0.9)). Overall, the continent was more speciose than Galápagos (Wilcoxon-test, W=1, $p<0.008$; Fig. 4B). Population sizes, however, did not differ significantly between the mainland and Galápagos (Wilcoxon-test, W=18, $p=0.839$; Fig. 4A). The largest and densest populations were found at Pelado and Santa Elena Peninsula and Galápagos (Fig. 4A and B), and consisted primarily of *M. plantaginea* and *L. alba*.

Population sizes and size-frequencies of species that were common on both mainland and in the Galápagos, and were found to define clusters in the heatmap (Fig. 3B), namely *M. plantaginea* (cluster C1) *L. alba* (cluster C3) and *H. hickmani* (clusters C1–C3), were further investigated.

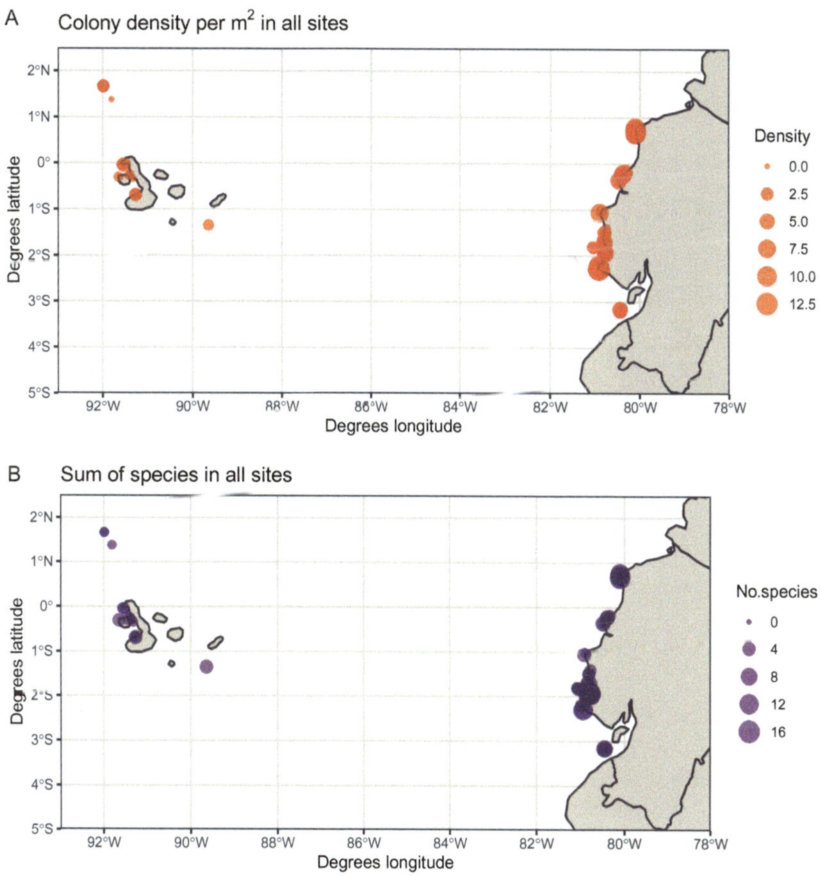

Fig. 4 (A) Sum of colony densities (Sum_dens=number of colonies m^{-2}) in all species and (B) sum of species (Sum_spec) present in each site (including a category "unidentified" and two colour morphs listed separately for *M. plantaginea*), as indicator of population size in the sample sites.

All three species had comparably as dense populations in the Galápagos as on the continent. On the continent and Galápagos (cluster C1), *L. alba* were skewed towards smaller specimens, which was not so clearly the case at Pelado and Santa Elena peninsula (cluster C2) nor at Esmeraldas (cluster C3), which had the sparsest population (Fig. 5A and B). *M. plantaginea*, also had the smallest and least dense populations in Esmeraldas (cluster C3) but

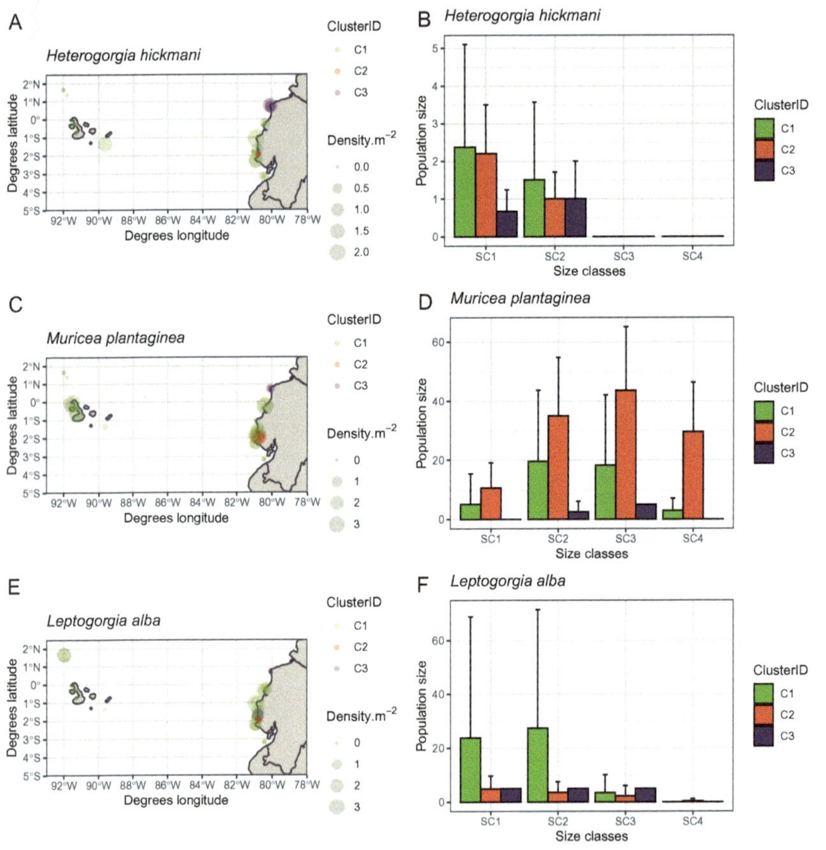

Fig. 5 Characterization of populations of some common octocoral species and defined assemblages on the mainland coast of Ecuador and the Galápagos Islands (A,C, and E) denoting the size of populations, coded by cluster membership (B,D, and F) and the size distribution of the population within each cluster (size classes SC1 $<100\,cm^2$, SC2 $100–399\,cm^2$, SC3 $400–899\,cm^2$ and SC4 $\geq 900\,cm^2$). Cluster ID refers to the clusters identified in Fig. 2A and B, where the latter also shows the contribution of species to cluster formation. Note that location of all sample sites is indicated, even if empty (smallest circle, see legend). Cluster ID refers to the assemblage clusters (same colour-code) obtained from Fig. 3. HHI in legend is *Heterogorgia hickmani*, MPI is *Muricea plantaginea*, LAL is *Leptogorgia alba*.

was similarly dense on the mainland coast and Galápagos (cluster C1) and at Pelado and the Santa Elena peninsula (cluster C2). The populations in all three clusters were overall clearly skewed towards the larger size-classes (Fig. 5C and D). *H. hickmani* was distributed at similar density in clusters C1 and C2 but had low density in cluster C2 (Pelado). It was densest in western Galápagos (especially at Isabela, cluster C2).

Only one species was widespread in Galápagos and was not found on the mainland coast, *Pacifigorgia darwinii*. The species was common in the colder region of Galápagos on the west-coasts of Isabela and Fernandina. It was also encountered, albeit more rarely, on the other islands. Most of the population was encompassed within the smallest two size-classes, and few larger colonies were encountered (Fig. 6A and B).

Among some of the assemblage-defining species that only occurred on the continent, site differences in population sizes and colony size distributions were observed. *Leptogorgia manabiensis* (Fig. 5C and D) was widespread and dense on the continental coast and populations consisted mainly of size class 2 (100–399 cm^2). At Pelado (cluster C2), the population consisted of less dense, but larger colonies than at the other sites. No very small individuals were encountered in Esmeraldas (cluster C3). Three *Muricea* species were dense and assemblage-defining on the mainland coast, but population structures differed. *M. fruticosa* had the numerically largest population, which was heavily skewed towards the smallest size class at Pelado and Santa Elena peninsula (cluster C2). This was not the case in the other numerically smaller populations of this species on the mainland (Fig. 7A and B). Also, in *Muricea austera,* the Pelado and Santa Elena peninsula population was skewed towards the smallest size-class, while the middle (SC2) and large (SC3) size-classes characterized the other mainland sites in this species (Fig. 7C and D). *Muricea squarrosa* was relatively dense and had larger colonies in Esmeraldas (cluster C3) compared to other sites.

3.3 Connectivity patterns

Based on biophysical larval flow models under assumption of a variably long (30, 60, 90-day) PLD, octocoral communities along the South American mainland appeared well-connected, and connectivity also existed from the continent to the Galápagos Islands in all three PLD scenaria (Fig. 8). In Galápagos, sites within the southern (all islands except Darwin and Wolf) and northern archipelago (Darwin, Wolf) were modelled as well connected, and connectivity existed between the two parts of the archipelago, but no direct connection back to the mainland existed in the model.

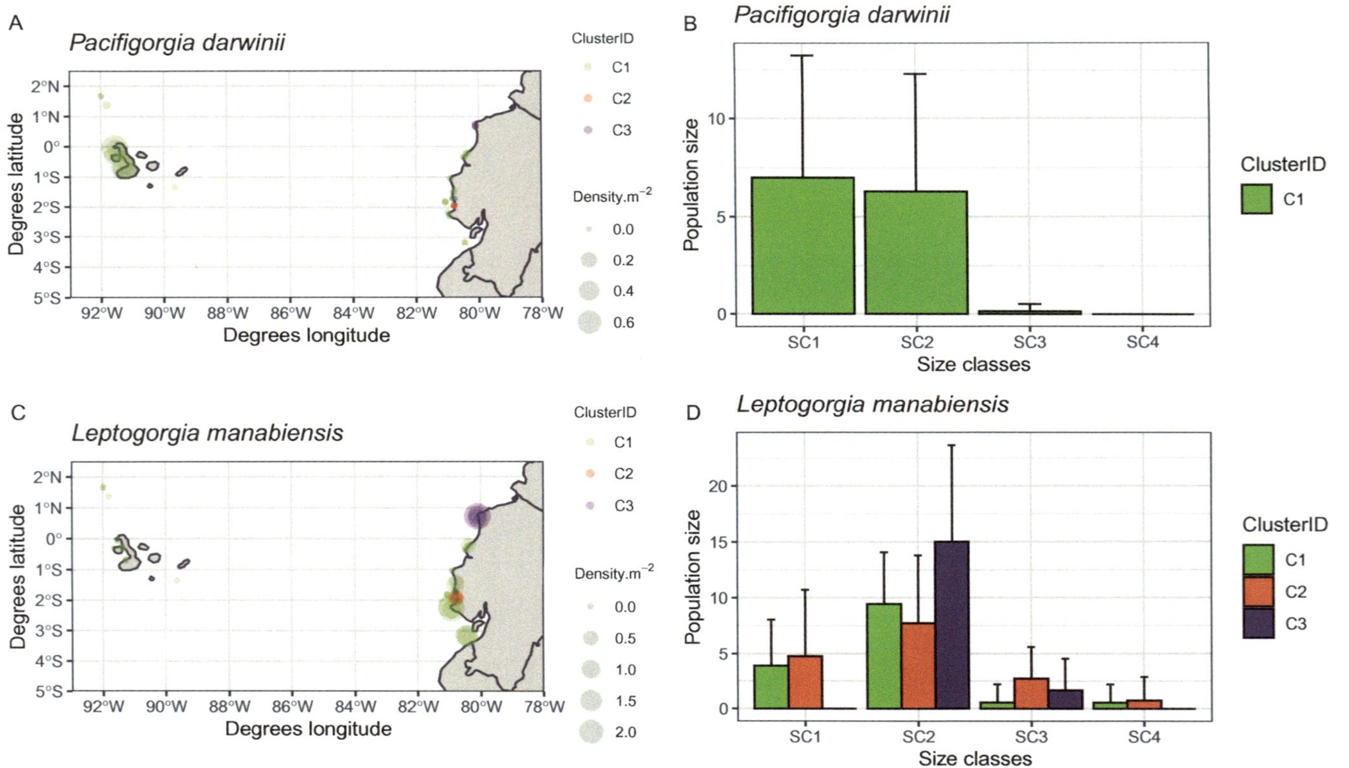

Fig. 6 (A and B) Locally restricted populations of *Pacifigorgia darwinii* were exclusively encountered in Galápagos (A) shows the size of the population (B) shows the size distribution of the population (size classes SC1 $<100\,cm^2$, SC2 $100–399\,cm^2$, SC3 $400–899\,cm^2$ and SC4 $\geq 900\,cm^2$). Cluster ID refers to the clusters identified in Fig. 2A and B, where the latter also shows the contribution of species to cluster formation. *Pacifigorgia darwinii* occurred only in Isabela (Galápagos), which was contained in cluster C1. (C and D) Colonies of *Leptogorgia manabiensis* were widespread on the continent, but absent in Galápagos (same size classes as B). Note that position of all sample sites is indicated, even if empty (smallest circle, see legend). Cluster ID refers to the assemblage clusters (same colour-code) obtained in Fig. 3. PAD in legend is *Pacifigorgia darwinii*, LMA *Leptogorgia manabiensis*.

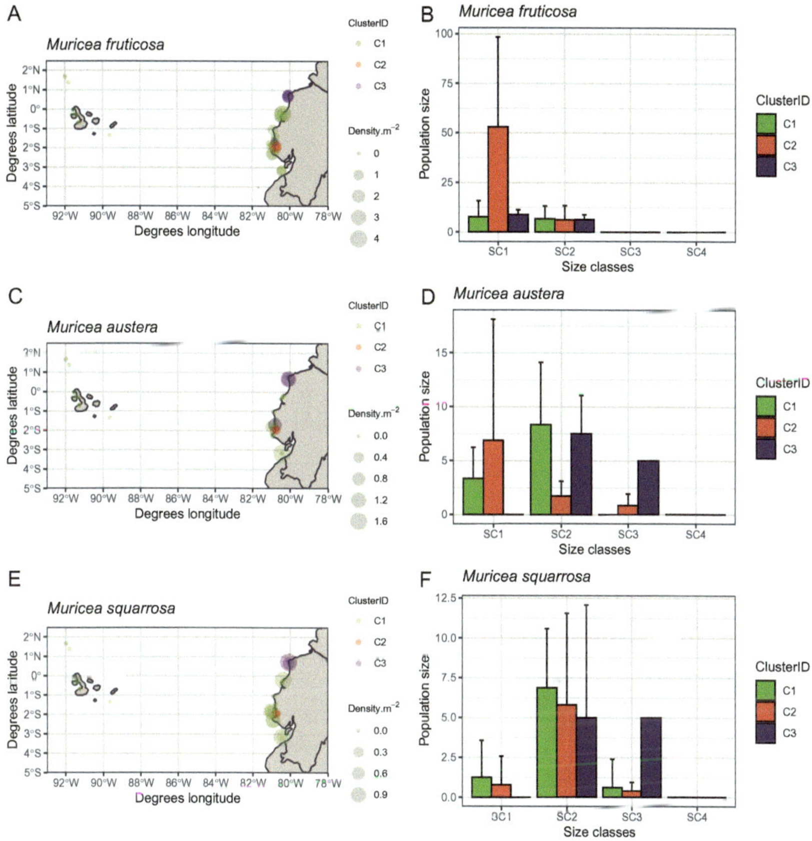

Fig. 7 Characterization of populations of some important octocoral species that defined octocoral assemblages on the Ecuador mainland only (A,C,E, and G) show the size of the population, coded by cluster membership (B,D,F, and H) show the size distribution of the population within each cluster (size classes SC1 $<100\,cm^2$, SC2 $100–399\,cm^2$, SC3 $400–899\,cm^2$ and SC4 $\geq 900\,cm^2$). Cluster ID refers to the clusters identified in Fig. 2A and B, where the latter also shows the contribution of species to cluster formation. Note that position of all sample sites is indicated, even if empty (smallest circle, see legend). Cluster ID refers to the assemblage clusters (same colour-code) obtained in Fig. 3. MUF in legend is *Muricea fruticosa*, MUA is *Muricea austera*, MUS is *Muricea squarrosa*.

Larvae from Galápagos were dispersed towards Cocos and Malpelo Islands and Mesoamerica. Connectivity from Galápagos back to the Ecuador mainland may exist primarily via Cocos and Malpelo Islands, which can be easily reached by larvae from Galápagos. Larvae released from these islands reached the continent. Some larval trajectories originating from the northern

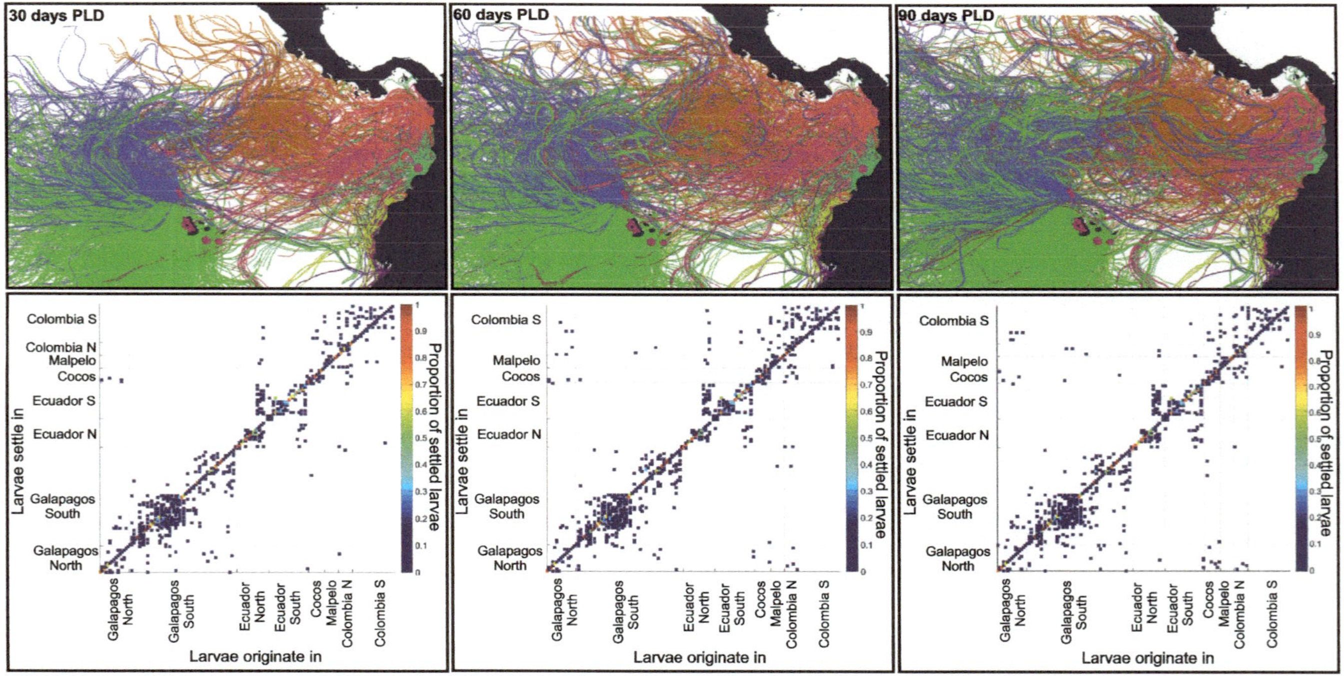

Fig. 8 Potential connectivity pattern of gorgonian corals in the Eastern Pacific under assumption of 30, 60, 90-day PLD. Upper Row: Larval trajectories are colour coded according to site-provenance (originating in southern Galápagos: bright green; northern Galápagos: slate; Cocos: orange; Malpelo: pink; Colombia: turquoise; northern Ecuador [Esmeraldas, Manabí]: gold; southern Ecuador [Santa Elena, El Oro]: purple). Lower row: Connectivity matrices between the four larval precincts shown in above row. The x-axis shows larval provenance (larvae originate in) and the y-axis shows destination (larvae settled in).

Galápagos reached the continent but did not encounter one of the model-defined settlement areas, so were presumed lost. If more suitable settlement areas than assumed in the model exist along the continent (especially in Colombia), then direct two-way connectivity between the continent, Galápagos, and back may indeed exist.

4. Discussion

Of the 26 octocoral species so far reported for continental Ecuador (Breedy et al., 2009, Breedy and Guzman, 2002, 2005, 2007, 2011, 2015, 2016; Figueroa, 2015; Hickman Jr., 2008; Soler-Hurtado et al., 2016; Soler–Hurtado and López–Gonzáles, 2012), 23 were identified in our surveys, which thus covered a representative species spectrum. In Galápagos only three of the 10–15 species recorded (exact number pending taxonomic classification, see Breedy et al., 2009 ; Hickman, 2008; Williams and Breedy, 2004) were identified, and many sites without octocorals were encountered. This points to a patchier distribution of octocorals across the archipelago and to possible impacts of the El Niño–Southern Oscillation (ENSO), such as that of 1982/83, after which substantial mortality was observed, following tissue-sloughing in *P. darwinii*, *P. douglasii* (accepted but doubtful species status) and *Muricea* spp., in the Galápagos (Robinson, 1985). *Heterogorgia hickmani, M. plantaginea*, and *L. alba* were the only species co-occurring on the mainland and in Galápagos. The only other species so far reported from the mainland and Galápagos were *M. fruticosa* and *Eugorgia daniana* (Breedy et al., 2009; Hickman, 2008), the former was only observed on the mainland, the latter not at all.

Our large-scale structural analysis of assemblages in Ecuador suggested a broad pattern separating the northern and southern mainland provinces of Esmeraldas and Santa Elena, from a larger group of continental and Galápagos sites. The numerical dominance of *M. plantaginea, M. fruticosa* and *L. alba,* previously observed at Pelado (Santa Elena, Steiner et al., 2018a, 2018b) was mirrored in the present, geographically much broader, study. This is not to say that local distributional patterns among the uncommon species in this study, which achieved comparatively higher local densities (i.e. *M. purpurea, Leptogorgia pumila*), would not be detectible with a finer-scaled survey. For example, the majority of uncommon and rare species were found in the more species-rich areas (Santa Elena, Esmeraldas), while the species-poor sites were more homogeneous in species composition. However, the low species and colony counts encountered at many sites may be a sign of under-sampling and undocumented previous mortality events, suggesting that more detailed surveys could yet reveal hitherto unseen patterns.

Colony size distributions and, most notably, maximum sizes which relate to the potential for microhabitat creation varied among the 23 species. *Muricea plantaginea* had the tallest colonies, reaching ~130 cm (Breedy and Guzman, 2016), followed by *Psammogorgia arbuscula,* rare in our study. When occurring in high densities, *M. plantaginea* fits the definition of an ecosystem engineer sensu Cuddington (2012) at specific continental locations (Pelado and the Santa Elena Peninsula) and at Isabela (Galápagos). In contrast, *H. hickmani* is a smaller species (Breedy and Guzman, 2005), with a pronounced preference for cryptic settlement spots, limiting extensive multi-colony assemblages. Except for *M. fruticosa,* seldom >15 cm, the other *Muricea* spp. also reached only 20–30 cm height (see Breedy and Guzman, 2016) yet formed the vertically most complex and species-rich assemblages. *Leptogorgia* and *Pacifigorgia* (Gorgoniidae) influenced benthic environments less, either because of their moderate size (*L. alba,*) or due to their scarcity (*P. irene*). However, the depth limits of this study restricted observations to turbulent sites where colony detachment is more frequent. Species-richer, taller and denser octocoral assemblages, visually better associated with animal forests (Rossi et al., 2017), are often found deeper than 20 m (Steiner et al., 2018a, 2018b; Rivera, pers. obs.) and were underrepresented. Such assemblages may serve as octocoral refuges from ENSO surface warming.

A biophysical larval flow model showed a clear N–S connectivity along the Ecuadorian mainland (approx. 600 km between northern and southernmost study site), a notable influx from the Colombian coast to Ecuador, and a pronounced mainly unidirectional E to W connectivity from the continent towards Galápagos (>1000 km). The N–S connectivity may be further enhanced by larvae drifting shorter distances between suitable coastal habitats that connect the wetter and warmer north in Esmeraldas with the drier and cooler south in Santa Elena and El Oro (Gálvez and Regalado, 2007). Mainland assemblages potentially could consist of panmictic populations, but panmixia is not necessarily common (Selkoe et al., 2014).

The E to W connectivity from continental Ecuador towards Galápagos, suggested in the model, also provides further links to Cocos, Malpelo and to continental Colombia (Lequeux et al., 2018; Riegl et al., 2019b; Vanegas and Borrero-Pérez, 2020; Wood et al., 2016). Three mainland species were reported in Galápagos, two of those (*M. plantaginea* and *L. alba*) had the densest and largest populations in our surveys. This may indicate that a certain (yet unquantified) threshold in population size must be attained for greater geographical dispersion (Edmunds and Lasker, 2019; Privitera-Johnson et al., 2015; Santangelo et al., 2003) and also to replenish the Galápagos populations. Species with smaller populations may simply not produce enough planulae to compensate for pelagic mortality on the drift to Galápagos. Alternatively, co-occurring species might best benefit from the surface currents between January and May, based on their reproductive timing (Carson et al., 2010; Gomez et al., 2018), and/or could be broadcast spawners, which in Scleractinia have longer pelagic distribution potential than brooders (Harrison, 2011; Kahng et al., 2011).

This study provides hints at the overall structure of octocoral populations in the equatorial ETP. Except for *C. riisei*, all are limited to the Eastern Pacific, forming potentially well-mixed and highly inter-connected populations of occasionally dense patches along the Colombian and Ecuadorian continental coasts. The latter is also the more species-rich area in this region (Cortés et al., 2017). Oceanic archipelagoes and islands appeared biogeographically more isolated in our models (Fig. 8) and harboured fewer species. The Galápagos could receive larvae from the continent and serve as a source for dispersal to Cocos, at least for *L. alba* (Cortés et al., 2017), from where Malpelo (Sánchez et al., 2011) can be settled (Lequeux et al., 2018; Riegl et al., 2019b; Wood et al., 2016). Larvae from Malpelo and Cocos seem to reach the continent (Fig. 8). Our models showed a largely unidirectional (clockwise) circular connectivity and thus local "upstream" disturbances or extinctions could interrupt "downstream" replenishment. On islands with low local recruitment and/or small populations, such scenarios could cause a cascading shrinking effect on populations. To date, few species are reported for either Cocos (6 spp.), or Malpelo (4 spp.), with three co-occurring species (*P. curta, L. alba* and *C. riisei*; Cortés et al., 2017) or Galápagos (10–15 spp., see Breedy et al., 2009; Cortés et al., 2017). Octocorals in the ETP may thus follow the classical continent-island pattern of species richness (MacArthur and Wilson, 1967). Such a pattern can also be observed when comparing Malpelo to species rich Panama (64 species, in Cortés et al., 2017) and exemplify isolation by distance. The low number of species and co-occurrences among the islands, despite good connectivity, may suggest previous mortality events and/or that the reproductive output of most ETP populations cannot overcome the open ocean distribution barrier (bottleneck).

The broad ETP distribution of *L. alba* may be rooted in connectivity suggested by the biophysical model. This species is reported from every country lining the Pacific from Mexico to Peru, as well as Galápagos,

Cocos, Malpelo, Gorgona (see Cortés et al., 2017; Sánchez et al., 2014) and from northern headlands in Colombia (Vanegas and Borrero-Pérez, 2020). The widespread distribution of *L. alba* may respond to its dispersal ability, due to the high survivorship of its planulae and a high recruitment rate of this broadcast spawner (Gomez et al., 2018). Gomez et al. (2014) also observed recruitment patterns suggestive of an *r*-selected species in Panama. In our surveys, *L. alba* populations primarily comprised small colonies, possibly attributed to ongoing recruitment or an artefact of reduced survival of tall colonies in our shallow turbulent sites. However, this species has also been observed intertidally and occasionally exposed (pers. obs. J. Cortés).

Carijoa riisei is the only species in this study with an Indo-Pacific distribution (Grigg, 2003; Kahng and Grigg, 2005). Kahng et al. (2008) determined that it is a gonochoric broadcast spawner with negatively buoyant eggs in Hawaii, which does not fit a long-distance current-driven distribution well. However, it is a fast-growing invasive Pacific species with vegetative propagation. Interpreting the distribution of *C. riisei* with ocean connectivity is further complicated by its hull fouling capabilities (Concepcion et al., 2010), which provides an alternate mean of dispersal, also suggesting a high risk of introduction to the Galápagos (Keith and Martinez, 2017).

4.1 Conclusion

Ecuador's octocoral fauna was richer on the continent than on its islands. Strong connectivity was suggested for continental sites, with a sharp drop observed towards the islands. Primarily species with large continental populations had dispersed to Galápagos and from there to the other equatorial ETP islands. The Ecuadorian octocoral fauna appears therefore of Eastern Pacific-wide relevance given its species richness and location in ETP connectivity ratchets. The populations currently best fit for monitoring susceptibility of octocorals to disturbance are those of *M. plantaginea*, *H. hickmani* and *L. alba*, due to their co-occurrence on the mainland and Galápagos, regional distribution, and because of the habitat-forming sizes and density of *M. plantaginea*. North - South connectivity along the Colombo-Ecuadorian coasts and the potential structural role of arborescent *Muricea* spp. on the mainland justify closer inspection of their regional population differentiations as baseline for the detection of disturbances. Connectivity between Galápagos, Cocos, Malpelo and Colombian headlands make *L. alba* a candidate for wider ETP population analysis. Our understanding of regional population dynamics relevant to cross-boundary conservation of this and other taxa shaping tropical marine ecosystems would greatly benefit from additional studies covering broader depth ranges, sampling efforts, and reproductive traits. The greatest challenge herein is the recent speed of habitat alterations rooted in phase shifts (Dudgeon et al., 2010; Vergés et al., 2014), invasive species (Steiner and Willette, 2015a) and the subsequent transformations of benthic organismal landscapes (Lasker et al., 2020a, in this volume; Steiner and Willette, 2015b).

Acknowledgements

S. Steiner, P. Martinez, F. Rivera surveyed continental sites and B. Riegl and S. Steiner surveyed Galápagos sites. All surveyors contributed personal funding for the fieldwork. SS was also assisted by the Prometeo Program (B 2015–0296), Ecuadorian National Secretariat of Higher Education, Science and Technology and granted the research permit (MAE013–15, IC–FAU–DPSE–MA) by Ministry of the Environment. PM and FR were supported by Nazca under grants of Conservation International, Ecuador and Fauna & Flora International. K. Collins and J. Mallinson (University of Southampton, UK) funded ship costs in Galápagos under Charles Darwin Foundation research permit number: PC20-18, granted by the Galápagos National Parks Directorate. This publication is listed as contribution 2352 of the Charles Darwin Foundation for the Galápagos Islands. O. Breedy (University of Costa Rica) provided guidance in the identification of species. Three anonymous reviewers shared constructive critiques to the mansucript.

References

Abeytia, R., Guzmán, H.M., Breedy, O., 2013. Species composition and bathymetric distribution of gorgonians (Anthozoa: Octocorallia) on the southern Mexican Pacific coast. Rev. Biol. Trop. 61 (3), 157–1166.

Alvarez-Filip, L., Dulvy, N.K., Gill, J.A., Côté, I.M., Watkinson, A.R., 2009. Flattening of Caribbean coral reefs: region-wide declines in architectural complexity. Proc. R. Soc. B 276, 3019–3025. https://doi.org/10.1098/rspb.2009.0339.

Baird, A.H., Guest, J.R., Willis, B.L., 2009. Systematic and biogeographical patterns in the reproductive biology of scleractinian corals. Annu. Rev. Ecol. Evol. Syst. 40, 551–571.

Bayer, F.M., 1961. The Shallow Water Octocorallia of the West Indian Region. Martinus Nijhoff, The Hague. https://doi.org/10.1039/b821918b.

Bayer, F.M., 1981. Status of knowledge of octocorals of world seas. Seminários de Biologia Marinha. Academia Brasileira de Ciéncias, Rio de Janeiro, pp. 3–11.

Borcard, D., Gillet, F., Legendre, P., 2011. Numerical Ecology With R. Springer.

Breedy, O., Guzman, H.M., 2002. A revision of the genus *Pacifigorgia* (Coelenterata: Octocorallia: Gorgoniidae). P. Biol. Soc. Wash. 115, 782–839.

Breedy, O., Guzman, H.M., 2005. A new species of alcyonacean octocoral from the Galápagos archipelago. J. Mar. Biol. Assoc. UK 85, 801–807.

Breedy, O., Guzman, H.M., 2007. A revision of the genus *Leptogorgia* Milne Edwards and Haime, 1857 (Coelenterata: Octocoralia: Gorgonidae) in the eastern Pacific. Zootaxa 1419, 1–90.

Breedy, O., Guzman, H.M., 2011. A revision of the genus *Heterogorgia* Verrill, 1868 (Anthozoa: Octocorallia: Plexauridae). Zootaxa 2995, 27–44.

Breedy, O., Guzman, H.M., 2015. A revisión of the genus *Muricea* Lamouroux, 1821 (Anthozoa, Octocorallia) in the eastern Pacific. Part I: *Eumuricea* Verrill, 1869 revisited. ZooKeys 537, 1–32.

Breedy, O., Guzman, H.M., 2016. A revision of the genus *Muricea* Lamouroux, 1821 (Anthozoa, Octocorallia) in the eastern Pacific. Part II. ZooKeys 581, 1–69.

Breedy, O., Hickman, C. P, Williams, G. C., 2009. Octocorals in the Galapagos Islands. Galapagos Res 66, Research Article 27.

Cantera, J.R., Prahl, V.H., Neira, O.R., 1987. Moluscos, crustaceos y equinodermos asociados a la gorgonia *Leptogorgia alba* Duchassaing y Michelotti 1864, en la isla de Gorgona. Colombia. Bol. Ecotropica. 17, 3–23.

Carson, H.S., Lopez-Duarte, P.C., Rasmussen, L., Wang, D.X., Levin, L.A., 2010. Reproductive timing alters population connectivity in marine metapopulations. Curr. Biol. 20, 1926–1931. https://doi.org/10.1016/j.cub.2010.09.057.

Cary, L.R., 1918. The Gorgonacea as a factor in the formation of coral reefs. Carnegie Inst. Wash. Pop. Tortugas Lab, Publ 213 (9), 341–362.

Cerrano, C., Bavestrello, G., Bianchi, C.N., Cattaneo-Vietti, R., Bava, S., Morganti, S., Morri, C., Picco, P., Sara, G., Schiaparelli, S., Siccardi, A., Sponga, F., 2000. A catastrophic mass-mortality episode of gorgonians and other organisms in the Ligurian Sea (northwestern Mediterranean), summer 1999. Ecol. Lett. 3, 284–293. https://doi.org/10.1046/j.1461-0248.2000.00152.x.

Coelho, M. A. G., Lasker, H. R., 2016. Larval behavior and settlement dynamics of a ubiquitous Caribbean octocoral and its implications for dispersal. Mar. Ecol. Prog. Ser. 561, 109–121. https://doi.org/10.3354/meps11941

Concepcion, G.T., Kahng, S.E., Crepeau, M.W., Franklin, E.C., Coles, S.L., Toonen, R.J., 2010. Resolving natural ranges and marine invasions in a globally distributed octocoral (genus *Carijoa*). Mar. Ecol. Prog. Ser. 401, 113–127. https://doi.org/10.3354/meps08364.

Cortés, J., Enochs, I. C., Sibaja-Cordero J., Hernández, L., Alvarado, J. J, et al., 2017. Marine biodiversity of eastern tropical Pacific, coral reefs, in Glynn et al. (Eds.), Coral Reefs of the Eastern Tropical Pacific, Coral Reefs of the World 8, Springer, Dordrecht.

Cuddington, K., Hastings, A., 2012. Ecosystem engineers. In: Gross, L.J. (Ed.), Encyclopedia of Theoretical Ecology. University of California Press, pp. 230–234.

Dahan, M., Benayahu, Y., 1998. Embryogenesis, planulae longevity, and competence in the octocoral *Dendronephthya hemprichi*. Invertebr. Biol. 117, 271–280.

Dudgeon, S., Aronson R., Richard, Bruno, J., Precht, W. 2010. Phase shifts and stable states on coral reefs. Mar. Ecol. Prog. Ser. 413, 201–216.

Edmunds, P.J., Lasker, H.R., 2019. Regulation of population size of arborescent octocorals on shallow Caribbean reefs. Mar. Ecol. Prog. Ser. 615, 1–14. https://doi.org/10.3354/meps12907.

Elzinga, C.L., Salzer, D.W., Willoghby, J.W., Gibbs, J.P., 2001. Monitoring Plant and Animal Populations. Blackwell. 359 pp.

Enochs, I.C., Manzello, D.P., Wirshing, H.H., Carlton, R., Serafy, J., 2016. Micro-CT analysis of the Caribbean octocoral Eunicea flexuosa subjected to elevated pCO(2). Ices J. Mar. Sci. 73, 910–919. https://doi.org/10.1093/icesjms/fsv159.

Etnoyer, P., Warernchuck, J., 2007. A catshark nursery in deep gorgonian fields in the Mississippi canyon, Gulf of Mexico. Bull. Mar. Sci. 88 (3), 553–559.

Fabricius, K., Alderslade, P., 2001. Soft Corals and Sea Fans: A Comprehensive Guide to the Tropical Shallow Water Genera of the Central–West Pacific, the Indian Ocean and the Red Sea. Australian Institute of Marine Science, Townsville, Australia.

Fiedler, P., 1992. Climatologies and Variability of Eastern Tropical Pacific Surface Waters. NOAA Tech Rpt. NMFS 109, National Oceanographic and Atmospheric Administration, US Department of Commerce.

Fiedler, P.C., Lavine, M.F., 2016. Oceanographic conditions of the eastern tropical Pacific. In: Glynn, P.W., Manzello, D., Enochs, I.C. (Eds.), Coral Reefs of the Eastern Tropical Pacific. Springer, pp. 59–83.

Figueroa, J., 2015. Contribución al conocimiento de la fauna de octocorales (Octocorallia: Gorgoniidae) en el arrecife rocos del islote Los Ahorcados, Manabí. Universidad Laica Eloy Alfaro de Manabí, Manabí Ecuador, Ecuador. Tesis de grado.

Figueroa-Pico, J., Tortosa, F.S., Carpio, A.J., 2020. Coral fracture by derelict fishing gear affects the sustainability of the marginal reefs of Ecuador. Coral Reefs. 39, 819–827. https://doi.org/10.1007/s00338-020-01926-6.

Fréchette, M., Butman, C.A., Geyer, W.R., 1989. The importance of boundary–layer flows in supplying phytoplankton to the benthic suspension feeder, *Mytilus edulis*. Limnol. Oceanogr. 34, 19–36.

Gálvez, H., Regalado, J., 2007. Características de las precipitaciones, la temperatura del aire y los vientos en la costa ecuatoriana. Acta Oceanográfica del Pacífico 14 (1), 201–205.

Glynn, P.W., Alvarado, J.J., Banks, S., et al., 2017. Eastern Pacific coral reef provinces, coral community structure and composition. An overview. In: Glynn, P.W., Manzello, D., Enochs, I.C. (Eds.), Coral Reefs of the Eastern Tropical Pacific. Springer, pp. 107–176.

Glynn, P.W., Mones, A.B., Podestá, G.T., Colwert, A., Colgan, M.W., 2017. El Niño - southern oscillation: effect on eastern Pacific coral reefs and associated biota. In: Glynn, P.W., Manzello, D.P., Enochs, I.C. (Eds.), Coral Reefs of the Eastern Tropcal Pacific, Coral Reefs of the World 8. Springer, Dordrecht, pp. 251–290.

Glynn, P., Feingold, J.S., Baker, A., Banks, S., Baums, I., et al., 2018. State of corals and coral reefs of the Galapagos Islands (Ecuador): past, present and future. Mar. Pollut. Bull. 133, 717–733.

Glynn, P.W., Coffman, B., Promov, K., Renegar, D.A., Gross, J., Blackwelder, P., Martinez, N., Dominguez, J., Vanderwoude, J., Riegl, B.M., 2019. Benthic ctenophore (Order Platyctenida) reproduction, recruitment, and seasonality in South Florida. Invertebr. Biol. 138, e12256. https://doi.org/10.1111/ivb.12256

Gomez, C.G., Guzman, H.M., Gonzalez, A., Breedy, O., 2014. Survival, growth, and recruitment of octocoral species (Coelenterata: Octocorallia) in Coiba National Park, Pacific Panama. Bull. Mar. Sci. 90 (2), 623–650. https://doi.org/10.5343/bms.2012.1092.

Gomez, C.G., Guzman, H.M., Gonzalez, A., 2015. Population decline and the effects of disturbances on the structure and recovery of octocoral communities (Coelenterata: Octocorallia) in Pacific Panama. J Mar Biol Assoc U K. 95 (01), 81–90. https://doi.org/10.1017/S0025315414000915.

Gomez, C.G., Gonzalez, A., Guzman, H.M., 2018. Reproductive traits and their relationship with water temperature in three common octocoral (Anthozoa: Octocoralia) species from the tropical eastern Pacific. Bull Mar Sci. 94 (4), 1527–1541. https://doi.org/10.5343/bms.2017.1051.

González-Ortiz, V., Egea, L.G., Jiménez-Ramos, R., et al., 2014. Interactions between seagrass complexity, hydrodynamic F1 and biomixing Alter food availability for associated filter–feeding organisms. PLoS One 9 (8), e104949. https://doi.org/10.1371/journal.pone.0104949.

Government of Ecuador, 2009. Ministerial Agreement 162. Ministry of Agriculture, Livestock, Aquaculture and Fisheries.

Graham, N., Nash, K., 2013. The importance of structural complexity in coral reef ecosystems. Coral Reefs. 32, 315–326. https://doi.org/10.1007/s00338-012-0984-y.

Grigg, R.W., 1977. Population dynamics of two gorgonian corals. Ecology 58, 278–290. https://doi.org/10.2307/1935603.

Grigg, R.W., 2003. Invasion of a deep black coral bed by an alien species, *Carijoa riisei*, off Maui, Hawaii. Coral Reefs 22, 121–122. https://doi.org/10.1007/s00338-003-0306-5.

Harrison, P.L., 2011. Sexual reproduction of Scleractinian corals. In: Dubinsky, Z., Stambler, N. (Eds.), Coral Reefs: An Ecosystem in Transition. Springer, Dordrecht.
Hickman Jr., C.P., 2008. A Field Guide to Corals and Other Radiates of. Sugar Spring Press, Lexington, VA, USA.
Hughes, T.P., et al., 2017. Global warming and recurrent mass bleaching of corals. Nature 543, 373–377. https://doi.org/10.1038/nature21707.
Irlandi, E.A., 1996. The effects of seagrass patch size and energy regime on growth of a suspension-feeding bivalve. J. Mar. Res. 54 (1), 161–185.
Johnston, M.W., Akins, J.L., 2016. The non-native royal damsel (*Neopomacentrus cyanomos*) in the southern Gulf of Mexico; an invasion risk? Mar. Biol. 163 (1), 1–14.
Kahng, S.E., Grigg, W., 2005. Impact of an alien octocoral, *Carijoa riisei,* on black corals in Hawaii. Coral Reefs. 24, 556–562. https://doi.org/10.1007/s00338-005-0026-0.
Kahng, S.E., Benayahu, Y., Wagner, D., Rothe, N., 2008. Sexual reproduction in the invasive octo-coral *Carijoa riisei* in Hawaii. Bull. Mar. Sci. 82, 1–17.
Kahng, S.E., Benayahu, Y., Lasker, H.R., 2011. Sexual reproduction in octocorals. Mar. Ecol. Prog. Ser. 443, 265. https://doi.org/10.3354/meps09414.
Keith, I., Martinez, P.C., 2017. The danger of *Carijoa riisei* invading the Galápagos Marine Reserve. In: GNPD, GCREG, CDF, GC, (Eds.), Galápagos Report 2015–2016. Puerto Ayora, Galápagos, Ecuador, pp. 118–124.
Konishi, K., 1981. Alcyonarian Spiculite: limestone of Soft Corals. Proc 4th Int Coral Reef Symp 1, pp. 643–649.
Lasker, H., 1985. Prey preferences and browsing pressure of the butterflyfich *Chaetodon capistratus* on Caribbean gorgonians. Mar. Ecol. Prog. Ser. 21, 213–220.
Lasker, H.R., Coffroth, M.A., 1988. Temporal and spatial variability among grazers: variability in the distribution of the gastropod *Cyphoma gibbosum* on octocorals. Mar. Ecol. Prog. Ser. 3, 285–295. https://doi.org/10.3354/meps043285.
Lasker, H.R., Bramanti, L., Tsounis, G., Edmunds, P.J., 2020a. The rise of octocoral forests on Caribbean reefs. In: Riegl, B. (Ed.), Population Dynamics of the Reef Crisis, vol. 87. Advances in Marine Biology. Elsevier, pp. 361–410 (in this volume). https://doi.org/10.1016/bs.amb.2020.08.009.
Lasker, H.R., Martínez-Quintana, Á., Bramanti, L., Edmunds, P.J., 2020b. Resilience of octocoral forests to catastrophic storms. Sci. Rep. 10, 4286.
Legendre, P., Legendre, L.F.J., 2012. Numerical Ecology. vol. 24. Elsevier.
Lenz, E.A., Bramanti, L., Lasker, H.R., Edmunds, P.J., 2015. Long-term variation of octocoral populations in St. John, US Virgin Islands. Coral Reefs 34, 1099. https://doi.org/10.1007/s00338-015-1315-x.
Lequeux, B.D., Ahumada-Sempoal, M.-A., Lopez-Perez, A., Reyes-Hernandez, C., 2018. Coral connectivity between equatorial eastern Pacific marine protected areas: a biophysical modelling approach. PLoS One 13 (8), e0202995. https://doi.org/10.1371/journal.pone.0202995.
Loya, Y., Sakai, K., Yamazato, K., Nakano, Y., Sambali, H., van Woesik, R., 2001. Coral bleaching: the winners and the losers. Ecol. Lett. 4, 122–131. https://doi.org/10.1046/j.1461-0248.2001.00203.x.
MacArthur, R.H., Wilson, E.O., 1967. The Theory of Island Biogeography. Princeton University Press, Princeton, N.J.
Metaxas, A., Lacharité, M., de Mendonça, S.N., 2019. Hydrodynamic connectivity of habitats of deep-water corals in Corsair Canyon, Northwest Atlantic: a case for cross-boundary conservation. Front. Mar. Sci. 6. https://doi.org/10.3389/fmars.2019.00159.
Mosher, C., Watling, L., 2009. Partners for life: a brittle star and its octocoral host. Mar. Ecol. Prog. Ser. 397, 81–88. https://doi.org/10.3354/meps08113.
Mouillot, D., Bellwood, D.R., Baraloto, C., Chave, J., Galzin, R., et al., 2013. Rare species support vulnerable functions in high-diversity ecosystems. PLoS Biol. 11 (5), e1001569. https://doi.org/10.1371/journal.pbio.1001569.

Oksanen, J., Blanchet, F.G., Friendly, M., et al., 2017. Vegan: Community Ecology Package. R Package Version 2.4-2. https://CRAN.R-project.org/package=vegan.

Patton, W.K., 1972. Studies on the animal symbionts of the gorgonian coral, *Leptogorgia virgulata* (Lamarck). Bull. Mar. Sci. 22, 419–431.

Plass-Johnson, J.G., Taylor, M.H., Husain, A., Teichberg, M.C., Ferse, S.C., 2016. Non-random variability in functional composition of coral reef fish communities along an environmental gradient. PLoS One 11 (4), e0154014. Published 2016 Apr 21. https://doi.org/10.1371/journal.pone.0154014.

Privitera-Johnson, K., Lenz, E.A., Edmunds, P.J., 2015. Density-associated recruitment in octocoral communities in St. John, US Virgin Islands. J. Exp. Mar. Biol. Ecol. 473, 103–109.

R Development Core Team, 2012. R: a language and environment for statistical computing. R Foundation for Statistical Computing, Vienna, Austria. ISBN: 3-900051-07-0. Available from http://www.R-project.org/.

Ramos, G.E., 1995. *Neopontonides henryvonprahli*, una nueva especie de camarón pontoníno del Pacífico de Colombia (Decapoda: Palaemonidae) simbionte de las gorgonias *Muricea robusta* y *Leptogorgia alba*. Rev. Biol. Trop. 43, 231–237.

Richardson, L.E., Graham, N.A.J., Hoey, A.S., 2020. Coral species composition drives key ecosystem function on coral reefs. Proc. R. Soc. B287, 20192214.

Riegl, B., Glynn, P.W., Banks, S., Keith, I., Rivera, F., Vera-Zambrano, M., d'Angelo, C., Wiedenmann, J., 2019a. Heat attenuation and nutrient delivery by localized upwelling avoided coral bleaching mortality in northern Galapagos during 2015/16 ENSO. Coral Reefs 38, 773–785. https://doi.org/10.1007/s00338-019-01787-8.

Riegl, B., Johnston, M., Glynn, P.W., Keith, I., Rivera, F., Vera-Zambrano, M., Banks, S., Feingold, J., Glynn, P., 2019b. Some environmental and biological determinants of coral richness, resilience and reef building in Galápagos (Ecuador). Sci. Rep. 9, 10322. www.nature.com/articles/s41598-019-46607-9.

Robinson, G., 1985. Influence of the 1982–83 El Niño on Galapagos Marine Life. In: Robinson, G., del Pino, E.M. (Eds.), El Niño in the Galápagos Islands. The 1982–1983 Event. Publication of the Charles Darwin Foundation for the Galápagos Islands, Quito Ecuador, pp. 153–190.

Rossi, S., Bramanti, L., Gori, A., Orejas, C., 2017. An overview of the animal forests of the world. In: Rossi, S., Bramanti, L., Gori, A., Orejas, C. (Eds.), Marine Animal Forests. The Ecology of Benthic Biodiversity Hotspots. Springer, Chan. 465 pp.

Ruzicka, R.R., Colella, M.A., Porter, J.W., Morrison, J.M., Kidney, J.A., Brinkhuis, V., Lunz, K.S., Macaulay, K.A., Bartlett, L.A., Meyers, M.K., Colee, J., 2013. Temporal changes in benthic assemblages on Florida keys reefs 11 years after the 1997/1998 El Niño. Mar. Ecol. Prog. Ser. 489, 125–141.

Sammarco, P., Coll, J., 1992. Chemical adaptations in the Octocorallia: evolutionary considerations. Mar. Ecol. Prog. Ser. 88, 93–104.

Sánchez, J.A., Ballesteros, D., 2014. The invasive snowflake coral (*Carijoa riisei*) in the tropical eastern Pacific. Colomb. Rev. Biol. Trop. 62 (Suppl. 1), 197–207.

Sánchez, J.A., Gómez, C.E., Escobar, D., Dueñas, L.F., 2011. Diversidad, abundancia y amenazas de los octocorales de la isla Malpelo, Pacífico Oriental Tropical, Colombia. Boletín de Investigaciones Marinas y Costeras 40, 139–154.

Sánchez, J.A., Ardila, N.E., Andrade, J., Dueñas, L.F., Navas, R., Ballesteros, D., 2014. Octocoral densities and mortalities in Gorgona Island, Colombia, tropical eastern Pacific. Rev. Biol. Trop. 62 (1), 209–219.

Sánchez, J.A., Gómez-Corrales, M., Gutierrez-Cala, L., Vergara, D.C., Roa, P., González-Zapata, F.L., Gnecco, M., Puerto, N., Neira, L., Sarmiento, A., 2019. Steady decline of corals and other benthic organisms in the SeaFlower biosphere reserve (southwestern Caribbean). Front. Mar. Sci. 6, 73.

Santangelo, G., Carletti, E., Maggi, E., Bramanti, L., 2003. Reproduction and population sexual structure of the overexploited Mediterranean red coral Corallium rubrum. Mar. Ecol. Progr. Ser. 248, 99–108. https://doi.org/10.3354/meps248099.

Selkoe, K.A., Gaggiotti, O.E., Bowen, B.W., Toonen, R.J., ToBo, L., 2014. Emergent patterns of population genetic structure for a coral reef community. Mol. Ecol. 23, 3064–3079.

Simpson, A., 2009. Reproduction in Octocorals (Subclass Octocorallia): A Review of Published Literature. Version 16 July 2009. In Deep-Sea Corals Portal. http://www.ucs.louisiana.edu/~scf4101/Bambooweb/repro_AS.html/, accessed 8 April 2020.

Soler–Hurtado, M.M., López–Gonzáles, P.J., 2012. Two new gorgonian species (Anthozoa: octocorallia: Gorgoniidae) from Ecuador (Eastern Pacific). Mar. Biol. Res. 8, 380–387.

Soler-Hurtado, M.M., Machordon, A., Munus, J., López-Golnzáles, P.J., 2016. New Records of the genera *Leptogorgia*, *Pacofigorgia* and *Eugorgia* (Octocorallia: Gorgoniidae) from Ecuador, with a description of a new species. Sci. Mar. 80 (3). https://doi.org/10.3989/scimar.04392.14s.

Spalding, M.D., Ravilious, C., Green, E.P., 2001. World atlas of coral reefs. UNEP World Conservation Monitoring Center. University of California Press, Berkeley, USA, 424 p.

Steiner, S.C.C., Willette, D.A., 2015a. Dimming sand halos in Dominica and the expansion of the invasive seagrass *Halophila stipulacaea*. Reef Encounter 30, 43–44.

Steiner, S.C.C., Willette, D.A., 2015b. The expansion of *Halophila stipulacea* (Hydrocahritaceae, Angiospermae) is changing the seagrass landscape in the Commonwealth of Dominica, Lesser Antilles. Caribbean Naturalist 22, 1–19.

Steiner, S.C.C., Lavorato, A., Rodríguez, J., 2018a. A Manual on Documenting Benthic Marine Invertebrate Communities in the Marine Reserve El Pelado, Santa Elena. Ecuador. Escuela Superior Politécnica del Litoral, Guayaquil, Ecuador, p. 88.

Steiner, S.C.C., Riegl, B., Lavorato, A., Rodriguez, J., 2018b. Community structure of shallow water Alcyonacea (Anthozoa: Octocorallia) from the southern tropical eastern Pacific. Ecol. Res. 33, 457–469. https://doi.org/10.1007/s11284-018-1567-3.

Strub, P.T., Mesias, J.M., Montecino, V., Rutllant, J., Salinas, S.H., 1998. Coastal ocean circulation off western South America: 273-313. In: Robinson, A.R., Brink, K.H. (Eds.), The Sea. In: The Global Coastal Ocean: Regional Studies and Syntheses, vol. 11. Wiley, New York, pp. 273–313.

Taylor, M.L., Yesson, C., Angew, D.J., Mitchell, R.E., Rogers, A.D., 2013. Using fisheries by-catch data to predict octocoral habitat suitability around South Georgia. J. Biogeogr. 40, 1600–1701.

Tsounis, G., Edmunds, P.J., 2017. Three decades of coral reef community dynamics in St. John, USVI: a contrast of scleractinians and octocorals. Ecosphere 8 (1), e01646. https://doi.org/10.1002/ecs2.1646.

Tsounis, G., Martinez, L., Bramanti, L., Viladrich, N., Gil, J.-M., Martinez, A., Rossi, S., 2012. Anthropogenic effects on reproductive effort and allocation of energy reserves in the Mediterranean octocoral *Paramuricea clavata*. Mar. Ecol. Prog. Ser. 449, 161–193.

Valisano, L., Notari, F., Mori, M., Cerrano, C., 2016. Temporal variability of sedimentation rates and mobile fauna inside and outside a gorgonian garden. Mar. Ecology-an Evolut. Perspect. 37, 1303–1314. https://doi.org/10.1111/maec.12328.

Van Oppen, M.J.H., Miegog, J.C., Sanchez, C.A., Fabricius, K.E., 2005. Diversity of algal endosymbionts (zooxanthellae) in octocorals: the roles of geography and host relationships. Mol. Ecol. 14, 2403–2417. https://doi.org/10.1111/j.1365–294X.2005.02545.x.

Vanegas, G.M.J., Borrero-Pérez, G.H., 2020. First records and new information on the associations of echinoderms with other phyla in the rocky reefs of northern Chocó, Colombian Pacific. ZooKeys 921, 1–22. https://doi.org/10.3897/zookeys.921.32802.

Vergés, A., Steinberg, P.D., Hay, M.E., Poore, A.G.P., Campbell, A.H., et al., 2014. The tropicalization of temperate marine ecosystems: climate-mediated changes in herbivory and community phase shifts. Proc. R. Soc. B 281, 0846.

Vogel, S., 1994. Life in Moving Fluids. Princeton Univ Press.
Vreeland, H.V., Lasker, H.R., 1989. Selective feeding of the polychaeta *Hermodice carunculata* Pallas on Caribbean gorgonians. J. Exp. Mar. Biol. Ecol. 129, 265–267.
Wainwright, S.A., Biggs, W.D., Currey, J.D., Gosline, J.M., 1976. Mechanical Design in Organisms. Princeton Univ. Press.
Wanatabe, S., Metaxas, A., Sameoto, J., Lawton, P., 2009. Patterns in abundance and size of two deep-water gorgonian octocorals, in relation to depth and substrate features off Nova Scotia. Deep sea research part 1, Oceanogr. Res. Pap. 56, 2235–2248.
Williams, G.C., Breedy, O., 2004. The Panamic gorgonian genus *Pacifigorgia* (Octocorallia: Gorgoniidae) in the Galápagos Archipelago, with descriptions of three new species. Proc. Calif. Acad. Sci. 55, 55–88.
Wood, S., Baums, I.B., Paris, C.B., Ridgwell, A., Kessler, W.S., Hendy, E.J., 2016. El Niño and coral larval dispersal across the eastern Pacific marine barrier. Nature Comms 7, 12571. https://doi.org/10.1038/ncomms12571.

CHAPTER FIFTEEN

A tropical eastern Pacific invasive brittle star species (Echinodermata: Ophiuroidea) reaches southeastern Florida

Peter W. Glynn[a,*], Renata Alitto[b], Joshua Dominguez[c], Ana B. Christensen[d], Phillip Gillette[a], Nicolas Martinez[c], Bernhard M. Riegl[e], and Kyle Dettloff[f]

[a]Department of Marine Biology and Ecology, Rosenstiel School of Marine and Atmospheric Science, University of Miami, Miami, FL, United States
[b]Instituto de Biologia, Universidade Estadual de Campinas, São Paulo, Brazil
[c]Undergraduate Marine and Atmospheric Science Program, Rosenstiel School of Marine and Atmospheric Science, University of Miami, Coral Gables, FL, United States
[d]Biology Department, Lamar University, Beaumont, TX, United States
[e]Department of Marine and Environmental Sciences, Halmos College of Arts and Sciences, Nova Southeastern University, Dania Beach, FL, United States
[f]National Oceanographic and Atmospheric Administration, National Marine Fisheries Service, Southeast Fisheries Science Center, Miami, FL, United States
[*]Corresponding author: e-mail address: pglynn@rsmas.miami.edu

Contents

Advances in Marine Biology, Volume 87
ISSN 0065-2881
https://doi.org/10.1016/bs.amb.2020.08.010

Abstract

The invasive brittle star *Ophiothela mirabilis* (family Ophiotrichidae), a tropical Indo-Pacific endemic species, first reported in Atlantic waters off southern Brazil in 2000, has extended its range northward to the Caribbean Sea, to the Lesser Antilles in 2011, and was first reported in south Florida in January 2019. Its occurrence in southeast Florida extends along nearly 70 km of coastline, from near the Port of Miami, Miami-Dade County, northward to Deerfield Beach, Broward County. It occurs abundantly as an epizoite on octocorals, attaining population densities of 25 individuals and more per 10-cm long octocoral stem. The surface texture of octocoral hosts (rough, smooth) did not affect the densities of the ophiuroid epizoites, and there were significantly greater abundances on octocorals during two winter sampling periods than in the summer. Beige and orange-coloured morphs are sometimes present on the same octocoral stem. Gut content analysis supported a suspension feeding mode, revealing essentially identical ingested items in both colour morphs with a preponderance of amorphous detritus and filamentous algae. Molecular genetic evidence (COI & 16s) has established the identity of *O. mirabilis* and its relationship to invasive Brazilian populations. The orange and beige morphs form two distinct, but closely related lineages that may represent two separate introductions. The orange morph shares haplotypes with Brazilian and Caribbean specimens suggesting a further range expansion of the 'original' invasion. The beige morph, however, shares haplotypes with specimens from the Mexican Pacific and Peru and potentially represents a secondary introduction. Traits promoting dispersal and establishment of this species in new habitats are manifold: vagility and ability to cling tightly to diverse host taxa (e.g. sponges, cnidarians, bryozoans, and echinoderms), frequent asexual reproduction (fissiparity), suspension feeding, including a wide range of dietary items, possession of integument-covered ossicles and arm spines offering protection from predators, and an effective competitive edge over associated microbiota for substrate space.

1. Introduction

Releases of non-indigenous, exotic species that eventually may become invasive and take on major modifying roles in their new ecosystems are a major component of the changes forced by humankind upon many natural ecosystems, among them coral reefs. Relatively little is known about the dynamics of such invasions. The tropical Indo-Pacific brittle star, *Ophiothela mirabilis* Verrill, 1867, described from specimens collected off the Pacific coast of Panama, was first reported in the Atlantic Ocean from the south coast of Brazil in 2000 (Hendler et al., 2012). This occurrence and subsequent records have attracted much interest due to the prodigious numbers observed and its rapid and presumed far-reaching dispersal in the western Atlantic region. In less than two decades, *O. mirabilis* has lengthened its latitudinal range and now extends ~5800 km along the coastal western Atlantic region. This invasive ophiuroid is now known from Brazil, Santa Catarina State (Lawley et al., 2018) north to the Amazon River mouth (Araújo et al., 2018; Mantelatto et al., 2016), from French Guiana, Trinidad and Tobago, to Saint Vincent, Saint Kitts, and the Grenadines (Hendler et al., 2012; Lawley et al., 2018; Rich et al., 2020) in the Caribbean, and most recently off the southeastern coast of Florida, USA (Glynn et al., 2019).

Ophiothela mirabilis is primarily an epizoite, associating closely with sponges, scleractinian corals, echinoids, and especially octocorals on which it wraps its arms around the host's branches. Due to the high abundance and diversity of potential octocoral hosts on western Atlantic hard grounds and coral reefs (Banks et al., 2008; Goldberg, 1973; Jaap et al., 2008; Johnson and Hallock, 2020; Sánchez, 2016), researchers have expressed concern over the rapid spread and potential impacts of this alien brittle star species (e.g. Granja-Fernández and López-Pérez, 2011; Mantelatto et al., 2016). The often-high density of individuals of *O. mirabilis* in contact with polyps might compromise their host's ability to obtain nutrients (Araújo et al., 2018) and engage in suspension feeding. Additionally, the multitude of epizoites that utilize octocorals as an essential habitat—Goh et al. (1999) reported 7 phyla and 30 species associates in Singapore, and Bayer (1961) 7 phyla and 14+ species in the western Atlantic—could be threatened directly or indirectly by the presence of this invasive ophiuroid.

In this study, we confirm the identity and occurrence of *O. mirabilis* as epizoites of octocorals in southeast Florida. Octocoral host species and the character of their epidermis and polyp armature are examined in relation to brittle star densities. Three colour morphs were present in south Florida; their locations, densities and relative proportions on individual hosts are examined. The presence or absence of individuals of *O. mirabilis* along 235 km of the southeastern Florida coast, including the Florida Keys, is noted in relation to major ship ports, which are considered potentially important in the introduction and establishment of *O. mirabilis* (Carlton and Geller, 1993; Lawley et al., 2018; Roy and Sponer, 2002). Gut analyses were performed on the two ophiuroid colour morphs to determine their diets and possible differences.

2. Materials and methods

2.1 Survey sites and sampling

Survey sites along the Florida coast, from 24° 40′ N (south of Vaca Key) to 26° 20′ N (Deerfield Beach), were sampled by surface diving and scuba during 2019 and early 2020 (Fig. 1A–E; Table 1). These sites were at shallow depths (5–8 m), exposed to the open sea, and supported species-rich octocoral communities with numerous species of associated biota, such as macroalgae, sponges, and diverse cnidarian taxa including zooxanthellate scleractinian corals (Banks et al., 2008; Jaap et al., 2008). At Dome Reef, Biscayne National Park, Jaap et al. (2008) listed the occurrence of 20 species of octocorals, 10 species of scleractinian corals, and one hydrozoan species. The ophiuroid populations at the collection sites are not continuous because the octocoral hardgrounds are separated by sand patches of varying distances (Banks et al., 2008). These populations may best be regarded as members of a metapopulation. The regional setting and subtropical environmental conditions along the southeastern coast of Florida are described in Jaap et al. (2008) and Banks et al. (2008).

Plexaurid octocorals, particularly members of the genera *Eunicea* and *Muricea*, were preferentially sampled since species in these taxa are among the commonly observed hosts of *O. mirabilis*. These plexaurid octocorals also commonly host the benthic ctenophore *Coeloplana waltoni*, the subject of an earlier study in which individuals of *O. mirabilis* were first reported in Florida (Glynn et al., 2019). Surveys in Florida have not revealed the presence of the invasive brittle star on scleractinian corals, bryozoans, ascidians or other benthic taxa as has been reported in Brazil (Lawley et al., 2018; Mantelatto et al., 2016) and the eastern Pacific (Granja-Fernández and López-Pérez, 2011; Granja-Fernández et al., 2014). All plexaurid octocorals sampled were $\geq$0.5 m in height with each colony bearing several elongate, dichotomous branches. Approximately 10-cm long side branch tips were cut from octocorals with scissors, secured in polyethylene bags then transferred to an insulated water cooler at the surface for transport to the laboratory. Brittle stars were collected at all sites for laboratory study except in the Florida Keys National Marine Sanctuary, Key Largo (Fig. 1D) where only visual searches were performed. Collections are prohibited in this marine protected area. Approximately 50 octocoral colonies at each of two sites were inspected for ophiuroids in the sanctuary. Octocoral branches ranging

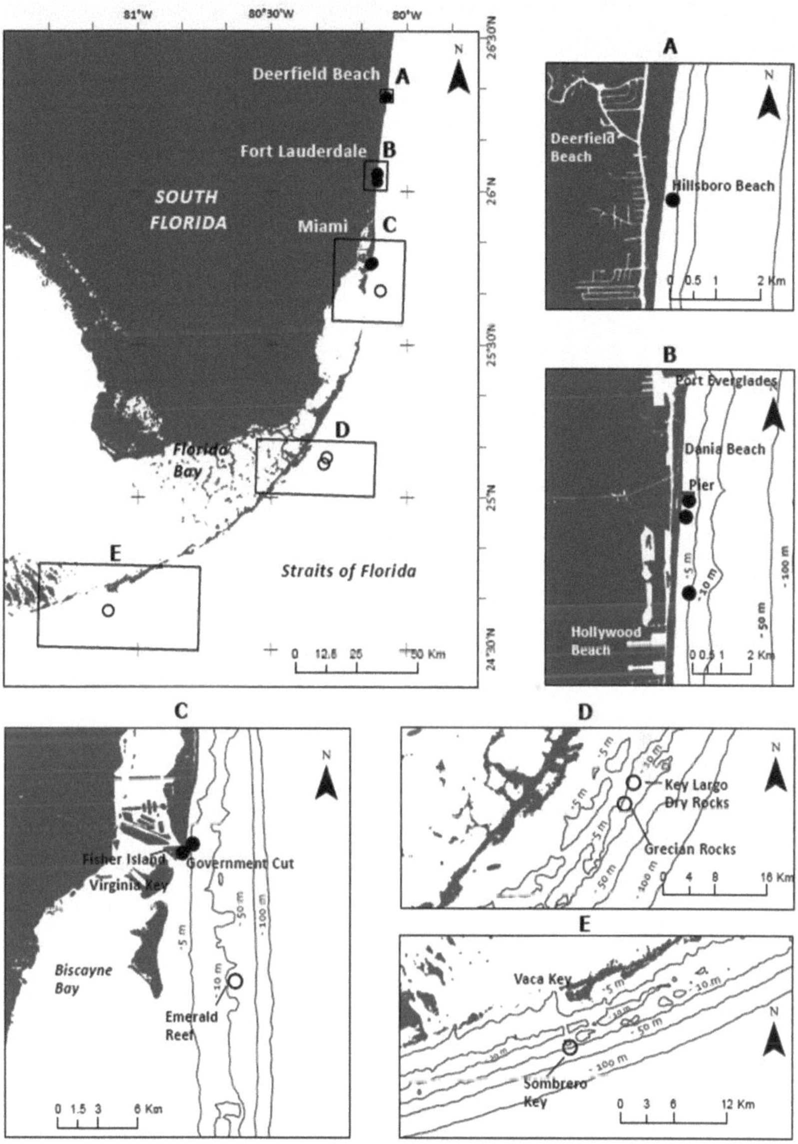

Fig. 1 Sampling sites for *Ophiothela mirabilis* along southeast Florida from Hillsboro Beach to Vaca Key, Florida Keys. Dark circles, *O. mirabilis* present on octocorals; clear circles, *O. mirabilis* not observed on octocorals or other benthos. (A) Hillsboro Beach, 1 site; (B) Dania/Hollywood beaches, 3 sites; (C) Government Cut/Key Biscayne, 2 sites; (D) Key Largo area, 2 sites; (E) Vaca Key, 1 site. All sites in panels (A) and (B), and Government Cut in (C), octocoral hosts present on flat, pavement-like substrates. Emerald Reef in panel (C), and all sites in (D) and (E), octocoral hosts were associated with structural coral reefs.

Table 1 Sample densities and occurrence of colour morphs of *Ophiothela mirabilis* present on octocoral hosts on flat, pavement-like substrates at localities off the southeastern Florida coast.

Locality	Date	Number of octocorals sampled[a]	Mean number per 10 cm octocoral branch (SD, *n*[b])	Range	Colour morphs[c]
Dania Beach study site	5/1/2019	16	27.0 (12.6, 432)	0–57	Beige
	27/2/2019	21	14.0 (10.5, 294)	0–42	Beige
	17/3/2019	20	14.0 (7.4, 279)	0–33	Beige
	21/4/2019	20	7.0 (5.8, 140)	0–19	Beige
	8/5/2019	30	2.5 (1.7, 76)	0–5	Beige
	7/6/2019	24	8.8 (6.4, 210)	0–28	Beige
	29/7/2019	25	3.4 (2.3, 85)	0–7	Beige
	18/8/2019	30	10.4 (12.0, 313)	0–50	Beige
	27/9/2019	27	15.8 (7.2, 422)	5–37	Beige
	27/10/2019	27	2.1 (5.2, 142)	0–12	Beige
	8/11/2019	26	18.2 (10.5, 455)	5–38	Beige
	5/12/2019	30	19.5 (9.5, 606)	4–73	Orange, 19.1% Beige, 80.9%
	7/1/2020	30	30.4 (13.1, 875)	7.5–58.0	Beige
Fisher Island	8/8/2019	25	0.04 (0.20, 1)	0–1	Orange
	7/12/2019	25	23.2 (18.6, 747)	3–85	Orange, 100%

Table 1 Sample densities and occurrence of colour morphs of *Ophiothela mirabilis* present on octocoral hosts on flat, pavement-like substrates at localities off the southeastern Florida coast.—cont'd

Locality	Date	Number of octocorals sampled[a]	Mean number per 10 cm octocoral branch (SD, n^b)	Range	Colour morphs[c]
Miami Beach	7/12/2019	25	1.8 (2.0, 110)	0–6	Orange, 94.5% Beige, 5.5%
Hollywood Beach	29/8/2019	25	11.2 (5.7, 280)	3–26	Orange, 66.4% Beige, 33.6%
Deerfield Beach	28/9/2019	7	3.2 (1.1, 25)	1.8–5.3	Beige
Pompano Beach/ Dania Beach[d]	6/2/20	17	20.6 (11.2, 337)	4–39	Faint yellow-orange, 1.5% Beige, 98.5%

[a]Assorted species of *Eunicea* spp., *Muricea elongata, Plexaurella dichotoma, Pseudoplexaura acerosa,* and *Gorgonia ventalina*. See Table 2.
[b]*n* refers to total number of unstandardized brittle stars sampled.
[c]Colours are for disc and arms; orange morphs were uniformly bright orange; beige, pale to greyish yellow with dark-brown bands segmenting the length of the arms.
[d]Mixed sites: Pompano Beach—80° 05.002′; 26° 13.268′ and 80° 05.094′; 26° 12.509′. Dania Beach—80° 06.344′; 26° 03.506′. 15 of 17 octocorals collected off Pompano Beach.

from 8 to 12 cm in length were identified and measured in the laboratory, their condition noted, and all ophiuroid epizoites counted and their respective colours recorded. Ophiuroid population densities were obtained by adjusting counts to numbers of individuals per 10-cm octocoral branch length.

2.2 Species

Identification of octocoral species was performed in the laboratory. The six species of plexaurid and three species of gorgoniid octocorals were identified with reference to Bayer (1961), and voucher specimens in the U.S. National Museum (Smithsonian Institution), the University of Miami Invertebrate

Museum, and collections in the Guy Harvey Oceanographic Center, Nova Southeastern University, Dania Beach, Florida. The scientific names of the eight species, the character of their branch epidermis, and respective museum catalogue numbers, are listed in Table 2.

The surface texture of octocorals and relative density of polyps were determined by noting the character and location of sclerites and their degree of projection. Calyces were not well developed in some species, which was noted. Based on these criteria, octocoral species were classified as potentially 'rough' or 'smooth' host substrates.

Ophiothela mirabilis was initially identified by Gordon Hendler, with reference to Verrill's (1867) description of specimens collected at the Pearl Islands, Gulf of Panama, Republic of Panama. See also Alitto et al. (2020) for Brazilian character sets. Representative specimens from the present study were deposited in the U.S. National Museum, Smithsonian Institution, Accession No. 2087409: USNM 1592554 (orange morph), USNM 1592555 (beige morph). Specimens analysed genetically are also to be deposited in the collections of the USNM.

2.3 Behavioural observations

Macroinvertebrate epizoites (ophiuroids, caprellids, and gastropods) were observed visually on octocoral hosts in the field utilizing scuba during late morning (0900–1200) daylight hours. Observations in the laboratory, including macroinvertebrate interactions with individuals of the minute ctenophore *Coeloplana waltoni* (Glynn et al., 2014), were made microscopically (8–35 ×) under indirect, subdued lighting on recently collected octocoral branches with their natural complement of epizoites. Visual observations of ophiuroids on their octocoral hosts were also made during non-feeding and feeding periods in the laboratory.

2.4 Gut analysis

Gut contents of beige colour morphs were examined irregularly at a few study sites during 2019 and 2020. For a critical comparison of possible differences between beige and orange colour morphs, gut analysis was performed on 20 individuals each of orange and beige individuals collected at the Hollywood Beach study site on 29 May 2020. The two colour morphs were collected from the same octocoral host colonies at the same time of day (mid-morning) and were fixed in 70% ethanol in the field immediately after

Table 2 Plexaurid and gorgoniid octocoral species observed hosting individuals of *Ophiothela mirabilis* in southeast Florida, 2018–19.

Family/species[a]	Character of stem epidermis	Museum catalogue numbers[c]
Plexauridae		
Eunicea asperula Milne Edwards & Haime, 1857	Rough, with upturned calyx lip; sharp calyces in some colonies	USNM 1516217, 1516219 UMML 7.287
Eunicea clavigera Bayer, 1961	Smooth, calyces well separated, moderately prominent	UMML 7.627
Eunicea flexuosa (Lamouroux, 1821)[b]	Smooth, apertures with inconspicuous lower lip	USNM 1516214, 1516215, 1516216 UMML 7.286
Eunicea tourneforti Milne Edwards & Haime, 1857	Rough, calyces crowded with well-developed lower lip	UMML 7.277
Muricea elongata Lamouroux, 1821	Rough, calyces crowded, sharply pointed and directed upward	UMML 7.1272
Plexaurella dichotoma (Esper, 1791)	Smooth, surface raised around apertures, calyces moderately separated	USNM 1516218 UMML 7.285
Gorgoniidae		
Pterogorgia guadalupensis Duchassaing & Michelin, 1846	Smooth, low profile calyces, widely separated	UMML 7.158
Pseudopterogorgia acerosa (Pallas, 1766)	Smooth, colony branching closely and regularly pinnate; calyces indistinct or absent	UMML 7.261, 7.647
Gorgonia ventalina Linnaeus, 1758	Smooth, network of interconnected branches	UMML 7.241

[a]Octocoral species identified according to colony height and branching morphology, disposition and spacing of polyps, morphology of calyces, and nature of sclerites.
[b]Formerly *Plexaura flexuosa*, see Grajales et al. (2007).
[c]USNM, United States National Museum, Washington DC; UMML, University of Miami Marine Laboratory, Miami, Florida.
Character of branch surfaces and ornamentation after Bayer (1961) and our observations of locally-collected colonies.

collection. The aboral epidermal disc covering the stomach was removed and the stomach contents were washed gently onto a slide with 70% ethanol in preparation for counts and identifications at 100 × magnification. Bengal Rose stain was applied occasionally to distinguish between inorganic and organic matter; counts included the latter only. All items thus exposed, generally ranging between about 30 and 400 μm, were enumerated on a 24 × 60 mm grid at 100 × magnification. Seven categories were created based on the items ingested: amorphous detritus (ad), filamentous algae (fa), crustaceans (cr), macroalgae (ma), nematode worms (ne), diatoms (di), and miscellania (mo). The mo category included faecal pellets, nematocysts, non-nematode worms, dinoflagellates, tardigrades, eggs, algal cells, and unidentified fragments. The methods of Rosi-Marshall et al. (2015) were generally followed except that gut contents were examined directly in alcohol and not concentrated on filters.

2.5 DNA extraction, amplification, and sequencing

Total genomic DNA was isolated from specimens using a Chelex protocol. A small portion of an arm (3–5 mm length) was allowed to dry overnight to allow the ethanol to evaporate and then was gently ground with a pestle in 500 μL of TE buffer. After grinding, 200 μL of a 10% Chelex solution (Biorad) was added and the mixture incubated at 57°C for 30 min followed by boiling for 8 min. After cooling to room temperature, the solution was gently inverted several times and centrifuged at 14,000 × *g* for 2 min. The supernatant was transferred to a clean microtube and refrigerated until use in PCR.

A fragment of the mitochondrial 16S gene was PCR-amplified using the forward primer 16S Sofi F (5′-CAG TAC TCT GAC TGT GCA A-3′) and the reverse primer 16S Sofi R (5′-GGA AAC TAT GAT CCA ACA TC-3′) (Pérez-Portela et al., 2013). A fragment of the mitochondrial cytochrome oxidase subunit 1 (COI) gene was amplified using the forward primer OphiF (5′-ATA ATG ATA GGA GGA TTT GGA AA-3′) (Bribiesca-Contreras et al., 2013) or COIceF (5′-ACT GCC CAC GCC CTA GTA ATG ATA TTT TTT ATG GTN ATG CC-3′) and the reverse primer COIceR (5′-TCG TGT GTC TAC GTC CAT TCC TAC TGT RAA CAT RTG-3′) (Hoareau and Boissin, 2010) as well as degenerate forward primer Oph-CO1-F (5′-TTT CAA CTA ATC AYA AGG AYA TWG G-3′) and reverse primer Oph-CO1-R (5′-CTT CAG GRT GWC CRA ARA AYC A-3′) (O'Hara et al., 2014). Total reaction volumes were 20 μL: 2 μL 10× standard buffer, 0.5 μL dNTPs (10 mM), 0.5 μL each primer (10 μM), 0.1 μL Taq polymerase (2.5 units) (New England Biolabs), 4 μL DNA template, and 12.4 μL molecular grade water. The amplification program for 16S fragment included an initial cycle of 95° C for 4 min followed by 35 cycles of 95°C for 45 s, 50°C for 1 min, 72°C for 45 s, and a final cycle of 72°C for 3 min; this was carried out on a SimpliAmp thermocycler. The amplification program for COI differed as it consisted of 40 cycles and annealing temperatures of 48–50°C. Amplification was confirmed on 1% agarose gels stained with ethidium bromide. PCR products were then purified using GenCatch PCR Cleanup kit (Epoch Life Science Inc.) as per manufacturer's instructions and then sent to MCLAB (San Francisco, CA) for sequencing.

DNA fragments of 353 bp for 16S and 623 bp for COI were obtained from 76 individuals. All the sequences are deposited in GenBank (www.ncbi.nlm.nih.gov/Genbank) (Table S1 in Supplementary Material in the online version at https://doi.org/10.1016/bs.amb.2020.08.010). The nucleotide sequences were edited using BIOEDIT Sequence Alignment Editor v. 7.0.1 (Hall, 1999) and GeneStudio 2.2.0.0 (Genestudio, Suwanee, GA, USA).

2.6 Phylogenetic analyses

Three mitochondrial datasets were considered: (i) 16S, (ii) COI, and (iii) concatenated (16s + COI). All sequences were aligned with GeneStudio 2.2.0.0 via Clustal W 1.83 (GeneStudio, Suwanee, GA, USA) providing a unified matrix. Additional sequences from other Ophiotrichidae (16S, COI) available in GenBank were added to the matrix for comparison. All samples used are listed in Table S1 in Supplementary Material in the online version at https://doi.org/10.1016/bs.amb.2020.08.010.

Phylogenetic reconstructions were made based on two different optimality criteria, Bayesian Inference (BI), and Maximum Likelihood (ML), to assess whether there were any differences in the trees recovered with regard to the method used.

The BI was performed with MrBayes version 3.2 (Ronquist et al., 2012). The best-fitting model of molecular evolution for concatenated datasets were SYM + G, chosen based on the Akaike Information Criterion (AIC), and General Time Reversible (GTR), chosen based on the Hierarchical Likelihood Ratio Tests (hLRTs), according to the estimation by MR.MODELTEST v.2.3 (Nylander, 2004). The BI analyses used one cold and three incrementally heated Monte Carlo Markov chains

(MCMC) on two simultaneous runs. The standard deviation of the split frequencies between the two runs reached a value lower than c. 0.001 at one million generations, with one tree sampled every 100th generation, each using a random tree as a starting point and a temperature parameter value of 0.2 (the default in MrBayes). The first 25% of the total sampled trees were discarded as 'burnin' to achieve the MCMC log-likelihoods that had become stationary and converged. The analyses resulted in similar likelihood scores, with ESS > 200, as verified using TRACER. The phylogenetic trees were visualized and edited in FIGTREE v1.4.3 (http://tree.bio.ed.ac.uk/software/figtree/).

The ML was performed with the MEGA v. X (Kumar et al., 2018) using the Kimura 2-parameter model (Kimura, 1980). The statistical support was obtained with a bootstrap function using 1000 replicates.

2.7 Genetic distances

The genetic distances between and within the different groups were estimated by *P*-distance using MEGA X v. 10.1.7 (Kumar et al., 2018), ignoring the alignment gaps in pairwise comparisons. Following Hart and Podolsky (2005) and Pérez-Portela et al. (2013), pairwise genetic *P*-distances of >3% for 16S and >15% for COI were considered interspecific variations as most of the species already described have been differentiated by this minimum. DnaSP v. 6.12.03 (Rozas et al., 2017) was used to generate multi-sequence aligned haplotype data, and Pop ART v 1.7 (Leigh and Bryant, 2015) was used to draw haplotype networks using the TCS method based on the haplotypes generated by DnaSP.

2.8 Statistical testing

Two multinomial logistic models were fit to counts of brittle star gut contents using the R package *nnet* (Venables and Ripley, 2002). Using colour morph as a categorical predictor (20 brittle stars per group), the first model included item counts of all seven individual food classes as the response, while the second model classified the response into two categories of interest (therefore reducing to a binomial model); amorphous detritus + filamentous algae (considered lower quality food) and the sum of the remaining categories (considered higher quality food). To adjust for extra-multinomial variation among the grouped counts, the ratio of the Pearson Chi-square goodness of fit statistic to the residual degrees of freedom for each model was used as an estimate of the overdispersion parameter σ^2 (McCullagh and Nelder, 1989). Test statistics were scaled by σ^2 accordingly to ensure appropriate *P*-values were obtained from subsequent hypothesis tests.

To test the effect of octocoral surface texture (rough, smooth) on ophiuroid densities, a generalized linear mixed model with a negative binomial response structure ($\theta = 3.16$) was fit to *Ophiothela mirabilis* densities per 10 cm length of octocoral branch using the R package lme4 (Bates et al., 2015). Survey date was treated as a random effect to account for the non-independence of observations taken on the same survey.

3. Results

3.1 Site occurrences and abundances

The population densities of individuals of *Ophiothela mirabilis* were highly variable within and among six sites quantitatively sampled from Miami to Deerfield Beach (Fig. 1). Of 13 surveys conducted at Dania Beach, from January to December 2019 and January 2020, mean monthly densities ranged from 2.1 to 30.4 individuals 10 cm^{-1} octocoral branch length with no evident temporal trends in abundance over the 13-month period of sampling (Table 1). Disc diameters of individuals in the February sample, with a mean density of 14.0 individuals 10 cm^{-1}, ranged from 0.9 to 1.6 mm ($n = 18$). Some smaller individuals possessed only three long arms and three short arms in the process of asexual reproduction (fissiparity). In other surveys, individuals of *O. mirabilis* were not observed on their usual octocoral hosts (e.g. *Eunicea* spp., *Muricea elongata*, and *Plexaurella dichotoma*) or other hosts along the Florida Keys at Vaca Key, Key Largo or Key Biscayne, sites located from 12 to 200 km south of Miami (Fig. 1). The first sample collected off Fisher Island contained only a single ophiuroid, equivalent to 0.04 individuals 10 cm^{-1}. Four months later, a mean density of 23.2 ophiuroid individuals per host branch was observed at this site; a maximum density of 85 individuals 10 cm^{-1} was present on one octocoral host (Table 1). Surveys at Miami and Hollywood beaches exhibited mean ophiuroid densities of 1.8 and 11.2 individuals 10 cm^{-1} host branches, respectively (Table 1). Several sampled branches were without ophiuroids at three sites (Dania Beach, Fisher Island, and Miami Beach), reflecting the high patchiness of occurrence.

Three colour patterns occurred in the collections: (1) 'beige', (2) orange, and (3) mottled (a mixture of orange and beige colouring); the 'beige' morph is actually a striking pattern of white, tan, and yellow when viewed under magnification, but appears a more uniform beige colour in the field (contrast Figs 2 and 3). Only beige colour morphs were present in the earliest and most recent collections at Dania Beach, from January to November 2019 and January 2020 (Fig. 2A and B; Table 1). The December 2019 collection contained 19.1% orange individuals (Fig. 2C), with 80.9% classified as beige. However, eight of the beige individuals exhibited mixed colours. Five of the beige individuals revealed predominantly orange discs and three others bore one to three mostly orange-coloured arms each. Unfortunately, these mixed colour morphs were not preserved or found in subsequent collections after intensive searching. Close-up views of the adoral discs demonstrate the degree of colour variability, especially among beige colour morphs (Fig. 3A, C–G, I–N, P). All individuals bore a centrally located white patch of various shapes. Yellow pigmentation was faint to pronounced with straight, looped or circular patterns. Orange morphs exhibited radially distributed short, white lines more or less pronounced and 12 in number (Fig. 3B, H and O). Spherical granules were scattered over the discs of both colour morphs. Orange-coloured morphs of *Ophiothela mirabilis* were first observed in August 2019 at Fisher Island and Hollywood Beach. The single ophiuroid collected at Fisher Island was orange and 66.4% of 280 individuals were orange in the Hollywood Beach collection. Orange-coloured morphs predominated in samples collected in December 2019 at Fisher Island (100%, $n = 747$) and Miami Beach (94.5%, $n = 110$). The orange-coloured individuals were readily visible, not matching their octocoral hosts, which generally ranged from grey to brown.

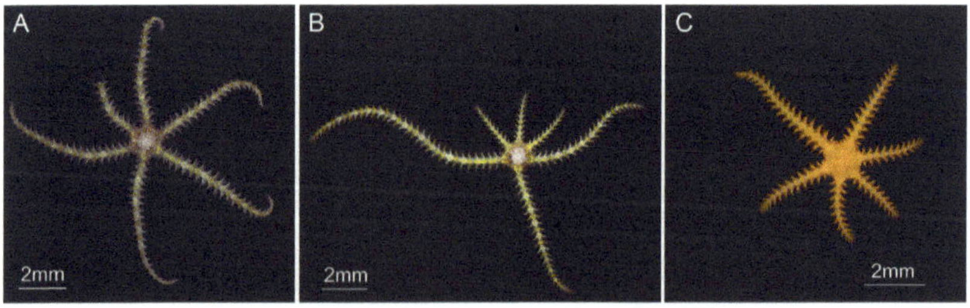

Fig. 2 Live colour adoral morph images of individuals of *Ophiothela mirabilis* collected in southeast Florida. (A) Beige morph, Dania Beach study site 20/02/2020; (B) beige morph regenerating following fission, Dania Beach study site, 07/02/2020; (C) orange morph, Hollywood Beach study site, 29/05/2020.

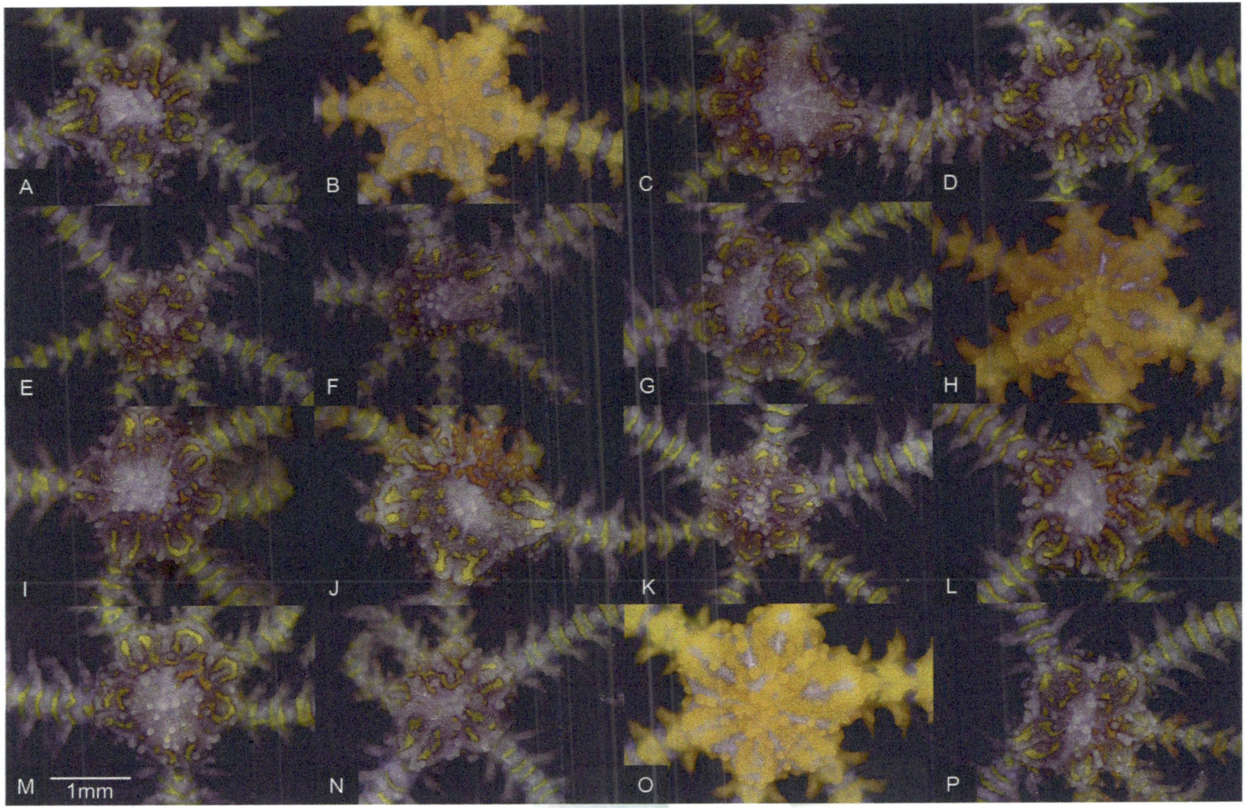

Fig. 3 Enlarged images of 16 live individuals of *Ophiothela mirabilis*, adoral disc views, beige morphs, (A, C–G, I–N, P); orange morphs, (B, H, O). Collected at Hollywood Beach, 29 May 2020. Viewed with Zeiss Stemi 508, 50 × magnification, Canon 7D Mark II camera.

3.2 Ophiuroid hosts

Numerous individuals of *Ophiothela mirabilis* occurred as epizoites on six species of octocorals in the family Plexauridae and on three octocoral species in the family Gorgoniidae (Table 2). No consistent trend in the densities of *O. mirabilis* as epizoites on octocoral hosts with differing epidermal textures was exhibited ($p = 0.74$) (Fig. 4). Model selection by AIC revealed evidence of a second order time effect across the seasonal series of observations (conditional $R^2 = 0.497$), with lowest densities of *O mirabilis* occurring in the summer months and highest densities in the winter months ($p = 0.006$).

Known non-octocoral hosts included a red alga, five species of sponges, and a cnidarian hydrocoral (Table 3). Only two sponge species, *Aplysina* sp. and *Cliona varians*, and the hydrocoral *Millepora alcicornis* hosted numerous individuals of *Ophiothela mirabilis*. The ophiuroid densities per host surfaces were not quantified but appeared to be decidedly lower than those living as epizoites on octocorals.

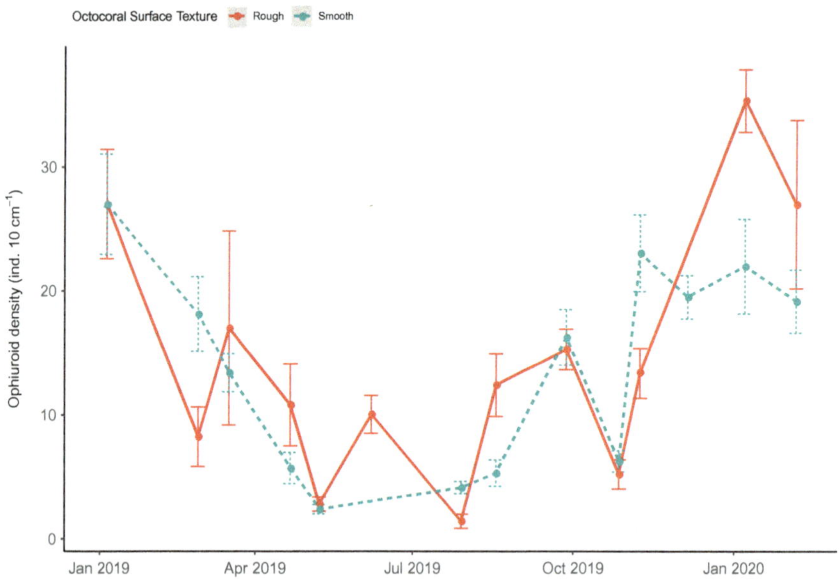

Fig. 4 Relationship between octocoral epidermal surface texture (rough, smooth) and density of individuals of *Ophiothela mirabilis* over a 1 year sampling period at Dania Beach, Florida, 5 January 2019–6 February 2020. Dots are mean values, error bars represent +/− standard errors. Thirteen collections each for rough and smooth octocoral hosts; sample size of rough and smooth octocorals in each collection ranged from 16 to 30. Rough octocoral surfaces consisted of crowded, sharply ornamented calyces (*Muricea elongata*, *Eunicea asperula*, and *Eunicea tournefortí*); smooth surfaced colonies with relatively smooth calyces or simple epidermal slits (*Eunicea clavigera*, *Eunicea flexuosa*, *Plexaurella dichotoma*, *Pseudopterogorgia acerosa*).

Table 3 Non-octocoral invertebrate hosts of invasive *Ophiothela mirabilis* with locations and dates when first observed, Southeast Florida.

Taxa	Location/date	Relative abundance[a]
Algae: Rhodophyta		
Jania adhaerens J.V. Lamouroux, 1816	Fisher Island 7/12/2019	Few individuals
Porifera: Demospongiae		
Anthosigmella varians (Duchassaing & Michelotti, 1864)	Dania Beach 27/09/2019	Single individual
Aplysina sp.		Several individuals
Haliclona rubens sensu (Duchassaing & Michelotti, 1864)		Few individuals
Cliona varians (Duchassaing & Michelotti, 1864)	Dania Beach 8/11/2019	Several individuals
Desmapsamma anchorata (Carter, 1882)	Dania Beach 7/1/2020	Few individuals
Cnidaria: Hydroida		
Millepora alcicornis Linnaeus, 1758	Dania Beach 27/09/2019	Several individuals

[a]Relative abundance refers to approximate number of individuals of *O. mirabilis* present on single host species.

3.3 Behaviour

Field observations. Individuals of *O. mirabilis* clung tenaciously to their octocoral hosts, wrapping all six arms over and around available branch surfaces (Fig. 5). Ophiuroid arms frequently contacted octocoral polyps, which usually partially or fully retracted into their calices. If contact was brief, polyps often remained expanded. Ophiuroids were usually evenly distributed on octocoral stems, but sometimes clumped, forming dense aggregations with individuals in frequent physical contact. Brittle stars were not present on dead, branched segments devoid of live octocoral tissue.

Laboratory observations. When ophiuroids were dislodged from their hosts in the laboratory they quickly remounted when offered the opportunity. When ophiuroids encountered individuals of a benthic ctenophore (*Coeloplana waltoni*), the latter would usually contract in size and adhere firmly to the octocoral or less often flee. Escape was usually short-lived

Fig. 5 Octocortal hosts harbouring individuals of *Ophiothela mirabilis.* (A) *Gorgonia ventalina*, Dania Beach study site, beige colour morphs, 6 m depth, 25 June 2019. (B) *Pseudopterogorgia acerosa*, mixed beige and orange colour morphs present. (C) *Eunicea* sp., orange colour morph. (D) *Muricea* sp., orange colour morph. (B)–(D), Hollywood Beach, 7 m depth, 29 August 2019.

because the ophiuroids moved more quickly than the ctenophores. Ophiuroids were never observed preying on ctenophores or other octocoral epizoites such as caprellids, or isopods. Caprellid epizoites occasionally attacked ophiuroids but were unsuccessful in penetrating the former's armoured epidermis. However, caprellids were frequently successful in attacking and quickly mutilating soft-bodied ctenophores.

Individuals of *O. mirabilis* were observed actively fleeing from the ovulid gastropod *Cyphoma gibbosum*. This gastropod, a common predator on octocorals, denudes coral stems of polyps and coenosarc in its grazing activities. The brittle stars moved up the octocoral branch ahead of the gastropod, moving away quicker from the approaching gastropod, and eventually falling off the branch tip.

Arm waving typically occurred when ophiuroids engaged in suspension feeding. We observed this behaviour in the field and laboratory. Prey capture was facilitated by the ophiuroid's arm spines, hooks and tube feet with arm tips often curling around adhering food items.

Table 4 Gut contents of *Ophiothela mirabilis*, mean proportions and standard errors of colour morphs by category.

	Colour morph[a]	
Category	**Orange**	**Beige**
Amorphous detritus	0.80 ± 0.024	0.75 ± 0.022
Filamentous algae	0.13 ± 0.022	0.19 ± 0.019
Miscellaneous[b]	0.028 ± 0.0043	0.027 ± 0.0057
Crustacea[c]	0.016 ± 0.0018	0.011 ± 0.0014
Macroalgae	0.0096 ± 0.0019	0.0091 ± 0.0011
Nematoda	0.0045 ± 0.0016	0.0055 ± 0.0012
Diatoms	0.0029 ± 0.0012	0.0027 ± 0.001

[a]Twenty individuals examined in each colour morph.
[b]Includes tardigrades, non-polychaete and non-nematode worms, dinoflagellates, eggs, nematocysts, faecal pellets, minute algal cells, and unidentified organic matter.
[c]Exoskeletal fragments.

3.4 Gut contents

The predominant gut contents of both colour morphs were amorphous detritus, which amounted overall to 80% and 75% respectively in orange and beige individuals (Table 4). Filamentous algae were the second most frequently observed food item with the remaining five categories (miscellanea through diatoms) in similar descending numerical order, each contributing <3% to <0.3% of pooled samples overall.

A likelihood ratio test found evidence of a difference among the proportions of overall food categories present between orange and beige morphs (scaled $\chi^2 = 16.6$, df = 6, $p = 0.011$). Differences in the likelihood of observing each individual food category relative to amorphous detritus (the most abundant item) between the two colour morphs were examined using Wald tests on estimated coefficients and scaled standard errors from the fitted model. Filamentous algae was the only category for which there was evidence of a difference ($p = 0.0001$), all other $p > 0.47$). This suggests that beige morphs tend to consume a greater proportion of filamentous algae relative to amorphous detritus compared with orange morphs. However, when food item counts were summed to only two categories (high vs. low nutrition items), there was no evidence of a difference in gut contents between the colour morphs at this level (scaled $\chi^2 = 0.339$, df = 1, $p = 0.56$).

3.5 Phylogenetic analyses

The phylogenetic cladograms obtained in the Bayesian (BI) analysis of the concatenated dataset (16S + COI, Fig. 6) show two main lineages with high support values, which were referred to as Clade A and Clade B. For Maximum Likelihood (ML), high support value occurred only for Clade B. Trees generated in the individual gene COI analyses resulted in the same clades with high support values for BI and ML (Clade B) and low support value for ML (Clade A) (Fig. S1 in Supplementary Material in the online version at https://doi.org/10.1016/bs.amb.2020.08.010). In relation to 16S, analyses resulted in only one clade with low support value (Clade B) for BI (Fig. S2 in Supplementary Material in the online version at https://doi.org/10.1016/bs.amb.2020.08.010). These differences are most likely due to lack of genetic data for both 16s and COI genes for all individuals used in the analyses, leading to the low support topologies in the concatenated dataset. Clade A is formed by the orange morph from Florida and Brazil as well as *O. mirabilis* from French Guyana and St. Vincent and *O. danae* from Australia. Clade B is formed by the beige morph from Florida, *O. mirabilis* from Peru and Mexico Pacific (Fig. 6). Unfortunately, specimens of the mottled colour morph were not preserved and were unavailable for inclusion in the genetic analyses.

3.6 Genetic distances

The genetic distances observed within orange and beige morphs were low (from 0% to 0.2%). In contrast, slightly higher genetic distances were observed between orange and beige morphs from Florida as well as between Clade A and B (1.3%).

3.7 Haplotype network

The haplotype network indicated the existence of two most common haplotypes: Hap_1 formed by Floridian and Brazilian orange morph along with *O. mirabilis* from French Guyana and St. Vincent; Hap_2 formed by Floridian beige morph along with *O. mirabilis* from Mexico Pacific and Peru. The divergence among these two haplotypes was caused by three mutations. The most distant haplotypes are from *O. danae* (Hap_6), which is separated from Hap_1 by six mutational steps, but is connected by the intermediate haplotypes (Fig. 6).

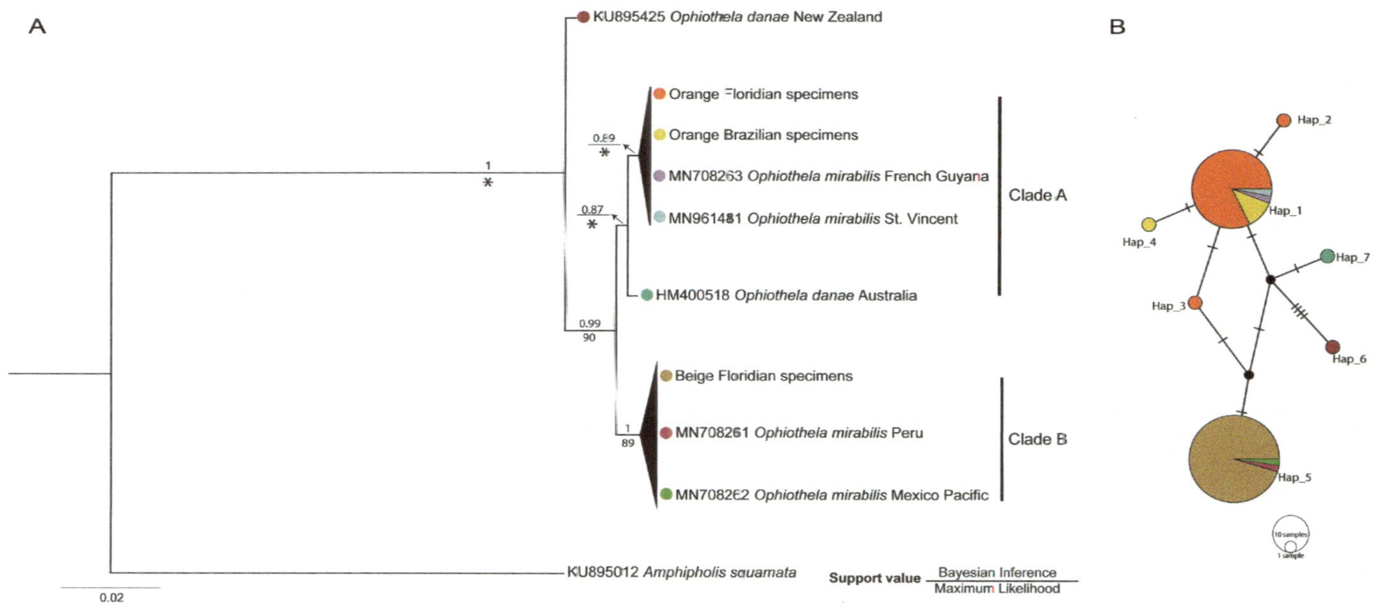

Fig. 6 Phylogenetic and genetic diversity analyses based on 16S + COI dataset. (A) Phylogenetic tree inferred from the Bayesian analysis (BI). The numbers above the branches represent posterior probabilities. For the cladelines also inferred in the Maximum Likelihood analysis (ML), bootstrap values (%) are provided below the branches. Asterisks (*) indicate that the branches were not recovered in ML analysis. Branches are identified by individual codes (Table 1 in Supplementary Material in the online version at https://doi.org/10.1016/bs.amb.2020.08.010). (B) Median-joining haplotype network. Each circle represents one haplotype and their sizes are proportional to the number of individuals sampled. The black circles represent possible intermediate haplotypes and crossing bars mutational steps.

4. Discussion

It is often difficult to determine exactly when a non-indigenous species arrives and becomes established in a new region. Some 19 years have passed since the first report of individuals of *O. mirabilis* in southern Brazil and its first reported occurrence in Florida. *Ophiothela mirabilis* is an eastern Pacific endemic species described from the Pearl Islands, Panama, from specimens occurring as epizoites on sponges and gorgonians (Verrill, 1867). It also has been reported from Pacific Colombia, at several sites along Middle America, from Mexico, often associated with corals and coral reefs, and as far north as Baja California (Granja-Fernández and López-Pérez, 2011; Granja-Fernández et al., 2014). It is possible that individuals of *O. mirabilis* were present on octocorals at our collection sites in 2018, or as early as 2017, since small numbers of similar appearing six-armed brittle stars were occasionally observed in earlier surveys. However, the identity of *O. mirabilis* in Florida was not confirmed until 2019 (Glynn et al., 2019). Thus, we tentatively conclude that the occurrence of *O. mirabilis* in southeastern Florida is a recent introduction, probably not present earlier than 2017 or 2018.

Fofonoff et al. (2003) summarized evidence supporting shipping as the predominant means of introduction of invasive species to shallow, North American waters. The presence of *O. mirabilis* near Brazilian and Caribbean ports prompted several workers to hypothesize that the introduction and spread of this ophiuroid could have occurred by shipping (e.g. Araújo et al., 2018; Bumbeer and Rocha, 2016; Hendler and Brugneaux, 2013; Hendler et al., 2012; Lawley et al., 2018; Mantelatto et al., 2016). High abundances of *Ophiothela mirabilis* were found in our study near sea ports with high volumes of ship traffic, namely $\leq$1 km from Government Cut at Fisher Island and Miami Beach (Table 1) near the entrance to the Port of Miami (Fig. 1C), and at Dania Beach (Table 1), 4 km south of Port Everglades (Fig. 1B). The two principal subvectors of shipping are hull fouling communities and ballast water. In recent decades, the inter-oceanic movement of oil drilling platforms (often supporting diverse fouling species), which are slower moving than ships, would offer perhaps a more effective means of transport. The slower forward motion would reduce the force of water flow and thus the dislodgment of fouling organisms. Indeed, several workers have recently noted the apparent introduction of alien invasive species via offshore oil platforms (Page et al., 2006; Pajuelo et al., 2016; Zaouali et al., 2008). The purported invasive species have involved sessile and motile species introductions, including crustaceans and fishes. The ability of individuals of *Ophiothela mirabilis* to hold on tenaciously to a variety of host species, and to reproduce by fissiparity, would advantage these species in dispersal.

The relatively rapid colonization and establishment of *Ophiothela mirabilis* in southern Brazil at the Tamboretes Archipelago in Santa Catarina state, and at several coastal localities northwards to St. Vincent in the Caribbean Windward Islands, spanning approximately 38.4° of latitude in 20 years, has been variously attributed to shipping and passive dispersal by ocean currents (Araújo et al., 2018; Hendler and Brugneaux, 2013; Lawley et al., 2018). The Northeastern Brazil Current (NBC), originating as a branch of the South Equatorial Current, flows northwest off Pernambuco state and enters the Caribbean Sea in the vicinity of Tobago at about 11° N. Populations of *O. mirabilis* in the vicinity of Pernambuco (ca. 8.1° S) could potentially contribute to the NBC planktonic larvae or adults that could be dispersed northwards in flotsam, perhaps eventually to join the Gulf Stream that flows by southeastern Florida. If dispersal in the tropical western Atlantic occurred by ocean current transport, one would expect reports of *O. mirabilis* at some Caribbean sites, such as the northern Lesser Antilles, the Greater Antilles or continental areas of northern South America and Central America bordering N and NW current flow. No such records have yet been forthcoming.

In contrast with the eastern Pacific (Granja-Fernández et al., 2014) and Brazil (Mantelatto et al., 2016), where individuals of *O. mirabilis* have been reported to associate with diverse hosts other than octocorals, e.g., algae, sponges, scleractinian corals, bryozoans, echinoderms, and ascidians, in Florida the brittle stars to date have been observed as epizoites on a single algal and hydrocoral species but predominantly on sponges and octocorals. Among the Octocorallia, however, they occur commonly on several species of *Eunicea* (four species), on *Muricea elongata*, *Plexaurella dichotoma*, *Gorgonia ventalina*, *Pterogorgia guadalupensis* and *Pseudopterogorgia acerosa*. Comprehensive searches of other benthos in Florida are necessary. It is likely that other non-octocoral taxa will serve as hosts as the brittle star populations increase in size in octocoral and coral reef communities.

The overall colour of individuals of *O. mirabilis*, including disc and arms, present along the Brazilian coast, at sites off northeastern South America, and in the Lesser Antilles, is yellow-orange (Alitto et al., 2016; Hendler and Brugneaux, 2013; Hendler et al., 2012; Mantelatto et al., 2016). Ophiuroids in most of our collections off southeast Florida were beige, i.e., light-greyish yellowish brown or pale to greyish yellow (Fig. 2A and B). Yellow-orange colour morphs (Fig. 2C), similar to those described from Brazil and the southern Caribbean Sea, were first collected at four sites from August to December 2019, and contributed importantly (≥66%) to samples from Fisher Island, Miami Beach, and Hollywood Beach. In contrast with the yellow-orange individuals of *O. mirabilis* in Brazil and the southern Caribbean region, Granja-Fernández et al. (2014) noted that this indigenous eastern tropical Pacific species displays a great variety of colours (purple rosaceous, creamy, burgundy, and yellow). The consistent identical colour of individuals of *O. mirabilis* from Brazil and the southern Caribbean prompted Hendler and Brugneaux (2013) to suggest that they represent a single lineage derived from a single introduction from the eastern Pacific to the southwestern Atlantic.

Differences in the Florida specimens extend beyond the variation in colour pattern. The phylogenetic and genetic diversity results indicate two distinct lineages, which may represent at least two separate invasion events. One lineage is formed by Floridian specimens sharing haplotypes with those from Brazil, French Guiana and the West Indies, representing the Atlantic Ocean. The majority of the specimens from Brazil and Florida share a common haplotype (Hap_1) with the other Caribbean specimens lending support to the hypothesis of Hendler and Brugneaux (2013) that one lineage of *Ophiothela* is spreading through the greater Caribbean region. However, the presence of three additional haplotypes, one in Brazil (Hap_4) and two in Florida (Hap_2 and Hap_3) suggest additional introductions, either by different clones in the same event or different events. The presence of additional colour morphs (purple) in Brazil (Alitto et al., 2020) points towards multiple introductions.

The second lineage is formed by beige Floridian specimens, closely related to the orange morphs, but sharing haplotypes with individuals from Mexico and Peru, both from the eastern Pacific region. There is only a single haplotype (Hap_5) for this colour morph, indicating that the introduction was by a single clonal type that has spread rapidly throughout the collection sites. Also, as the beige morph was the first detected at any of the sites, thus suggesting that the initial introduction of *O. mirabilis* into Florida may not have been through the Caribbean, but directly by a Pacific vector. While the beige colour morph has not been reported in Mexico (M.R. Granja-Fernández, Personal Communication), Vanegas González and Borrero-Pérez (2020) reported the presence of a similar colour morph from the Pacific coast of Colombia.

Ophiothela mirabilis exhibited some unexpected and intriguing variations in abundance in this study. A plausible explanation for the low abundances of individuals of *O. mirabilis* in the summer of 2019 is not readily apparent. The lowest densities of ophiuroids observed in May and July occurred before Hurricane Dorian in August, a destructive category 5 cyclone with 265 km h^{-1} winds, centred 170 km from the Dania Beach study site. Another possible factor is the seasonal availability of plankton. The scant data available (Calypso LNG Deepwater Port License Application, 2008) indicate highest abundances of planktonic crustaceans near Dania Beach in May (during declining ophiuroid abundances) and lowest abundances in July and September (during low ophiuroid abundances). Clearly, the factors influencing these fluctuations in abundance, if indeed they can be shown to be recurrent seasonal trends, need further study.

In the eastern Pacific, individual *O. mirabilis* colours frequently matched the colours of the gorgonian host species from which they were collected (Granja-Fernández et al., 2014). This was not the case in the southern Atlantic or in our southeast Florida study. Some beige-coloured individuals were sometimes difficult to detect on octocoral hosts with crowded and highly sculptured calyces, such as *Eunicea asperula*, *E. tourneforti*, and *Muricea elongata*. When not feeding or when disturbed, brittle stars would often lower their discs and arms between calyces, becoming difficult to detect. Orange morphs contrasted markedly from their usually grey, blue or brown-coloured hosts, highly visible under natural conditions.

Currently, no other invasive species affecting tropical octocorals have been reported, although non-native turf algae (Linares et al., 2012) and macroalgae (Cebrian et al., 2012) have been shown to impact octocorals in the Mediterranean Sea. An invasive Indo-Pacific soft coral (*Xenia*) has been observed overgrowing zooxanthellate scleractinian corals in Venezuela (Ruiz Allais et al., 2014) but has not been reported overgrowing octocorals. Marine metazoan symbioses are generally host specific, influenced by various factors such as potency of cnidae (stinging cells), mucus production, noxious secondary metabolites, protection from predators, and provision of surfaces from which to feed. Symbiotic partnerships need more study to understand the vulnerability of host species and to better exercise measures necessary to safeguard these vital coral reef engineers.

In Florida, *O. mirabilis* inhabits octocorals, which have become the dominant cnidarians on Caribbean reefs (Lasker et al., 2020, in this volume). It is possible that this invasion was facilitated by the ecological upheaval wrought by the replacement of dominant benthic taxa (Stachowicz, 2001). The speed and apparent success of this invasion could have been favoured as a secondary effect of the well-documented decline of the Scleractinia and ascendency of octocorals (Lasker et al., 2020, in this volume) and might not have been so successful otherwise. Neither the level of potential facilitation, nor the effects of the invader on the facilitating host fauna are known. The brittle stars possibly have a deleterious influence on some other epizoites inhabiting octocoral branches, such as caprellids amphipods and the platyctenid ctenophore *Coeloplana waltoni* (Glynn et al., 2014). While it is unknown whether these latter species play a positive role in the life-history of the octocoral host, the brittle star invasion seems to lead at the very least to a reduction of local biodiversity associated with octocorals. If high densities of brittle stars should be demonstrated to interfere with octocoral life processes, such as feeding, growth and reproduction, then this newly introduced species would have the potential to cause further declines of reefal fauna in the Caribbean. Scleractinian abundances have declined due to a combination of epizootics, heat anomalies, and competition with other benthos (particularly algae; Dudgeon et al., 2010; Precht et al., 2020, in this volume). Hopefully, the incidence of pathogenic infections or increases in other octocoral mortality factors will not be stimulated by the presence of the invading brittle star *O. mirabilis*, nor that the brittle star adversely affects the survivorship of the rich epibiont community. Of additional concern is that this aggressively increasing new component of Caribbean and Brazilian reef faunas is not a harbinger of further declines of the region's coral reefs.

5. Conclusions

With the occurrence of *O. mirabilis* in southeastern Florida, seemingly firmly established, attention should now turn to studying the effects of this ophiuroid species on its octocoral hosts and native octocoral epizoites such as the benthic ctenophore *Coeloplana waltoni*, diverse polychaete worm species, crustaceans, and mollusks. Conditions triggering fission should also be investigated as this trait appears to be one of the keys to the success of *O. mirabilis* as an invasive species. Additional genetic data from Pacific, Brazilian, and Caribbean populations are needed to further determine the source of origin and trace the path of spreading. Such knowledge may help to predict the potential impacts of the non-native ophiuroid as well as when and where such events are likely to occur.

Acknowledgements

We thank Gordon Hendler for the initial identification of *Ophiothela*, Herman Wirshing and Charles Messing for identifying octocoral host species, Cristina Diaz for identifying sponges and Larry E. Brand for verifying the identification of *Jania*. Jorge Quintela kindly created the map in Fig. 1. Amanda Robinson and Christopher Meyer kindly catalogued specimens into the USNM collections. Claudia Alvarez provided field observations following Hurricane Dorian, and David E. Newhard (Enzo) the use of his boat and help with collections. Help with the literature and insightful discussions were generously offered by Walter Goldberg and Howard Lasker. The NSU (Nova Southeastern University) scuba team—William L. Bullock (Landis), Brooke Enright and Rachel L. Ionata—and Brian Coffman also helped with sample collections. All collections were made pursuant to the terms of the Special Activity License SAL-18-1105-E, Florida Fish and Wildlife Conservation Commission, Division of Marine Fisheries Management. Laboratory and facilities support were provided to the P. Glynn team by the Department of Marine Biology and Ecology, Rosenstiel School of Marina and Atmospheric Science, University of Miami, and to the A. Christensen group by the Biology Department, Lamar University. R. Alitto's contributions were supported by grant 2018/10313-0, São Paulo Research Foundation (FAPESP).

Data availability statement

Sequence data of *Ophiothela mirabilis* that support the findings of this study have been deposited in GenBank with the accession codes: 16S, MT337300-MT337373; COI, MT340982-MT341052. DNA fragments of 353 bp for 16S and 623 bp for COI were obtained from 76 individuals. All the sequences are deposited in GenBank (www.ncbi.nlm.nih.gov/Genbank) (Table S1 in Supplementary Material in the online version at https://doi.org/10.1016/bs.amb.2020.08.010). The nucleotide sequences were edited using BIOEDIT Sequence Alignment Editor v. 7.0.1 (Hall, 1999) and GeneStudio 2.2.0.0 (Genestudio, Suwanee, GA, USA).

References

Alitto, R.A.S., Bueno, M.L., Domenico, M.D., Borger, M., 2016. Annotated checklist of echinoderms from Araca Bay, southeastern Brazil. Check List 12 (1), 1836. https://doi.org/10.15560/12.1.1836.

Alitto, R.A.S., Granadier, G., Christensen, A.B., O'Hara, T., Di Domenico, M., Borges, M., 2020. Unravelling the taxonomic identity of *Ophiothela* Verrill, 1867 (Ophiuroidea) along the Brazilian coast. J. Mar. Biol. Assoc. UK. 1–14. https://doi.org/10.1017/S002531542000034X.

Araújo, J.T., de Oliveira Soares, M., Matthews-Cascon, H., Monteiro, F.A.C., 2018. The invasive brittle star *Ophiothela mirabilis* Verrill, 1867 (Echinodermata, Ophiuroidea) in the southwestern Atlantic: filling gaps of distribution, with comments on an octocoral host. Lat. Am. J. Aquat. Res. 46, 1123–1127. https://doi.org/10.3856/vol46-issue5fulltext-25.

Banks, K.W., Riegl, B.M., Richards, V.P., Walker, B.K., Helmle, K.P., et al., 2008. The reef tract of continental Southeast Florida (Miami-Dade, Broward and Palm Beach Counties, USA). In: Riegl, B.M., Dodge, R.E. (Eds.), Coral Reefs of the USA. Coral Reefs of the World I. Springer Science + Business Media B.V, pp. 175–220.

Bates, D., Mächler, M., Bolker, B., Walker, S., 2015. Fitting linear mixed-effects models using lme4. J. Stat. Softw. 67 (1), 1–48. https://doi.org/10.18637/jss.v067.i01.

Bayer, F.M., 1961. The shallow-water Octocorallia of the West Indian region. A manual for marine biologists. In: Stud Fauna Curaçao Caribb Islands. Martinus Nijhoff, The Hague, p. 373.

Bribiesca-Contreras, G., Solis-Marin, F.A., Laguarda-Figueras, A., Zaldivar-Riveron, A., 2013. Identification of echinoderms (Echinodermata) from an anchialine cave in Cozumel Island, Mexico, using DNA barcodes. Mol. Ecol. Resour. 13, 1137–1145. https://doi.org/10.1111/1755-0998.12098.

Bumbeer, J., Rocha, R.M., 2016. Invading the natural marine substrates: a case study with invertebrates in South Brazil. Zoologia (Curitiba) 33 (3), e20150211. https://doi.org/10.1590/S1984-4689zool-20150211.

Calypso LNG Deepwater Port License Application. 2008. Docket No. USCG-2006-26009 1, 39–43.

Carlton, J.T., Geller, J.B., 1993. Ecological roulette: the global transport of non-indigenous marine organisms. Science 261, 78–82.

Cebrian, E., Linares, C., Marschal, C., Garrabou, J., 2012. Exploring the effects of invasive algae on the persistence of gorgonian populations. Biol. Invasions 14, 2647–2656.

Dudgeon, S.R., Aronson, R.B., Bruno, J.F., Precht, W.F., 2010. Phase shifts and stable states on coral reefs. Mar. Ecol. Prog. Ser. 413, 201–216.

Fofonoff, P.W., Ruiz, G.M., Steves, B., Carlton, J.T., 2003. In ships or on ships? Mechanisms of transfer and invasion for nonnative species to the coasts of North America. In: Ruiz, G.M., Carlton, J.T. (Eds.), Invasive Species: Vectors and Management Strategies. Island Press, Washington DC, pp. 152–182.

Glynn, P.W., Bayer, F.M., Renegar, D.A., 2014. *Coeloplana waltoni*, a new species of minute benthic ctenophore (Ctenophora: Platyctenida) from South Florida. Proc. Biol. Soc. Wash. 127, 423–436.

Glynn, P.W., Coffman, B., Primov, K., Renegar, D.A., Gross, J., et al., 2019. Benthic ctenophores (Order Platyctenida) reproduction, recruitment, and seasonality in South Florida. Invertebr. Biol. 2019, e12256. https://doi.org/10.1111/ivb.12256.

Goh, N.K.C., Ng, P.K.L., Chou, L.M., 1999. Notes on the shallow water gorgonian-associated fauna on coral reefs in Singapore. Bull. Mar. Sci. 65 (1), 259–282.

Goldberg, W., 1973. The ecology of the coral-octocoral communities off the Southeast Florida coast: geomorphology, species composition, and zonation. Bull. Mar. Sci. 23, 465–488.

Grajales, A., Aguilar, C., Sánchez, J.A., 2007. Phylogenetic reconstruction using secondary structures of internal transcribed spacer 2 (ITS2, rDNA): finding the molecular and morphological gap in Caribbean gorgonian corals. BMC Evol. Biol. 7, 90. https://doi.org/10.1186/1471-2148-7-90.

Granja-Fernández, M.R., López-Pérez, R.A., 2011. Nuevos registros de ofiuroideos (Echinodermata: Ophiuroidea) para localidades de Zihuatanejo (Guerrero) y Puerto Escondido (Oaxaca), Pacífico mexicano. Rev. Mex. Biodivers. 82, 1320–1323.

Granja-Fernández, R., Herrero-Pérezrul, M.D., López-Pérez, R.A., Hernández, L., Rodríguez-Zaragoza, F.A., et al., 2014. Ophiuroidea (Echinodermata) from coral reefs in the Mexican Pacific. Zookeys 406, 101–145.

Hall, T.A., 1999. BioEdit: a user-friendly biological sequence alignment editor and analysis program for Windows 95/98/NT. Nucleic Acids Symp. Ser. 41, 95–98.

Hart, M.W., Podolsky, R.D., 2005. Mitochondrial DNA phylogeny and rates of larval evolution in *Macrophiothrix* brittlestars. Mol. Phylogenet. Evol. 34, 438–447.

Hendler, G., Brugneaux, S.J., 2013. New records of brittle stars from French Guiana: *Ophiactis savignyi* and the alien species *Ophiothela mirabilis* (Echinodermata: Ophiuroidea). Mar. Biodivers. Rec. 6, e113. https://doi.org/10.1017/s1755267213000845.

Hendler, G., Migotto, A.E., Ventura, C.R.R., Wilk, L., 2012. Epizoic *Ophiothela* brittle stars have invaded the Atlantic. Coral Reefs 31, 1005. https://doi.org/10.1007/s00338-012-0936-6.

Hoareau, T.B., Boissin, E., 2010. Design of phylum-specific hybrid primers for DNA barcoding: addressing the need for efficient COI amplification in the Echinodermata. Mol. Ecol. Resour. 10, 960–967.

Jaap, W.C., Szmant, A., Jaap, K., Dupont, J., Clarke, R., et al., 2008. A perspective on the biology of Florida Keys coral reefs. In: Riegl, B.M., Dodge, R.E. (Eds.), Coral Reefs of the USA. Coral Reefs of the World I. Springer Science + Business Media B.V, pp. 75–125.

Johnson, S.K., Hallock, P., 2020. A review of symbiotic gorgonian research in the Western Atlantic and Caribbean with recommendations for future work. Coral Reefs 39, 239–258.

Kimura, M., 1980. A simple method for estimating evolutionary rates of base substitutions through comparative studies of nucleotide sequences. J. Mol. Evol. 16, 111–120.

Kumar, S., Stecher, G., Li, M., Knyaz, C., Tamura, K., 2018. MEGA X: molecular evolutionary genetics analysis across computing platforms. Mol. Biol. Evol. 35, 1547–1549.

Lasker, H.R., Bramanti, L., Tsounis, G., Edmunds, P.J., 2020. The rise of octocoral forests on Caribbean reefs. In: Riegl, B. (Ed.), Population Dynamics of the Reef Crisis, vol. 87 Advances in Marine Biology. Elsevier, pp. 361–410 (in this volume). https://doi.org/10.1016/bs.amb.2020.08.009.

Lawley, J.W., Fonseca, A.C., Faria Júnior, E., Lindner, A., 2018. Occurrence of the non-indigenous brittle star *Ophiothela* cf. *mirabilis* Verrill, 1867 (Echinodermata: Ophiuroidea) in natural and anthropogenic habitats off Santa Catarina, Brazil. Check List 14, 453–459. https://doi.org/10.15560/14.2.453.

Leigh, J.W., Bryant, D., 2015. Popart: full-feature software for haplotype network construction. Methods Ecol. Evol. 6, 1110–1116.

Linares, C., Cebrian, E., Coma, R., 2012. Effects of turf algae on recruitment and juvenile survival of gorgonian corals. Mar. Ecol. Prog. Ser. 452, 81–88.

Mantelatto, M.C., Vidon, L.F., Silveira, R.B., Menegola, C., da Rocha, R.M., Creed, J.C., 2016. Host species of the non-indigenous brittle star *Ophiothela mirabilis* (Echinodermata: Ophiuroidea): an invasive generalist in Brazil? Mar. Biodivers. Rec. 9, 8. https://doi.org/10.1186/s41200-016-0013-x.

McCullagh, P., Nelder, J.A., 1989. Generalized Linear Models, second ed. Chapman and Hall, London.

Nylander, J.A.A., 2004. MrModeltest.pl. In: Program distributed by the author. Evolutionary Biology Centre, Uppsala University, Uppsala.

O'Hara, T.D., Hugall, A.F., Thuy, B., Moussalli, A., 2014. Phylogenomic resolution of the class Ophiuroidea unlocks a global microfossil record. Curr. Biol. 24, 1874–1879.

Page, H.M., Dugan, J.E., Culver, C.S., Hoesterey, J.C., 2006. Exotic invertebrate species on offshore oil platforms. Mar. Ecol. Prog. Ser. 325, 101–107.

Pajuelo, J.G., González, J.A., Triay-Portella, R., Martín, J.A., Ruiz-Díaz, R., Lorenzo, J.M., Luque, Á., 2016. Introduction of non-native marine fish species to the Canary Islands waters through oil platforms as vectors. J. Marine Syst. 163, 23–30.

Pérez-Portela, R., Almada, V., Turon, X., 2013. Cryptic speciation and genetic structure of widely distributed brittle stars (Ophiuroidea) in Europe. Zool. Scr. 42, 151–169.

Precht, W.F., Aronson, R.B., Gardner, T.A., Gill, J.A., Hawkins, J.P., Hernández-Delgado, E.A., Jaap, W.C., Mcclanahan, T.R., Mcfield, M.D., Murdoch, T.J.T., Nugues, M.M., Roberts, C.M., Schelten, C.K., Watkinson, A.R., Côté, I.M., 2020. The timing and causality of ecological shifts on Caribbean reefs. In: Riegl, B. (Ed.), Population Dynamics of the Reef Crisis, vol. 87. Advances in Marine Biology. Elsevier, pp. 331–360 (in this volume). https://doi.org/10.1016/bs.amb.2020.08.008.

Rich, L.P., Dennis, M.M., Freeman, M.A., 2020. New record of the non-native *Ophiothela mirabilis* (Verill 1867) in St. Kitts, West Indies. Adv. Oceanogr. Mar. Biol. 2 (1), 1–4. https://doi.org/10.33552/AOMB.2020.02.000526.

Ronquist, F., Teslenko, M., van der Mark, P., Ayres, D.L., Darling, A., Höhna, S., Larget, B., Liu, L., Suchard, M.A., Huelsenbeck, J.P., 2012. MrBayes 3.2: efficient Bayesian phylogenetic inference and model choice across a large model space. Syst. Biol. 61, 539–542. https://doi.org/10.1093/sysbio/sys029.

Rosi-Marshall, E.J., Wellard Kelly, H.A., Hall Jr., R.O., Vallis, K.A., 2015. Methods for quantifying aquatic macroinvertebrate diets. Freshw. Sci. 35 (1), 229–236.

Roy, M.S., Sponer, R., 2002. Evidence of a human-mediated invasion of the tropical western Atlantic by the 'world's most common brittlestar. Proc. R. Soc. Lond. B. 269, 1017–1023. https://doi.org/10.1098/rspb.2002.1977.

Rozas, J., Ferrer-Mata, A., Sánchez-DelBarrio, J.C., Guirao-Rico, S., Librado, P., Ramos-Onsins, S.E., Sánchez-Gracia, A., 2017. DnaSP 6:DNA sequence polymorphism analysis of large data sets. Mol. Biol. Evol. 34, 3299–3302. https://doi.org/10.1093/molbev/msx248.

Ruiz Allais, J.P., Amaro, M.E., McFadden, C.S., Halász, A., Benayahu, Y., 2014. The first incidence of an alien soft coral of the family Xeniidae in the Caribbean, an invasion in eastern Venezuelan coral communities. Coral Reefs 33, 287.

Sánchez, J.A., 2016. Diversity and evolution of octocoral animal forests at both sides of tropical America. In: Rossi, S., Bramanti, L., Gori, A., Drejas Saco del Valle, C. (Eds.), Marine Animal Forests. Springer, Cham, pp. 1–33.

Stachowicz, J.J., 2001. Mutualism, facilitation, and the structure of ecological communities: positive interactions play a critical, but underappreciated, role in ecological communities by reducing physical or biotic stresses in existing habitats and by creating new habitats on which many species depend. Bioscience 51 (3), 235–246.

Vanegas González, M.J., Borrero-Pérez, G.H., 2020. First records and new information on the associations of echinoderms with other phyla in the rocky reefs of northern Chocó, Colombian Pacific. Zookeys 921, 1–22.

Venables, W.N., Ripley, B.D., 2002. Random and mixed effects. In: Modern Applied Statistics With S. Springer, New York, pp. 271–300.

Verrill, A.E., 1867. Notes on Radiata in the museum of Yale College with descriptions of new genera and species. No. 2. Notes on the echinoderms of Panama and the west coast of America, with descriptions of a new genus. Trans. Connecticut Acad. Arts Sci. 1 (2), 251–322.

Zaouali, J., Souissi, J., Galil, B.S., d'Acoz, C.D.U., Abdullah, A., 2008. Grapsoid crabs (Crustacea: Decapoda: Brachyura) new to the Sirte Basin, southern Mediterranean Sea—the roles of vessel traffic and climate change. J. Mar. Biol. Assoc. Biodivers. Rec. 5770, 4. https://doi.org/10.1017/S1755267207007701.

FSC
www.fsc.org
MIX
Papier | Fördert
gute Waldnutzung
FSC® C083411

Kolibri 360 GmbH
Ferdinand-Jühlke-Straße 7
99095 Erfurt, Deutschland
produktsicherheit@kolibri360.de